Algebra

Exponents and Radicals

$$a^m \cdot a^n = a^{m+n}$$

$$\frac{a^m}{a^n} = a^{m-n} \quad a \neq 0$$

$$(a^m)^n = a^{mn}$$

$$(ab)^n = a^n b^n$$

$$\left(\frac{a}{b}\right)^n = \frac{a^n}{b^n} \quad b \neq 0$$

$$a^0 = 1 \quad a \neq 0$$

$$a^{-n} = \frac{1}{a^n} \quad a \neq 0$$

$$a^{m/n} = \sqrt[n]{a^m} = (\sqrt[n]{a})^m$$

$$\sqrt[m]{\sqrt[n]{a}} = \sqrt[mn]{a}$$

$$\sqrt[n]{a}\sqrt[n]{b} = \sqrt[n]{ab}$$

$$\frac{\sqrt[n]{a}}{\sqrt[n]{b}} = \sqrt[n]{\frac{a}{b}} \quad b \neq 0$$

Special Products

$$a(x + y) = ax + ay$$

$$(x + y)(x - y) = x^2 - y^2$$

$$(x + y)^2 = x^2 + 2xy + y^2$$

$$(x - y)^2 = x^2 - 2xy + y^2$$

$$(x + y)^3 = x^3 + 3x^2y + 3xy^2 + y^3$$

$$(x - y)^3 = x^3 - 3x^2y + 3xy^2 - y^3$$

$$(x^3 - y^3) = (x - y)(x^2 + xy + y^2)$$

$$(x^3 + y^3) = (x + y)(x^2 - xy + y^2)$$

Binomial Series

$$(1 + x)^n = 1 + nx + \frac{n(n - 1)}{2!}x^2 + \frac{n(n - 1)(n - 2)}{3!}x^3 + \cdots$$

$$\tan \theta = \frac{m_2 - m_1}{1 + (m_1 m_2)}$$

Quadratic Equation and Formula

$$ax^2 + bx + c = 0$$

$$x = \frac{-b \pm \sqrt{b^2 - 4ac}}{2a}$$

Properties of Logarithms

$$\log_b x + \log_b y = \log_b xy$$

$$\log_b x - \log_b y = \log_b\left(\frac{x}{y}\right)$$

$$n \log_b x = \log_b(x^n)$$

Geometric Formulas

Triangle

$$A = \frac{1}{2}bh$$

Circle

$$A = \pi r^2$$
$$c = 2\pi r$$

Cylinder

$$V = \pi r^2 h$$
$$A = 2\pi rh + 2\pi r^2$$

Cone

$$V = \frac{1}{3}\pi r^2 h$$

Sphere

$$V = \frac{4}{3}\pi r^3$$
$$A = 4\pi r^2$$

Technical Calculus with Analytic Geometry

Third Edition

Other Benjamin/Cummings Titles of Related Interest

Technical Calculus with Analytic Geometry

Third Edition

Allyn J. Washington

Dutchess Community College

The Benjamin/Cummings Publishing Company
Menlo Park, California • Reading, Massachusetts
Don Mills, Ontario • Wokingham, U.K. • Amsterdam
Sydney • Singapore • Tokyo • Mexico City
Bogota • Santiago • San Juan

Sponsoring editor: Craig Bartholomew

Production editor: Cici Oremland

Production service: Stacey Sawyer, San Francisco

Book and cover designer: Gary Head

Copyeditor: Charlie Hibbard

Technical art: Carl Brown

Composition: Typothetae

Library of Congress Cataloging-in-Publication Data

Washington, Allyn J.
   Technical calculus with analytic geometry.

   Bibliography: p.
   Includes index.
   1. Calculus. 2. Geometry, Analytic.   I. Title
QA303.W38   1986   515'.15   85-23002
ISBN 0-8053-9512-1

ABCDEFGHIJ-DO-89876

The Benjamin/Cummings Publishing Company, Inc.
2727 Sand Hill Road
Menlo Park, California 94025

Preface

Scope of the Book

This book is intended primarily for students taking technical programs at two-year and four-year colleges and technical institutes. The text emphasizes elementary topics in calculus and certain selected topics from the more advanced areas. They have been selected because of their application in the technologies, especially electrical and mechanical technologies. The topics are developed in a nonrigorous and intuitive manner, with stress being placed on the interpretation and application of the material presented.

It is assumed that students using this text will have had courses covering basic algebra and trigonometry. However, several sections on algebra and trigonometry are included for review and reference purposes.

The general topics covered in the text are basic analytic geometry, differentiation and integration of algebraic and elementary transcendental functions, an introduction to partial derivatives and double integrals, expansion of functions in series, and differential equations.

The analytic geometry is developed primarily for use in the calculus, although numerous direct applications of curves such as the conic sections are discussed to show the importance of graphical methods. A chapter on polar coordinates demonstrates the use of different coordinate systems. Also, empirical curve fitting is included to show how a curve can be fitted to data. In the chapter dealing with partial derivatives and double integrals, a section is devoted to solid analytic geometry. Also, the graphs of the trigonometric, inverse trigonometric, exponential, and logarithmic functions are reviewed.

The calculus will give the student a mathematical understanding of many topics that arise in other courses. The standard methods of differentiation and integration are covered, along with certain numerical methods, applicable for calculator and computer use.

The chapter on partial derivatives and double integrals is included for use in appropriate courses. Applications are shown for such fields as electronics, thermodynamics, and mechanics.

The chapter on expansion of functions in series will give students an understanding of various types of series and operations using them. The section on Fourier series is appropriate for curricula in electrical technology.

Differential equations provide one of the richest areas of mathematical applications in technical areas, especially in electrical and mechanical technologies. Here a mathematical basis can be established for many concepts and formulas that students previously have had to accept without basis. For this reason, considerable space is devoted to elementary differential equations.

Laplace transforms are introduced to show methods and terminology for electrical and mechanical technology students. Numerical methods of solving differential equations are appropriate for use with calculators and computers—familiarity with which will make the technician more effective in modern industry.

New Features

This third edition of **Technical Calculus with Analytic Geometry** includes all the basic features of the first two editions. However, *most sections have been rewritten* to some degree to include additional explanatory material, examples, and exercises. Specifically, among the new features of the third edition are the following:

- **New Chapter**

  Because four new sections on methods of integration have been added, this chapter has been split into two chapters. Chapter 8 covers *Integration by Standard Forms,* and Chapter 9 covers *Methods of Integration.* The new sections are as follows:

  1. *Integration by Substitution*—Section 9-2 develops a basic method of integration by algebraic substitution.
  2. *Integration by Partial Fractions: Nonrepeated Linear Factors*—Section 9-4 now introduces the topic of integration by partial fractions.
  3. *Integration of Partial Fractions: Other Cases*—Section 9-5 covers integration using other types of partial fractions.
  4. *Improper Integrals*—Section 9-7 introduces improper integrals.

- **New Sections**

  Other sections of new material are as follows:

  1. *Higher Derivatives*—Section 2-10 is now devoted solely to the introduction of higher derivatives.
  2. *Newton's Method for Solving Equations*—Section 3-2 now covers Newton's method.
  3. *Sequences and Series*—Section 13-2 is a new introductory section for the chapter on *Expansion of Functions in Series.*

- **Additional Coverage**

  Many additional examples illustrating the *evaluation of derivatives* have been included in Chapters 2, 6, and 7. Also, *Fourier Series* with a period of 2L are discussed.

- **Computer Programs in BASIC**

  A new appendix gives 20 selected computer programs in BASIC. These programs are keyed to text material by use of a computer symbol (shown to the left).

- **Common Error Margin Symbol**

  Concepts that students often handle incorrectly or have difficulty handling are identified with a special margin symbol (shown to the left) and light color shading.

- **Special Example Comments**

  In many examples, special comments in color are included to emphasize certain points. Also, specific parts of examples are frequently identified by the use of color. Arrows are used to clearly indicate the part of the example to which reference is made.

- **Figures**

  There are many more figures in this edition—more than 425 figures, an increase of over 50% from the second edition. These new figures assist the student in the examples and exercises.

- **Examples**

  The text includes more than 575 worked examples. This is an increase of about 15% from the second edition.

- **Exercises**

  There are now more than 3700 exercises—an increase of more than 20% from the second edition. Each exercise group is designated by the number of the section that it follows. Most exercises are grouped such that there is an even-numbered exercise equivalent to each odd-numbered exercise.

Other Features

- **Applications**

  The text illustrates the numerous applications of calculus to all fields of technology. It includes definitions of physical quantities, the relations among displacement, velocity, and acceleration, areas and volumes, centroids and moments of inertia, and work and liquid pressure. Many of the new exercises are related to modern technology.

- **Calculator Use**

  A number of exercises are specifically designed for solution using a calculator. These exercises can be used in developing certain basic concepts such as a limit. It is assumed that students will use a scientific calculator for most of their calculating work.

- **Review Exercises**

  Each chapter is followed by review exercises. These can be used either as additional problems or as review assignments.

- **Important Formulas**

  These formulas are set off in color so that they are easily located and used.

- **Units of Measurement**

  Both U.S. customary units and metric units are used. The text includes an appendix on units.

- **Graphical Methods**

  In addition to specific topics from analytic geometry, graphical techniques and interpretations are included throughout the book.

- **Examples**

  Many of the examples are used advantageously to introduce concepts, as well as to clarify and illustrate points made in the text.

- **Order of Topics**

  The order of coverage can be changed in several places without loss of continuity. Any omissions or changes will, of course, depend on the type of course and completeness required.

- **Supplements**

  Supplements to this text include an *Instructor's Guide* and a *Solutions Manual*. The *Solutions Manual* includes worked-out solutions to all odd-numbered exercises.

Acknowledgments

The author wishes to acknowledge the valuable comments and suggestions of current reviewers and many of those who used the earlier editions. These include the members of the mathematics department at Dutchess Community College, as well as the following people:

Amad Abu Said
Southern Institute of Technology

Dennis Adams
Knoxville State Technical Institute

William D. Brower
New Jersey Institute of
  Technology

Donald De Bonee
University of Hartford

Richard Gregg
OMI College of Applied Sciences

George R. Mach
California Polytechnic State
  University
San Luis Obispo

F. P. Mathur
California Polytechnic State
  University, Pomona

Don Sibrel
Nashville State Technical Institute

V. M. Smith
California Polytechnic State
   University, Pomona

Thomas J. Stark
Cincinnati Technical College

Jeff Zacharakis-Jutz
Casper College

In addition, I wish to thank Robert E. Seaver of Loraine County Community College, who reviewed all of the computer programs and wrote the introductions to these programs. Finally, the assistance, cooperation, and diligent efforts of the Benjamin/Cummings staff during the production of this text are also greatly appreciated.

A. J. W.

Contents

Technical Calculus with Analytic Geometry

Third Edition

CHAPTER 1

Plane Analytic Geometry

The problems which can be solved by the methods of algebra and trigonometry are numerous. There are, however, a great many problems which arise in the various fields of technology which require for their solution methods beyond those available from algebra and trigonometry. The use of these traditional topics remains of definite importance, but it is necessary to develop additional methods of analyzing and solving problems.

One very important type of problem which arises involves the rate of change of one quantity with respect to another. Examples of rates of change are velocity (the rate of change of distance with respect to time), the rate of change of the length of a metal rod with respect to temperature, the rate of change of light intensity with respect to the distance from the source, the rate of change of electric current with respect to time, and many other similar physical situations. The methods of **differential calculus,** which we shall start developing in Chapter 2, will enable us to solve problems involving these quantities.

Another principal type of problem which calculus allows us to solve is that of finding a function when its rate of change is known. This is **integral calculus,** a study of which starts in Chapter 4. One of the principal applications comes from electricity, where current is the time rate of change of electric charge. Integral calculus also leads to the solution of a great many other apparently unrelated problems, which include plane areas, volumes, and the physical concepts of work and pressure.

Before starting a study of the calculus, we will first develop some of the basic concepts of **analytic geometry.** This branch of mathematics deals with the relationship between algebra and geometry. The definitions and methods developed here, along with certain basic curves and their characteristics, will enable us to develop the calculus more readily. Also, many technical applications of analytic geometry itself are illustrated.

1–1 Rectangular Coordinates

From algebra we recall that *a **variable** is a quantity which may take on admissible values during a given discussion. A **constant** is a quantity which remains fixed during the discussion.* Generally letters toward the end of the alphabet are used to denote variables, and letters near the beginning of the alphabet denote constants.

One of the most valuable ways of representing an equation in two variables is by means of a graph of the equation. By using graphs we are able to obtain a "picture" of the equation, and this picture enables us to learn a great deal about the equation.

To achieve this graphical representation we use the fact that we may represent numbers by points on a line. Since two variables are involved, it is necessary to use a line for each. This is done most conveniently by placing the lines perpendicular to each other.

Thus *we place one line horizontally and label it the **x**-axis. The other line we place vertically and label the **y**-axis. The point of intersection is called the **origin** and is designated by 0. This is called the **rectangular coordinate system.***

Starting from the origin, equal intervals are marked off along each of the axes, and the **positive integers** *(the numbers 1, 2, 3, . . . , which represent whole quantities)* are placed at these positions to the right of the origin. **Zero** *(which is an integer, but is neither positive nor negative)* is placed at the origin, and the **negative integers** are placed to the left of the origin. On the y-axis the positive integers are placed above the origin and the negative integers are placed below. *The four parts into which the plane is divided are called* **quadrants** (see Fig. 1–1).

The points between the integers represent either **rational numbers** [*numbers which can be represented by the division of one integer by another (not zero)*] or **irrational numbers** (*those numbers which cannot be represented by the division of one integer by another; π and $\sqrt{2}$ are such numbers*). In this way all **real numbers,** *which include the rational numbers (the integers are rational) and*

Figure 1–1   Figure 1–2

irrational numbers, may be represented as points on the axes. We shall use real numbers throughout this text unless otherwise noted. We will occasionally refer to and use **imaginary numbers,** *which is the name given to square roots of negative numbers.*

Example A

The number 7 is an integer, rational (since $7 = \frac{7}{1}$), and real (the real numbers include all the rational numbers); 3π is irrational and real; $\sqrt{5}$ is irrational and real; $\frac{1}{8}$ is rational and real; $7\sqrt{-1}$ is imaginary; $\frac{6}{3}$ is rational and real (it is an integer when we use the symbol ''2'' to represent it); $\pi/6$ is irrational and real; $\sqrt{-3}/2$ is imaginary.

A point P in the plane is designated by the pair of numbers (x, y). *The x-value, called the* **abscissa,** *is the perpendicular distance of P from the y-axis. The y-value, called the* **ordinate,** *is the perpendicular distance of P from the x-axis. The values x and y together, written as (x, y) are the* **coordinates** *of P.*

Example B

The positions of points $P(4, 5)$, $Q(-2, 3)$, $R(-1, -5)$, $S(4, -2)$ and $T(0, 3)$ are shown in Fig. 1–2. Note that this representation allows for *one point for any pair of values* (x, y).

Example C

Where are all the points whose ordinates are 2? All such points are two units above the x-axis; thus the answer can be stated as ''on a line 2 units above the x-axis.'' Note the dashed line in Fig. 1–2.

Another important mathematical concept we use in dealing with numbers is the **absolute value** of a number. By definition, *the absolute value of a positive number is the number itself, and the absolute value of a negative number is the corresponding positive number (obtained by changing its sign).* We may interpret the absolute value as being the number of units a given number is from the origin, regardless of direction. The absolute value is designated by $|\ |$ placed around the number.

Example D

The absolute value of 6 is 6, and the absolute value of -7 is 7. We designate this by $|6| = 6$ and $|-7| = 7$.

Other examples are $|-\pi| = \pi$, $|\frac{7}{5}| = \frac{7}{5}$, and $|-\sqrt{2}| = \sqrt{2}$.

On the x-axis, *if a first number is to the right of a second number, the first number is said to be* **greater than** *the second. If the first number is to the left of the second, it is* **less than** *the second.* On the y-axis, if one number is above a second, the first number is greater than the second. "Greater than" is designated by $>$, and "less than" is designated by $<$. *These are called* **signs of inequality.**

Example E $6 > 3, \quad 8 > -1, \quad 5 < 9, \quad 0 > -4, \quad -2 > -4, \quad -1 < 0$

Exercises 1–1

In Exercises 1 and 2, designate the given numbers as integers, rational, irrational, real, or imaginary. (More than one designation may be correct.)

1. (a) $\dfrac{5}{4}$ (b) $-\pi$ (c) $\sqrt{-4}$ (d) $6\sqrt{-1}$

2. (a) 3 (b) $6 + \sqrt{2}$ (c) $\dfrac{1}{3}\sqrt{7}$ (d) $\dfrac{7}{3}$

In Exercises 3 and 4, find the absolute values of the given numbers.

3. (a) 3 (b) $\dfrac{7}{2}$ (c) -4 4. (a) $-\left(\dfrac{\pi}{18}\right)$ (b) x (if $x < 0$)

In Exercises 5 and 6, insert the correct inequality signs ($>$ or $<$) between the numbers of the given pairs of numbers.

5. (a) $6 \quad\quad 8$ (b) $7 \quad\quad -5$ (c) $\pi \quad\quad -1$
6. (a) $-4 \quad\quad -3$ (b) $-\sqrt{2} \quad\quad -9$ (c) $0.2 \quad\quad 0.6$

In Exercises 7 through 10, plot the given points.

7. $A(2, 7), B(-1, -2), C(-4, 2)$ 8. $A\left(3, \dfrac{1}{2}\right), B(-6, 0), C\left(-\dfrac{5}{2}, -5\right)$

9. $A(1, 3), B(-1, -1), C(6, 13)$ Join these points by straight-line segments. What conclusion can you reach?

10. $A(-3, -2), B(-3, 5), C(3, 5)$ Join these points by straight-line segments. What conclusion can you reach?

In Exercises 11 through 24, locate the indicated points, and answer the questions.

11. Where are all the points whose abscissas are 1?

12. Where are all the points whose ordinates are -3?

13. What is the abscissa of all points on the y-axis?

14. What is the ordinate of all points on the x-axis?

15. Where are all the points (x, y) for which $x > 0$?

16. Where are all the points for which $y < 0$?

17. Locate all points for which $|y| > 2$.

18. Locate all points (x, y) for which $|x| < 1$.

19. Locate all points (x, y) for which $x < 0$ and $y > 1$.

20. Locate all points (x, y) for which $x > 0$ and $y < 0$.

21. In which quadrants is the ratio y/x positive?

22. In which quadrants is the ratio y/x negative?

23. Three vertices of a rectangle are $(5, 2)$, $(-1, 2)$, and $(-1, 4)$. What are the coordinates of the fourth vertex?

24. Two vertices of an equilateral triangle are $(7, 1)$ and $(2, 1)$. What is the abscissa of the third vertex?

In Exercises 25 and 26, determine the indicated types of numbers.

25. If a and b are positive integers and $b > a$, what type of number is represented by (a) $b - a$, (b) $a - b$, (c) $\dfrac{b - a}{b + a}$?

26. If a and b represent positive integers, what type of number is represented by (a) $a + b$, (b) a/b, (c) $a \times b$?

In Exercises 27 and 28, shade in the appropriate regions for the given inequalities. See Appendix B for the use of units of measurement, and the symbols used to represent them.

27. After 2 s, the current in a certain circuit is less than 3 A. Using t to represent time and i to represent current, this statement may be written as "for $t > 2$ s, $i < 3$ A." Using the x-axis for values of t and the y-axis for values of i, shade in the region for the above statement, assuming that $i > 0$.

28. Less than 3 ft from a light source the illuminance is greater than 8 lx (lux). Using x for the distance and y for the illuminance, shade in the appropriate region.

1–2 The Graph of an Equation

The *graph of an equation is the set of all points whose coordinates (x, y) satisfy the equation relating x and y.* In order to find sets of coordinates for the purpose of constructing the graph, we assume certain values of one of the variables, usually x, and then determine the values of the other variable by using the equation.

Since there is no limit to the possible number of points which can be chosen, we normally select a few values of x, obtain the corresponding values of y, and plot these points. The points are then connected by a *smooth* curve (not short straight lines from one point to the next), normally from left to right.

Example A

Graph the equation $y = 3x - 5$.

We now let x take on various values and determine the corresponding values of y. Note that once we choose a given value of x, we have no choice about the corresponding y-value, as it is determined by the equation. If $x = 0$ we find that $y = -5$. This means that the point $(0, -5)$ is on the graph of the equation $y = 3x - 5$. Choosing another value of x, for example, 1, we find that $y = -2$. This means that the point $(1, -2)$ is on the graph of the equation. Continuing to choose a few other values of x, we tabulate the results, as shown in Fig. 1–3. It is best to arrange the table so that the values of x increase; then there is no doubt how they are to be connected, for they are connected in the order shown. Finally, we connect the points as shown in Fig. 1–3, and see that the graph of the equation $y = 3x - 5$ is a straight line.

x	y
-1	-8
0	-5
1	-2
2	1
3	4
4	7

Figure 1–3

x	y
-2	-6
-1	-2
0	0
1	0
2	-2
3	-6

Figure 1–4

There are some special points with regard to certain curves which should be noted. Each of the next four examples illustrates and explains a particular point to be considered when plotting the graph of an equation.

Example B

Graph the equation $y = x - x^2$.

First we determine the values in the table as shown with Fig. 1–4. We must be careful with negative values of x. For $x = -1$, we have $y = (-1) - (-1)^2 = -1 - (+1) = -1 - 1 = -2$. Once all the values have been found and plotted, we note that $y = 0$ for both $x = 0$ and $x = 1$. The question arises: What happens between these values? Trying $x = \frac{1}{2}$, we find that $y = \frac{1}{4}$. Using this point completes the necessary information. Note that in plotting these graphs we do not stop the graph with the last point determined, but indicate that the curve continues.

Example C

Graph the equation $y = 2x^2 - 4$.

    We first tabulate the values as shown. In constructing the table we note that the same value of y appears for both $+3$ and -3. Also the same value of y appears for $+2$ and for -2, and again for $+1$ and for -1. Since $(-x)^2 = x^2$ and $(+x)^2 = x^2$ we can understand why this happens. The graph of this equation (see Fig. 1–5) is seen to be **symmetrical** to the y-axis. This can be thought of as meaning that the right half of the curve is the reflection of the left half, and conversely. As in this case, *any time $-x$ can replace x in an equation without changing the equation, its graph will be symmetrical to the y-axis.* In the same way, *if $-y$ can replace y in an equation without changing the equation, its graph is symmetrical to the x-axis.* Many of the curves to be discussed in the coming sections will demonstrate the property of symmetry.

Example D

Graph the equation $y = 1 + \dfrac{1}{x}$.

    In finding points on this graph (see Fig. 1–6) we note that y is not defined if we try $x = 0$, since *division by zero is not defined*. Although x cannot *equal* zero, it can take on any value near zero. Thus, by choosing values of x between -1 and $+1$ we obtain the points necessary to determine the shape of the graph.

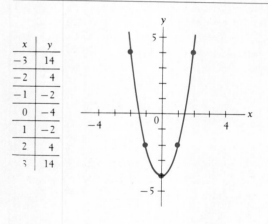

x	y
-3	14
-2	4
-1	-2
0	-4
1	-2
2	4
3	14

Figure 1–5

x	y
-4	$3/4$
-3	$2/3$
-2	$1/2$
-1	0
$-1/2$	-1
$-1/3$	-2
$1/3$	4
$1/2$	3
1	2
2	$3/2$
3	$4/3$
4	$5/4$

Figure 1–6

In examining the possible values in example D, we note that as the values of x get closer and closer to zero the values of y become numerically larger and larger. That is, as x *approaches* zero, y becomes numerically unbounded. In this case we say that $x = 0$ (the y-axis) is an **asymptote** of the curve. More generally, *an asymptote is a line which the graph of an equation approaches*. We can find the asymptotes by determining those values of x which lead to division by zero. In this way we determine those asymptotes which are vertical, although we must realize that the graph of any given equation may have asymptotes which are horizontal or oblique. We shall not consider these other types at this time. Asymptotes are important in a number of graphs. In trigonometry the graph of $y = \tan x$ has vertical asymptotes for $x = \frac{1}{2}\pi, \frac{3}{2}\pi, \ldots$, and we shall find asymptotes of importance when discussing the hyperbola.

Example E

Graph the equation $y = \sqrt{x + 1}$.

When finding the points for the graph, we may not let x equal any negative value less than -1. All such values lead to imaginary values for y, and *only real values of the variables are permissible*. Also, *since we have indicated the positive square root, all values of y are positive or zero* (see Fig 1–7). Note that the graph starts at the point $(-1, 0)$.

Graphs and graphical methods will be used throughout this text. It will be seen that a great deal can be learned about an equation by analyzing its graph. Also, graphs can often demonstrate a given property more clearly than can other descriptions. The following example illustrates an application of graphing an equation, along with conclusions that can easily be seen from the graph.

x	y
−1	0
0	1
1	1.4
2	1.7
3	2
4	2.2
5	2.4
6	2.6
7	2.8
8	3

Figure 1–7

Figure 1–8

Example F

The power of a certain voltage source in terms of the load resistance is given by

$$P = \frac{100R}{(0.5 + R)^2}$$

where P is measured in watts (W) and R in ohms (Ω). Plot the graph, using the y-direction for power and the x-direction for resistance.

Since a negative value for the resistance has no physical significance, we need not plot P for negative values of R. The following table of values is obtained.

R (ohms)	0.0	0.25	0.50	1.0	2.0	3.0	4.0	5.0	10.0
P (watts)	0.0	44.4	50.0	44.4	32.0	24.5	19.8	16.5	9.1

The values 0.25 and 0.50 are used for R when it is found that P is less for $R = 2$ than for $R = 1$. In this way a smoother curve is obtained (see Fig. 1–8).

From the graph we see that P changes quite rapidly for small values of R, with a peak value of about 50 W when $R = 0.50\ \Omega$ (actually this is exact for this equation). Also, as R becomes larger, P becomes smaller, but much less rapidly. This shows some of the conclusions that can be drawn from a graph.

Also note that the scale on the P-axis is different from that on the R-axis. This reflects the different magnitudes and ranges of values used for each of the variables. Different scales are normally used in such cases.

Equations of a particular type have graphs which are of a specific form, and many of them have been named. We noted that the graph of the function in Example A is a **straight line. Parabolas** are illustrated in Examples B and C. The graph of the function in Example D is a **hyperbola.** Other types of graphs are found in the exercises, and a more detailed analysis of several of these curves is found later in this chapter. The use of graphs is extensive in mathematics and in nearly all areas of application.

Exercises 1–2

In Exercises 1 through 32, graph the given equations.

1. $y = 3x$

2. $y = -2x$

3. $y = 2x - 4$

4. $y = 3x + 5$

5. $y = 7 - 2x$

6. $y = 5 - 3x$

7. $y = \frac{1}{2}x - 2$

8. $y = 6 - \frac{1}{3}x$

9. $y = x^2$

10. $y = -2x^2$

11. $y = 3 - x^2$

12. $y = x^2 - 3$

13. $y = \frac{1}{2}x^2 + 2$

14. $y = 2x^2 + 1$

15. $y = x^2 + 2x$

16. $y = 2x - x^2$ 17. $y = x^2 - 3x + 1$ 18. $y = 2 + 3x + x^2$

19. $y = x^3$ 20. $y = -2x^3$ 21. $y = x^3 - x^2$

22. $y = 3x - x^3$ 23. $y = x^4 - 4x^2$ 24. $y = x^3 - x^4$

25. $y = \dfrac{1}{x}$ 26. $y = \dfrac{1}{x + 2}$ 27. $y = \dfrac{1}{x^2}$

28. $y = \dfrac{1}{x^2 + 1}$ 29. $y = \sqrt{x}$ 30. $y = \sqrt{4 - x}$

31. $y = \sqrt{16 - x^2}$ 32. $y = \sqrt{x^2 - 16}$

In Exercises 33 and 34, determine whether or not the graphs of the equations of the indicated exercises have symmetry to the x-axis or to the y-axis.

33. (a) Exercise 12   (b) Exercise 16   (c) Exercise 29

34. (a) Exercise 14   (b) Exercise 17   (c) Exercise 32

In Exercises 35 and 36, determine whether or not the graphs of the equations of the indicated exercises have vertical asymptotes.

35. (a) Exercise 19   (b) Exercise 25   (c) Exercise 28

36. (a) Exercise 22   (b) Exercise 27   (c) Exercise 26

In Exercises 37 through 50, graph the given equations. In each case plot the first mentioned variable as the ordinate and the second variable as the abscissa.

37. The velocity v (in feet per second) of an object under the influence of gravity, in terms of the time t (in seconds), is given by $v = 100 - 32t$. If the object strikes the ground after 4 s, graph v vs. t.

38. If \$1000 is placed in an account earning 6% simple interest, the amount A in the account after t years is given by the equation $A = 1000(1 + 0.06t)$. If the money is withdrawn from the account after 6 years, plot A vs. t.

39. The consumption c of fuel, in liters per hour (L/h), of a certain engine is determined in terms of the rotation rate r, in revolutions per minute (r/min), of the engine to be $c = 0.011r + 4.0$. This formula is valid from 500 r/min to 3000 r/min. Plot c vs. r.

40. The heat capacity (in joules per kilogram) of an organic liquid is related to the temperature (in degrees Celsius) by the equation $c_p = 2320 + 4.73T$ for the temperature range of $-40°C$ to $120°C$. Graphically show that this equation does not satisfy experimental data above $120°C$. Plot the graph of the equation and the following data points.

c_p	3030	3220	3350
T	140	160	200

41. The maximum speed v, in miles per hour, at which a car can safely travel around a circular turn of radius r, in feet, is given by $r = 0.42v^2$. Plot r vs. v.

42. The energy in the electric field around an inductor is given by $E = \frac{1}{2}LI^2$, where I is the current in the inductor and L is the inductance. Plot E vs. I (a) if $L = 1$ unit, (b) if $L = 0.1$ unit, (c) if the E-axis is marked off in units of L.

43. If a rectangular tract of land has a perimeter of 600 m, its area in terms of its width is $A = 300w - w^2$. Plot A vs. w.

44. A formula used to determine the number N of board feet of lumber which can be cut from a 4-ft section of a log of diameter d, in inches, is $N = 0.22d^2 - 0.71d$. Plot N vs. d for values of d from 10 in. to 40 in.

45. The illuminance (in lumens per square meter) of a certain source of light in terms of the distance (in meters) from the source is given by $I = 400/r^2$. Plot I vs. r.

46. The value V, in dollars, of a certain model of a car depreciates according to its age t, in years, according to the formula $V = \dfrac{12{,}000}{t + 1}$. Plot V vs. t for $t < 9$.

47. The total profit P, in dollars, a manufacturer makes in producing x units of a certain product is given by $P = 4x^3 - 18x^2 - 120x - 200$. Plot P vs. x for values to $x = 10$.

48. The deflection y of a beam at a horizontal distance x from one end is given by $y = -k(x^4 - 30x^2 + 1000x)$, where k is a constant. Plot the deflection (in terms of k) vs. x (in feet) if there are 10 ft between the end supports of the beam.

49. Plot the graphs of $y = x$ and $y = |x|$ on the same coordinate system. Note how the graphs differ.

50. Plot the graphs of $y = 2 - x$ and $y = |2 - x|$ on the same coordinate system. Note how the graphs differ.

1–3 Some Basic Definitions

We earlier noted that analytic geometry deals with the relationship between algebra and geometry. In the sections which follow we will see that proper recognition of certain important equations will lead immediately to the graph of that equation. In this section we will develop certain basic concepts which are needed to establish the proper relationships between equation and curve.

The first of these concepts involves the distance between any two points in the coordinate plane. If these two points lie on a line parallel to the x-axis, the **directed distance** from the first point $A(x_1, y)$ to the second point $B(x_2, y)$ is denoted as $\overline{AB}$ and is defined as $x_2 - x_1$. We can see that $\overline{AB}$ *is positive if B is to the right of A, and it is negative if B is to the left of A.* Similarly, the directed distance between two points $C(x, y_1)$ and $D(x, y_2)$ on a line parallel to the y-axis is $y_2 - y_1$. We see that $\overline{CD}$ *is positive if D is above C, and it is negative if D is below C.*

Example A

Figure 1–9

The line segment joining $A(-1, 5)$ and $B(-4, 5)$ in Fig. 1–9 is parallel to the x-axis. Therefore, the directed distance

$$\overline{AB} = -4 - (-1) = -3$$

and the directed distance

$$\overline{BA} = -1 - (-4) = 3$$

Also in Fig. 1–9, the line segment joining $C(2, -3)$ and $D(2, 6)$ is parallel to the y-axis. The directed distance

$$\overline{CD} = 6 - (-3) = 9$$

and the directed distance

$$\overline{DC} = -3 - 6 = -9$$

Figure 1–10

We now wish to find the length of a line segment joining any two points in the plane. If these points are on a line which is not parallel to either of the axes (Fig. 1–10), we must use the Pythagorean theorem to find the distance between them. By making a right triangle with the line segment joining the two points as the hypotenuse, and line segments parallel to the axes as the legs, we have the formula which gives the distance between any two points in the plane. This formula, called *the* **distance formula,** *is*

$$d = \sqrt{(x_2 - x_1)^2 + (y_2 - y_1)^2} \qquad (1\text{–}1)$$

Here we choose the positive square root since we are concerned only with the magnitude of the length of the line segment.

Example B

The distance between $(3, -1)$ and $(-2, -5)$ is given by

$$d = \sqrt{[(-2) - 3]^2 + [(-5) - (-1)]^2}$$
$$= \sqrt{(-5)^2 + (-4)^2} = \sqrt{25 + 16} = \sqrt{41}$$

It makes no difference which point is chosen as (x_1, y_1) and which is chosen as (x_2, y_2), since the difference in the x-coordinates (and y-coordinates) is squared. We also obtain $\sqrt{41}$ if we set up the distance as

$$d = \sqrt{[3 - (-2)]^2 + [(-1) - (-5)]^2}$$

Another important quantity which is defined for a line is its **slope.** *The slope gives a measure of the direction of a line, and is defined as the vertical directed distance from one point to another on the same straight line, divided by the horizontal directed distance from the first point to the second.* Thus, using the letter *m* to represent slope, we have

$$m = \frac{y_2 - y_1}{x_2 - x_1}$$

(1–2)

Example C

Figure 1–11

The slope of the line joining $(3, -5)$ and $(-2, -6)$ is

$$m = \frac{-6 - (-5)}{-2 - 3} = \frac{-6 + 5}{-5} = \frac{1}{5}$$

See Fig. 1–11. Again we may interpret either of the points as (x_1, y_1) and the other as (x_2, y_2). We can also obtain the slope of this same line from

$$m = \frac{-5 - (-6)}{3 - (-2)} = \frac{1}{5}$$

The larger the numerical value of the slope of a line, the more nearly vertical is the line. Also, *a line rising to the right has a positive slope, and a line falling to the right has a negative slope.*

Example D

Figure 1–12

The line through the two points in Example C has a positive slope, which is numerically small. From Fig. 1–11 it can be seen that the line rises slightly to the right.

The line joining $(3, 4)$ and $(4, -6)$ has a slope of

$$m = \frac{4 - (-6)}{3 - 4} = -10$$

This line falls sharply to the right (see Fig. 1–12).

Figure 1–13

From the definition of slope, we may conclude that the slope of a line parallel to the y-axis cannot be defined (the x-coordinates of any two points on the line would be the same, which would then necessitate division by zero). This, however, does not prove to be of any trouble.

If a given line is extended indefinitely in either direction, it must cross the x-axis at some point unless it is parallel to the x-axis. *The angle measured from the x-axis in a positive direction to the line is called the* **inclination** *of the line* (see Fig. 1–13). The inclination of a line parallel to the x-axis is defined to be zero. *An alternative definition of slope, in terms of the inclination, is*

$$m = \tan \alpha, \qquad 0° \le \alpha < 180° \tag{1–3}$$

where α is the inclination. This can be seen from the fact that the slope can be defined in terms of any two points on the line. Thus, if we choose as one of these points that point where the line crosses the x-axis and any other point, we see from the definition of the tangent of an angle that Eq. (1–3) is in agreement with Eq. (1–2).

Example E

Figure 1–14

The slope of a line with an inclination of 45° is

$$m = \tan 45° = 1.000$$

The slope of a line having an inclination of 120° is

$$m = \tan 120° = -\sqrt{3} = -1.732$$

See Fig. 1–14.

Figure 1–15

Any two parallel lines crossing the x-axis will have the same inclination. Therefore, *the slopes of parallel lines are equal.* This can be stated as

$$m_1 = m_2 \qquad \text{for} \parallel \text{lines} \tag{1–4}$$

If two lines are perpendicular, this means that there must be 90° between their inclinations (Fig. 1–15). The relation between their inclinations is

$$\alpha_2 = \alpha_1 + 90°$$

which can be written as

$$90° - \alpha_2 = -\alpha_1$$

Taking the tangent of each of the angles in this last relation, we have

$$\tan(90° - \alpha_2) = \tan(-\alpha_1)$$

or

$$\cot \alpha_2 = -\tan \alpha_1$$

since a function of the complement of an angle is equal to the cofunction of that angle, and since $\tan(-\alpha) = -\tan \alpha$. But $\cot \alpha = 1/\tan \alpha$, which means that $1/\tan \alpha_2 = -\tan \alpha_1$. Using the inclination definition of slope, we may write, as *the relation between slopes of perpendicular lines.*

$$m_2 = -\frac{1}{m_1} \quad \text{or} \quad m_1 m_2 = -1 \qquad \text{for } \perp \text{ lines} \qquad (1\text{-}5)$$

Example F

Figure 1–16

The line through $(3, -5)$ and $(2, -7)$ has a slope of

$$m_1 = \frac{-5 + 7}{3 - 2} = 2$$

The line through $(4, -6)$ and $(2, -5)$ has a slope of

$$m_2 = \frac{-6 - (-5)}{4 - 2} = -\frac{1}{2}$$

These lines are perpendicular. See Fig. 1–16.

Using the formulas for distance and slope, we can show certain basic geometric relations. The following examples illustrate the use of the formulas, and thus show the use of algebra in solving problems which are basically geometric. This illustrates the methods of analytic geometry.

Example G

Figure 1–17

Show that line segments joining $A(-5, 3)$, $B(6, 0)$, and $C(5, 5)$ form a right triangle (see Fig. 1–17).

If these points are vertices of a right triangle, the slopes of two of the sides must be negative reciprocals. This would show perpendicularity. Thus, we find the slopes of the three lines to be

$$m_{AB} = \frac{3 - 0}{-5 - 6} = -\frac{3}{11}, \qquad m_{AC} = \frac{3 - 5}{-5 - 5} = \frac{1}{5}, \qquad m_{BC} = \frac{0 - 5}{6 - 5} = -5$$

We see that the slopes of AC and BC are negative reciprocals, which means that $AC \perp BC$. From this we can conclude that the triangle is a right triangle.

Example H

Figure 1–17

Find the area of the triangle in Example G.

Since the right angle is at C, the legs of the triangle are AC and BC. The area is one-half the product of the lengths of the legs of a right triangle. The lengths of the legs are

$$d_{AC} = \sqrt{(-5 - 5)^2 + (3 - 5)^2} = \sqrt{104} = 2\sqrt{26}$$

and

$$d_{BC} = \sqrt{(6 - 5)^2 + (0 - 5)^2} = \sqrt{26}$$

Therefore, the area is

$$A = \frac{1}{2}(2\sqrt{26})(\sqrt{26}) = 26$$

Exercises 1–3

In Exercises 1 through 10, find the distances between the given pairs of points.

1. $(3, 8)$ and $(-1, -2)$
2. $(-1, 3)$ and $(-8, -4)$
3. $(4, -5)$ and $(4, -8)$
4. $(-3, 7)$ and $(2, 10)$
5. $(-1, 0)$ and $(5, -7)$
6. $(15, -1)$ and $(-11, 1)$
7. $(-4, -3)$ and $(3, -3)$
8. $(-2, 5)$ and $(-2, -2)$
9. $(1.22, -3.45)$ and $(-1.07, -5.16)$
10. $(-5.6, 2.3)$ and $(8.2, -7.5)$

In Exercises 11 through 20, find the slopes of the lines through the points in Exercises 1 through 10.

In Exercises 21 through 24, find the slopes of the lines with the given inclinations.

21. $30°$
22. $60°$
23. $150°$
24. $135°$

In Exercises 25 through 28, find the inclinations of the lines with the given slopes.

25. 0.3640
26. 0.8243
27. -6.691
28. -1.428

In Exercises 29 through 32, determine whether the lines through the two pairs of points are parallel or perpendicular.

29. $(6, -1)$ and $(4, 3)$, and $(-5, 2)$ and $(-7, 6)$
30. $(-3, 9)$ and $(4, 4)$, and $(9, -1)$ and $(4, -8)$
31. $(-1, -4)$ and $(2, 3)$, and $(-5, 2)$ and $(-19, 8)$
32. $(-1, -2)$ and $(3, 6)$, and $(2, -6)$ and $(5, 0)$

In Exercises 33 through 36, determine the value of k.

33. The distance between $(-1, 3)$ and $(11, k)$ is 13.
34. The distance between $(k, 0)$ and $(0, 2k)$ is 10.
35. The points $(6, -1)$, $(3, k)$, and $(-3, -7)$ are all on the same line.
36. The points in Exercise 35 are the vertices of a right triangle, with the right angle at $(3, k)$.

In Exercises 37 through 40, show that the given points are vertices of the indicated geometric figures.

37. Show that the points (2, 3), (4, 9), and $(-2, 7)$ are the vertices of an isosceles triangle.
38. Show that $(-1, 3)$, (3, 5), and (5, 1) are the vertices of a right triangle.
39. Show that (3, 2), (7, 3), $(-1, -3)$, and $(3, -2)$ are the vertices of a parallelogram.
40. Show that $(-5, 6)$, (0, 8), $(-3, 1)$, and (2, 3) are the vertices of a square.

In Exercises 41 and 42, find the indicated areas.

41. Find the area of the triangle of Exercise 38.
42. Find the area of the square of Exercise 40.

In Exercises 43 and 44, use the following information to find the midpoints between the given points on a straight line.

The *midpoint* between points (x_1, y_1) and (x_2, y_2) on a straight line is the point

$$\left(\frac{x_1 + x_2}{2}, \frac{y_1 + y_2}{2} \right)$$

43. $(-4, 9)$ and (6, 1)   44. $(-1, 6)$ and $(-13, -8)$

1–4 The Straight Line

Figure 1–18

Using the definition of slope, we can derive the general type of equation which always represents a straight line. This is another basic method of analytic geometry. That is, equations of a particular form can be shown to represent a particular type of curve. When we recognize the form of the equation, we know the kind of curve it represents. This can be of great assistance in sketching the graph.

A straight line can be defined as a "curve" with constant slope. By this we mean that for any two different points chosen on a given line, if the slope is calculated, the same value is always found. Thus, if we consider one point (x_1, y_1) on a line to be fixed (Fig. 1–18), and another point $P(x, y)$ which can *represent* any other point on the line, we have

$$m = \frac{y - y_1}{x - x_1}$$

which can be written as

$$\underline{\hspace{3cm}}$$
$$y - y_1 = m(x - x_1) \qquad\qquad (1\text{–}6)$$

Equation (1–6) is known as the **point-slope form** *of the equation of a straight line.* It is useful when we know the slope of a line and some point through which the line passes. Direct substitutions can then give us the equation of the line. Such information is often available about a given line.

Example A

Find the equation of the line which passes through $(-4, 1)$ with a slope of -2.
    By using Eq. (1–6), we find that

$$y - 1 = (-2)(x + 4)$$

which can be simplified to

$$y + 2x + 7 = 0$$

This line is shown in Fig. 1–19.

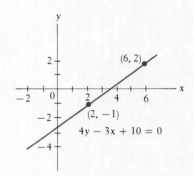

$y + 2x + 7 = 0$

$4y - 3x + 10 = 0$

Figure 1–19   **Figure 1–20**

Example B

Find the equation of the line through $(2, -1)$ and $(6, 2)$.
    We first find the slope of the line through these points:

$$m = \frac{2 + 1}{6 - 2} = \frac{3}{4}$$

Then, by using either of the two known points and Eq. (1–6), we can find the equation of the line:

$$y + 1 = \frac{3}{4}(x - 2)$$

or

$$4y + 4 = 3x - 6$$

or

$$4y - 3x + 10 = 0$$

This line is shown in Fig. 1–20.

Equation (1–6) can be used for any line except one parallel to the y-axis. Such a line has an undefined slope. However, it does have the property that all points have the same x-coordinate, regardless of the y-coordinate. *We represent a line parallel to the y-axis as*

$$x = a \qquad (1-7)$$

A line parallel to the x-axis has a slope of 0. From Eq. (1–6), we can find its equation to be $y = y_1$. To keep the same form as Eq. (1–7), we normally write this as

$$y = b \qquad (1-8)$$

Example C

The line $x = 2$ is a line parallel to the y-axis and two units to the right of it. This line is shown in Fig. 1–21.

Figure 1–21   Figure 1–22

Example D

The line $y = -4$ is a line parallel to the x-axis and 4 units below it. This line is shown in Fig. 1–22.

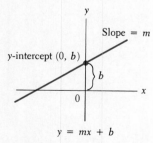

$y = mx + b$

Figure 1–23

If we choose the special point $(0, b)$, which is the y-intercept of the line, as the point to use in Eq. (1–6), we have

$$y - b = m(x - 0)$$

or

$$y = mx + b \qquad (1-9)$$

Equation (1–9) is the **slope-intercept form** *of the equation of a straight line.* Its primary usefulness lies in the fact that once we find the equation of a line and then write it in slope-intercept form, we know that the slope of the line is the coefficient of the x-term and that it crosses the y-axis with the coordinate indicated by the constant term. See Fig. 1–23.

Example E

Figure 1–24

Find the slope and the y-intercept of the straight line whose equation is $2y + 4x - 5 = 0$.

We write this equation in slope-intercept form:

$$2y = -4x + 5$$

$$y = -2x + \frac{5}{2}$$

Since the coefficient of x in this form is -2, the slope of the line is -2. The constant on the right is $\frac{5}{2}$, which means that the y-intercept is $\frac{5}{2}$. See Fig. 1–24.

From Eqs. (1–6) and (1–9), and from the examples of this section, we see that the equation of the straight line has certain characteristics: we have a term in y, a term in x, and a constant term if we simplify as much as possible. *This form is represented by the equation*

$$Ax + By + C = 0 \qquad\qquad (1\text{–}10)$$

which is known as the **general form** *of the equation of the straight line.* We have seen this form before in Section 1–2. Now we have shown why it represents a straight line.

Example F

Find the general form of the equation of the line which is parallel to the line $3x + 2y - 6 = 0$ and which passes through the point $(-1, 2)$.

Since the line whose equation we want is parallel to the line $3x + 2y - 6 = 0$, it must have the same slope. Therefore, writing $3x + 2y - 6 = 0$ in slope-intercept form, we have

$$3x + 2y - 6 = 0$$

$$2y = -3x + 6$$

$$y = -\frac{3}{2}x + 3$$

Therefore, the slope of $3x + 2y - 6 = 0$ is $-\frac{3}{2}$, which means that the slope of the required line is also $-\frac{3}{2}$. We use $m = -\frac{3}{2}$, the given point $(-1, 2)$, and the point-slope form to find the equation.

$$y - 2 = -\frac{3}{2}(x + 1)$$

$$2y - 4 = -3(x + 1)$$

$$2y - 4 = -3x - 3$$

$$3x + 2y - 1 = 0$$

Figure 1–25

This last form is the general form of the equation of the line. This line and the line $3x + 2y - 6 = 0$ are shown in Fig. 1–25.

In many physical situations a linear relationship exists between variables. A few examples of situations where such relationships exist are between (1) the distance traveled by an object and the elapsed time, when the velocity is constant, (2) the amount a spring stretches and the force applied, (3) the change in electric resistance and the change in temperature, (4) the force applied to an object and the resulting acceleration, and (5) the pressure at a certain point within a liquid and the depth of the point. The following example illustrates the use of a straight line in dealing with an applied problem.

Example G

Under the condition of constant acceleration, the velocity of an object varies linearly with the time. If after 1 s a certain object has a velocity of 40 ft/s, and 3 s later it has a velocity of 55 ft/s, find the equation relating the velocity and time, and graph this equation. From the graph determine the initial velocity (the velocity when $t = 0$) and the velocity after 6 s.

If we use v for velocity and t for time, the slope of the straight line is

$$m = \frac{v_2 - v_1}{t_2 - t_1}$$

Using the information given in the problem, we have

$$m = \frac{55 - 40}{4 - 1} = 5$$

Then, using the point-slope form of the equation of a straight line, we have $v - 40 = 5(t - 1)$, or $v = 5t + 35$, which is the required equation (see Fig. 1–26). For purposes of graphing the line, the values given are sufficient. Of course, there is no need to include negative values of t, since these have no physical meaning. From the graph we see that the line crosses the v-axis at 35. This means that the initial velocity is 35 ft/s. Also, when $t = 6$ we see that $v = 65$ ft/s.

Figure 1–26

Exercises 1–4

In Exercises 1 through 20, find the equation of each of the lines with the given properties.

1. Passes through $(-3, 8)$ with a slope of 4.
2. Passes through $(-2, -1)$ with a slope of -2.
3. Passes through $(-2, -5)$ and $(4, 2)$.
4. Passes through $(-3, -5)$ and $(-2, 3)$.
5. Passes through $(1, 3)$ and has an inclination of $45°$.
6. Has a y-intercept of -2 and an inclination of $120°$.
7. Passes through $(6, -3)$ and is parallel to the x-axis.
8. Passes through $(-4, -2)$ and is perpendicular to the x-axis.
9. Is parallel to the y-axis and is 3 units to the left of it.
10. Is parallel to the x-axis and is 5 units below it.

11. Has an x-intercept of 4 and a y-intercept of -6.

12. Has an x-intercept of -3 and a slope of 2.

13. Is perpendicular to a line with a slope of 3 and passes through $(1, -2)$.

14. Is perpendicular to a line with a slope of -4 and has a y-intercept of 3.

15. Is parallel to a line with a slope of $-\frac{1}{2}$ and has an x-intercept of 4.

16. Is parallel to a line through $(-1, 2)$ and $(3, 1)$ and passes through $(1, 2)$.

17. Is parallel to a line through $(7, -1)$ and $(4, 3)$ and has a y-intercept of -2.

18. Is perpendicular to the line joining $(4, 2)$ and $(3, -5)$ and passes through $(4, 2)$.

19. Is perpendicular to the line $5x - 2y - 3 = 0$ and passes through $(3, -4)$.

20. Is parallel to the line $2y - 6x - 5 = 0$ and passes through $(-4, -5)$.

In Exercises 21 through 24, draw the lines with the given equations.

See Appendix C for a
computer program for
graphing a straight line.

21. $4x - y = 8$ 22. $2x - 3y - 6 = 0$

23. $3x + 5y - 10 = 0$ 24. $4y = 6x - 9$

In Exercises 25 through 28, reduce the equations to slope-intercept form and determine the slope and y-intercept.

25. $3x - 2y - 1 = 0$ 26. $4x + 2y - 5 = 0$

27. $5x - 2y + 5 = 0$ 28. $6x - 3y - 4 = 0$

In Exercises 29 through 32, determine the value of k.

29. What is the value of k if the lines $4x - ky = 6$ and $6x + 3y + 2 = 0$ are to be parallel?

30. What must k equal in Exercise 29 if the given lines are to be perpendicular?

31. What must be the value of k if the lines $3x - y = 9$ and $kx + 3y = 5$ are to be perpendicular?

32. What must k equal in Exercise 31 if the given lines are to be parallel?

In Exercises 33 and 34, show that the given lines are parallel. In Exercises 35 and 36, show that the given lines are perpendicular.

33. $3x - 2y + 5 = 0$ and $4y = 6x - 1$

34. $3y - 2x = 4$ and $6x - 9y = 5$

35. $6x - 3y - 2 = 0$ and $x + 2y - 4 = 0$

36. $4x - y + 2 = 0$ and $2x + 8y - 1 = 0$

In Exercises 37 through 40, find the equations of the given lines.

37. The line which has an x-intercept of 4 and a y-intercept the same as the line $2y - 3x - 4 = 0$.

38. The line which is perpendicular to the line $8x + 2y - 3 = 0$ and has the same x-intercept as this line.

39. The line with a slope of -3 which also passes through the intersection of the lines $5x - y = 6$ and $x + y = 12$.

40. The line which passes through the point of intersection of $2x + y - 3 = 0$ and $x - y - 3 = 0$ and through the point $(4, -3)$.

In Exercises 41 through 52, some applications and methods involving straight lines are shown.

41. The length l of a pulley belt is 8 cm longer than twice the circumference c of one of the pulley wheels. Express l in terms of c.

42. An acid solution is made from x liters of a 20% solution and y liters of a 30% solution. If the solution contains a total of 20 L of acid, find the equation relating x and y.

43. One microcomputer printer can print x characters per second and a second printer can print y characters per second. If the first prints for 50 s and the second for 60 s and they print a total of 12,200 characters, find the equation relating x and y.

44. A light ray travels (from the left) along the line $x + y = 1$ and reflects off the x-axis. Write the equation for the reflected ray.

45. The average velocity of an object is defined as the change in displacement s divided by the corresponding change in time t. Find the equation relating the displacement s and time t for an object for which the average velocity is 50 m/s and $s = 10$ m when $t = 0$ s.

46. A wall is 15 cm thick. At the outside, the temperature is 3°C and at the inside, it is 23°C. If the temperature changes at a constant rate through the wall, write an equation of the temperature T in the wall in terms of the distance x from the outside to the inside of the wall.

47. Within certain limits, the amount which a spring stretches varies linearly with the amount of force applied. If a spring whose natural length is 15 in. stretches 2.0 in. when 3.0 lb of force are applied, find the equation relating the length of the spring and the applied force.

48. The electric resistance of a certain resistor increases by 0.005 Ω for every increase of 1°C. Given that its resistance is 2.000 Ω at 0°C, find the equation relating the resistance and temperature. From the equation find the resistance when the temperature is 50°C.

49. The amount of heat required to raise the temperature of water varies linearly with the increase in temperature. However, a certain quantity of heat is required to change ice (at 0°C) into water without a change in temperature. An experiment is performed with 10 g of ice at 0°C. It is found that ice requires 4.19 kJ to change it into water at 20°C. Another 1.26 kJ is required to warm the water to 50°C. How many kilojoules are required to melt the ice at 0°C into water at 0°C?

50. The pressure at a point below the surface of a body of water varies linearly with the depth of the water. If the pressure at a depth of 10.0 m is 199 kPa and the pressure at a depth of 30.0 m is 395 kPa, what is the pressure at the surface? (This is the atmospheric pressure of the air above the water.)

51. A survey of the traffic on a particular highway showed that the number of cars passing a particular point each minute varied linearly from 6:30 A.M. to 8:30 A.M. on workday mornings. The study showed that an average of 45 cars passed the point in one minute at 7 A.M. and that 115 cars passed in one minute at 8 A.M. If n is the number of cars passing the point in one minute and t is the number of minutes after 6:30 A.M., find the equation relating n and t, and graph the equation. From the graph, determine n at 6:30 A.M. and at 8:30 A.M.

52. The total fixed cost for a company to operate a certain plant is $400 per day. It also costs $2 for each unit produced in the plant. Find the equation relating the total cost C of operating the plant and the number of units n produced each day. Graph the equation, assuming that $n \leq 300$.

x	x^2	y
0	0	2
1	1	5
2	4	14
3	9	29
4	16	50
5	25	77

Figure 1–27

In Exercises 53 through 56, treat the given nonlinear equations as linear equations in order to sketch their graphs. As an example, $y = 2 + 3x^2$ can be sketched as a straight line by graphing y in terms of x^2. A table of values for this graph is shown along with the corresponding graph in Fig. 1–27.

53. Sketch the graph of the equation $y = 1 + \sqrt{x}$ by graphing y vs. $\sqrt{x}$.

54. Using this method, sketch the equation $C = 10/E$ as a linear equation. This is the equation relating the voltage across a capacitor and the capacitance, assuming the charge on the capacitor is constant.

55. A spacecraft is launched such that its altitude h, in kilometers, is given by $h = 300 + 2t^{3/2}$ for $0 \le t < 100$ s. Sketch this as a linear equation.

56. The current i, in amperes, in a certain electric circuit is given by $i = 6(1 - e^{-t})$. Sketch this as a linear equation, using $e = 2.72$.

Another method of treating a nonlinear equation as a linear equation is to use logarithms (see Section 7–1). Using logarithmic or semilogarithmic paper to plot x or y, distances corresponding to $\log y$ or $\log x$ are plotted automatically if the scale is logarithmic in the respective direction. Therefore, the graphs of many nonlinear equations are linear when using this paper. In Exercises 57 through 60, plot the given equations on semilogarithmic or logarithmic paper as indicated.

57. An equation of the form $y = ax^n$ is a straight line when plotted on logarithmic paper, because $\log y = \log a + n \log x$ is in the form of a straight line. The variables are $\log y$ and $\log x$; the slope can be found from $(\log y - \log a)/\log x = n$, and the intercept is a. (To get the slope from the graph, it is necessary to measure vertical and horizontal distances between two points. The $\log y$ intercept is found where $\log x = 0$, and this occurs when $x = 1$.) Plot $y = 3x^4$ on logarithmic paper to verify this analysis.

58. An equation of the form $y = a(b^x)$ is a straight line on semilogarithmic paper, since $\log y = \log a + x \log b$ is in the form of a straight line. The variables are $\log y$ and x, the slope is $\log b$, and the intercept is a. [To get the slope from the graph, we must calculate $(\log y - \log a)/x$ for some set of values x and y. The intercept is read directly off the graph where $x = 0$.] Plot $y = 3(2^x)$ on semilogarithmic paper to verify this analysis.

59. If experimental data are plotted on logarithmic paper and the points lie on a straight line, it is possible to determine the equation (see Exercise 57). The following data come from an experiment to determine the relationship between the tension in a string and the velocity of a wave in the string. From the graph on logarithmic paper, determine v in terms of T.

V (meters per second)	16.0	32.0	45.3	55.4	64.0
T (newtons)	0.100	0.400	0.800	1.20	1.60

60. If experimental data are plotted on semilogarithmic paper, and the points lie on a straight line, it is possible to determine the equation (see Exercise 58). The following data come from an experiment designed to determine the relationship between the voltage across an inductor and the time, after the switch is opened. Determine v in terms of t.

v (volts)	40	15	5.6	2.2	0.8
t (milliseconds)	0.0	20	40	60	80

1–5 The Circle

We have found that we can obtain a general equation which represents a straight line by considering a fixed point on the line and then a general point $P(x, y)$ which can represent any other point on the same line. Mathematically we can state this as "the line is the **locus** of a point $P(x, y)$ which *moves* along the line." That is, the point $P(x, y)$ can be considered as a variable point which moves along the line.

In this way we can define a number of important curves. *A* **circle** *is defined as the locus of a point P(x, y) which moves so that it is always equidistant from a fixed point. We call this fixed distance the* **radius,** *and we call the fixed point the* **center** *of the circle.* Thus, using this definition, calling the fixed point (h, k) and the radius r, we have

$$\sqrt{(x - h)^2 + (y - k)^2} = r$$

or, by squaring both sides, we have

$$(x - h)^2 + (y - k)^2 = r^2 \qquad (1\text{–}11)$$

Figure 1–28

Equation (1–11) is called the **standard equation** *of a circle with center at (h, k) and radius r* (Fig. 1–28).

Example A

Figure 1–29

The equation $(x - 1)^2 + (y + 2)^2 = 16$ represents a circle with center at $(1, -2)$ and a radius of 4. We determine these values by considering the equation of this circle to be in the form of Eq. (1–11) as

$$(x - 1)^2 + [y - (-2)]^2 = 4^2$$
$$\uparrow \qquad\qquad \uparrow$$
$$\text{Form requires } - \text{ signs}$$

Note carefully the way in which we find the y-coordinate of the center. This circle is shown in Fig. 1–29.

If the center of the circle is at the origin, which means that the coordinates of the center are $(0, 0)$, the equation of the circle becomes

$$x^2 + y^2 = r^2 \qquad (1\text{–}12)$$

A circle of this type clearly exhibits the important property of **symmetry,** which we first discussed in Section 1–2. We can see that if we replace y by $-y$ the equation is unchanged, since $x^2 + (-y)^2 = r^2$ can be simplified to

$x^2 + y^2 = r^2$. Also, the equation remains unchanged if we replace x by $-x$. Therefore, it is symmetrical to the x-axis and to the y-axis.

This type of circle is symmetrical to the origin as well as being symmetrical to both axes. The meaning of symmetry to the origin is that the origin is the midpoint of any two points (x, y) and $(-x, -y)$ which are on the curve. Thus, *if $-x$ can replace x, and $-y$ replace y at the same time, without changing the equation, the graph of the equation is **symmetrical to the origin**.*

Example B

Figure 1–30

The equation of the circle with its center at the origin and with a radius of 6 is $x^2 + y^2 = 36$.

The symmetry of this circle can be shown analytically by the substitutions mentioned above. Replacing x by $-x$, we obtain $(-x)^2 + y^2 = 36$. Since $(-x)^2 = x^2$, this equation can be rewritten as $x^2 + y^2 = 36$. Since this substitution did not change the equation, the graph is symmetrical to the y-axis.

Replacing y by $-y$, we obtain $x^2 + (-y)^2 = 36$, which is the same as $x^2 + y^2 = 36$. This means that the curve is symmetrical to the x-axis.

Replacing x by $-x$, and simultaneously replacing y by $-y$, we obtain $(-x)^2 + (-y)^2 = 36$, which is the same as $x^2 + y^2 = 36$. This means that the curve is symmetrical to the origin. The circle is shown in Fig. 1–30.

Example C

Figure 1–31

Find the equation of the circle with center at $(2, 1)$ and which passes through $(4, 8)$.

In Eq. (1–11), we can determine the equation if we can find h, k, and r for this circle. From the given information, $h = 2$ and $k = 1$. To find r, we use the fact that *all points on the circle must satisfy the equation of the circle.* The point $(4, 8)$ must satisfy Eq. (1–11), with $h = 2$ and $k = 1$. Thus, $(4 - 2)^2 + (8 - 1)^2 = r^2$. From this relation we find $r^2 = 53$. The equation of the circle is

$$(x - 2)^2 + (y - 1)^2 = 53$$

This circle is shown in Fig. 1–31.

Example D

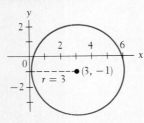

Figure 1–32

Find the equation of the circle with center at $(3, -1)$ and which is tangent to the y-axis.

We know that $h = 3$ and $k = -1$. Since the circle is tangent to the y-axis, the radius is the distance from the center to the y-axis. Therefore, $r = 3$, since the center is 3 units to the right of the y-axis ($h = 3$). Therefore the equation of the circle is

$$(x - 3)^2 + (y + 1)^2 = 9$$

The circle is shown in Fig. 1–32.

If we multiply out each of the terms in Eq. (1–11), we may combine the resulting terms to obtain

$$x^2 - 2hx + h^2 + y^2 - 2ky + k^2 = r^2$$

$$x^2 + y^2 - 2hx - 2ky + (h^2 + k^2 - r^2) = 0 \qquad \textbf{(1–13)}$$

Since each of h, k, and r is constant for any given circle, the coefficients of x and y and the term within parentheses in Eq. (1–13) are constants. Equation (1–13) can then be written as

$$x^2 + y^2 + Dx + Ey + F = 0 \qquad \textbf{(1–14)}$$

Equation (1–14) is called the **general equation** *of the circle.* It tells us that any equation which can be written in that form will represent a circle.

Example E

▶

Figure 1–33

Find the center and radius of the circle

$$x^2 + y^2 - 6x + 8y - 24 = 0$$

We can find this information if we write the given equation in standard form. To do so, we must complete the square in the x-terms and also in the y-terms. This is done by first writing the equation in the form

$$(x^2 - 6x \qquad) + (y^2 + 8y \qquad) = 24$$

To complete the square of the x-terms, we take half of 6, which is 3, square it, and add the result, 9, to each side of the equation. In the same way, we complete the square of the y-terms by adding 16 to each side of the equation, which gives

$$(x^2 - 6x + 9) + (y^2 + 8y + 16) = 24 + 9 + 16$$

$$(x - 3)^2 + (y + 4)^2 = 49 \quad \uparrow \quad \uparrow$$

Add to both sides

Thus, the center is $(3, -4)$, and the radius is 7 (see Fig. 1–33).

Example F

The equation $3x^2 + 3y^2 + 6x - 20 = 0$ can be seen to represent a circle by writing it in general form. This is done by dividing through by 3. In this way we have $x^2 + y^2 + 2x - \frac{20}{3} = 0$. To determine the center and radius, we write the equation in standard form and complete the necessary squares. This leads to

$$(x^2 + 2x + 1) + y^2 = \frac{20}{3} + 1$$

$$(x + 1)^2 + y^2 = \frac{23}{3}$$

Figure 1–34

Thus, the center is $(-1, 0)$; the radius $\sqrt{\frac{23}{3}} = 2.77$ (see Fig. 1–34).

We can also see that this circle is symmetrical to the x-axis, but it is not symmetrical to the y-axis or to the origin. If we replace y by $-y$, the equation does not change, but if we replace x by $-x$, the term $6x$ in the original equation becomes negative, and the equation *does* change.

Exercises 1–5

In Exercises 1 through 4, determine the center and radius of each circle.

1. $(x - 2)^2 + (y - 1)^2 = 25$ 2. $(x - 3)^2 + (y + 4)^2 = 49$
3. $(x + 1)^2 + y^2 = 4$ 4. $x^2 + (y + 6)^2 = 64$

In Exercises 5 through 20, find the equation of each of the circles from the given information.

5. Center at $(0, 0)$, radius 3   6. Center at $(0, 0)$, radius 1
7. Center at $(2, 2)$, radius 4   8. Center at $(0, 2)$, radius 2
9. Center at $(-2, 5)$, radius $\sqrt{5}$ 10. Center at $(-3, -5)$, radius 8
11. Center at $(-3, 0)$, radius $\frac{1}{2}$ 12. Center at $(\frac{3}{2}, -2)$, radius $\frac{5}{2}$
13. Center at $(2, 1)$, passes through $(4, -1)$.
14. Center at $(-1, 4)$, passes through $(-2, 3)$.
15. Center at $(-3, 5)$, tangent to the x-axis.
16. Center at $(2, -4)$, tangent to the y-axis.
17. Tangent to both axes and the lines $y = 4$ and $x = 4$.
18. Tangent to both axes, radius 4, and in the second quadrant.
19. Center on the line $5x = 2y$, radius 5, tangent to the x-axis.
20. The points $(3, 8)$ and $(-3, 0)$ are the ends of a diameter.

In Exercises 21 through 32, determine the center and radius of each circle. Sketch each circle.

21. $x^2 + (y - 3)^2 = 4$ 22. $(x - 2)^2 + (y + 3)^2 = 49$
23. $4(x + 1)^2 + 4(y - 5)^2 = 81$ 24. $2(x + 4)^2 + 2(y + 3)^2 = 25$
25. $x^2 + y^2 - 25 = 0$ 26. $x^2 + y^2 - 9 = 0$
27. $x^2 + y^2 - 2x - 8 = 0$ 28. $x^2 + y^2 - 4x - 6y - 12 = 0$
29. $x^2 + y^2 + 8x - 10y - 8 = 0$ 30. $x^2 + y^2 + 8x + 6y = 0$
31. $2x^2 + 2y^2 - 4x - 8y - 1 = 0$ 32. $3x^2 + 3y^2 - 12x + 4 = 0$

In Exercises 33 through 36, determine whether the circles with the given equations are symmetrical to either axis or the origin.

33. $x^2 + y^2 = 100$ 34. $x^2 + y^2 - 4x - 5 = 0$
35. $x^2 + y^2 + 8y - 9 = 0$ 36. $x^2 + y^2 - 2x + 4y - 3 = 0$

In Exercises 37 through 44, find the indicated quantities.

37. Determine where the circle $x^2 - 6x + y^2 - 7 = 0$ crosses the x-axis.
38. Find the points of intersection of the circle $x^2 + y^2 - x - 3y = 0$ and the line $y = x - 1$.
39. Find the locus of a point $P(x, y)$ which moves so that its distance from $(2, 4)$ is twice its distance from $(0, 0)$. Describe the locus.

40. Find the equation of the locus of a point $P(x, y)$ which moves so that the line joining it and $(2, 0)$ is always perpendicular to the line joining it and $(-2, 0)$. Describe the locus.

41. If an electrically charged particle enters a magnetic field of flux density B with a velocity v at right angles to the field, the path of the particle is a circle. The radius of the path is given by $R = mv/Bq$, where m is the mass of the particle and q is its charge. If a proton ($m = 1.67 \times 10^{-27}$ kg, $q = 1.60 \times 10^{-19}$ C) enters a magnetic field of flux density 1.50 T (tesla: V · s/m²) with a velocity of 2.40×10^8 m/s, find the equation of the path of the proton. (The fact that a charged particle travels in a circular path in a magnetic field is an important basis for the construction of a cyclotron.)

42. The relationship of the impedance Z, resistance R, capacitive reactance X_C, and inductive reactance X_L in an electric circuit is $Z = \sqrt{(X_L - X_C)^2 + R^2}$. Sketch the graph of R vs. X_L if Z and X_C are constant.

43. A cylindrical oil tank, 4.00 ft in diameter, is on its side. A circular hole, with center 18.0 in. below the center of the end of the tank, has been drilled in the end of the tank. If the radius of the hole is 3.00 in., what is the equation of the circle of the end of the tank and the equation of the circle of the hole? Use the center of the end of the tank as the origin.

44. A student is drawing a friction drive in which two circular disks are in contact with each other. They are represented by circles in the drawing. The first has a radius of 10.0 cm and the second has a radius of 12.0 cm. What is the equation of each circle if the origin is at the center of the first circle and the x-axis passes through the center of the second circle?

1-6 The Parabola

Another important curve is the parabola, which we have come across several times in the past. In this section we shall find the form of the equation of the parabola and see that it is a familiar form.

A **parabola** *is defined as the locus of a point P(x, y) which moves so that it is always equidistant from a given line and a given point. The given line is called* the **directrix,** *and the given point is called the* **focus.** The line through the focus which is perpendicular to the directrix is called the **axis** of the parabola. The point midway between the directrix and focus is the **vertex** of the parabola. Using the definition, we shall find the equation of the parabola for which the focus is the point $(p, 0)$ and the directrix is the line $x = -p$. By choosing the focus and directrix in this manner, we shall be able to find a general representation of the equation of a parabola with its vertex at the origin.

According to the definition of the parabola, the distance from a point $P(x, y)$ on the parabola to the focus $(p, 0)$ must equal the distance from $P(x, y)$ to the directrix $x = -p$. The distance from P to the focus can be found by use of the distance formula. The distance from P to the directrix is the perpendicular distance, and this can be found as the distance between two points on a line parallel to the *x*-axis. These distances are indicated in Fig. 1–35.

Thus, we have

$$\sqrt{(x - p)^2 + (y - 0)^2} = x + p$$

Squaring both sides of this equation, we have

$$(x - p)^2 + y^2 = (x + p)^2$$

or

$$x^2 - 2px + p^2 + y^2 = x^2 + 2px + p^2$$

Simplifying, we obtain

$$y^2 = 4px \qquad\qquad (1\text{–}15)$$

Equation (1–15) is called the **standard form** *of the equation of a parabola with its axis along the x-axis and the vertex at the origin.* Its symmetry to the *x*-axis can be proven since $(-y)^2 = 4px$ is the same as $y^2 = 4px$.

Figure 1–35

Figure 1–36

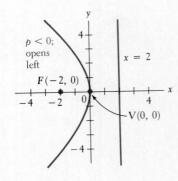

Figure 1–37

Example A   Find the coordinates of the focus, the equation of the directrix, and sketch the graph of the parabola $y^2 = 12x$.

From the form of the equation, we know that the vertex is at the origin (Fig. 1–36). The coefficient 12 tells us that the focus is (3, 0), since $p = \frac{12}{4}$ [note that the coefficient of *x* in Eq. (1–15) is $4p$]. Also, this means that the directrix is the line $x = -3$. Since *p* is positive, the parabola opens to the right.

Example B

If the focus is to the left of the origin, with the directrix an equal distance to the right, the coefficient of the x-term will be negative. This tells us that the parabola opens to the left, rather than to the right, as is the case when the focus is to the right of the origin. For example, the parabola $y^2 = -8x$ has its vertex at the origin, its focus at $(-2, 0)$, and the line $x = 2$ as its directrix. This is consistent with Eq. (1–15), in that $4p = -8$, or $p = -2$. The parabola opens to the left as shown in Fig. 1–37.

If we chose the focus as the point $(0, p)$ and the directrix as the line $y = -p$, we would find that the resulting equation is

$$x^2 = 4py \qquad\qquad (1\text{–}16)$$

This is the standard form of the equation of a parabola with the y-axis as its axis and the vertex at the origin. Its symmetry to the y-axis can be proven, since $(-x)^2 = 4py$ is the same as $x^2 = 4py$. We note that the difference between this equation and Eq. (1–15) is that x is squared and y appears to the first power in Eq. (1–16), rather than the reverse, as in Eq. (1–15).

Example C

The parabola $x^2 = 4y$ has its vertex at the origin, focus at the point $(0, 1)$, and its directrix the line $y = -1$. The graph is shown in Fig. 1–38, and we see in this case that the parabola opens up.

Figure 1–38

Figure 1–39

Example D

The parabola $x^2 = -6y$ has its vertex at the origin, its focus at the point $(0, -\frac{3}{2})$, and its directrix the line $y = \frac{3}{2}$. This parabola opens down, as Fig. 1–39 shows.

Example E

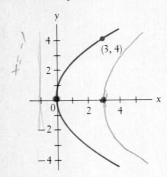

Figure 1–40

Find the equation of the parabola with vertex at the origin, axis along the x-axis, and which passes through $(3, 4)$.

Since the parabola has its vertex at the origin and axis along the x-axis, its general form is the equation given by Eq. (1–15), or $y^2 = 4px$. We can find the value of p by using the fact that it passes through $(3, 4)$. These coordinates must satisfy the equation. This means that

$$4^2 = 4p(3), \quad \text{or} \quad p = \frac{4}{3}$$

Therefore, the equation is $y^2 = \frac{16}{3}x$. The parabola is shown in Fig. 1–40.

Equations (1–15) and (1–16) give us the general form of the equation of a parabola with its vertex at the origin and its focus on one of the coordinate axes. The following example illustrates the use of the definition of the parabola to find the equation of a parabola which has its vertex at a point other than at the origin.

Example F

Using the definition of the parabola, determine the equation of the parabola with its focus at $(2, 3)$ and its directrix the line $y = -1$ (see Fig. 1–41).

Figure 1–41

Choosing a general point $P(x, y)$ on the parabola, and equating the distances from this point to $(2, 3)$ and to the line $y = -1$, we have

$$\sqrt{(x - 2)^2 + (y - 3)^2} = y + 1$$

Squaring both sides of this equation and simplifying, we have

$$(x - 2)^2 + (y - 3)^2 = (y + 1)^2$$
$$x^2 - 4x + 4 + y^2 - 6y + 9 = y^2 + 2y + 1$$

or

$$8y = 12 - 4x + x^2$$

The x-term and the constant (12 in this case) are characteristic of a parabola which does not have its vertex at the origin if the directrix is parallel to the x-axis.

Thus we can conclude that *a parabola is characterized by the presence of the square of either (but not both) x or y, and a first-power term in the other.* In this way we can recognize a parabola by inspecting the equation.

Exercises 1–6

In Exercises 1 through 12, determine the coordinates of the given parabolas. Sketch each curve.

1. $y^2 = 4x$ 2. $y^2 = 16x$ 3. $y^2 = -4x$ 4. $y^2 = -16x$

5. $x^2 = 8y$ 6. $x^2 = 10y$ 7. $x^2 = -4y$ 8. $x^2 = -12y$

9. $y^2 = 2x$ 10. $x^2 = 14y$ 11. $y = x^2$ 12. $x = 3y^2$

In Exercises 13 through 20, find the equations of the parabolas satisfying the given conditions.

13. Focus (3, 0), directrix $x = -3$ 14. Focus $(-2, 0)$, directrix $x = 2$

15. Focus (0, 4), vertex (0, 0) 16. Focus $(-3, 0)$, vertex (0, 0)

17. Vertex (0, 0), directrix $y = -1$ 18. Vertex (0, 0), directrix $y = \dfrac{1}{2}$

19. Vertex at (0, 0), axis along the y-axis, passes through $(-1, 8)$.

20. Vertex at (0, 0), axis along the x-axis, passes through $(2, -1)$.

In Exercises 21 through 24, use the definition of the parabola to find the equations of the parabolas satisfying the given conditions. Sketch each curve.

21. Focus (6, 1), directrix $x = 0$ 22. Focus $(1, -4)$, directrix $x = 2$

23. Focus (1, 1), vertex (1, 3) 24. Vertex $(-2, -4)$, directrix $x = 3$

In Exercises 25 through 34, find the indicated quantities.

25. In calculus it can be shown that a light ray coming from the focus of a parabola will be reflected off parallel to the axis of the parabola. Suppose that a light ray from the focus of a parabolic reflector described by the equation $y^2 = 16x$ strikes the reflector at the point (1, 4). Along what lines does the incident ray move, and along what line does the reflected ray move?

Figure 1–42

Figure 1–43

26. A radio wave reflector has a parabolic cross section, and a wave that comes in parallel to its axis will be reflected through the focus. A particular reflector is designed for the incoming and reflected waves, as shown in Fig. 1–42 (measurements in meters). What is the equation of the parabola?

27. A linear solar reflector has a parabolic cross section, which is shown in Fig. 1–43. From the given dimensions, what is the focal length (vertex to focus) of the reflector?

28. The rate of development of heat H (measured in watts) in a resistor of resistance R (measured in ohms) of an electric circuit is given by $H = Ri^2$, where i is the current (measured in amperes) in the resistor. Sketch the graph of H versus i, if $R = 6.0 \ \Omega$.

29. Under certain conditions, a cable which hangs between two supports can be closely approximated as being parabolic. Assuming that this cable hangs in the shape of a parabola, find its equation if a point 10 ft horizontally from its lowest point is 1.0 ft above its lowest point. Choose the lowest point as the origin of the coordinate system.

30. A wire is fastened 36.0 ft up on each of two telephone poles which are 200 ft apart. Halfway between the poles the wire is 30.0 ft above the ground. Assuming the wire is parabolic, find the height of the wire 50.0 ft from either pole.

31. The period T (measured in seconds) of the oscillation for resonance in an electric circuit is given by $T = 2\pi\sqrt{LC}$, where L is the inductance (in henrys) and C is the capacitance (in farads). Sketch the graph of the period versus capacitance for a constant inductance of 1 H. Assume values of C from 1 μF to 250 μF, for this is a common range of values.

32. The velocity v of a jet of water flowing from an opening in the side of a certain container is given by $v = 8\sqrt{h}$, where h is the depth of the opening. Sketch a graph of v (in feet per second) versus h (in feet).

33. A small island is 4 km from a straight shoreline. A ship channel is equidistant between the island and the shoreline. Write an equation for the channel.

34. Under certain circumstances, the maximum power P in an electric circuit varies as the square of the voltage of the source E_0 and inversely as the internal resistance R_i of the source. If 10 W is the maximum power for a source of 2.0 V and internal resistance of 0.10 Ω, sketch the graph of P versus E_0, if R_i remains constant.

1–7 The Ellipse

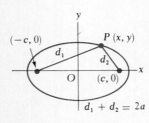

Figure 1–44

The next important curve we shall discuss is the ellipse. *The **ellipse** is defined as the locus of a point P(x, y) which moves so that the sum of its distances from two fixed points is constant. These two fixed points are called the **foci** of the ellipse.* We will let this sum of distances be $2a$ and the foci be the points $(c, 0)$ and $(-c, 0)$. These choices set up the ellipse with its center at the origin such that c is the length of the line segment from the center to a focus. Also, by choosing $2a$ as the sum of distances, we will see that the length a has a special meaning.

From the definition of an ellipse, we have (see Fig. 1–44)

$$\sqrt{(x - c)^2 + y^2} + \sqrt{(x + c)^2 + y^2} = 2a$$

When solving equations involving radicals, we find that we should remove one of the radicals to the right side, and then square each side. Thus we have the following steps:

$$\sqrt{(x + c)^2 + y^2} = 2a - \sqrt{(x - c)^2 + y^2}$$

$$(x + c)^2 + y^2 = 4a^2 - 4a\sqrt{(x - c)^2 + y^2} + (\sqrt{(x - c)^2 + y^2})^2$$

$$x^2 + 2cx + c^2 + y^2 = 4a^2 - 4a\sqrt{(x - c)^2 + y^2} + x^2 - 2cx + c^2 + y^2$$

$$4a\sqrt{(x - c)^2 + y^2} = 4a^2 - 4cx$$

$$a\sqrt{(x - c)^2 + y^2} = a^2 - cx$$

$$a^2(x^2 - 2cx + c^2 + y^2) = a^4 - 2a^2cx + c^2x^2$$

$$(a^2 - c^2)x^2 + a^2y^2 = a^2(a^2 - c^2)$$

At this point we define (and the usefulness of this definition will be shown presently) $a^2 - c^2 = b^2$. This substitution gives us

$$b^2x^2 + a^2y^2 = a^2b^2$$

Dividing through by a^2b^2, we have

$$\frac{x^2}{a^2} + \frac{y^2}{b^2} = 1 \tag{1-17}$$

If we let $y = 0$, we find that the x-intercepts are $(-a, 0)$ and $(a, 0)$. We see that the distance $2a$, chosen as the sum of the distances in the derivation of Eq. (1–17), is the distance between the x-intercepts. *These two points are known as the **vertices** of the ellipse, and the line between them is known as the **major axis** (Fig. 1–45(a)) of the ellipse. Thus, a is called the **semimajor axis.***

If we now let $x = 0$, we find the y-intercepts to be $(0, -b)$ and $(0, b)$. *The line joining these two intercepts is called the **minor axis** of the ellipse (Fig. 1–45(b)) which means b is called the **semiminor axis.*** The point $(0, b)$ is on the ellipse, and is also equidistant from $(-c, 0)$ and $(c, 0)$. Since the sum of the distances from these points to $(0, b)$ equals $2a$, the distance from $(c, 0)$ to $(0, b)$ must be a. Thus, we have a right triangle formed by line segments of lengths a, b, and c, with a as the hypotenuse (Fig. 1–45(c)). From this *we have*

$$a^2 = b^2 + c^2 \tag{1-18}$$

as the relation between the distances a, b, and c. This equation also shows us why b was defined as it was in the derivation.

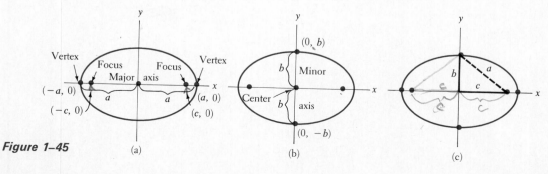

Figure 1–45

(a)   (b)   (c)

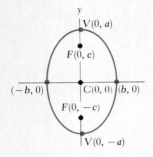

Figure 1–46

Equation (1–17) is called the **standard equation** *of the ellipse with its major axis along the x-axis and its center at the origin.*

If we choose points on the y-axis as the foci, *the standard equation of the ellipse, with its center at the origin and its major axis along the y-axis, is*

$$\frac{y^2}{a^2} + \frac{x^2}{b^2} = 1 \qquad\qquad (1\text{–}19)$$

In this case the vertices are $(0, a)$ and $(0, -a)$, the foci are $(0, c)$ and $(0, -c)$, and the ends of the minor axis are $(b, 0)$ and $(-b, 0)$. See Fig. 1–46.

The symmetry of the ellipses given by Eqs. (1–17) and (1–19) to each of the axes and the origin can be proven, since each of x and y can be replaced by its negative in these equations without changing the equations.

Example A

The ellipse

$$\frac{x^2}{25} + \frac{y^2}{9} = 1$$

has vertices at $(5, 0)$ and $(-5, 0)$. Its minor axis extends from $(0, 3)$ to $(0, -3)$, as we see in Fig. 1–47. This information is directly obtainable from this form of the equation, since the 25 tells us that $a^2 = 25$, or $a = 5$. In the same way we have $b = 3$. Since we know both a^2 and b^2, we can find c^2 from the relation $c^2 = a^2 - b^2$. Thus, $c^2 = 16$, or $c = 4$. Thus, the foci are $(4, 0)$ and $(-4, 0)$.

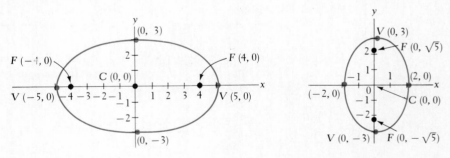

Figure 1–47   **Figure 1–48**

Example B

The ellipse

$$\frac{x^2}{4} + \frac{y^2}{9} = 1$$

has vertices at $(0, 3)$ and $(0, -3)$. The minor axis extends from $(2, 0)$ to $(-2, 0)$. This we can find directly from the equation, since $a^2 = 9$ and $b^2 = 4$. One might ask how we chose $a^2 = 9$ in this example, and $a^2 = 25$ in the preceding example. Since $a^2 = b^2 + c^2$, *a* is always larger than *b*. Thus, we can tell which axis (x or y) the major axis is along by seeing which number (in the denominator) is larger, when the equation is written in standard form. The larger one stands for a^2. In this example, the foci are $(0, \sqrt{5})$ and $(0, -\sqrt{5})$. See Fig. 1–48.

Example C

Find the coordinates of the vertices, the ends of the minor axis, and the foci of the ellipse $4x^2 + 16y^2 = 64$.

This equation must be put in standard form first, which we do by dividing through by 64. When this is done, we obtain

$$\frac{x^2}{16} + \frac{y^2}{4} = 1 \leftarrow$$

Form requires
+ and 1

Thus, $a = 4$, $b = 2$, and $c = 2\sqrt{3}$. The vertices are at $(4, 0)$ and $(-4, 0)$. The ends of the minor axis are $(0, 2)$ and $(0, -2)$, and the foci are $(2\sqrt{3}, 0)$ and $(-2\sqrt{3}, 0)$. See Fig. 1–49.

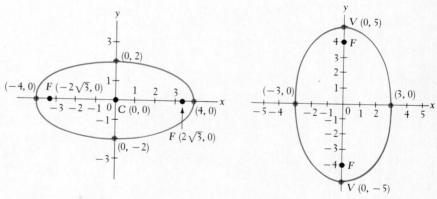

Figure 1–49   Figure 1–50

Example D

Find the equation of the ellipse for which the center is at the origin, the major axis is along the y-axis, the minor axis is 6 units long, and there are 8 units between foci.

Directly we know that $2b = 6$, or $b = 3$. Also $2c = 8$, or $c = 4$. Thus, we find that $a = 5$. Since the major axis is along the y-axis, we have the equation

$$\frac{y^2}{25} + \frac{x^2}{9} = 1$$

This ellipse is shown in Fig. 1–50.

Example E

Find the equation of the ellipse with its center at the origin, an end of its minor axis at (2, 0), and which passes through $(-1, \sqrt{6})$.

Since the center is at the origin and an end of the minor axis is at (2, 0), we know that the ellipse is of the form of Eq. (1–19), and that $b = 2$. Thus, we have

$$\frac{y^2}{a^2} + \frac{x^2}{2^2} = 1$$

In order to find a^2 we use the fact that the ellipse passes through $(-1, \sqrt{6})$. This means that these coordinates satisfy the equation of the ellipse. This gives us

$$\frac{(\sqrt{6})^2}{a^2} + \frac{(-1)^2}{4} = 1$$

$$\frac{6}{a^2} + \frac{1}{4} = 1$$

$$\frac{6}{a^2} = \frac{3}{4}$$

$$a^2 = 8$$

Thus, the equation of the ellipse is

$$\frac{y^2}{8} + \frac{x^2}{4} = 1$$

The ellipse is shown in Fig. 1–51.

Figure 1–51

Equations (1–17) and (1–19) give us the standard form of the equation of an ellipse with its center at the origin and its foci on one of the coordinate axes. The following example illustrates the use of the definition of the ellipse to find the equation of an ellipse with its center at a point other than the origin.

Example F

Using the definition, find the equation of the ellipse with foci at (1, 3) and (9, 3), with a major axis of 10.

Using the same method as in the derivation of Eq. (1–17), we have the following steps. (Remember, the sum of distances in the definition equals the length of the major axis.)

$$\sqrt{(x-1)^2 + (y-3)^2} + \sqrt{(x-9)^2 + (y-3)^2} = 10$$

$$\sqrt{(x-1)^2 + (y-3)^2} = 10 - \sqrt{(x-9)^2 + (y-3)^2}$$

$$(x-1)^2 + (y-3)^2 = 100 - 20\sqrt{(x-9)^2 + (y-3)^2} + (\sqrt{(x-9)^2 + (y-3)^2})^2$$

$$x^2 - 2x + 1 + y^2 - 6y + 9 = 100 - 20\sqrt{(x-9)^2 + (y-3)^2} + x^2 - 18x + 81 + y^2 - 6y + 9$$

$$20\sqrt{(x - 9)^2 + (y - 3)^2} = 180 - 16x$$
$$5\sqrt{(x - 9)^2 + (y - 3)^2} = 45 - 4x$$
$$25(x^2 - 18x + 81 + y^2 - 6y + 9) = 2025 - 360x + 16x^2$$
$$25x^2 - 450x + 2250 + 25y^2 - 150y = 2025 - 360x + 16x^2$$
$$9x^2 - 90x + 25y^2 - 150y + 225 = 0$$

The additional x- and y-terms are characteristic of the equation of an ellipse whose center is not at the origin (see Fig. 1–52).

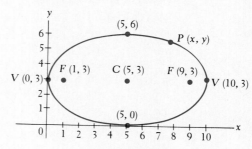

Figure 1–52

We can conclude that *the equation of an ellipse is characterized by the presence of both an x^2- and a y^2-term, having different coefficients (in value but not in sign).* The difference between the equation of an ellipse and that of a circle is that the coefficients of the squared terms in the equation of the circle are the same, whereas those of the ellipse differ.

Exercises 1–7

In Exercises 1 through 12, find the coordinates of the vertices and foci of the given ellipses. Sketch each curve.

1. $\dfrac{x^2}{4} + \dfrac{y^2}{1} = 1$

2. $\dfrac{x^2}{100} + \dfrac{y^2}{64} = 1$

3. $\dfrac{x^2}{25} + \dfrac{y^2}{36} = 1$

4. $\dfrac{x^2}{49} + \dfrac{y^2}{81} = 1$

5. $4x^2 + 9y^2 = 36$

6. $x^2 + 36y^2 = 144$

7. $49x^2 + 4y^2 = 196$

8. $25x^2 + y^2 = 25$

9. $8x^2 + y^2 = 16$

10. $2x^2 + 3y^2 = 6$

11. $4x^2 + 25y^2 = 25$

12. $9x^2 + 4y^2 = 9$

In Exercises 13 through 20, find the equations of the ellipses satisfying the given conditions. The center of each is at the origin.

13. Vertex $(15, 0)$, focus $(9, 0)$

14. Minor axis 8, vertex $(0, -5)$

15. Focus $(0, 2)$, major axis 6

16. Semiminor axis 2, focus $(3, 0)$

17. Vertex $(8, 0)$, passes through $(2, 3)$.

18. Focus $(0, 2)$, passes through $(-1, \sqrt{3})$.

19. Passes through $(2, 2)$ and $(1, 4)$.

20. Passes through $(-2, 2)$ and $(1, \sqrt{6})$.

In Exercises 21 through 24, find the equations of the ellipses with the given properties by use of the definition of an ellipse.

21. Foci at $(-2, 1)$ and $(4, 1)$, a major axis of 10.

22. Foci at $(-3, -2)$ and $(-3, 8)$, major axis 26.

23. Vertices at $(1, 5)$ and $(1, -1)$, foci at $(1, 4)$ and $(1, 0)$.

24. Vertices at $(-2, 1)$ and $(-2, 5)$, foci at $(-2, 2)$ and $(-2, 4)$.

In Exercises 25 through 34, solve the given problems.

25. Show that the ellipse $2x^2 + 3y^2 - 8x - 4 = 0$ is symmetrical to the x-axis.

26. Show that the ellipse $5x^2 + y^2 - 3y - 7 = 0$ is symmetrical to the y-axis.

27. The arch of a bridge across a stream is in the form of half an ellipse above the water level. If the span of the arch at water level is 100 ft and the maximum height of the arch above water level is 30 ft, what is the equation of the arch? Choose the origin of the coordinate system at the most convenient point.

28. An elliptical gear (Fig. 1–53) which rotates about its center is kept continually in mesh with a circular gear which is free to move horizontally. If the equation of the ellipse of the gear (with the origin of the coordinate system at its center) in its present position is $3x^2 + 7y^2 = 20$, how far does the center of the circular gear move going from one extreme position to the other? (Assume the units are centimeters.)

Figure 1–53

29. A satellite to study the earth's atmosphere has a minimum altitude of 500 mi and a maximum altitude of 2000 mi. If the path of the satellite about the earth is an ellipse with the center of the earth at one focus, what is the equation of its path? (Assume the radius of the earth is 4000 mi.)

30. An electric circuit is caused to flow in a loop of wire rotating in a magnetic field. In a study of this phenomenon, a piece of wire is cut into two pieces, one of which is bent into a circle and the other into a square. If the sum of areas of the circle and square is always π units, find the relation between the radius r of the circle and the side x of the square. Sketch the graph of r versus x.

31. A vertical pipe 6.0 inches in diameter is to pass through a roof inclined at 45°. What are the dimensions of the elliptical hole which must be cut in the roof for the pipe?

32. The ends of a horizontal tank 20.0 ft long are ellipses, which can be described by the equation $9x^2 + 20y^2 = 180$, where x and y are measured in feet. The area of an ellipse is $A = \pi ab$. Find the volume of the tank.

33. A draftsman draws a series of triangles with a base from $(-3, 0)$ to $(3, 0)$ and a perimeter of 14 cm (all measurements in centimeters). Find the equation of the curve on which all of the third vertices of the triangles are located.

34. A drainage trough 1.50 m wide, 0.50 m deep, and 8.00 m long has a semielliptical cross section. At capacity what volume does it hold? (See Exercise 32.)

1–8 The Hyperbola

The final curve we shall discuss in detail is the hyperbola. *The* **hyperbola** *is defined as the locus of a point P(x, y) which moves so that the difference of the distances from two fixed points is a constant. These fixed points are the* **foci** *of the hyperbola.* We choose the foci of the hyperbola as the points $(c, 0)$ and

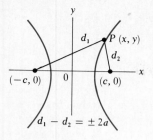

Figure 1–54

$(-c, 0)$ (see Fig. 1–54) and the constant difference to be $2a$. As with the ellipse, these choices make c the length of the line segment from the center to a focus and a (as we will see) the length of the line segment from the center to a vertex. Therefore, we have

$$\sqrt{(x + c)^2 + y^2} - \sqrt{(x - c)^2 + y^2} = 2a$$

Following the same procedure as in the preceding section, we find the equation of the hyperbola to be

$$\frac{x^2}{a^2} - \frac{y^2}{b^2} = 1 \tag{1–20}$$

When we derive this equation, we have a definition of the relation between a, b, and c which is different from that for the ellipse. This relation is

$$c^2 = a^2 + b^2 \tag{1–21}$$

An analysis of Eq. (1–20) will reveal the significance of a and b. First, by letting $y = 0$, we find that the x-intercepts are $(a, 0)$ and $(-a, 0)$, just as they are for the ellipse. *These points are called the* **vertices** *of the hyperbola.* By letting $x = 0$, we find that we have imaginary solutions for y, which means that there are no points on the curve which correspond to a value of $x = 0$.

To find the significance of b, we shall solve Eq. (1–20) for y in a particular form. First we have

$$\frac{y^2}{b^2} = \frac{x^2}{a^2} - 1$$

$$= \frac{x^2}{a^2} - \frac{a^2 x^2}{a^2 x^2}$$

$$= \frac{x^2}{a^2}\left(1 - \frac{a^2}{x^2}\right)$$

Multiplying through by b^2 and then taking the square root of each side, we have

$$y^2 = \frac{b^2 x^2}{a^2}\left(1 - \frac{a^2}{x^2}\right)$$

$$y = \pm\frac{bx}{a}\sqrt{1 - \frac{a^2}{x^2}} \tag{1–22}$$

We note that, if large values of x are assumed in Eq. (1–22), the quantity under the radical becomes approximately 1. In fact, the larger x becomes, the nearer 1 this expression becomes, since the x^2 in the denominator of a^2/x^2 makes this term nearly zero. Thus, for large values of x, Eq. (1–22) is approximately

$$y = \pm \frac{bx}{a} \tag{1-23}$$

Equation (1–23) can be seen to represent the equations for two straight lines, each of which passes through the origin. One has a slope of b/a and the other a slope of $-b/a$. *These lines are called the* **asymptotes** *of the hyperbola.* The concept of an asymptote was introduced in Section 1–2. The fact that the hyperbola *approaches* the asymptote as x becomes large without bound (*approaches* infinity) is designated by

$$y \to \frac{bx}{a} \quad \text{as} \quad x \to \infty$$

Since straight lines are easily sketched, the easiest way to sketch a hyperbola is to draw its asymptotes and then to draw the hyperbola so that it comes closer and closer to these lines as x becomes larger numerically. To draw in the asymptotes, the usual procedure is to first draw a small rectangle, $2a$ by $2b$, with the origin in the center. Then straight lines are drawn through opposite vertices. These lines are the asymptotes (see Fig. 1–55). Thus we see that the significance of the value of b lies in the slope of the asymptotes of the hyperbola.

Equation (1–20) is called the **standard equation** *of the hyperbola with its center at the origin. It has a* **transverse axis** *of length 2a along the x-axis and a* **conjugate axis** *of length 2b along the y-axis.* This means that a represents the length of the semitransverse axis, and b represents the length of the semiconjugate axis (see Fig. 1–56). From the definition of c we know that c is the length of the line segment from the center to a focus. Also, c is the length of the semidiagonal as shown in Fig. 1–56. This shows us the geometric meaning of the relation among a, b, and c which is given in Eq. (1–21).

Figure 1–55

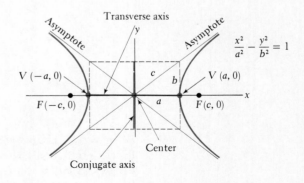

Figure 1–56

If the transverse axis is along the y-axis and the conjugate axis is along the x-axis, the equation of a hyperbola with its center at the origin is

$$\frac{y^2}{a^2} - \frac{x^2}{b^2} = 1 \qquad (1\text{--}24)$$

The symmetry of the hyperbolas given by Eqs. (1–20) and (1–24) to each of the axes and to the origin can be proven since each of x and y can be replaced by its negative in these equations without changing the equations.

Example A

The hyperbola

$$\frac{x^2}{16} - \frac{y^2}{9} = 1$$

has vertices at $(4, 0)$ and $(-4, 0)$. Its transverse axis extends from one vertex to the other. Its conjugate axis extends from $(0, 3)$ to $(0, -3)$. Since $c^2 = a^2 + b^2$, we find that $c = 5$, which means the foci are the points $(5, 0)$ and $(-5, 0)$. Drawing in the rectangle and then the asymptotes (Fig. 1–57), we draw the hyperbola from each vertex toward the asymptotes.

Figure 1–57

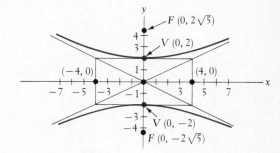

Figure 1–58

Example B

The hyperbola

$$\frac{y^2}{4} - \frac{x^2}{16} = 1$$

has vertices at $(0, 2)$ and $(0, -2)$. Its conjugate axis extends from $(4, 0)$ to $(-4, 0)$. The foci are $(0, 2\sqrt{5})$ and $(0, -2\sqrt{5})$. Since, with 1 on the right side of the equation, the y^2-term is the positive term, this hyperbola is in the form of Eq. (1–24). [Example A illustrates a hyperbola of the type of Eq. (1–20), since the x^2-term is the positive term.] Since $2a$ extends along the y-axis, we see that the equations of the asymptotes are $y = \pm(a/b)x$. This is not a contradiction of Eq. (1–23), but an extension of it for the case of a hyperbola with its transverse axis along the y-axis. The ratio a/b simply expresses the slope of the asymptote (see Fig. 1–58).

Example C

Determine the coordinates of the vertices and foci of the hyperbola $4x^2 - 9y^2 = 36$.

First, by dividing through by 36, we can put this equation in standard form. Thus, we have

$$\frac{x^2}{9} - \frac{y^2}{4} = 1 \longleftarrow$$

Form requires
— and 1 —

The transverse axis is along the x-axis with vertices at $(3, 0)$ and $(-3, 0)$. Since $c^2 = a^2 + b^2$, $c = \sqrt{13}$. This means that the foci are at $(\sqrt{13}, 0)$ and $(-\sqrt{13}, 0)$ (see Fig. 1–59).

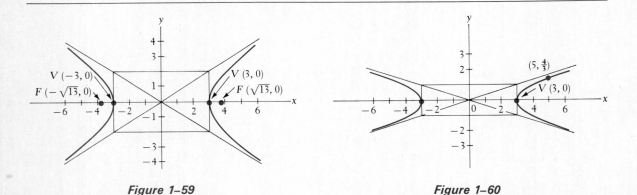

Figure 1–59   Figure 1–60

Example D

Find the equation of the hyperbola with its center at the origin, a vertex at $(3, 0)$, and which passes through $(5, \frac{4}{3})$.

When we know that the hyperbola has its center at the origin and a vertex at $(3, 0)$, we know that the standard form of its equation is given by Eq. (1–20). This also tells us that $a = 3$. By using the fact that the coordinates of the point $(5, \frac{4}{3})$ must satisfy the equation, we are able to find the value of b^2. Substituting, we have

$$\frac{25}{9} - \frac{\left(\frac{16}{9}\right)}{b^2} = 1$$

from which we find $b^2 = 1$. Therefore, the equation of the hyperbola is

$$\frac{x^2}{9} - \frac{y^2}{1} = 1 \quad \text{or} \quad x^2 - 9y^2 = 9$$

This hyperbola is shown in Fig. 1–60.

Equations (1–20) and (1–24) give us the standard form of the equation of the hyperbola with its center at the origin and its foci on one of the coordinate axes. There is one other important equation form which represents a hyperbola, and that is

$$xy = c \tag{1–25}$$

The asymptotes of this hyperbola are the coordinate axes, and the foci are on the line $y = x$, or on the line $y = -x$, if c is negative. This hyperbola is symmetrical to the origin, for if $-x$ replaces x, and $-y$ replaces y at the same time, we have $(-x)(-y) = c$, or $xy = c$. The equation remains unchanged. However, if either $-x$ replaces x, or if $-y$ replaces y, but not both, the sign on the left is changed. Therefore, it is not symmetrical to either axis. The c here represents a constant and is not related to the focus. The following example illustrates this type of hyperbola.

Example E

Plot the graph of the equation $xy = 4$.

We find the values of the table below, and then plot the appropriate points. Here it is permissible to use a limited number of points, since we know that the equation represents a hyperbola (Fig. 1–61). Thus, using $y = 4/x$, we obtain the values in the table.

x	-8	-4	-1	$-\frac{1}{2}$	$\frac{1}{2}$	1	4	8
y	$-\frac{1}{2}$	-1	-4	-8	8	4	1	$\frac{1}{2}$

Figure 1–61

We conclude that the equation of a hyperbola is characterized by the presence of both an x^2- and a y^2-term, having different signs, or by the presence of an xy-term with no squared terms.

Exercises 1–8

In Exercises 1 through 12, find the coordinates of the vertices and the foci of the given hyperbolas. Sketch each curve.

1. $\dfrac{x^2}{25} - \dfrac{y^2}{144} = 1$ 2. $\dfrac{x^2}{16} - \dfrac{y^2}{4} = 1$ 3. $\dfrac{y^2}{9} - \dfrac{x^2}{1} = 1$

4. $\dfrac{y^2}{2} - \dfrac{x^2}{2} = 1$ 5. $2x^2 - y^2 = 4$ 6. $3x^2 - y^2 = 9$

7. $2y^2 - 5x^2 = 10$ 8. $3y^2 - 2x^2 = 6$ 9. $4x^2 - y^2 + 4 = 0$

10. $9x^2 - y^2 - 9 = 0$ 11. $4x^2 - 9y^2 = 16$ 12. $y^2 - 9x^2 = 25$

In Exercises 13 through 20, find the equations of the hyperbolas satisfying the given conditions. The center of each is at the origin.

13. Vertex (3, 0), focus (5, 0)

14. Vertex (0, 1), focus $(0, \sqrt{3})$

15. Conjugate axis = 12, vertex (0, 10)

16. Focus (8, 0), transverse axis = 4

17. Passes through (2, 3), focus (2, 0).

18. Passes through $(8, \sqrt{3})$, vertex (4, 0).

19. Passes through (5, 4) and $(3, \frac{4}{5}\sqrt{5})$.

20. Passes through (1, 2) and $(2, 2\sqrt{2})$.

In Exercises 21 through 24, sketch the graphs of the hyperbolas given.

21. $xy = 2$ 22. $xy = 10$ 23. $xy = -2$ 24. $xy = -4$

In Exercises 25 through 28, find the equations of the hyperbolas with the given properties by use of the definition of the hyperbola.

25. Foci at (1, 2) and (11, 2), with a transverse axis of 8.

26. Vertices $(-2, 4)$ and $(-2, -2)$, with conjugate axis of 4.

27. Center at (2, 0), vertex at (3, 0), conjugate axis of 6.

28. Center at $(1, -1)$, focus at (1, 4), vertex at (1, 2).

In Exercises 29 through 34, solve the given problems.

29. Sketch the graph of impedance versus resistance if the reactance $X_L - X_C$ of a given electric circuit is constant at 60 Ω (see Exercise 42 of Section 1–5).

30. One statement of Boyle's law is that the product of the pressure and volume, for constant temperature, remains a constant for a perfect gas. If one set of values for a perfect gas under the condition of constant temperature is that the pressure is 300 kPa for a volume of 8.0 L, sketch a graph of pressure versus volume.

31. The relationship between the frequency f, wavelength λ, and the velocity v of a wave is given by $v = f\lambda$. The velocity of light is a constant, being 3.0×10^{10} cm/s. The visible spectrum ranges in wavelength from about 4.0×10^{-5} cm (violet) to about 7.0×10^{-5} cm (red). Sketch a graph of frequency f (in Hertz) as a function of wavelength for the visible spectrum.

32. Wavelengths of gamma rays vary from about 10^{-8} cm to about 10^{-13} cm, but the gamma rays have the same velocity as light. On logarithmic paper, plot the graph of frequency versus wavelength for gamma rays. What type of curve results when this type of hyperbola is plotted on logarithmic paper? (See Exercises 57–60 of Section 1–4.)

33. An electronic instrument located at point P records the sound of a rifle shot and the sound of the bullet striking the target at the same instant. Show that P lies on a branch of a hyperbola.

34. Two concentric hyperbolas are called conjugate hyperbolas if the transverse and conjugate axes of one are respectively the conjugate and transverse axes of the other. What is the equation of the hyperbola conjugate to the hyperbola of Exercise 14?

1–9 Translation of Axes

Until now, except by direct use of the definition, the equations considered for the parabola, the ellipse, and the hyperbola have been restricted to the particular cases in which the vertex of the parabola is at the origin and the center of the ellipse or hyperbola is at the origin. In this section we shall consider, without specific use of the definition, the equations of these curves for the cases in which the axis of the curve is parallel to one of the coordinate axes. This is done by **translation of axes.**

We choose a point (h, k) in the xy-coordinate plane and let this point be the origin of another coordinate system, the $x'y'$-coordinate system. The x'-axis is parallel to the x-axis and the y'-axis is parallel to the y-axis. Every point in the plane now has two sets of coordinates associated with it, (x, y) and (x', y'). From Fig. 1–62 we see that

$$x = x' + h \quad \text{and} \quad y = y' + k \tag{1–26}$$

Equations (1–26) can also be written in the form

$$x' = x - h \quad \text{and} \quad y' = y - k \tag{1–27}$$

The following examples illustrate the use of Eqs. (1–27) in the analysis of equations of the parabola, ellipse, and hyperbola.

Figure 1–62

Example A

Describe the locus of the equation

$$\frac{(x - 3)^2}{25} + \frac{(y + 2)^2}{9} = 1$$

In this equation, given that $h = 3$ and $k = -2$, we have $x' = x - 3$ and $y' = y + 2$. In terms of x' and y', the equation is

$$\frac{(x')^2}{25} + \frac{(y')^2}{9} = 1$$

We recognize this equation as that of an ellipse (see Fig. 1–63) with a semimajor axis of 5 and a semiminor axis of 3. The center of the ellipse is at $(3, -2)$ since this was the choice of h and k to make the equation fit a standard form.

Figure 1–63

Example B

Find the equation of the parabola with vertex at $(2, 4)$ and focus at $(4, 4)$.

If we let the origin of the $x'y'$-coordinate system be the point $(2, 4)$, the point $(4, 4)$ would be the point $(2, 0)$ in the $x'y'$-system. This means that $p = 2$ and $4p = 8$ (Fig. 1–64). In the $x'y'$-system, the equation is

$$(y')^2 = 8(x')$$

Since $(2, 4)$ is the origin of the $x'y'$-system, this means that $h = 2$ and $k = 4$. Using Eq. (1–27), we have

$$(y - 4)^2 = 8(x - 2)$$

as the equation of the parabola in the xy-coordinate system. If this equation is multiplied out, and like terms are combined, we obtain

$$y^2 - 8x - 8y + 32 = 0$$

Figure 1–64

Example C

Find the center of the hyperbola $2x^2 - y^2 - 4x - 4y - 4 = 0$.

To analyze this curve, we first complete the square in the x-terms and in the y-terms. This will allow us to recognize properly the choice of h and k.

$$2x^2 - 4x - y^2 - 4y = 4$$

$$2(x^2 - 2x \qquad) - (y^2 + 4y \qquad) = 4 \qquad \overset{\text{Subtract 4}}{\downarrow}$$

$$2(x^2 - 2x + 1) - (y^2 + 4y + 4) = 4 + 2 - 4$$

We note here that when we added 1 to complete the square of the x-terms within the parentheses, we were actually adding 2 to the left side. Thus, we added 2 to the right side. Similarly, when we added 4 to the y-terms within the parentheses, we were actually subtracting 4 from the left side. Continuing, we have

$$2(x - 1)^2 - (y + 2)^2 = 2$$

$$\frac{(x - 1)^2}{1} - \frac{(y + 2)^2}{2} = 1$$

Figure 1–65

Thus, if we let $h = 1$ and $k = -2$, the equation in the $x'y'$-system becomes

$$\frac{(x')^2}{1} - \frac{(y')^2}{2} = 1$$

This means that the center is at $(1, -2)$, since this point corresponds to the origin of the $x'y'$-coordinate system (see Fig. 1–65).

Example D

Figure 1–66

Find the vertex of the parabola $2x^2 - 12x - 3y + 15 = 0$.

First, we complete the square in the x-terms, placing all other resulting terms on the right. By factoring the resulting expression on the right, we may determine the values of h and k.

$$2x^2 - 12x = 3y - 15$$
$$2(x^2 - 6x\ \ \ \ ) = 3y - 15$$
$$2(x^2 - 6x + 9) = 3y - 15 + 18$$
$$2(x - 3)^2 = 3(y + 1)$$
$$(x - 3)^2 = \frac{3}{2}(y + 1)$$
$$x'^2 = \frac{3}{2}y'$$

Therefore, the vertex is at $(3, -1)$, since this point corresponds to the origin of the $x'y'$-coordinate system. See Fig. 1–66.

Example E

Glass beakers are to be made with a height of 3 in. Express the surface area of the beakers in terms of the radius of the base. Sketch the graph of area versus radius.

The total surface area of a beaker is the sum of the area of the base and the lateral surface area of the side. In general, this surface area S in terms of the radius r of the base and the height h of the side is

$$S = \pi r^2 + 2\pi rh$$

Since h is constant at 3 in., we have

$$S = \pi r^2 + 6\pi r$$

which is the desired relationship.

For the purposes of sketching the graph of S and r, we now complete the square of the r terms:

$$S = \pi(r^2 + 6r)$$
$$S + 9\pi = \pi(r^2 + 6r + 9)$$
$$S + 9\pi = \pi(r + 3)^2$$
$$(r + 3)^2 = \frac{1}{\pi}(S + 9\pi)$$

(continued on next page)

We note that this equation represents a parabola with vertex at $(-3, -9\pi)$ for its coordinates (r, S). Since $4p = \frac{1}{\pi}$, $p = \frac{1}{4\pi}$ and the focus of this parabola is at $(-3, \frac{1}{4\pi} - 9\pi)$. Only positive values for S and r have meaning, and therefore the part of the graph for negative r is shown as a dashed curve (see Fig. 1–67).

Figure 1–67

Exercises 1–9

In Exercises 1 through 8, describe the locus of each of the given equations. Identify the type of curve and its center (vertex if it is a parabola). Sketch each curve.

1. $(y - 2)^2 = 4(x + 1)$

2. $\dfrac{(x + 4)^2}{4} + \dfrac{(y - 1)^2}{1} = 1$

3. $\dfrac{(x - 1)^2}{4} - \dfrac{(y - 2)^2}{9} = 1$

4. $(y + 5)^2 = -8(x - 2)$

5. $\dfrac{(x + 1)^2}{1} + \dfrac{y^2}{9} = 1$

6. $\dfrac{(y - 4)^2}{16} - \dfrac{(x + 2)^2}{4} = 1$

7. $(x + 3)^2 = -12(y - 1)$

8. $\dfrac{x^2}{16} + \dfrac{(y + 1)^2}{1} = 1$

In Exercises 9 through 20, find the equation of each of the curves described by the given information.

9. Parabola: vertex $(-1, 3)$, $p = 4$, axis parallel to x-axis.

10. Parabola: vertex $(2, -1)$, directrix $y = 3$.

11. Parabola: vertex $(-3, 2)$, focus $(-3, 3)$.

12. Parabola: focus $(2, 4)$, directrix $x = 6$.

13. Ellipse: center $(-2, 2)$, focus $(-5, 2)$, vertex $(-7, 2)$.

14. Ellipse: center $(0, 3)$, focus $(12, 3)$, major axis 26 units.

15. Ellipse: vertices $(-2, -3)$ and $(-2, 5)$, end of minor axis $(0, 1)$.

16. Ellipse: foci $(1, -2)$ and $(1, 10)$, minor axis 5 units.

17. Hyperbola: vertex $(-1, 1)$, focus $(-1, 4)$, center $(-1, 2)$.

18. Hyperbola: foci $(2, 1)$ and $(8, 1)$, conjugate axis 6 units.

19. Hyperbola: vertices $(2, 1)$ and $(-4, 1)$, focus $(-6, 1)$.
20. Hyperbola: center $(1, -4)$, focus $(1, 1)$, transverse axis 8 units.

In Exercises 21 through 28, determine the center (or vertex if the curve is a parabola) of the given curves. Sketch each curve.

21. $x^2 + 2x - 4y - 3 = 0$ 22. $y^2 - 2x - 2y - 9 = 0$
23. $4x^2 + 9y^2 + 24x = 0$ 24. $2x^2 + 9y^2 + 8x - 72y + 134 = 0$
25. $9x^2 - y^2 + 8y - 7 = 0$ 26. $5x^2 - 4y^2 + 20x + 8y = 4$
27. $2x^2 - 4x = 9y - 2$ 28. $4x^2 + 16x = y - 20$

In Exercises 29 through 32, find the required equations.

29. Find the equation of the hyperbola with asymptotes $x - y = -1$ and $x + y = -3$, and vertex $(3, -1)$

30. The circle $x^2 + y^2 + 4x - 5 = 0$ passes through the foci and the ends of the minor axis of an ellipse which has its major axis along the x-axis. Find the equation of the ellipse.

31. A first parabola has its vertex at the focus of a second parabola, and its focus at the vertex of the second parabola. If the equation of the second parabola is $y^2 = 4x$, find the equation of the first parabola.

32. Verify each of the following equations as being the standard form as indicated. Parabola, vertex at (h, k), axis parallel to the x-axis:

$$(y - k)^2 = 4p(x - h)$$

Parabola, vertex at (h, k), axis parallel to the y-axis:

$$(x - h)^2 = 4p(y - k)$$

Ellipse, center at (h, k), major axis parallel to the x-axis:

$$\frac{(x - h)^2}{a^2} + \frac{(y - k)^2}{b^2} = 1$$

Ellipse, center at (h, k), major axis parallel to the y-axis:

$$\frac{(y - k)^2}{a^2} + \frac{(x - h)^2}{b^2} = 1$$

Hyperbola, center at (h, k), transverse axis parallel to the x-axis:

$$\frac{(x - h)^2}{a^2} - \frac{(y - k)^2}{b^2} = 1$$

Hyperbola, center at (h, k), transverse axis parallel to the y-axis:

$$\frac{(y - k)^2}{a^2} - \frac{(x - h)^2}{b^2} = 1$$

In Exercises 33 through 36, solve the given problems.

33. The power supplied to a circuit by a battery with a voltage E and an internal resistance r is given by $P = EI - rI^2$, where P is the power (in watts) and I is the current (in amperes). Sketch the graph of P vs. I for a 6.0 V battery with an internal resistance of 0.30 Ω.

34. A calculator company determined that its total income I from the sale of a particular type of calculator is given by $I = 100x - x^2$, where x is the selling price of the calculator. Sketch a graph of I vs. x.

35. The planet Pluto moves about the sun in an elliptical orbit, with the sun at one focus. The closest that Pluto approaches the sun is 2.8 billion miles, and the farthest it gets from the sun is 4.6 billion miles. If the sun is at the origin of a coordinate system and the other focus is on the positive x-axis, what is the equation of the path of Pluto?

36. A rectangular tract of land is to have a perimeter of 800 m. Express the area in terms of its width and sketch the graph.

1–10 The Second-Degree Equation

The equations of the circle, parabola, ellipse, and hyperbola are all special cases of the same general equation. In this section we discuss this equation and how to identify the particular form it takes when it represents a specific type of curve.

Each of these curves can be represented by a **second-degree equation** *of the form*

$$Ax^2 + Bxy + Cy^2 + Dx + Ey + F = 0 \tag{1–28}$$

The coefficients of the second-degree terms determine the type of curve which results. Recalling the discussions of the general forms of the equations of the circle, parabola, ellipse, and hyperbola from the previous sections of this chapter, we have the following results.

Equation (1–28) represents the indicated curve for the given conditions for A, B, and C.

1. If $A = C$, $B = 0$, a circle.
2. If $A \neq C$ (but they have the same sign), $B = 0$, an ellipse.
3. If A and C have different signs, $B = 0$, a hyperbola.
4. If $A = 0$, $C = 0$, $B \neq 0$, a hyperbola.
5. If either $A = 0$ or $C = 0$ (but not both), $B = 0$, a parabola.
   (Special cases, such as a single point or no real locus, can also result.)

Another conclusion about Eq. (1–28) is that, if either $D \neq 0$ or $E \neq 0$ (or both), the center (or vertex of a parabola) of the curve is not at the origin. If $B \neq 0$, the axis of the curve has been rotated. We have considered only one such case (the hyperbola $xy = c$) in this chapter.

Example A

For reference, Eq. (1–12) is $x^2 + y^2 = r^2$.

The equation $2x^2 = 3 - 2y^2$ represents a circle. This can be seen by putting it in the form of Eq. (1–28). This form is

$$2x^2 + 2y^2 - 3 = 0$$

Here $A = C$ and, since there is no xy-term, $B = 0$. This means that it is a circle. If we write the equation as $x^2 + y^2 = \frac{3}{2}$ it fits the form of Eq. (1–12).

Example B

The equation $3x^2 = 6x - y^2 + 3$ represents an ellipse. Before we analyze the equation, we should put it in the form of Eq. (1–28). For the given equation, this form is

$$3x^2 + y^2 - 6x - 3 = 0$$

Here we see that $B = 0$, A and C have the same sign, and $A \neq C$. Therefore, the curve is an ellipse. The $-6x$ term indicates that the center of the ellipse is not at the origin.

Example C

The equation $2(x + 3)^2 = y^2 + 2x^2$ represents a parabola. Putting it in the form of Eq. (1–28), we have

$$2(x^2 + 6x + 9) = y^2 + 2x^2$$
$$2x^2 + 12x + 18 = y^2 + 2x^2$$
$$y^2 - 12x - 18 = 0$$

We now note that $A = 0$, $B = 0$, and $C \neq 0$. This indicates that the equation represents a parabola. Here, the -18 term indicates that the vertex is not at the origin.

Example D

Identify the curve represented by the equation $2x^2 + 12x = y^2 - 14$. Determine the appropriate important quantities associated with the curve, and sketch the graph.

Writing this equation in the form of Eq. (1–28), we have

$$2x^2 - y^2 + 12x + 14 = 0$$

In this form we identify the equation as representing a hyperbola, since A and C have different signs, and $B = 0$. We now write it in the standard form of a hyperbola.

$$2x^2 + 12x - y^2 = -14$$
$$2(x^2 + 6x \quad) - y^2 = -14$$
$$2(x^2 + 6x + 9) - y^2 = -14 + 18$$
$$2(x + 3)^2 - y^2 = 4$$
$$\frac{(x + 3)^2}{2} - \frac{y^2}{4} = 1$$
$$\frac{x'^2}{2} - \frac{y'^2}{4} = 1$$

Figure 1–68

Thus, we see that the center (h, k) of the hyperbola is the point $(-3, 0)$. Also, $a = \sqrt{2}$ and $b = 2$. This means that the vertices are $(-3 + \sqrt{2}, 0)$ and $(-3 - \sqrt{2}, 0)$, and the conjugate axis extends from $(-3, 2)$ to $(-3, -2)$. Also, $c^2 = 2 + 4 = 6$, which means that $c = \sqrt{6}$. The foci are $(-3 + \sqrt{6}, 0)$ and $(-3 - \sqrt{6}, 0)$. The graph is shown in Fig. 1–68.

Example E

Identify the curve represented by the equation $4y^2 - 23 = 4(4x + 3y)$. Determine the appropriate important quantities associated with the curve, and sketch the graph.

Writing this equation in the form of Eq. (1–28), we have

$$4y^2 - 23 = 16x + 12y$$
$$4y^2 - 16x - 12y - 23 = 0$$

Therefore, we recognize the equation as representing a parabola, since $A = 0$ and $B = 0$. Now writing the equation in the standard form of a parabola, we have

$$4y^2 - 12y = 16x + 23$$
$$4(y^2 - 3y \quad) = 16x + 23$$
$$4\left(y^2 - 3y + \frac{9}{4}\right) = 16x + 23 + 9$$
$$4\left(y - \frac{3}{2}\right)^2 = 16(x + 2)$$
$$\left(y - \frac{3}{2}\right)^2 = 4(x + 2)$$
$$y'^2 = 4x'$$

Figure 1–69

We now note that the vertex is the point $(-2, \frac{3}{2})$ and that $p = 1$. Also, it is symmetric to the x'-axis. Therefore, the focus is $(-1, \frac{3}{2})$ and the directrix is $x = -3$. The graph is shown in Fig. 1–69.

The circle, parabola, ellipse, and hyperbola are usually referred to as **conic sections.** If a plane is passed through a cone, the intersection of the plane and the cone results in one of these curves; the curve formed depends on the angle of the plane with respect to the axis of the cone. This is indicated in Fig. 1–70.

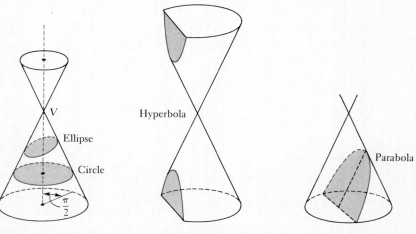

Figure 1–70

Exercises 1–10

In Exercises 1 through 20, identify each of the equations as representing either a circle, parabola, ellipse, or hyperbola.

1. $x^2 + 2y^2 - 2 = 0$

2. $x^2 - y = 0$

3. $2x^2 - y^2 - 1 = 0$

4. $3x^2 + 3y^2 - 1 = 0$

5. $2x^2 + 2y^2 - 3y - 1 = 0$

6. $x^2 - 2y^2 - 3x - 1 = 0$

7. $2x^2 - x - y = 1$

8. $2x^2 + 4y^2 - y - 2x = 4$

9. $x^2 = y^2 - 1$

10. $3x^2 = 2y - 4y^2$

11. $x^2 = y - y^2$

12. $y = 3 - 6x^2$

13. $x(y + 3x) = x^2 + xy - y^2 + 1$

14. $x(2 - x) = y^2$

15. $2xy + x - 3y = 6$

16. $(y + 1)^2 = x^2 + y^2 - 1$

17. $2x(x - y) = y(3 - y - 2x)$

18. $2x^2 = x(x - 1) + 4y^2$

19. $y(3 - 2y) = 2(x^2 - y^2)$

20. $4x(x - 1) = 2x^2 - 2y^2 + 3$

In Exercises 21 through 28, identify the curve represented by each of the given equations. Determine the appropriate important quantities associated with the curve, and sketch the graph.

21. $x^2 = 8(y - x - 2)$

22. $x^2 = 6x - 4y^2 - 1$

23. $y^2 = 2(x^2 - 2x - 2y)$

24. $4x^2 + 4 = 9 - 8x - 4y^2$

25. $y^2 + 42 = 2x(10 - x)$

26. $x^2 - 4y = y^2 + 4(1 - x)$

27. $4(y^2 - 4x - 2) = 5(4y - 5)$

28. $2(2x^2 - y) = 8 - y^2$

In Exercises 29 through 32, use the given values to determine the type of curve represented.

29. For the equation $x^2 + ky^2 = a^2$, what type of curve is represented if (a) $k = 1$, (b) $k < 0$, and (c) $k > 0$ $(k \neq 1)$?

30. For the equation $\dfrac{x^2}{4 - C} - \dfrac{y^2}{C} = 1$, what type of curve is represented if (a) $C < 0$, (b) $0 < C < 4$? (For $C > 4$, see Exercise 32.)

31. In Eq. (1–28), if $A = C \neq 0$ and $B = D = E = F = 0$, what locus is described?

32. For the equation in Exercise 30, what type of locus is described if $C > 4$?

In Exercises 33 through 36, set up the necessary equation and then determine the type of curve it represents.

33. For a given alternating-current circuit the resistance and capacitive reactance are constant. What type of curve is represented by the equation relating impedance and inductive reactance? (See Exercise 42 of Section 1–5.)

34. A room is 8.0 ft high, and the length is 6.5 ft longer than the width. Express the volume V of the (rectangular) room in terms of the width w. What type of curve is represented by the equation?

35. The sides of a rectangle are $2x$ and y, and the diagonal is $x + 5$. What type of curve is represented by the equation relating x and y?

36. The legs of a right triangle are x and y and the hypotenuse is $x + 2$. What type of curve is represented by the equation relating x and y?

1–11 Review Exercises for Chapter 1

In Exercises 1 through 12, find the equation of the indicated curve subject to the given conditions. Sketch each curve.

1. Straight line: passes through $(1, -7)$ with a slope of 4.
2. Straight line: passes through $(-1, 5)$ and $(-2, -3)$.
3. Straight line: perpendicular to $3x - 2y + 8 = 0$ and has a y-intercept of -1.
4. Straight line: parallel to $2x - 5y + 1 = 0$ and has an x-intercept of 2.
5. Circle: center at $(1, -2)$, passes through $(4, -3)$.
6. Circle: tangent to the line $x = 3$, center at $(5, 1)$.
7. Parabola: focus $(3, 0)$, vertex $(0, 0)$.
8. Parabola: directrix $y = -5$, vertex $(0, 0)$.
9. Ellipse: vertex $(10, 0)$, focus $(8, 0)$, center $(0, 0)$.
10. Ellipse: center $(0, 0)$, passes through $(0, 3)$ and $(2, 1)$.
11. Hyperbola: vertex $(0, 13)$, center $(0, 0)$, conjugate axis of 24.
12. Hyperbola: foci $(0, 10)$ and $(0, -10)$, vertex $(0, 8)$.

In Exercises 13 through 24, find the indicated quantities for each of the given equations. Sketch each curve.

13. $x^2 + y^2 + 6x - 7 = 0$, center and radius
14. $x^2 + y^2 - 4x + 2y - 20 = 0$, center and radius
15. $x^2 = -20y$, focus and directrix
16. $y^2 = 24x$, focus and directrix
17. $16x^2 + y^2 = 16$, vertices and foci
18. $2y^2 - 9x^2 = 18$, vertices and foci
19. $2x^2 - 5y^2 = 8$, vertices and foci
20. $2x^2 + 25y^2 = 50$, vertices and foci
21. $x^2 - 8x - 4y - 16 = 0$, vertex and focus
22. $y^2 - 4x + 4y + 24 = 0$, vertex and directrix
23. $4x^2 + y^2 - 16x + 2y + 13 = 0$, center
24. $x^2 - 2y^2 + 4x + 4y + 6 = 0$, center

In Exercises 25 through 28, determine the number of real solutions to the given systems of equations by sketching the curves and finding the number of points of intersection.

25. $x^2 + y^2 = 9$
    $4x^2 + y^2 = 16$

26. $x - 2y - 1 = 0$
    $x^2 - y^2 = 1$

27. $x^2 + y^2 - 4y - 5 = 0$
    $y^2 - 4x^2 - 4 = 0$

28. $x^2 - 4y^2 + 2x - 3 = 0$
    $y^2 - 4x - 4 = 0$

In Exercises 29 through 56, solve the given problems.

29. Find the points of intersection of the ellipses $25x^2 + 4y^2 = 100$ and $4x^2 + 9y^2 = 36$.
30. Find the points of intersection of the hyperbola $y^2 - x^2 = 1$ and the ellipse $x^2 + 25y^2 = 25$.

31. In two ways show that the line segments joining $(-3, 11)$, $(2, -1)$, and $(14, 4)$ form a right triangle.

32. Show that the altitudes of the triangle with vertices $(2, -4)$, $(3, -1)$, and $(-2, 5)$ meet at a single point.

33. By means of the definition of a parabola, find the equation of the parabola with focus at $(3, 1)$ and directrix the line $y = -3$. Find the same equation by the method of translation of axes.

34. Repeat the instructions of Exercise 33 for the parabola with focus at $(0, 2)$ and directrix the line $y = 4$.

35. A piece x centimeters long is cut from a board 250 cm long. Sketch the graph of the remaining length y as a function of x.

36. The acceleration of an object is defined as the change in velocity v divided by the corresponding change in time t. Find the equation relating the velocity v and time t for an object for which the acceleration is 20 ft/s² and $v = 5.0$ ft/s when $t = 0$ s.

37. In an electric circuit the voltage V equals the product of the current I and the resistance R. Sketch the graph of V (in volts) vs. I (in amperes) for a circuit in which $R = 3\ \Omega$.

38. An airplane touches down when landing at 100 mi/h. Its velocity v while coming to a stop is given by $v = 100 - 20{,}000t$, where t is the time in hours. Sketch the graph of v vs. t.

39. In a certain electric circuit, two resistors having resistances R_1 and R_2, respectively, act as a voltage divider. The relation between them is $\alpha = R_1/(R_1 + R_2)$, where α is a constant less than 1. Sketch the curve of R_1 vs. R_2 (both in ohms), if $\alpha = \frac{1}{2}$.

40. Let C and F denote corresponding Celsius and Fahrenheit temperature readings. If the equation relating the two is linear, determine this equation given that $C = 0$ when $F = 32$ and $C = 100$ when $F = 212$.

41. A solar panel reflector is circular and has a circumference of 4.50 m. Taking the origin of a coordinate system at the center of the reflector, what is the equation of the circumference?

42. A friction disk drive wheel is tangent to another wheel, as shown in Fig. 1–71. The radii of the wheels are 2 in. and 6 in. What is the equation of the circumference of each wheel if the coordinate system is located at the center of the drive wheel as shown?

Figure 1–71

43. The arch of a small bridge across a stream is parabolic. If, at water level, the span of the arch is 80 ft and the maximum height of the span above water level is 20 ft, what is the equation of the arch? Choose the most convenient point for the origin of the coordinate system.

44. If a source of light is placed at the focus of a parabolic reflector, the reflected rays are parallel. Where is the focus of a parabolic reflector which is 8 cm across and 6 cm deep?

45. Sketch the curve of the total surface area of a right circular cylinder as a function of its radius, if the height is always 10 units.

46. At very low temperatures certain metals have an electric resistance of zero. This phenomenon is called superconductivity. A magnetic field also affects the superconductivity. A certain level of magnetic field, the threshold field, is related to the

temperature T by

$$\frac{H_T}{H_0} = 1 - \left(\frac{T}{T_0}\right)^2$$

where H_0 and T_0 are specifically defined values of magnetic field and temperature. Sketch H_T/H_0 vs. T/T_0.

47. The vertical position of a projectile is given by $y = 120t - 16t^2$, and its horizontal position is given by $x = 60t$. By eliminating the time t, determine the path of the projectile. Sketch this path for the length of time it would need to strike the ground, assuming level terrain.

48. The rate r in grams per second at which a substance is formed in a chemical reaction is given by $r = 10.0 - 0.010t^2$. Sketch the graph of r vs. t.

49. The inside of the top of an arch is a semiellipse with an equation of $x^2 + 4y^2 = 100$. The outside of the arch is also a semiellipse with an equation $2x^2 + 5y^2 = 500$. If measurements are in meters, what is the thickness of the arch (a) at each base and (b) at the center? See Fig. 1–72, where t_b and t_c represent these thicknesses.

$2x^2 + 5y^2 = 500$

t_c

$x^2 + 4y^2 = 100$

t_b t_b

Figure 1–72

50. The earth moves about the sun in an elliptical path. If the closest the earth gets to the sun is 90.5 million miles, and the farthest it gets from the sun is 93.5 million miles, find the equation of the path of the earth. The sun is at one focus of the ellipse. Assume the major axis to be along the x-axis and that the center of the ellipse is at the origin.

51. Soon after reaching the vicinity of the moon, Apollo 11 (the first spacecraft to land a man on the moon) went into an elliptical lunar orbit. The closest the craft was to the moon in this orbit was 70 mi, and the farthest it was from the moon was 190 mi. What was the equation of the path if the moon was at one of the foci of the ellipse? Assume the major axis to be along the x-axis and that the center of the ellipse is at the origin. The radius of the moon is 1080 mi.

52. A machine-part designer wishes to make a model for an elliptical cam by placing two pins in her design board, putting a loop of string over the pins, and marking off the outline by keeping the string taut. (Note that she is using the definition of the ellipse.) If the cam is to measure 10 cm by 6 cm, how long should the loop of string be and how far apart should the pins be?

53. Under certain conditions the work W done on a wire by increasing the tension from T_1 to T_2 is given by $W = k(T_2^2 - T_1^2)$, where k is a constant depending on the properties of the wire. If the work is constant for various sets of initial and final tensions, sketch a graph of T_2 vs. T_1.

2 ft -------- 1 ft --------

|—————— 10 ft ——————|

Figure 1–73

54. In order to study the relationship between velocity and pressure of a stream of water, a pipe is constructed such that its cross section is a hyperbola. If the pipe is 10 ft long, 1 ft in diameter at the narrow point (middle), and 2 ft in diameter at the ends, what is the equation of the cross section of the pipe? (In physics it is shown that where the velocity of a fluid is greatest, the pressure is least.) See Fig. 1–73.

55. A hallway 16 ft wide has a ceiling whose cross section is a semiellipse. The ceiling is 10 ft high at the walls and 14 ft high at the center. Find the height of the ceiling 4 ft from each wall.

Length of rope = 60 ft

10 ft

h

4 ft

x

Figure 1–74

56. A 60-ft rope passes over a pulley 10 ft above the ground and a crate on the ground is attached at one end. A man holds the other end at a level of 4 ft above the ground. If the man walks away from the pulley, express the height of the crate above the ground in terms of the distance the man is from directly below the crate. Sketch the graph of distance vs. height. See Fig. 1–74. (Neglect the thickness of the crate.)

CHAPTER 2

The Derivative

In this chapter we shall start developing the methods of **differential calculus.** As we mentioned on p. 1, this subject deals with the very important problem involving the *rate of change* of one quantity with respect to another. For example, velocity is the rate of change of distance with respect to time.

Isaac Newton (1642–1727), an English mathematician and physicist, and Gottfried Wilhelm Leibniz (1646–1716), a German mathematician and philosopher, are credited with the creation of calculus. In order to solve problems in fields such as geometry and astronomy, each independently developed the basic methods of calculus.

Calculus has numerous applications in many different fields of study. It is fundamental for the study of higher mathematics and has extensive applications in the physical sciences, social sciences, and engineering. We will illustrate many of these applications throughout the book.

Before dealing directly with the concept of rate of change, we will review the meaning of a *function* in Section 2–1 and introduce the concept of a *limit* in Section 2–2.

2–1 Algebraic Functions

In Chapter 1 we discussed a number of equations and their associated graphs. In each case discussed, the equation expressed a relationship between the two variables. *If the equation is solved for one of the variables, say y, in terms of the other, x, then we say that y is a* **function** *of x. Here y is called the* **dependent variable** *and x is called the* **independent variable.**

Example A

The equation of a straight line written in slope-intercept form, $y = mx + b$, expresses y, the dependent variable, as a function of x, the independent variable.

Relationships which can be expressed as functions exist between various quantities in all fields of technology. An experiment to determine whether or not a relationship exists between the distance an object falls and the time of fall should indicate (approximately) that $s = 16t^2$, where s is the distance in feet and t is the time in seconds. Electrical measurements of current and voltage with respect to a particular resistor would show that $V = kI$, where V is the voltage, I is the current, and k is a constant.

Example B

The power developed in a certain resistor by a current I is given by $P = 4I^2$. Here P is a function of I. The dependent variable is P, and the independent variable is I.

As we have mentioned, the rate of change of a quantity is important in many physical situations. Since relationships between the appropriate quantities can be expressed as functions, and since we can determine the rate of change of a function through differential calculus, we need to review various operations on functions.

We shall now make our definition of a function more explicit. *Whenever a relationship exists between two variables such that for every permissible value of the first there is only one corresponding value of the second, then the second variable is a* **function** *of the first variable.*

In this definition, *the first variable is the independent variable* and *the second variable is the dependent variable.* Since we will be dealing only with functions whose values are real numbers, it is possible that there are restrictions on the permissible values of the variables.

There are many ways to express functions. Formulas, such as those mentioned above, define functions. Functions can also be expressed by means of tables, charts, and graphs.

Example C

We can demonstrate an important aspect of a function by inspecting the graphs shown in Figures 2–1(a) and 2–1(b).

In Fig. 2–1(a) there is only one value of y for each value of x. In Fig. 2–1(b) we can see that there are two values of y for some values of x. Therefore, the graph in Fig. 2–1(a) defines a function, whereas the graph in Fig. 2–1(b) does *not* define a function. There must be only one value of the dependent variable for each value of the independent variable for a relationship to be a function.

Figure 2–1   (a)   (b)

Example D

For the equation $y = 2x^2 - 6x$, we say that y is a function of x. The dependent variable is y, and the independent variable is x. Some of the values of y corresponding to chosen values of x are given in the following table.

x	-2	-1	0	$\frac{1}{2}$	1	2	3	π	10
y	20	8	0	$-\frac{5}{2}$	-4	-4	0	$2\pi^2 - 6\pi$	140

For convenience of notation,

the phrase "function of x" is written as $f(x)$.

This, in turn, means that the statement "y is a function of x" may be written as $y = f(x)$. (Note that $f(x)$ does **not** mean f times x.)

Example E

If $y = 6x^3 - 5x$, we may say that y is a function of x, where this function $f(x)$ is $6x^3 - 5x$. We may also write $f(x) = 6x^3 - 5x$. It is common to write functions in this form, rather than in the form $y = 6x^3 - 5x$. However, y and $f(x)$ represent the same expression.

One of the most important uses of this notation is to designate the value of a function for a particular value of the independent variable. That is, *the value of the function $f(x)$ when $x = a$ is written as $f(a)$.*

Example F

For the function $f(x) = 3x^2 - 5$, the value of $f(x)$ for $x = 2$ may be represented as $f(2)$. Thus, substituting 2 for x, we have

$$f(2) = 3(2^2) - 5 = 7$$

In the same way, the value of $f(x)$ for $x = -1$ is

$$f(-1) = 3(-1)^2 - 5 = -2$$

The value of $f(x)$ for $x = a + 2$ is

$$f(a + 2) = 3(a + 2)^2 - 5 = 3a^2 + 12a + 7$$

where we have substituted an expression involving a literal number.

In certain instances we need to define more than one function of x. Then we use different symbols to denote the functions. For example, $f(x)$ and $g(x)$ may represent different functions of x, such as $f(x) = 5x^2 - 3$ and $g(x) = 6x - 7$. Special functions are represented by particular symbols. For example, in trigonometry we shall come across the "sine of the angle ϕ," where the sine is a function of ϕ. This is designated by sin ϕ.

Example G

If $f(x) = \sqrt{3x} + x$ and $g(x) = ax^4 - 5x$, then

$$f(3) = \sqrt{3(3)} + 3 = 3 + 3 = 6$$

and

$$g(3) = a(3^4) - 5(3) = 81a - 15$$

A function may be looked upon as a set of instructions. These instructions tell us how to obtain the value of the dependent variable for a particular value of the independent variable, even if the set of instructions is expressed in literal symbols.

Example H

The function $f(x) = x^2 - 3x$ tells us to "square the value of the independent variable, multiply the value of the independent variable by 3, and subtract the second result from the first." An analogy would be a computer which was programmed so that when a number was entered into the computer, it would square the number, then multiply the number by 3, and finally subtract the second result from the first. This is represented in diagram form in Fig. 2–2.

Figure 2–2

The functions $f(t) = t^2 - 3t$ and $f(n) = n^2 - 3n$ are the same as the function $f(x) = x^2 - 3x$, since the operations performed on the independent variable are the same. Although different literal symbols appear, this does not change the function.

As we mentioned earlier, the values of the independent variable and the dependent variable must be real numbers. This means that the values chosen for the independent variable must give real number values for the dependent variable. Therefore, *values of the independent variable that lead to division by zero or to imaginary values of the dependent variable may not be used*. The following example illustrates this point.

Example I

The function $f(u) = \dfrac{u}{u - 4}$ is not defined if $u = 4$, since this would require division by zero. Therefore, the values of u are restricted to values other than 4.

The function $g(s) = \sqrt{3 - s}$ is not defined for real numbers if s is greater than 3, since such values would result in imaginary values for $g(s)$. Thus, the values of s are restricted to values equal to or less than 3.

Practical physical considerations may also restrict the values of the independent variable. For example, the surface area of a cube, as a function of its edge, is given by $A = 6e^2$. However, since the edge of a cube cannot be negative and there is no cube if $e = 0$, the values of e must be positive.

The functions which are studied in mathematics fall into two general classes: **algebraic functions** and **transcendental functions.** In this chapter we shall be concerned with the calculus associated with algebraic functions. Transcendental functions, such as the trigonometric functions and the logarithmic function, will be taken up in later chapters.

There are various types of algebraic functions. Those which are of primary concern here are as follows.

Polynomials *are of the form*

$$P(x) = a_0 x^n + a_1 x^{n-1} + \cdots + a_{n-1} x + a_n \tag{2-1}$$

where the a's are constants and n is a nonnegative integer.

Rational functions *are of the form*

$$F(x) = \frac{f(x)}{g(x)} \tag{2-2}$$

where f(x) and g(x) are polynomials.

Roots *of rational functions are of the form*

$$I(x) = \sqrt[n]{F(x)} \tag{2-3}$$

where $F(x)$ is a rational function. (A polynomial is also a rational function.) Combinations of these functions will also be encountered frequently.

Example J

The current I in an electric circuit equals the voltage E divided by the resistance R. If the voltage and resistance are functions of time, given by

$$E(t) = 2t^2 + t + 5 \quad \text{and} \quad R(t) = 3t + 20$$

we see from Eq. (2-1) that both are polynomials. (Also, both are rational functions, since we can express each with a denominator of 1, and 1 is a polynomial with $a_0 = 1$ and $n = 0$.)

The current is given by

$$I(t) = \frac{E(t)}{R(t)} = \frac{2t^2 + t + 5}{3t + 20}$$

which is a rational function.

Example K

If $f(x) = 2x^2 - 3$ and $g(x) = \sqrt{x + 2}$, then

$$f(x) + g(x) = 2x^2 - 3 + \sqrt{x \cdot + 2}$$

Also, we may write

$$g[f(x)] = \sqrt{(2x^2 - 3) + 2}$$
$$= \sqrt{2x^2 - 1}$$

Here we replace x in the function $g(x)$ with the expression for $f(x)$. This is the meaning of $g[f(x)]$, following functional notation.

Since we will be dealing with exponents and radicals, we shall now review their basic properties. The basic laws of exponents are as follows.

$$a^m \cdot a^n = a^{m+n} \tag{2-4}$$

$$\frac{a^m}{a^n} = a^{m-n}, \qquad a \neq 0 \tag{2-5}$$

$$(a^m)^n = a^{mn} \tag{2-6}$$

$$(ab)^n = a^n b^n, \qquad \left(\frac{a}{b}\right)^n = \frac{a^n}{b^n}, \qquad b \neq 0 \tag{2-7}$$

$$a^0 = 1, \qquad a \neq 0 \tag{2-8}$$

$$a^{-n} = \frac{1}{a^n}, \qquad a \neq 0 \tag{2-9}$$

We shall often find it convenient to express radicals in the form of fractional exponents. The relationship between fractional exponents and radicals, along with the basic operations with radicals, are listed below.

$$a^{m/n} = \sqrt[n]{a^m} \qquad (m \text{ and } n \text{ integers}, \quad n > 0) \tag{2-10}$$

$$\sqrt[n]{a^n} = a \tag{2-11}$$

$$\sqrt[m]{\sqrt[n]{a}} = \sqrt[mn]{a} \tag{2-12}$$

$$\sqrt[n]{a} \cdot \sqrt[n]{b} = \sqrt[n]{ab} \tag{2-13}$$

$$\frac{\sqrt[n]{a}}{\sqrt[n]{b}} = \sqrt[n]{\frac{a}{b}}, \qquad b \neq 0 \tag{2-14}$$

As we have noted, Eqs. (2–10) through (2–14) are valid with m and n as integers, $n > 0$. This also means that $m > 0$ in Eq. (2–12) but not necessarily in Eq. (2–10). Also, these properties hold in general if $a > 0$ and $b > 0$, but they are not necessarily valid for negative values of a and b. For example, a^m may be negative in Eq. (2–10) if n is odd, but not if n is even. The following examples illustrate the use of these laws of exponents and radicals with functions.

Example L

For the function $f(x) = 2x^2 - 1$, we have:

(a) $(2x^2 - 1)^2(2x^2 - 1)^5 = (2x^2 - 1)^{2+5} = (2x^2 - 1)^7$

(b) $[(2x^2 - 1)^3]^4 = (2x^2 - 1)^{3 \cdot 4} = (2x^2 - 1)^{12}$

(c) $(2x^2 - 1)^{-2} = 1/(2x^2 - 1)^2$

(d) $(2x^2 - 1)^{2/3} = \sqrt[3]{(2x^2 - 1)^2}$

(e) $\sqrt[4]{\sqrt{2x^2 - 1}} = \sqrt[4 \cdot 2]{2x^2 - 1} = \sqrt[8]{2x^2 - 1}$

Example M

Simplify the expression for the function

$$f(x) = \frac{(3x^2 - 1)^{1/3}(2x) - (2x^3)(3x^2 - 1)^{-2/3}}{(3x^2 - 1)^{2/3}}$$

We multiply the numerator and denominator of this expression by $(3x^2 - 1)^{2/3}$ to clear the fractional exponents in the numerator. Applying the laws of exponents, we then proceed with the simplification.

$$f(x) = \frac{(3x^2 - 1)^{1/3}(2x) - (2x^3)(3x^2 - 1)^{-2/3}}{(3x^2 - 1)^{2/3}} \cdot \frac{(3x^2 - 1)^{2/3}}{(3x^2 - 1)^{2/3}}$$

$$= \frac{(3x^2 - 1)^{1/3}(3x^2 - 1)^{2/3}(2x) - (2x^3)(3x^2 - 1)^{-2/3}(3x^2 - 1)^{2/3}}{[(3x^2 - 1)^{2/3}]^2}$$

$$= \frac{(3x^2 - 1)(2x) - (2x^3)(3x^2 - 1)^0}{(3x^2 - 1)^{4/3}}$$

$$= \frac{6x^3 - 2x - 2x^3}{(3x^2 - 1)^{4/3}} = \frac{4x^3 - 2x}{(3x^2 - 1)^{4/3}}$$

Such algebraic simplifications are commonly found in calculus problems.

Exercises 2–1

In Exercises 1 through 12, determine the required functions.

1. Express the area A of a square as a function of its side s.

2. Express the circumference c of a circle as a function of its radius r.

3. Express the volume V of a right circular cone of height 8 as a function of the radius r of the base.

4. Express the perimeter p of a square as a function of its area A.

5. Express the distance d traveled in t hours at 55 mi/h.

6. A rocket weighs 3000 tons at lift-off. If the first-stage engines burn fuel at the rate of 10 tons/s, express the weight w of the rocket as a function of the time t during which the first-stage engines operate.

7. A certain computer part costs \$3 to produce and distribute. Express the profit p made by selling 100 of these parts at a price of c dollars each.

8. A taxi fare is 55¢ plus 10¢ for every $\frac{1}{5}$ mile traveled. Express the fare F as a function of the distance s traveled.

9. Express the simple interest I on \$200 at 12% per year as a function of the number of years t.

10. A chemist adds x liters of a solution which is 50% alcohol to 100 L of a solution which is 70% alcohol. Express the number n of liters of alcohol in the final solution as a function of x.

11. A particular factory uses two kinds of fuel. To keep air pollutants below a specified level, it is determined that the number of tons n_1 of the first fuel and the number of tons n_2 of the second fuel which may be used weekly is determined by the equation $6n_1 + 9n_2 = 12{,}000$. Express n_2 as a function of n_1.

12. The distance p of an object from a particular lens is related to the distance q of the image from the lens by the equation $p = 5q/(q - 5)$. Express q as a function of p.

In Exercises 13 through 20, find the indicated values of the given functions.

13. $f(x) = x^2 - 9x$. Find $f(3)$ and $f(-5)$.

14. $f(x) = 2x^3 - 7x$. Find $f(1)$ and $f(\frac{1}{2})$.

15. $g(x) = 6\sqrt{x} - 3$. Find $g(9)$ and $g(\frac{1}{4})$.

16. $H(t) = t^4 - t + 3$. Find $H(0)$ and $H(-2)$.

17. $f(x) = ax^2 - a^2x$. Find $f(a)$ and $f(2a)$.

18. $K(s) = 3s^2 - s + 6$. Find $K(2s)$ and $K(s - 1)$.

19. $T(t) = 5t + 7$. Find $T(-t)$ and $T[T(t)]$.

20. $X(f) = \dfrac{f + 5}{1 - f}$. Find $X(f - 2)$ and $X[X(-f)]$.

In Exercises 21 through 24, use the functions $F(x) = \sqrt{x^2 + 4}$ and $G(x) = x - 1$.

21. Find $F[G(x)]$.

22. Find $[G(x)]^3$.

23. Find $G(x)/F(x)$.

24. Find $\{G[F(x)]\}^2$.

In Exercises 25 through 28, sketch the graph of the given functions. Algebraically each can be shown to be a familiar form. However, be careful to show only the appropriate portion.

25. $f(x) = 6 - 3x$

26. $f(x) = 4x^2 - 8$

27. $f(x) = \sqrt{9 - x^2}$

28. $f(x) = \sqrt{x^2 - 9}$

In Exercises 29 through 36, perform the indicated operations and simplify.

29. $\sqrt{\sqrt[3]{x^6 + 1}}$

30. $(\sqrt{x} - 1)(\sqrt{x^2 + x + 1})$

31. $\dfrac{(4x - 5)^4}{(4x - 5)^{1/2}}$

32. $\dfrac{3(x^2 + 1)^0}{(x^2 + 4)^{-1/2}}$

33. $(2x + 1)^{1/2} + (x + 3)(2x + 1)^{-1/2}$

34. $(3x - 1)^{-2/3}(1 - x) - (3x - 1)^{1/3}$

35. $\dfrac{(x^2 + 1)^{1/2} - x^2(x^2 + 1)^{-1/2}}{x^2 + 1}$

36. $\dfrac{(1 - 2x^2)^{1/4}(2x) + 4x^3(1 - 2x^2)^{-3/4}}{(1 - 2x^2)^{1/2}}$

In Exercises 37 through 40, answer the given questions involving functions.

37. Sketch the graph of $y^2 = x$. Does this equation express y as a function of x?

38. Sketch the graph of $y = 4x$. Does this equation express x as a function of y?

39. Does the table in Fig. 2–3(a) define y as a function of x? x as a function of y?

40. Does the graph in Fig. 2–3(b) define y as a function of x? x as a function of y?

x	0	1	2	3	4	5
y	3	5	5	2	0	1

(a)

(b)

Figure 2–3

In Exercises 41 through 44, state any restrictions that might exist on the values of the independent variable.

41. $Y(y) = \dfrac{y + 1}{y - 1}$

42. $G(z) = \dfrac{1}{(z - 4)(z + 2)}$

43. $F(y) = \sqrt{y - 1}$

44. $X(x) = \dfrac{6}{\sqrt{1 - x}}$

A function may be defined differently for different intervals of the independent variable. In Exercises 45 through 48, use the following functions.

$$f(x) = \begin{cases} x + 1 & \text{for } x < 1 \\ \sqrt{x + 3} & \text{for } x \geq 1 \end{cases} \qquad g(x) = \begin{cases} \dfrac{1}{x} & \text{for } x \neq 0 \\ 0 & \text{for } x = 0 \end{cases}$$

45. Find (a) $f(-2)$ and (b) $f(2)$.

46. Find (a) $g(-2)$ and (b) $g(0)$.

47. Sketch the graph of $f(x)$.

48. Sketch the graph of $g(x)$.

In Exercises 49 through 56, solve the given problems involving functions.

49. If the temperature within a certain refrigerator is maintained at 273 K (0°C), its *coefficient of performance p* as a function of the external temperature T (in kelvins) is

$$p = \frac{273}{T - 273}$$

What is its coefficient of performance at 308 K (35°C)?

50. The net profit P made on selling 20 items, if each costs $15, as a function of the price p is $P = 20(p - 15)$. What is the profit if the price is $28? $12?

51. The distance s that a freely falling body travels as a function of the time t is given by $s = 16t^2$, where s is measured in feet and t is measured in seconds. How far does an object fall in $t + 2$ seconds?

52. The electrical resistance of a certain ammeter as a function of the resistance of the coil of the meter is

$$R = \frac{10R_c}{10 + R_c}$$

Find the function which represents the resistance of the meter if the resistance of the coil is doubled.

53. A piece of wire 60 in. long is cut into two pieces, one of which is bent into a circle and the other into a square. Express the total area of the two figures as a function of the perimeter of the square.

54. A window is in the form of a rectangle surmounted by a semicircle. If the perimeter of the window is 20 ft, express the area of the window as a function of the vertical side of the rectangle.

55. Express the perimeter of a rectangular field in terms of one of its dimensions if the area of the field is A.

56. The power of an electric circuit varies jointly as the resistance and the square of the current. Given that the power is 10 W when the current is 0.5 A and the resistance is 40 Ω, express the power as a function of the current for this 40-Ω resistor.

2–2 Limits

Another important concept for the development of differential calculus is that of a **limit.** Reasoning similar to that which is necessary for establishing the meaning of a limit was encountered when we discussed the asymptotes of a hyperbola. We shall now introduce the necessary terminology in order to establish the meaning of a limit.

In order to help understand and develop the concept of a limit, we consider briefly the **continuity** of a function. *For a function to be* **continuous at a point,** *the function must exist at the point, and any small change in x must produce only a small change in f(x).* In fact, the change in $f(x)$ can be made as small as we wish by restricting the change in x sufficiently, if the function is continuous. Also, *a function is said to be* **continuous over an interval** *if it is continuous at each point in the interval.*

Example A

The function $f(x) = 3x^2$ is continuous for all values of x. That is, $f(x)$ is defined for all values of x, and a small change in x for any given value produces only a small change in $f(x)$. If we choose $x = 2$ and then let x change by 0.1, 0.01, and so on, we obtain the values in the following table by use of a calculator.

x	2	2.1	2.01	2.001
$f(x)$	12	13.23	12.1203	12.012003

We can also see that the change in $f(x)$ is made smaller by the smaller changes in x. This shows that $f(x)$ is continuous at $x = 2$. Since this type of result would be obtained for any other x we may choose, we see that $f(x)$ is continuous for all values, and therefore it is continuous over the interval of all values of x.

Example B

The function $f(x) = \dfrac{1}{x - 2}$ is not continuous at $x = 2$. When we substitute 2 for x in the function, we have division by zero. This means that the function is not defined. Therefore, the condition that the function must exist is not satisfied.

Considering continuity from a graphical point of view, a function which is continuous over an interval will have no "breaks" in its graph over that interval. If a function is not continuous, either the graph of the function has a break in it or it does not exist for the values of x.

Example C

The graph of the function $f(x) = 3x^2$, which we determined to be continuous at all values of x in Example A, is shown in Fig. 2–4. We see that there are no breaks in the graph.

The graph of $f(x) = \dfrac{1}{x-2}$, which we determined to be not continuous at $x = 2$ in Example B, is shown in Fig. 2–5. We see that there is a break in the graph for $x = 2$, and this shows the fact that $f(x)$ does not exist at $x = 2$.

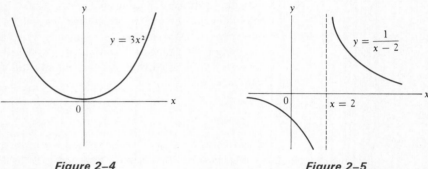

Figure 2–4   Figure 2–5

Example D

The function represented by the graph in Fig. 2–6 is continuous in the interval for which $x > 1$. The graph does not exist for values $x < 1$.

The function represented by the graph in Fig. 2–7 is not continuous at $x = 1$. The function is defined (by the solid circle point) for $x = 1$ (the open circle point indicates that point is not part of the graph). However, a small change from 1 may result in a change of at least 3 in $f(x)$, regardless of how small the change in x is made. Therefore, the small change condition is not satisfied.

The function represented by the graph in Fig. 2–8 is not continuous for $x = -2$. The open circle shows that point is not part of the graph, and therefore $f(x)$ is not defined for $x = -2$.

Figure 2–6

Figure 2–7

Figure 2–8

In our earlier discussions of the asymptotes of a hyperbola, we used the symbol →, which means "approaches." *When we say that x → 2, we mean that x may take on any value as close to 2 as desired, but it also distinctly means that x cannot be set equal to 2.*

Example E

Let us consider the behavior of the function $f(x) = 2x + 1$ as $x \to 2$.

Since we are not to use $x = 2$, we use a calculator to set up tables in order to determine values of $f(x)$, as x gets close to 2.

x	1.000	1.500	1.900	1.990	1.999
$f(x)$	3.000	4.000	4.800	4.980	4.998

x	3.000	2.500	2.100	2.010	2.001
$f(x)$	7.000	6.000	5.200	5.020	5.002

We can see that $f(x)$ approaches 5, as x approaches 2, from above 2 and from below 2.

In Example E, since $f(x) \to 5$ as $x \to 2$, the number 5 is called the limit of $f(x)$ as $x \to 2$. This leads us to the meaning of the limit of a function. In general, *the limit of a function $f(x)$ is that value which the function approaches as x approaches a given value a.* This is written as

$$\lim_{x \to a} f(x) = L \tag{2-15}$$

where L is the value of the limit of the function. Remember, in approaching a, x may come as arbitrarily close as desired to a, but it may not equal a.

An important conclusion can be drawn from the limit in Example E. The function $f(x)$ is a continuous function, and $f(2)$ equals the value of the limit as $x \to 2$. In general, it is true that *the limit of $f(x)$ as $x \to a$ is the same as $f(a)$, if the function is continuous at $x = a$.*

Although we can evaluate the limit for a continuous function as $x \to a$ by evaluating $f(a)$, it is possible that a function is not continuous at $x = a$, and that the limit exists and can be determined. Thus, we must be able to determine the value of a limit without finding $f(a)$. The following example illustrates the evaluation of such a limit.

Example F

Find

$$\lim_{x \to 2} \frac{2x^2 - 3x - 2}{x - 2}$$

We note immediately that the function is not continuous at $x = 2$, for division by zero is indicated. Thus, we cannot evaluate the limit by substituting $x = 2$ into

(*continued on next page*)

See Appendix C for a computer program for finding limits such as the one in this example.

the function. By setting up tables, we determine the value which $f(x)$ approaches, as x approaches 2.

x	1.000	1.500	1.900	1.990	1.999
$f(x)$	3.000	4.000	4.800	4.980	4.998

x	3.000	2.500	2.100	2.010	2.001
$f(x)$	7.000	6.000	5.200	5.020	5.002

We see that the values obtained are identical to those in Example E. Since $f(x) \rightarrow 5$ as $x \rightarrow 2$, we have

$$\lim_{x \to 2} \frac{2x^2 - 3x - 2}{x - 2} = 5$$

Thus, the limit exists as $x \rightarrow 2$, although the function does not exist at $x = 2$.

The reason that the functions in Examples E and F have the same limit is shown in the following example.

Example G

(a)

(b)

Figure 2–9

The function $\dfrac{2x^2 - 3x - 2}{x - 2}$ in Example F is the same as the function $2x + 1$ in Example E, except when $x = 2$. By factoring the numerator of the function of Example F, we have

$$\frac{2x^2 - 3x - 2}{x - 2} = \frac{(2x + 1)(x - 2)}{x - 2} = 2x + 1$$

The cancellation here is valid, as long as x does not equal 2, for we have division by zero at $x = 2$. Also, in finding the limit as $x \rightarrow 2$, we do not use the value $x = 2$. Therefore,

$$\lim_{x \to 2} \frac{2x^2 - 3x - 2}{x - 2} = \lim_{x \to 2} (2x + 1) = 5$$

The limits of the two functions are equal, since, again, in finding the limit, we do not let $x = 2$. The graphs of the two functions are shown in Fig. 2–9(a) and (b). We can see from the graphs that the limits are the same, although one of the functions is not continuous.

The limit of the function in Example F was determined by calculating values near $x = 2$ and by means of an algebraic change in the function. This illustrates that limits may be found through the meaning and definition, and through other procedures when the function is not continuous. The following example illustrates a function for which the limit does not exist as x approaches the indicated value.

Example H

In trying to find

$$\lim_{x \to 2} \frac{1}{x - 2}$$

we note that $f(x)$ is not defined for $x = 2$, since we would have division by zero. Therefore, we set up the following table to see how $f(x)$ behaves as $x \to 2$.

x	3	2.5	2.1	2.01	2.001
$f(x)$	1	2	10	100	1000

x	1	1.5	1.9	1.99	1.999
$f(x)$	-1	-2	-10	-100	-1000

We see that $f(x)$ gets larger as $x \to 2$ from above 2, and $f(x)$ gets smaller (large negative values) as $x \to 2$ from below 2. (This may be written as $f(x) \to +\infty$ as $x \to 2^+$ and $f(x) \to -\infty$ as $x \to 2^-$, but we must remember that ∞ is not a real number.) Therefore, the limit as $x \to 2$ does not exist. The graph of this function is shown in Fig. 2–5.

The following examples further illustrate the evaluation of limits.

Example I

Find $\lim_{x \to 4} (x^2 - 7)$.

Since the function $x^2 - 7$ is continuous at $x = 4$, we may evaluate this limit by substitution. For $f(x) = x^2 - 7$, we have $f(4) = 9$. This means that

$$\lim_{x \to 4} (x^2 - 7) = 9$$

Example J

Find

$$\lim_{x \to 2} \left(\frac{x^2 - 4}{x - 2} \right)$$

Since

$$\frac{x^2 - 4}{x - 2} = \frac{(x - 2)(x + 2)}{x - 2} = x + 2$$

is valid as long as $x \neq 2$, we find that

$$\lim_{x \to 2} \left(\frac{x^2 - 4}{x - 2} \right) = \lim_{x \to 2} (x + 2) = 4$$

Again, we do not have to concern ourselves with the fact that the cancellation is not valid for $x = 2$. In finding the limit we do not consider the value of $f(x)$ at $x = 2$.

Example K

Find $\lim_{x \to 0} (x\sqrt{x-3})$.

We see that $x\sqrt{x-3} = 0$ if $x = 0$, but this function does not have real values for values of x less than 3 other than $x = 0$. *Since x cannot approach 0, $f(x)$ does not approach 0 and the limit does not exist.* The point of this example is that even if $f(a)$ exists, we cannot evaluate the limit by finding $f(a)$, unless $f(x)$ is continuous at $x = a$. Here, $f(a)$ exists but the limit does not exist.

Limits as x approaches infinity are also of importance. However, when dealing with these limits, we must remember that ∞ does not represent a real number and that arithmetic operations may not be performed on it. Therefore, when we write $x \to \infty$, we know that we are to consider values of x which are becoming large without bound. Consider the following examples.

Example L

Figure 2–10

Find

$$\lim_{x \to \infty} \left(3 + \frac{1}{x^2}\right)$$

As x becomes larger and larger, $1/x^2$ becomes smaller and smaller and approaches zero. This means that $f(x) \to 3$ as $x \to \infty$. Thus,

$$\lim_{x \to \infty} \left(3 + \frac{1}{x^2}\right) = 3$$

We can verify our reasoning and the value of the limit by making a table of values for x and $f(x)$ as x becomes large.

x	1	10	100	1000
$f(x)$	4	3.01	3.0001	3.000001

Again we see that $f(x) \to 3$ as $x \to \infty$. See Fig. 2–10.

Example M

Figure 2–11

Find

$$\lim_{x \to \infty} \frac{x^2 + 1}{2x^2 + 3}$$

We note that as $x \to \infty$, both the numerator and denominator become large without bound. Therefore, we use a calculator to make a table to see how $f(x)$ behaves as x becomes very large.

x	1	10	100	1000
$f(x)$	0.4	0.4975369	0.4999750	0.4999998

From this table we see that $f(x) \to 0.5$ as $x \to \infty$. See Fig. 2–11.

This limit can also be found through algebraic operations and an examination of the resulting algebraic form. If we divide both the numerator and the denom-

▶ inator of the function by x^2, which is the largest power of x which appears in either the numerator or denominator, the function becomes

$$\frac{1 + \dfrac{1}{x^2}}{2 + \dfrac{3}{x^2}} \quad \begin{bmatrix} \text{Terms} \to 0 \\ \text{as } x \to \infty \end{bmatrix}$$

Here we see that $1/x^2$ and $3/x^2$ both approach zero as $x \to \infty$. This means that the numerator approaches 1 and the denominator approaches 2. Thus,

$$\lim_{x \to \infty} \frac{x^2 + 1}{2x^2 + 3} = \lim_{x \to \infty} \frac{1 + \dfrac{1}{x^2}}{2 + \dfrac{3}{x^2}} = \frac{1}{2}$$

which agrees with our previous result.

The definitions and development of continuity and of a limit presented in this section are not mathematically completely rigorous. However, the development is consistent with a more rigorous development, and the concept of a limit is the principal concern.

Exercises 2–2

In Exercises 1 through 6, determine the values of x for which the function is continuous. If the function is not continuous, determine the reason.

1. $f(x) = 3x - 2$ 2. $f(x) = 9 - x^2$ 3. $f(x) = \dfrac{1}{x + 3}$

4. $f(x) = \dfrac{2}{x^2 - x}$ 5. $f(x) = \dfrac{1}{\sqrt{x}}$ 6. $f(x) = \dfrac{\sqrt{x + 2}}{x}$

In Exercises 7 through 12, determine the values of x for which the function, as represented by the graph in Fig. 2–12, is continuous. If the function is not continuous, determine the reason.

7.

(a)

8.

(b)

9.

(c)

Figure 2–12

(d)   (e)   (f)

Figure 2–12

In Exercises 13 through 20, evaluate the given functions for the given values of x by use of a calculator. Do not change the form of the functions. Then, by observing the values obtained, find the indicated limits.

13. Evaluate $f(x) = 3x - 2$ for the following values of x: 4.000, 3.500, 3.100, 3.010, 3.001 and for 2.000, 2.500, 2.900, 2.990, 2.999. Find

$$\lim_{x \to 3} (3x - 2)$$

14. Evaluate $f(x) = 2x^2 - 4x$ for the following values of x: 0.4000, 0.4900, 0.4990, 0.4999 and for 0.6000, 0.5100, 0.5010, 0.5001. Find

$$\lim_{x \to 0.5} (2x^2 - 4x)$$

15. Evaluate $f(x) = \dfrac{x^3 - x}{x - 1}$ for the following values of x: 0.900, 0.990, 0.999 and for 1.100, 1.010, 1.001. Find

$$\lim_{x \to 1} \frac{x^3 - x}{x - 1}$$

16. Evaluate

$$f(x) = \frac{x^3 + 2x^2 - 2x + 3}{x + 3}$$

for the following values of x: -2.9, -2.99, -2.999 and -3.1, -3.01, -3.001. Find

$$\lim_{x \to -3} \frac{x^3 + 2x^2 - 2x + 3}{x + 3}$$

17. Evaluate

$$f(x) = \frac{2 - \sqrt{x + 2}}{x - 2}$$

for the following values of x: 2.1, 2.01, 2.001 and 1.9, 1.99, 1.999. Find

$$\lim_{x \to 2} \frac{2 - \sqrt{x + 2}}{x - 2}$$

18. Evaluate $f(x) = \dfrac{2 - \sqrt{x}}{4 - x}$ for the following values of x: 3.900, 3.990, 3.999 and for 4.100, 4.010, 4.001. Find $\displaystyle \lim_{x \to 4} \frac{2 - \sqrt{x}}{4 - x}$

19. Evaluate

$$f(x) = \frac{2x + 1}{5x - 3}$$

for the following values of x: 10, 100, 1000. Find

$$\lim_{x \to \infty} \frac{2x + 1}{5x - 3}$$

20. Evaluate

$$f(x) = \frac{1 - x^2}{8x^2 + 5}$$

for the following values of x: 10, 100, 1000. Find

$$\lim_{x \to \infty} \frac{1 - x^2}{8x^2 + 5}$$

In Exercises 21 through 40, evaluate the given limits by direct evaluation as in Examples I through M. Change the form of the function where necessary.

21. $\lim_{x \to 3} (x + 4)$

22. $\lim_{x \to -2} (2x - 1)$

23. $\lim_{x \to 2} \frac{x^2 - 1}{x + 1}$

24. $\lim_{x \to 5} \left(\frac{3}{x^2 + 2} \right)$

25. $\lim_{x \to 0} \frac{x^2 + x}{x}$

26. $\lim_{x \to 2} \frac{x^2 - 2x}{x - 2}$ $\frac{x(x-2)}{x-2}$

27. $\lim_{x \to -1} \frac{x^2 - 1}{x + 1}$

28. $\lim_{x \to 5} \frac{x - 5}{x^2 - 25}$

29. $\lim_{x \to 1} \frac{x^3 - x}{x - 1}$

30. $\lim_{x \to 1/3} \frac{3x - 1}{3x^2 + 5x - 2}$

31. $\lim_{x \to 3} \frac{x^2 - 2x - 3}{3 - x}$

32. $\lim_{x \to 0} \frac{(2 + x)^2 - 4}{x}$

33. $\lim_{x \to -1} \sqrt{x} (x + 1)$

34. $\lim_{x \to 1} (x - 1) \sqrt{x^2 - 4}$

35. $\lim_{x \to \infty} \frac{\frac{2}{x}}{1 - 2x}$

36. $\lim_{x \to \infty} \frac{6}{1 + \frac{2}{x^2}}$

37. $\lim_{x \to \infty} \frac{3x^2 + 5}{x^2 - 2}$

38. $\lim_{x \to \infty} \frac{x - 1}{7x + 4}$

39. $\lim_{x \to \infty} \frac{2x - 6}{x^2 - 9}$

40. $\lim_{x \to \infty} \frac{2 - x^2}{3x^3 - 1}$

In Exercises 41 and 42, use a calculator to evaluate the function at 0.1, 0.01, and 0.001 from both sides of the value which x approaches. In Exercises 43 and 44, evaluate the function for the values of x of 10, 100, and 1000. From these values determine the limit. Then, by using an appropriate change of algebraic form, evaluate the limit directly and compare values.

41. $\lim_{x \to 0} \frac{x^2 - 3x}{x}$

42. $\lim_{x \to 3} \frac{2x^2 - 6x}{x - 3}$

43. $\lim_{x \to \infty} \frac{2x^2 + x}{x^2 - 3}$

44. $\lim_{x \to \infty} \frac{x^2 + 6x + 5}{2x^2 + 1}$

In Exercises 45 through 48, solve the given problems involving limits.

45. The magnetic field B at the center of a loop of wire due to a current i in the wire is given by $B = ki/r$, where r is the radius of the loop and k is a constant. If $k = 6.30 \times 10^{-7}$ Wb/A-m and $r = 0.100$ m, find $\lim_{i \to 0.3} B$ (in teslas) by evaluating the function for the following values of i: 0.200, 0.290, 0.299, 0.301, 0.310, 0.400.

46. Velocity can be determined by dividing the displacement of an object by the time elapsed in moving through the displacement. In a certain experiment the following values were determined for the displacements and elapsed times for the motion of an object. Determine the limit of the velocity.

Displacement (centimeters)	0.480000	0.280000	0.029800	0.002998	0.00029998
Time (seconds)	0.200000	0.100000	0.010000	0.001000	0.00010000

47. A certain object, after being heated, cools at such a rate that its temperature T (in degrees Celsius) decreases 10% each minute. If the object is originally heated to 100°C find $\lim_{t \to 10} T$ and $\lim_{t \to \infty} T$, where t is the time (in minutes).

48. Show that $\lim_{x \to 0^+} 2^{1/x} \neq \lim_{x \to 0^-} 2^{1/x}$ where $\lim_{x \to 0^+}$ means to find the limit as x approaches zero through positive values only, and $\lim_{x \to 0^-}$ means to find the limit as x approaches zero through negative values only.

2-3 The Slope of a Tangent to a Curve

Figure 2-13

Figure 2-14

Having developed the basic operations with functions and the concept of a limit, we shall now turn our attention to a graphical interpretation of the rate of change of a function. This interpretation, basic to an understanding of the calculus, deals with the slope of a line tangent to the curve of a function.

Consider the points $P(x_1, y_1)$ and $Q(x_2, y_2)$ in Fig. 2-13. From Chapter 1 we know that the slope of the line through these points is given by

$$m = \frac{y_2 - y_1}{x_2 - x_1}$$

This, however, represents the slope of the line through P and Q and no other line. If we allow Q to be a point closer to P, the slope of PQ will more closely approximate the slope of a line drawn tangent to the curve at P (see Fig. 2-14). In fact, the closer Q is to P, the better this approximation becomes. It is not possible to allow Q to coincide with P, for then it would not be possible to define the slope of PQ in terms of two points. *The slope of the tangent line, often referred to as the slope of the curve, is the limiting value of the slope of PQ as Q approaches P.*

Example A

Figure 2–15

Find the slope of a line tangent to the curve $y = x^2 + 3x$ at the point $P(2, 10)$ by determining the limit of slopes of lines PQ as Q approaches P.

We shall let point Q have the x-values of 3.0, 2.5, 2.1, 2.01, and 2.001. Then, using a calculator, we tabulate the necessary values. Since P is the point $(2, 10)$, $x_1 = 2$ and $y_1 = 10$. Thus, using the values of x_2, we tabulate the values of y_2, $y_2 - 10$, $x_2 - 2$, and thereby the values of the slope m.

Point	Q_1	Q_2	Q_3	Q_4	Q_5	P
x_2	3.0	2.5	2.1	2.01	2.001	2
y_2	18.0	13.75	10.71	10.0701	10.007001	10
$y_2 - 10$	8.0	3.75	0.71	0.0701	0.007001	
$x_2 - 2$	1.0	0.5	0.1	0.01	0.001	
$m = \dfrac{y_2 - 10}{x_2 - 2}$	8.0	7.5	7.1	7.01	7.001	

We can see from the above table that the slope of PQ approaches the value of 7 as Q approaches P. Thus, the slope of the tangent line at $(2, 10)$ is 7. See Fig. 2–15.

With the proper notation, it is possible to express the coordinates of Q in terms of the coordinates of P. If we define the quantity Δx ("delta" x) by the equation

$$x_2 = x_1 + \Delta x \tag{2–16}$$

and Δy by the equation

$$y_2 = y_1 + \Delta y \tag{2–17}$$

the coordinates of Q become $(x_1 + \Delta x, y_1 + \Delta y)$. The quantities Δx and Δy represent the difference in the coordinates of P and Q. They are not to be thought of as "Δ times x," for the symbol Δ used here has no meaning by itself. *The name* **increment** *is given to the difference of the coordinates of two points, and therefore Δx and Δy are the increments in x and y, respectively.*

Using Eqs. (2–16) and (2–17), along with the definition of slope, we can express the slope of PQ as

$$m_{PQ} = \frac{(y_1 + \Delta y) - y_1}{(x_1 + \Delta x) - x_1} = \frac{\Delta y}{\Delta x} \tag{2–18}$$

By previous discussion, as Q approaches P, the slope of the tangent line is more *nearly* approximated by $\Delta y / \Delta x$.

Example B

Figure 2–16

Find the slope of a line tangent to the curve $y = x^2 + 3x$ at the point (2, 10) by the increment method indicated in Eq. (2–18). (This is the same slope as calculated in Example A.)

As in Example A, point P has the coordinates (2, 10). Thus the coordinates of any other point Q can be expressed as $(2 + \Delta x, 10 + \Delta y)$. See Fig. 2–16. The slope of PQ then becomes

$$m_{PQ} = \frac{(10 + \Delta y) - 10}{(2 + \Delta x) - 2} = \frac{\Delta y}{\Delta x}$$

This expression itself does not enable us to find the slope, since values of Δx and Δy are not known. If, however, we can express Δy in terms of Δx, we might derive more information. Both P and Q are on the curve of the function, which means that the coordinates of each must satisfy the function. Using the coordinates of Q, we have

$$(10 + \Delta y) = (2 + \Delta x)^2 + 3(2 + \Delta x)$$
$$= 4 + 4\Delta x + (\Delta x)^2 + 6 + 3\Delta x$$

Subtracting 10 from each side, we have

$$\Delta y = 7\Delta x + (\Delta x)^2$$

As Q approaches P, Δx becomes smaller and smaller. Calculating Δy and the ratio $\Delta y/\Delta x$ for increasingly small Δx, we have the following values

Δx	0.1	0.01	0.001
Δy	0.71	0.0701	0.007001
$\dfrac{\Delta y}{\Delta x}$	7.1	7.01	7.001

We note that the values of $\Delta y/\Delta x$ are the same as those for the slope for points Q_3, Q_4, and Q_5 in Example A. We also see that as $\Delta x \to 0$, $\Delta y/\Delta x \to 7$.

Now, substituting the expression for Δy into the expression for m_{PQ}, we have

$$m_{PQ} = \frac{7\Delta x + (\Delta x)^2}{\Delta x} = 7 + \Delta x$$

From this expression we can see directly that as $\Delta x \to 0$, $m_{PQ} \to 7$. Using this fact, we can see that the slope of the tangent line is

$$m_{\text{tan}} = \lim_{\Delta x \to 0} m_{PQ} = 7$$

We see that this result agrees with that found in Example A.

Example C

Find the slope of a line tangent to the curve $y = 4x - x^2$ at the point (x_1, y_1).

The points P and Q are $P(x_1, y_1)$ and $Q(x_1 + \Delta x, y_1 + \Delta y)$. Using the coordinates of Q in the function, we obtain

$$(y_1 + \Delta y) = 4(x_1 + \Delta x) - (x_1 + \Delta x)^2$$
$$= 4x_1 + 4\Delta x - x_1^2 - 2x_1\Delta x - (\Delta x)^2$$

Using the coordinates of P, we obtain

$$y_1 = 4x_1 - x_1^2$$

Subtracting the second of these expressions from the first to solve for Δy, we obtain

$$(y_1 + \Delta y) - y_1 = [4x_1 + 4\Delta x - x_1^2 - 2x_1\Delta x - (\Delta x)^2] - (4x_1 - x_1^2)$$

or

$$\Delta y = 4\Delta x - 2x_1\Delta x - (\Delta x)^2$$

Dividing through by Δx to obtain an expression for $\Delta y / \Delta x$, we obtain

$$\frac{\Delta y}{\Delta x} = 4 - 2x_1 - \Delta x$$

In this last equation, the desired expression is on the left, but all we can determine from $\Delta y / \Delta x$ itself is that the ratio will become one very small number divided by another very small number as $\Delta x \to 0$. The right side, however, approaches $4 - 2x_1$ as $\Delta x \to 0$. This indicates that the slope of a tangent at the point (x_1, y_1) is given by

$$m_{\text{tan}} = 4 - 2x_1$$

This method has an advantage over that of Example B—we now have a general expression for the slope of a tangent line for any value x_1. If $x_1 = -1$, $m_{\text{tan}} = 6$, and if $x_1 = 3$, $m_{\text{tan}} = -2$. These tangent lines are indicated in Fig. 2–17.

Figure 2–17

Example D

Find the expression for the slope of a line tangent to the curve $y = x^3 + 2$ at the general point (x_1, y_1), and use this expression to determine the slope when $x = \frac{1}{2}$.

Using the coordinates of the point $Q(x_1 + \Delta x, y_1 + \Delta y)$ in the function, we obtain

$$y_1 + \Delta y = (x_1 + \Delta x)^3 + 2 = x_1^3 + 3x_1^2\Delta x + 3x_1(\Delta x)^2 + (\Delta x)^3 + 2$$

and using the coordinates of the point $P(x_1, y_1)$, we obtain $y_1 = x_1^3 + 2$. Subtracting this second expression from the first, we have

$$\Delta y = 3x_1^2\Delta x + 3x_1(\Delta x)^2 + (\Delta x)^3$$

Dividing by Δx, we have

$$\frac{\Delta y}{\Delta x} = 3x_1^2 + 3x_1\Delta x + (\Delta x)^2$$

As $\Delta x \to 0$, the right side approaches the value $3x_1^2$. This means that

$$m_{\text{tan}} = 3x_1^2$$

When $x_1 = \frac{1}{2}$, we find that the slope of the tangent is $3(\frac{1}{4}) = \frac{3}{4}$. The curve and this tangent line are indicated in Fig. 2–18.

Figure 2–18

In interpreting this analysis as the rate of change of a function, we see that if $\Delta x = 1$ and the corresponding value of Δy is found, it can be said that y changes by the amount Δy as x changes one unit. If x changed by some lesser amount, we could still calculate a ratio for the amount of change in y for the given change in x. Therefore, as long as x changes at all, there will be a corresponding change in y. In this way *the ratio $\Delta y / \Delta x$ is the* **average rate of change** *of y, which is a function of x, with respect to x. As $\Delta x \to 0$ the limit of the ratio $\Delta y / \Delta x$ gives us the* **instantaneous rate of change** *of y with respect to x.*

Example E

In Example A let us consider points $P(2, 10)$ and $Q_2(2.5, 13.75)$. From P to Q_2, x changes by 0.5 unit and y by 3.75 units. This means that *the average change in y for a 1-unit change in x* is $3.75/0.5 = 7.5$ units. However, this is not the rate at which y is changing with respect to x **at** most points within this interval.

At point P, the slope of the tangent line of 7 tells us that y is changing 7 units for a one-unit change in x. However, *this is an instantaneous rate of change **at** point P*, and tells us the rate at which y is changing with respect to x **at P and only at P.**

Exercises 2–3

In Exercises 1 through 4, use the method of Example A to calculate the slope of a line tangent to the curve of each of the given functions. Let Q_1, Q_2, Q_3, and Q_4 have the indicated x-values. Sketch the curves and tangent lines.

1. $y = x^2$; P is $(2, 4)$; let Q have x-values of 1.5, 1.9, 1.99, 1.999.
2. $y = 1 - x^2$; P is $(2, -3)$; let Q have x-values of 1.5, 1.9, 1.99, 1.999.
3. $y = x^2 + 2x$; P is $(-3, 3)$; let Q have x-values of $-2.5, -2.9, -2.99, -2.999$.
4. $y = x^3 + 1$; P is $(-1, 0)$; let Q have x-values of $-0.5, -0.9, -0.99, -0.999$.

In Exercises 5 through 8, use the method of Example B (divide the expression for Δy by Δx and use the simplified expression for m_{PQ}) to calculate the slope of a line tangent to the curve of each of the given functions for the given points P. (These are the same functions and points P as for Exercises 1 through 4.)

5. $y = x^2$; P is $(2, 4)$.
6. $y = 1 - x^2$; P is $(2, -3)$.
7. $y = x^2 + 2x$; P is $(-3, 3)$.
8. $y = x^3 + 1$; P is $(-1, 0)$.

In Exercises 9 through 20, use the method of Example C to find a general expression for the slope of a tangent line to each of the indicated curves. Then find the slope for the given values of x. Sketch the curves and tangent lines.

9. $y = x^2$; $x = 2$, $x = -1$
10. $y = 1 - x^2$; $x = 2$, $x = -2$
11. $y = x^2 + 2x$; $x = -3$, $x = 1$
12. $y = 4 - 3x^2$; $x = 0$, $x = 2$
13. $y = x^2 + 4x + 5$; $x = -3$, $x = 2$
14. $y = 2x^2 - 4x$; $x = 1$, $x = 1.5$
15. $y = 6x - x^2$; $x = -2$, $x = 3$
16. $y = 2x - 3x^2$; $x = 0$, $x = 0.5$

17. $y = x^3 - 2x$; $x = -1$, $x = 0$, $x = 1$
18. $y = 3x - x^3$; $x = -2$, $x = 0$, $x = 2$
19. $y = x^4$; $x = 0$, $x = 0.5$, $x = 1$
20. $y = 1 - x^4$; $x = 0$, $x = 1$, $x = 2$

In Exercises 21 through 24, find the average rate of change of y with respect to x from P to Q. Then compare this with the instantaneous rate of change of y with respect to x at P by finding m_{tan} at P.

21. $y = x^2 + 2$, $P(2, 6)$, $Q(2.1, 6.41)$
22. $y = 1 - 2x^2$, $P(1, -1)$, $Q(1.1, -1.42)$
23. $y = 9 - x^3$, $P(2, 1)$, $Q(2.1, -0.261)$
24. $y = x^3 - 6x$, $P(3, 9)$, $Q(3.1, 11.191)$

2-4 The Derivative

Using the terminology and techniques developed in the earlier sections of this chapter, we are now ready to establish one of the fundamental definitions of calculus. Generalizing on the method of Example B of the preceding section, if we consider that the point $P(x, y)$ is held constant while the point $Q(x + \Delta x, y + \Delta y)$ approaches it, then both Δx and Δy approach zero. The points P and Q both lie on the curve of the function $f(x)$, which in turn means that the coordinates of each point must satisfy the function. For the point P this means that $y = f(x)$, and for the point Q this means that $y + \Delta y = f(x + \Delta x)$. Subtracting the first expression from the second, we have

$$(y + \Delta y) - y = f(x + \Delta x) - f(x)$$

$$\Delta y = f(x + \Delta x) - f(x) \tag{2-19}$$

From the discussion of the previous section, we saw that the slope of a line tangent to $f(x)$ at point $P(x, y)$ was found as the limiting value of the ratio $\Delta y / \Delta x$ as Δx approaches zero. Formally, *this limiting value of the ratio of $\Delta y / \Delta x$ is known as the* **derivative** *of the function*. Therefore, the derivative of a function $f(x)$ is defined by the relation

$$\lim_{\Delta x \to 0} \frac{\Delta y}{\Delta x} = \lim_{\Delta x \to 0} \frac{f(x + \Delta x) - f(x)}{\Delta x} \tag{2-20}$$

The process of finding the derivative of a function from its definition is called the delta-process. *The general process of finding a derivative is called* **differentiation.**

Example A

Find the derivative of the function $y = 2x^2 + 3x$ by the delta-process.

To find Δy, we must first derive the quantity $f(x + \Delta x) - f(x)$. Thus, we find $y + \Delta y = f(x + \Delta x)$ first:

$$y + \Delta y = 2(x + \Delta x)^2 + 3(x + \Delta x)$$

Then we subtract the original function:

$$y + \Delta y - y = 2(x + \Delta x)^2 + 3(x + \Delta x) - 2x^2 - 3x$$

Simplifying, we find that the result is

$$\Delta y = 4x\Delta x + 2(\Delta x)^2 + 3\Delta x$$

Next, dividing through by Δx, we obtain

$$\frac{\Delta y}{\Delta x} = 4x + 2\Delta x + 3$$

When we find the limit as Δx approaches zero, we see that the $2\Delta x$-term on the right approaches zero. Thus,

$$\lim_{\Delta x \to 0} \frac{\Delta y}{\Delta x} = 4x + 3$$

Therefore, we find that the derivative of the function $y = 2x^2 + 3x$ is the function $4x + 3$. This means that we can calculate the slope of a tangent line for any point on the graph of $y = 2x^2 + 3x$ by substituting the x-coordinate into the expression $4x + 3$. For example, the slope of a tangent line is 7 if $x = 1$.

Example B

Find the derivative of $y = 6x - 2x^3$ by the delta-process.

First we find $y + \Delta y = f(x + \Delta x)$.

$$\begin{aligned} y + \Delta y &= 6(x + \Delta x) - 2(x + \Delta x)^3 \\ &= 6x + 6(\Delta x) - 2[x^3 + 3x^2(\Delta x) + 3x(\Delta x)^2 + (\Delta x)^3] \\ &= 6x + 6(\Delta x) - 2x^3 - 6x^2(\Delta x) - 6x(\Delta x)^2 - 2(\Delta x)^3 \end{aligned}$$

Next we subtract the original function.

$$\begin{aligned} y + \Delta y - y &= [6x + 6(\Delta x) - 2x^3 - 6x^2(\Delta x) - 6x(\Delta x)^2 - 2(\Delta x)^3] \\ &\quad - (6x - 2x^3) \\ \Delta y &= 6(\Delta x) - 6x^2(\Delta x) - 6x(\Delta x)^2 - 2(\Delta x)^3 \end{aligned}$$

Dividing through by Δx is the next step.

$$\begin{aligned} \frac{\Delta y}{\Delta x} &= \frac{6(\Delta x) - 6x^2(\Delta x) - 6x(\Delta x)^2 - 2(\Delta x)^3}{\Delta x} \\ &= 6 - 6x^2 - 6x(\Delta x) - 2(\Delta x)^2 \end{aligned}$$

Now we find the limit as Δx approaches zero.

$$\lim_{\Delta x \to 0} \frac{\Delta y}{\Delta x} = \lim_{\Delta x \to 0} [6 - 6x^2 - 6x(\Delta x) - 2(\Delta x)^2]$$

$$= 6 - 6x^2$$

Therefore, the derivative of $y = 6x - 2x^3$ is $6 - 6x^2$.

Example C

Find the derivative of $y = \dfrac{1}{x}$ by the delta-process.

We write

$$y + \Delta y = \frac{1}{x + \Delta x}$$

Then we find that

$$y + \Delta y - y = \frac{1}{x + \Delta x} - \frac{1}{x}$$

$$\Delta y = \frac{x - (x + \Delta x)}{x(x + \Delta x)} = \frac{-\Delta x}{x(x + \Delta x)}$$

$$\frac{\Delta y}{\Delta x} = \frac{-1}{x(x + \Delta x)}$$

$$\lim_{\Delta x \to 0} \frac{\Delta y}{\Delta x} = \lim_{\Delta x \to 0} \frac{-1}{x(x + \Delta x)} = \frac{-1}{x^2}$$

Example D

Find the derivative of $y = x^2 + \dfrac{1}{x + 1}$ by the delta-process.

$$y + \Delta y = (x + \Delta x)^2 + \frac{1}{x + \Delta x + 1}$$

$$y + \Delta y - y = x^2 + 2x \Delta x + (\Delta x)^2 + \frac{1}{x + \Delta x + 1} - x^2 - \frac{1}{x + 1}$$

The problem is handled algebraically most easily if the fractions are combined and the other terms are handled separately:

$$\Delta y = 2x \Delta x + (\Delta x)^2 + \frac{1}{x + \Delta x + 1} - \frac{1}{x + 1}$$

$$= 2x \Delta x + (\Delta x)^2 + \frac{(x + 1) - (x + \Delta x + 1)}{(x + \Delta x + 1)(x + 1)}$$

$$= 2x \Delta x + (\Delta x)^2 - \frac{\Delta x}{(x + \Delta x + 1)(x + 1)}$$

$$\frac{\Delta y}{\Delta x} = 2x + \Delta x - \frac{1}{(x + \Delta x + 1)(x + 1)}$$

$$\lim_{\Delta x \to 0} \frac{\Delta y}{\Delta x} = 2x - \frac{1}{(x + 1)^2}$$

Several shorter notations are used for the derivative. The notation of the definition is clumsy for general use, so the notations

$$y', \quad D_x y, \quad f'(x), \quad \text{and} \quad \frac{dy}{dx}$$

are commonly used. The following example illustrates the use of basic notation with derivatives.

Example E

In Example B, $y = 6x - 2x^3$, and we found that the derivative is $6 - 6x^2$. Thus we may write

$$y' = 6 - 6x^2 \quad \text{or} \quad \frac{dy}{dx} = 6 - 6x^2$$

If we had written $f(x) = 6x - 2x^3$, we would then write $f'(x) = 6 - 6x^2$. If we wish to evaluate the derivative at some point, such as $(-2, 4)$, we write

$$\frac{dy}{dx} = 6 - 6x^2$$

$$\left. \frac{dy}{dx} \right|_{x=-2} = 6 - 6(-2)^2 = 6 - 24 = -18$$

We note that only the x-coordinate of the point was needed in the evaluation.

Example F

Find dy/dx of the function $y = \sqrt{x}$ by the delta-process.

We first square both sides of the equation, thus obtaining $y^2 = x$. (This is valid only for $y \geq 0$, since the original function $y = \sqrt{x}$ is not defined for $y < 0$.) Substituting the coordinates of the point $(x + \Delta x, y + \Delta y)$ into this equation, we have

$$(y + \Delta y)^2 = x + \Delta x$$

Expanding the left side and then subtracting the equation $y^2 = x$, we have

$$y^2 + 2y\,\Delta y + (\Delta y)^2 - y^2 = x + \Delta x - x$$

or

$$2y\,\Delta y + (\Delta y)^2 = \Delta x$$

Next, dividing through by Δx, we obtain

$$2y\,\frac{\Delta y}{\Delta x} + \Delta y\,\frac{\Delta y}{\Delta x} = 1$$

Solving for $\Delta y/\Delta x$, we have

$$\frac{\Delta y}{\Delta x} = \frac{1}{2y + \Delta y}$$

Finally, taking the limit as $\Delta x \to 0$, we obtain

$$\lim_{\Delta x \to 0} \frac{\Delta y}{\Delta x} = \lim_{\Delta x \to 0} \frac{1}{2y + \Delta y} = \frac{1}{2y}$$

The Δy is omitted, since $\Delta y \to 0$ as $\Delta x \to 0$. Now, substituting in the original function, we obtain the final result:

$$\frac{dy}{dx} = \frac{1}{2\sqrt{x}}$$

▶ One might ask why, when we are finding a derivative, we take a limit as Δx approaches zero and do not simply let Δx equal zero. If we did this, the ratio $\Delta y / \Delta x$ would be exactly $0/0$, which would then require division by zero. As we know, this is an undefined operation in mathematics, and therefore Δx *cannot equal* zero. However, it can equal any value as near zero as necessary. This idea is basic in the meaning of the word *limit*.

Exercises 2–4

In Exercises 1 through 24, find the derivative of each of the functions by using the delta-process.

1. $y = 3x - 1$
2. $y = 6x + 3$
3. $y = 1 - 2x$
4. $y = 2 - 5x$
5. $y = x^2 - 1$
6. $y = 4 - x^2$
7. $y = 5x^2$
8. $y = -6x^2$
9. $y = x^2 - 7x$
10. $y = x^2 + 4x$
11. $y = 8x - 2x^2$
12. $y = 3x - 5x^2$
13. $y = x^3 + 4x - 6$
14. $y = 2x - 4x^3$
15. $y = \dfrac{1}{x + 2}$
16. $y = \dfrac{1}{x + 1}$
17. $y = x + \dfrac{1}{x}$
18. $y = \dfrac{x}{x - 1}$
19. $y = \dfrac{2}{x^2}$
20. $y = \dfrac{1}{x^2 + 1}$
21. $y = x^4 + x^3 + x^2 + x$
22. $y = \frac{1}{3}x^3 + \frac{1}{2}x^2 + x$
23. $y = x^4 - \dfrac{2}{x}$
24. $y = \dfrac{1}{x} + \dfrac{1}{x^2}$

In Exercises 25 through 28, find the derivative of each of the given functions by using the delta-process. Then evaluate the derivative at the given point.

25. $y = 3x^2 - 2x$, $(-1, 5)$
26. $y = 9x - x^3$, $(2, 10)$
27. $y = \dfrac{6}{x + 3}$, $(3, 1)$
28. $y = x^2 - \dfrac{2}{x}$, $(-2, 5)$

In Exercises 29 through 32, find dy/dx for the given functions by the method of Example F.

29. $y = \sqrt{x + 1}$
30. $y = \sqrt{x - 2}$
31. $y = \sqrt{1 - 3x}$
32. $y = \sqrt{x^2 + 3}$

See Appendix C for a computer program to evaluate a derivative.

2–5 The Meaning of the Derivative

In Section 2–3, we saw that the slope of a curve was the limiting value of the slope of the line through points P and Q as Q approaches P. At the end of the section, we discussed the idea that $\Delta y/\Delta x$ indicated the rate of change of y with respect to x. In Section 2–4, we defined the limit of the ratio of $\Delta y/\Delta x$ as $\Delta x \to 0$ as the derivative. Thus, the derivative is a measure of the rate of change of y with respect to x at point P. However, P may represent any point, which means that the value of the derivative changes from one point on a curve to another point.

*We therefore interpret the derivative as the **instantaneous rate of change** of y with respect to x.*

Example A

In Examples A and B of Section 2–3, y is changing at the rate of 7 units for every change of 1 unit in x, *when x is 2, and only when x is 2.* In Example C of Section 2–3, y is increasing 6 units for every increase of 1 unit of x *when* $x = -1$. When $x = 3$, y is decreasing 2 units for 1 unit of increase in x.

If a functional relationship exists between any two variables, then one can be thought of as varying with respect to the other. There are many applications of this principle, one of them being the velocity of an object. We shall consider here the case of rectilinear motion, that is, motion along a straight line.

As we have seen, the velocity of an object is found by dividing the change in displacement by the time required for this change. This, however, gives a value only for the **average velocity** for the specified time interval. If the time interval considered becomes smaller and smaller, then the average velocity which is calculated more nearly approximates the **instantaneous velocity** at some particular time. In the limit, the value of the average velocity gives the value of the instantaneous velocity. Using the symbols as defined for the derivative, *the instantaneous velocity of an object moving in rectilinear motion at a particular time t is given by*

$$v = \lim_{\Delta t \to 0} \frac{\Delta s}{\Delta t} \tag{2–21}$$

where s is the displacement.

In Eq. (2–21) the derivative gives us the instantaneous rate of change of s with respect to t. Therefore, the units of this derivative would be in units of displacement divided by units of time. In general, *units of the derivative of $y = f(x)$ are in units of y divided by units of x.*

Example B

Find the instantaneous velocity when $t = 4$ s of an object for which the displacement s (in feet) is given by $s = 16t^2$ by calculating values of $\Delta s/\Delta t$ and determining the limit as Δt approaches zero.

Here we shall let t take on values of 3.5, 3.9, 3.99, and 3.999 s. When $t = 4$ s, $s = 256$ ft. Therefore, we calculate Δt by subtracting values of t from 4, and Δs is calculated by subtracting values from 256. The values of velocity are then calculated by using $v = \Delta s/\Delta t$.

t (seconds)	3.5	3.9	3.99	3.999
s (feet)	196.0	243.36	254.7216	255.872016
Δs (feet)	60.0	12.64	1.2784	0.127984
Δt (seconds)	0.5	0.1	0.01	0.001
v (feet per second)	120.0	126.4	127.84	127.984

We can see that the limit is approaching 128 ft/s, which is therefore the instantaneous velocity when $t = 4$ s.

Example C

By use of the delta-process, determine an expression for the instantaneous velocity of the object of Example B, for which the displacement s (in feet) is given by $s = 16t^2$ and t represents the time (in seconds). By use of the derived expression, determine the instantaneous velocity when $t = 2$ s and when $t = 4$ s.

Applying the delta-process to this function, we find the derivative of s with respect to t:

$$s + \Delta s = 16(t + \Delta t)^2 = 16t^2 + 32t\,\Delta t + 16(\Delta t)^2$$
$$\Delta s = 32t\,\Delta t + 16(\Delta t)^2$$
$$\frac{\Delta s}{\Delta t} = 32t + 16\,\Delta t$$
$$v = \lim_{\Delta t \to 0}\frac{\Delta s}{\Delta t} = \frac{ds}{dt} = 32t$$

Finding the instantaneous velocity for $t = 2$ s and $t = 4$ s, we have

$$\left.\frac{ds}{dt}\right|_{t=2} = 32(2) = 64 \text{ ft/s}$$

and

$$\left.\frac{ds}{dt}\right|_{t=4} = 32(4) = 128 \text{ ft/s}$$

We see that this second result agrees with that obtained in Example B.

By finding $\lim_{\Delta x \to 0} \Delta y / \Delta x$, we can find the instantaneous rate of change of y with respect to x. The expression $\lim_{\Delta t \to 0} \Delta s / \Delta t$ gives the velocity, or instantaneous rate of change of displacement with respect to time. Generalizing, we can say that

the derivative can be interpreted as instantaneous rate of change of the dependent variable with respect to the independent variable.

This is true no matter what the variables represent, so long as a function is defined between the variables.

Example D

Find an expression for the rate of change of the volume of a sphere with respect to its radius.

$$V = \frac{4}{3} \pi r^3$$

$$V + \Delta V = \frac{4\pi}{3} (r + \Delta r)^3$$

$$= \frac{4\pi}{3} [r^3 + 3r^2 \Delta r + 3r (\Delta r)^2 + (\Delta r)^3]$$

$$\Delta V = \frac{4\pi}{3} [3r^2 \Delta r + 3r (\Delta r)^2 + (\Delta r)^3]$$

$$\frac{\Delta V}{\Delta r} = \frac{4\pi}{3} [3r^2 + 3r \Delta r + (\Delta r)^2]$$

$$\lim_{\Delta r \to 0} \frac{\Delta V}{\Delta r} = 4\pi r^2$$

We can see that, as the radius increases, the rate of change of the volume with respect to the radius also increases. This should be expected, since a sphere changing from a radius of 1 to a radius of 2 has an increase in volume from $4\pi/3$ to $32\pi/3$, whereas a change in the radius from 2 to 3 involves a volume increase from $32\pi/3$ to $108\pi/3$.

Example E

The power P produced by an electric current i in a resistor varies directly as the square of the current. Given that 1.2 W of power are produced by a current of 0.5 A in a particular resistor, find an expression for the rate of change of power with respect to current.

We must first find the functional relationship between power and current. This we can do by solving the indicated problem in variation:

$$P = ki^2, \qquad 1.2 = k(0.5)^2, \qquad k = 4.8 \text{ W/A}^2, \qquad P = 4.8i^2$$

Now, knowing the function, we may determine the expression for the rate of change of P with respect to i by use of the delta-process:

$$P + \Delta P = 4.8(i + \Delta i)^2 = 4.8[i^2 + 2i(\Delta i) + (\Delta i)^2]$$

$$\Delta P = 4.8[2i(\Delta i) + (\Delta i)^2]$$

$$\frac{\Delta P}{\Delta i} = 4.8(2i + \Delta i)$$

$$\lim_{\Delta i \to 0} \frac{\Delta P}{\Delta i} = 9.6i$$

This last expression tells us that as the current increases, the power changes at a rate of $9.6i$ W/A. For example, if the current is 2 A, the power is changing at the rate of 19.2 W/A. The larger the current, the greater the increase in power. This should be expected, since the power varies as the square of the current.

Exercises 2–5

In Exercises 1 through 4, calculate the instantaneous velocity for the indicated value of the time (in seconds) of an object for which the displacement (in feet) is given by the indicated function. Use the method of Example B, and calculate values of $\Delta s / \Delta t$ for the given values of t, and determine the limit as Δt approaches zero.

1. $s = 4t + 10$; when $t = 3$; use values of t of 2.0, 2.5, 2.9, 2.99, 2.999.
2. $s = 20 - 8t$; when $t = 2$; use values of t of 1.0, 1.5, 1.9, 1.99, 1.999.
3. $s = 1 - 2t^2$; when $t = 4$; use values of t of 3.0, 3.5, 3.9, 3.99, 3.999.
4. $s = 120t - 16t^2$; when $t = 0.5$; use values of t of 0.4, 0.45, 0.49, 0.499, 0.4999.

In Exercises 5 through 8, use the delta-process to find an expression for the instantaneous velocity of an object moving with rectilinear motion according to the given functions (the same as those of Exercises 1 through 4) relating s (in feet) and t (in seconds). Then calculate the instantaneous velocity for the given value of t.

5. $s = 4t + 10$; $t = 3$ 6. $s = 20 - 8t$; $t = 2$
7. $s = 1 - 2t^2$; $t = 4$ 8. $s = 120t - 16t^2$; $t = 0.5$

In Exercises 9 through 12, use the delta-process to find an expression for the instantaneous velocity of an object moving with rectilinear motion according to the given functions relating s and t.

9. $s = 3t - \dfrac{2}{t}$ 10. $s = \dfrac{1}{2t + 3}$

11. $s = t^3 - 6t + 2$ 12. $s = s_0 + v_0 t - \frac{1}{2}at^2$ (s_0, v_0, and a are constants.)

In Exercises 13 through 16, use the delta-process to find an expression for the instantaneous acceleration a of an object moving with rectilinear motion according to the given functions. The instantaneous acceleration of an object is defined as the instantaneous rate of change of velocity with respect to time.

13. $v = 6t^2 - 4t + 2$ 14. $v = v_0 - gt$ (v_0 and g are constants.)
15. $s = 1 - 2t^2$ (Find v, then find a.)
16. $s = s_0 + v_0 t - \frac{1}{2}at^2$ (s_0, v_0, and a are constants.) (Find v, then find a.)

In Exercises 17 through 28, find the indicated rates of change.

17. The electric current i at a point in an electric circuit is the instantaneous rate of change of the electric charge q which passes the point, with respect to the time t. Find i in a circuit for which $q = 30 - 2t$.

18. The profit P in selling 100 items at price p, when the cost of each item is \$10, is given by $P = 100(p - 10)$. Find the expression for the instantaneous rate of change of P with respect to p.

19. Find a general expression for the instantaneous rate of change of the area of a circle with respect to its radius. Then find the instantaneous rate of change of A with respect to r, when $r = 3$.

20. A load L, in newtons, is distributed along a beam 8 m long such that $L = 4x - \frac{1}{2}x^2$, where x is the distance from one end of the beam. Find the instantaneous rate of change of L with respect to x.

21. The resistance of a certain resistor as a function of the temperature is given by the relation $R = 50(1 + 0.0053T + 0.00001T^2)$. Determine the instantaneous rate of change of resistance with respect to temperature (in ohms per degree Celsius) when the temperature is 20°C.

22. At very low temperatures, near absolute zero, the specific heat c of solids is given by the Debye equation $c = kT^3$, where T is the thermodynamic temperature and k is a constant dependent on the material. Determine the expression for the instantaneous rate of change for the specific heat with respect to temperature. The units of the specific heat are joules per kilogram kelvin and T is measured in kelvins.

23. The total solar radiation H, in watts per square meter, on a particular surface during an average clear day is given by $H = \dfrac{5000}{t^2 + 10}$, where $t\,(-6 \le t \le 6)$ is the number of hours from noon (6 A.M. is equivalent to $t = -6$ h). Find the instantaneous rate of change of H with respect to t at 3 P.M.

24. The value (in dollars) of a certain car is given by the function $V = \dfrac{6000}{t + 1}$, where t is measured in years. Find a general expression for the rate of change of V with respect to t, and then evaluate this expression when $t = 3$ years.

25. Oil in a certain machine is stored in a conical reservoir, for which the radius and height are both 4 cm. Find the instantaneous rate of change of the volume V of oil in the reservoir with respect to the depth d of the oil. See Fig. 2–19.

26. The total energy supplied to an inductor while the current increases from 0 to I is proportional to the square of the current I. If 8 J are supplied to a certain inductor as the current increases from 0 to 2 A, evaluate the expression for the instantaneous rate of change of energy with respect to I when $I = 1.5$ A.

27. The illuminance of a light source varies inversely as the square of the distance from the source. Given that the illuminance is 25 lx (lux) at a distance of 0.20 m, find the expression for the instantaneous rate of change of illuminance with respect to distance. Calculate this rate for distances of 0.01 m and 0.10 m.

28. The frequency of vibration of a wire varies directly as the square root of the tension on the wire. Find the expression for the instantaneous rate of change of frequency with respect to tension. Find the value of this rate of change for $T = 1$ N if the frequency is 400 Hz when $T = 1$ N.

Figure 2–19

2-6 Derivatives of Polynomials

The task of finding the derivative of a function can be considerably shortened from that involved in the direct use of the delta-process. We can use the delta-process to derive certain basic formulas for finding derivatives of particular types of functions. These formulas will then be used to find the derivatives. In this section, we shall derive the formulas for finding the derivatives of polynomial functions of the form $f(x) = a_0 x^n + a_1 x^{n-1} + \cdots + a_n$.

First we shall find the derivative of a constant. By letting $y = c$ and applying the delta-process to this function, we can obtain the desired result:

$$y = c, \qquad y + \Delta y = c, \qquad \Delta y = 0, \qquad \frac{\Delta y}{\Delta x} = 0, \qquad \lim_{\Delta x \to 0} \frac{\Delta y}{\Delta x} = 0$$

From this we conclude that *the derivative of a constant is zero*. No assumption was made as to the value of the constant, and thus this result holds for all constants. Therefore, if $y = c$, $dy/dx = 0$. This is more commonly written as

$$\frac{dc}{dx} = 0 \tag{2-22}$$

Graphically this means that for any function of the type $y = c$ the slope is always zero. From our knowledge of straight lines, we know that $y = c$ represents a straight line parallel to the x-axis. From the definition of slope, we know that any line parallel to the x-axis has a slope of zero. See Fig. 2-20. We can see that the two results are consistent.

Figure 2-20

Next we shall find the derivative of any integral power of x. If $y = x^n$, where n is an integer, by use of the binomial theorem we have the following:

$$(y + \Delta y) = (x + \Delta x)^n$$

$$= x^n + n x^{n-1} \Delta x + \frac{n(n-1)}{2} x^{n-2} (\Delta x)^2 + \cdots + (\Delta x)^n$$

$$\Delta y = n x^{n-1} \Delta x + \frac{n(n-1)}{2} x^{n-2} (\Delta x)^2 + \cdots + (\Delta x)^n$$

$$\frac{\Delta y}{\Delta x} = n x^{n-1} + \underbrace{\frac{n(n-1)}{2} x^{n-2} \Delta x + \cdots + (\Delta x)^{n-1}}$$

$$\lim_{\Delta x \to 0} \frac{\Delta y}{\Delta x} = n x^{n-1} \qquad \text{Each term} \to 0 \text{ as } \Delta x \to 0$$

Thus, *the derivative of the nth power of x is found to be*

$$\frac{dx^n}{dx} = n x^{n-1} \tag{2-23}$$

Example A   Find the derivative of the function $y = -5$.

Applying the result of Eq. (2–22), since -5 is a constant, we have

$$\frac{dy}{dx} = \frac{d(-5)}{dx} = 0$$

or

$$\frac{dy}{dx} = 0$$

Example B   Find the derivative of the function $y = x$.

Applying the result of Eq. (2–23) (here $n = 1$), we have

$$\frac{dy}{dx} = \frac{d(x)}{dx} = (1)x^{1-1} = 1$$

or simply

$$\frac{dy}{dx} = 1$$

Thus, the derivative of $y = x$ is 1, which means that the slope of the line $y = x$ is always 1. This is consistent with our previous discussion of the straight line.

Example C   Find the derivative of $y = x^3$.

We find that

$$\frac{dy}{dx} = \frac{d(x^3)}{dx}$$
$$= 3x^{3-1}$$
$$= 3x^2$$

We can see that this result is consistent with the results found in the examples and exercises of the previous sections in this chapter.

Example D   Find the derivative of $y = x^{10}$.

We find that

$$\frac{dy}{dx} = \frac{d(x^{10})}{dx}$$
$$= 10x^{10-1}$$
$$= 10x^9$$

Next we shall find the derivative of a constant times a function of x. If $y = cu$, where $u = f(x)$, we have the following result:

$$y + \Delta y = c(u + \Delta u)$$

(As x increases by Δx, u increases by Δu, since u is a function of x.) Then

$$\Delta y = c\,\Delta u, \qquad \frac{\Delta y}{\Delta x} = c\,\frac{\Delta u}{\Delta x}, \qquad \lim_{\Delta x \to 0} \frac{\Delta y}{\Delta x} = c \lim_{\Delta x \to 0} \frac{\Delta u}{\Delta x}$$

Therefore, *the derivative of the product of a constant and a function of x is the product of the constant and the derivative of the function of x.* This is written as

$$\frac{d(cu)}{dx} = c\,\frac{du}{dx} \tag{2–24}$$

Example E

Find the derivative of $y = 3x^2$.

In this case $c = 3$ and $u = x^2$. Thus, $du/dx = 2x$. Therefore,

$$y = 3x^2$$

$$\frac{dy}{dx} = 3\,\frac{dx^2}{dx} = 3(2x) = 6x$$

Occasionally the derivative of a constant times a function of x is confused with the derivative of a constant (alone). The distinction between a constant multiplying a function and an isolated constant must be made.

Finally, if the types of functions for which we have found derivatives are added, the result is a polynomial function with more than one term. The derivative of such a function is found by lettting $y = u + v$, where u and v are functions of x. Applying the delta-process, since u and v are functions of x, each has an increment corresponding to an increment in x. Thus, we have the following result:

$$y + \Delta y = (u + \Delta u) + (v + \Delta v)$$

$$\Delta y = \Delta u + \Delta v$$

$$\frac{\Delta y}{\Delta x} = \frac{\Delta u}{\Delta x} + \frac{\Delta v}{\Delta x}$$

$$\lim_{\Delta x \to 0} \frac{\Delta y}{\Delta x} = \lim_{\Delta x \to 0} \frac{\Delta u}{\Delta x} + \lim_{\Delta x \to 0} \frac{\Delta v}{\Delta x}$$

This tells us that *the derivative of the sum of functions of x is the sum of the derivatives of the functions.* This is written as

$$\frac{d(u + v)}{dx} = \frac{du}{dx} + \frac{dv}{dx} \tag{2–25}$$

Example F

Find the derivative of $y = 4x^2 + 5$.

Here $u = 4x^2$ and $v = 5$. Thus, $du/dx = 8x$ and $dv/dx = 0$. Hence we have

$$\frac{dy}{dx} = \frac{d(4x^2)}{dx} + \frac{d(5)}{dx}$$

$$= 8x + 0$$

$$= 8x$$

Example G

Evaluate the derivative of $y = 2x^4 - 6x^2 - 8x - 9$ at the point $(-2, 15)$.

First, finding the derivative, we have

$$\frac{dy}{dx} = \frac{d(2x^4)}{dx} - \frac{d(6x^2)}{dx} - \frac{d(8x)}{dx} - \frac{d(9)}{dx}$$

$$= 8x^3 - 12x - 8$$

Since the derivative is a function only of x, we now evaluate it for $x = -2$.

$$\left.\frac{dy}{dx}\right|_{x=-2} = 8(-2)^3 - 12(-2) - 8$$

$$= -48$$

Example H

Figure 2–21

Find the slope of a line which is tangent to the curve of the function $y = 4x^7 - x^4$ at the point $(1, 3)$. See Fig. 2–21.

We must find, and then evaluate, the derivative of the function for the value $x = 1$.

$$\frac{dy}{dx} = 28x^6 - 4x^3$$

Evaluating this, we have

$$\left.\frac{dy}{dx}\right|_{x=1} = 28(1) - 4(1)$$

$$= 24$$

Thus, the slope of the tangent line is 24. Again, we note that the substitution $x = 1$ must be made ***after*** the differentiation has been performed.

Example I

Find the instantaneous velocity when $t = 3$ s for an object moving with rectilinear motion according to the function $s = 8t - 2t^3$, where s is measured in centimeters.

In this example we must find the derivative of s with respect to t, ds/dt, and then evaluate it for $t = 3$. Thus,

$$s = 8t - 2t^3$$

$$\frac{ds}{dt} = 8 - 6t^2$$

$$\left.\frac{ds}{dt}\right|_{t=3} = 8 - 6(3^2) = 8 - 54 = -46 \text{ cm/s}$$

Therefore, the object is moving at -46 cm/s (it is moving in a negative direction) when $t = 3$ s.

Exercises 2–6

In Exercises 1 through 16, find the derivative of each of the given functions.

1. $y = x^5$
2. $y = x^{12}$
3. $y = 4x^9$
4. $y = -7x^6$

5. $y = x^4 - 6$
6. $y = 3x^5 - 1$
7. $y = x^2 + 2x$
8. $y = x^3 - 2x^2$

9. $y = 5x^3 - x - 1$
10. $y = 6x^2 - 6x + 5$

11. $y = x^8 - 4x^7 - x$
12. $y = 4x^4 - 2x + 9$

13. $y = 6x^7 - 5x^3 + 2$
14. $y = 13x^4 - 6x^3 - x - 1$

15. $y = \frac{1}{3}x^3 + \frac{1}{2}x^2$
16. $y = \frac{1}{4}x^8 - \frac{1}{2}x^4 - \sqrt{3}$

In Exercises 17 through 20, evaluate the derivatives of the given functions at the given points.

17. $y = 6x^2 - 8x + 1$ $(2, 9)$
18. $y = x^3 - 5x^2 + 4$ $(-1, -2)$

19. $y = 2x^3 + 9x - 7$ $(-2, -41)$
20. $y = x^4 - 9x^2 - 5x$ $(3, -15)$

In Exercises 21 through 24, find the slope of a tangent line to the curve of each of the given functions for the given values of x.

21. $y = 2x^6 - 6x^2$ $(x = 2)$
22. $y = 3x^3 - 9x$ $(x = 1)$

23. $y = 6x - 2x^4$ $(x = -1)$
24. $y = x^4 - \frac{1}{2}x^2 + 2$ $(x = -2)$

In Exercises 25 through 28, find an expression for the instantaneous velocity of objects moving according to the functions given, if s represents displacement in terms of time t.

25. $s = 6t^5 - 5t + 2$
26. $s = 4t^2 - 7t$

27. $s = 120t - 16t^2$
28. $s = s_0 + v_0 t + \frac{1}{2}at^2$

In Exercises 29 through 32, s represents the displacement and t represents the time for objects moving according to the given functions. Find the value of the instantaneous velocity for the given times.

29. $s = 2t^3 - 4t^2$, $t = 4$

30. $s = t^2 - 6t + 4$, $t = 2$

31. $s = 80t - 4.9t^2$, $t = 3$

32. $s = 8t^2 - 10t + 6$, $t = 5$

In Exercises 33 through 42, solve the given problems by finding the appropriate derivative.

33. For what value(s) of x is the tangent to the curve of $y = 3x^2 - 6x$ parallel to the x-axis? (That is, where is the slope zero?)

34. For what value(s) of x is the tangent to the curve of $y = 4x^3 - 12x^2$ parallel to the x-axis? (See Exercise 33.)

35. If the radius of a right circular cylinder always equals its height, find an expression for the instantaneous rate of change of volume with respect to the radius.

36. For pillars of a given length, the crushing load varies as the fourth power of the radius. If 18 tons will crush a certain pillar 5 in. in radius, find an expression which gives the instantaneous rate of change of crushing load with respect to the radius.

37. The voltage V produced by a certain thermocouple as a function of the temperature T is given by $V = 4.0\,T + 0.005T^2$. If the temperature changes slowly, find the instantaneous rate of change of voltage with respect to temperature when the temperature is 200°C.

38. The cost C (in dollars) to a firm producing x units is given by the equation $C = 0.02x^3 - 2.40x^2 + 100x$. Determine the instantaneous rate of change of C with respect to x for $x = 50$ units.

39. The altitude h (in meters) of an airplane as a function of the horizontal distance x (in kilometers) it has traveled is given by $h = 30x^2 - x^3$. Find the instantaneous rate of change of h with respect to x for $x = 10$ km.

40. The ends of a 10-ft beam are supported at different levels. The deflection y of the beam is given by $y = kx^2(x^3 + 450x - 3500)$, where x is the horizontal distance from one end and k is a constant. Determine the expression for the instantaneous rate of change of deflection with respect to x.

41. The force F, in newtons, exerted by a cam on a lever is given by $F = x^4 - 12x^3 + 46x^2 - 60x + 25$, where x ($1 \le x \le 5$) is the distance, in centimeters, from the center of rotation of the cam to the edge of the cam in contact with the lever (see Fig. 2–22). Find the instantaneous rate of change of F with respect to x when $x = 4$ cm.

42. A company determines that the cost C, in dollars, of producing x machine parts per day is given by $C = x^3 - 300x + 100$. Find the value of the instantaneous rate of change of C with respect to x when $x = 12$ parts.

Figure 2–22

2-7 Derivatives of Products and Quotients of Functions

The formulas developed in the previous section are valid for polynomial functions. However, many functions are not polynomial in form. Some functions can best be expressed as the product of two or more simpler functions, others are the quotient of two simpler functions, and some are expressed as powers of a function. In this section, we shall develop the formula for the derivative of a product of functions and the formula for the derivative of the quotient of two functions.

Example A

The functions $f(x) = x^2 + 2$ and $g(x) = 3 - 2x$ can be combined to form new functions which represent the types mentioned above. For example, the function $p(x) = f(x)g(x) = (x^2 + 2)(3 - 2x)$ is an example of a function expressed as the product of two simpler functions. The function

$$q(x) = \frac{g(x)}{f(x)} = \frac{3 - 2x}{x^2 + 2}$$

is an example of a rational function that is the quotient of two other functions. The function $F(x) = [g(x)]^3 = (3 - 2x)^3$ is an example of a power of a function.

If u and v both represent functions of x, the derivative of the product of u and v is found by applying the delta-process. This leads to the following result:

$$y = u \cdot v$$
$$y + \Delta y = (u + \Delta u)(v + \Delta v)$$

(Since u and v are functions of x, each has an increment corresponding to an increment in x.) Then we have

$$\Delta y = u\,\Delta v + v\,\Delta u + \Delta u\,\Delta v$$

$$\frac{\Delta y}{\Delta x} = u\,\frac{\Delta v}{\Delta x} + v\,\frac{\Delta u}{\Delta x} + \Delta u\,\frac{\Delta v}{\Delta x}$$

$$\lim_{\Delta x \to 0} \frac{\Delta y}{\Delta x} = u \lim_{\Delta x \to 0} \frac{\Delta v}{\Delta x} + v \lim_{\Delta x \to 0} \frac{\Delta u}{\Delta x} + \lim_{\Delta x \to 0}\left(\Delta u\,\frac{\Delta v}{\Delta x}\right)$$

(The functions u and v are not affected by Δx approaching zero, but Δu and Δv both approach zero as Δx approaches 0.) Thus,

$$\frac{dy}{dx} = u\,\frac{dv}{dx} + v\,\frac{du}{dx} + 0\,\frac{dv}{dx}$$

We conclude that *the derivative of the product of two functions equals the first function times the derivative of the second function plus the second function times the derivative of the first function.* This is written as

$$\frac{d(uv)}{dx} = u\,\frac{dv}{dx} + v\,\frac{du}{dx} \qquad (2\text{--}26)$$

Example B

For the product function in Example A, the derivative is found as follows:

$$y = (x^2 + 2)(3 - 2x), \qquad u = x^2 + 2, \qquad v = 3 - 2x$$

$$\frac{d(uv)}{dx} = \quad u \qquad \frac{dv}{dx} \quad + \quad v \qquad \frac{du}{dx}$$

$$\downarrow \qquad \downarrow \qquad \qquad \downarrow \qquad \downarrow$$

$$\frac{dy}{dx} = (x^2 + 2)(-2) + (3 - 2x)(2x) = -2x^2 - 4 + 6x - 4x^2$$

$$= -6x^2 + 6x - 4$$

Example C

Find the derivative of the function $y = (3 - x - 2x^2)(x^4 - x)$.
In this problem, $u = 3 - x - 2x^2$ and $v = x^4 - x$. Hence

$$\frac{dy}{dx} = (3 - x - 2x^2)(4x^3 - 1) + (x^4 - x)(-1 - 4x)$$

$$= 12x^3 - 3 - 4x^4 + x - 8x^5 + 2x^2 - x^4 - 4x^5 + x + 4x^2$$

$$= -12x^5 - 5x^4 + 12x^3 + 6x^2 + 2x - 3$$

In both of these examples, we could have multiplied the functions first and then taken the derivative as a polynomial. However, we shall soon meet functions for which this latter method would not be applicable.

We shall now find the derivative of the quotient of two functions. We apply the delta-process to the function $y = u/v$ and obtain the result as follows:

$$y + \Delta y = \frac{u + \Delta u}{v + \Delta v}$$

$$\Delta y = \frac{u + \Delta u}{v + \Delta v} - \frac{u}{v} = \frac{vu + v\,\Delta u - uv - u\,\Delta v}{v(v + \Delta v)}$$

$$\frac{\Delta y}{\Delta x} = \frac{v(\Delta u/\Delta x) - u(\Delta v/\Delta x)}{v(v + \Delta v)}$$

$$\lim_{\Delta x \to 0} \frac{\Delta y}{\Delta x} = \frac{v \lim\limits_{\Delta x \to 0}(\Delta u/\Delta x) - u \lim\limits_{\Delta x \to 0}(\Delta v/\Delta x)}{\lim\limits_{\Delta x \to 0} v(v + \Delta v)}$$

Thus

$$\frac{dy}{dx} = \frac{v\,(du/dx) - u\,(dv/dx)}{v^2}$$

This last result says that *the derivative of the quotient of two functions equals the denominator times the derivative of the numerator minus the numerator times the derivative of the denominator, all divided by the square of the denominator.* This is written as

$$\frac{d\dfrac{u}{v}}{dx} = \frac{v\dfrac{du}{dx} - u\dfrac{dv}{dx}}{v^2}$$

(2–27)

Example D

Find the derivative of the quotient indicated in Example A.

$$y = \frac{3 - 2x}{x^2 + 2}, \qquad u = 3 - 2x, \qquad v = x^2 + 2$$

$$\frac{dy}{dx} = \frac{\overset{v}{\overset{\downarrow}{(x^2 + 2)}}\overset{\frac{du}{dx}}{\overset{\downarrow}{(-2)}} - \overset{u}{\overset{\downarrow}{(3 - 2x)}}\overset{\frac{dv}{dx}}{\overset{\downarrow}{(2x)}}}{(x^2 + 2)^2} = \frac{-2x^2 - 4 - 6x + 4x^2}{(x^2 + 2)^2}$$

$$= \frac{2(x^2 - 3x - 2)}{(x^2 + 2)^2} \qquad \overset{\nwarrow}{v^2}$$

Example E

Find the derivative of

BoTToM × dirr ToP

– ToP × dirr BoTToM

diviDED by boTToM²

$$y = \frac{x^3 - x - 1}{1 - 2x^2}$$

$$\frac{dy}{dx} = \frac{(1 - 2x^2)(3x^2 - 1) - (x^3 - x - 1)(-4x)}{(1 - 2x^2)^2}$$

$$= \frac{3x^2 - 1 - 6x^4 + 2x^2 + 4x^4 - 4x^2 - 4x}{(1 - 2x^2)^2}$$

$$= \frac{-2x^4 + x^2 - 4x - 1}{(1 - 2x^2)^2}$$

Example F

Evaluate the derivative of $y = \dfrac{3x^2 + x}{1 - 4x}$ at $(2, -2)$.

We first find dy/dx and then find its value for $x = 2$.

$$\frac{dy}{dx} = \frac{(1 - 4x)(6x + 1) - (3x^2 + x)(-4)}{(1 - 4x)^2}$$

$$= \frac{6x + 1 - 24x^2 - 4x + 12x^2 + 4x}{(1 - 4x)^2}$$

$$= \frac{-12x^2 + 6x + 1}{(1 - 4x)^2}$$

$$\frac{dy}{dx}\bigg|_{x=2} = \frac{-12(2^2) + 6(2) + 1}{[1 - 4(2)]^2} = \frac{-48 + 12 + 1}{49} = \frac{-35}{49}$$

$$= -\frac{5}{7}$$

Exercises 2–7

In Exercises 1 through 8, find the derivative of each function by use of Eq. (2–26). Do not find the product before finding dy/dx.

1. $y = x^2(3x + 2)$ 2. $y = 3x(x^3 + 1)$
3. $y = 6x(3x^2 - 5x)$ 4. $y = 2x^3(3x^4 + x)$
5. $y = (x + 2)(2x - 5)$ 6. $y = (3x - 1)(x^2 + 1)$
7. $y = (x^4 - 3x^2 + 3)(1 - 2x^3)$ 8. $y = (x^3 - 6x)(2 - 4x^3)$

In Exercises 9 through 12, find the derivative of each function by use of Eq. (2–26). Then multiply out each function and find the derivative by treating it as a polynomial. Compare results.

9. $y = (2x - 7)(5 - 2x)$ 10. $y = (x^2 + 2)(2x^2 - 1)$
11. $y = (x^3 - 1)(2x^2 - x - 1)$ 12. $y = (3x^2 - 4x + 1)(5 - 6x^2)$

In Exercises 13 through 24, find the derivative of each function by Eq. (2–27).

13. $y = \dfrac{x}{2x + 3}$ 14. $y = \dfrac{2x}{x + 1}$ 15. $y = \dfrac{1}{x^2 + 1}$

16. $y = \dfrac{x + 2}{2x + 3}$ 17. $y = \dfrac{x^2}{3 - 2x}$ 18. $y = \dfrac{2}{3x^2 - 5x}$

19. $y = \dfrac{2x - 1}{3x^2 + 2}$ 20. $y = \dfrac{2x^3}{4 - x}$ 21. $y = \dfrac{x + 8}{x^2 + x + 2}$

22. $y = \dfrac{3x}{4x^5 - 3x - 4}$ 23. $y = \dfrac{2x^2 - x - 1}{x^3 + 2x^2}$ 24. $y = \dfrac{3x^3 - x}{2x^2 - 5x + 4}$

In Exercises 25 through 28, evaluate the derivatives of the given functions for the given values of x. In Exercises 25 and 26, find the derivative by use of Eq. (2–26).

25. $y = (3x - 1)(4 - 7x)$, $x = 3$ 26. $y = (3x^2 - 5)(2x^2 - 1)$, $x = -1$

27. $y = \dfrac{3x - 5}{2x + 3}$, $x = -2$ 28. $y = \dfrac{2x^2 - 5x}{3x + 2}$, $x = 2$

In Exercises 29 through 40, solve the given problems by finding the appropriate derivatives.

29. Find the derivative of $y = \dfrac{x^2(1 - 2x)}{3x - 7}$, but do not multiply out the numerator before taking the derivative.

30. Find the derivative of $y = 4x^2 - \dfrac{1}{x - 1}$, but do not combine the terms before finding the derivative.

31. Find the slope of a tangent line to the curve $y = (4x + 1)(x^4 - 1)$ at the point $(-1, 0)$. Do not multiply the factors together before taking the derivative.

32. Find the slope of a line tangent to the curve of the function $y = (3x + 4)(1 - 4x)$ at the point $(2, -70)$. Do not multiply before finding the derivative.

33. For what value(s) of x is the slope of a tangent to the curve $y = \dfrac{x}{x^2 + 1}$ equal to zero?

34. Determine the sign of the derivative of the function $y = \dfrac{2x - 1}{1 - x^2}$ for the following values of x: $-2, -1, 0, 1, 2$. Is the slope of a tangent to this curve ever negative?

35. If the distance (in feet) traveled by a particle in a given time is found from the expression $s = 2/t^2$, find the velocity after 4 s. (Use the quotient rule.)

36. During a chemical change the number n of grams of a compound being formed is given by $n = \dfrac{6t}{t + 2}$, where t is measured in seconds. How many grams per second are being formed after 3 s?

37. The voltage across a resistor in an electric circuit is the product of the resistance and the current. If the current I (in amperes) varies with time (in seconds) according to the relation $I = 5.00 + 0.01t^2$, and the resistance varies with time according to the relation $R = 15.00 - 0.10t$, find the time rate of change of the voltage when $t = 5$ s.

38. In thermodynamics, an equation relating pressure p, volume V, and temperature T is

$$\left(p + \frac{a}{V^2}\right)(V - b) = RT$$

where a, b, and R are constants. This equation is known as the van der Waals equation. Solve this equation for p, and then find the expression for the rate of change of p with respect to V, assuming a constant temperature.

39. The electric power produced by a certain source is given by

$$P = \frac{E^2 r}{R^2 + 2Rr + r^2}$$

where E is the voltage of the source, R is the resistance of the source, and r is the resistance in the circuit. Find the expression for the rate of change of power with respect to the resistance in the circuit, assuming that the other quantities remain constant.

40. In the study of simple harmonic motion, under certain conditions the tangent of the angle by which the displacement lags the impressed force is given by

$$\tan \phi = \frac{30\omega}{600 - \omega^2}$$

where ω (in rad/s) is the frequency of vibration. Find the rate of change of $\tan \phi$ with respect to ω when $\omega = 3$ rad/s.

2–8 The Derivative of a Power of a Function

In Example A of Section 2–7 we illustrated $y = (3 - 2x)^3$ as the power of a function of x. Here, $3 - 2x$ is the function of x, and it is being raised to the third power. If we let $u = 3 - 2x$, we can write

$$y = u^3 \quad \text{where } u = 3 - 2x$$

Writing it this way, y is a function of u and u is a function of x. This means *that y is a function of a function of x, referred to as a* **composite function.** However, y is still a function of x, since u is a function of x.

Since we will frequently need to find the derivative of the power of a function, we now develop the necessary formula. Considering $y = f(u)$ and $u = g(x)$, we can express the derivative dy/dx in terms of dy/du and du/dx. If Δx is the increment in x, then Δy and Δu are the corresponding increments in y and u, respectively. We may then write

$$\frac{\Delta y}{\Delta x} = \frac{\Delta y}{\Delta u} \cdot \frac{\Delta u}{\Delta x}$$

When Δx approaches zero, Δu and Δy both approach zero, for u and y are functions of x. Thus,

$$\lim_{\Delta x \to 0} \frac{\Delta y}{\Delta x} = \lim_{\Delta u \to 0} \frac{\Delta y}{\Delta u} \cdot \lim_{\Delta x \to 0} \frac{\Delta u}{\Delta x}$$

or

$$\frac{dy}{dx} = \frac{dy}{du} \cdot \frac{du}{dx} \tag{2–28}$$

(Here we have assumed $\Delta u \neq 0$, although it can be shown that this condition is not necessary.) *Equation (2–28) is known as the* **chain rule** *for derivatives.*

Using Eq. (2–28) for $y = u^n$, where u is a function of x, we have

$$\frac{dy}{dx} = \frac{d(u^n)}{du} \cdot \frac{du}{dx}$$

or

$$\frac{du^n}{dx} = nu^{n-1}\left(\frac{du}{dx}\right) \tag{2–29}$$

By use of Eq. (2–29) we may find the derivative of a power of a function of x.

Example A

Find the derivative of $y = (3 - 2x)^3$.

For this function, we identify $n = 3$ and $u = 3 - 2x$. This means that $du/dx = -2$, and therefore

$$\frac{du^n}{dx} = n \quad u^{n-1} \quad \left(\frac{du}{dx}\right)$$

$$\frac{dy}{dx} = 3(3 - 2x)^2(-2) \qquad = (9 - 6x)(-4)$$

$$= -6(3 - 2x)^2$$

▶ A common type of error in finding this type of derivative is to omit the du/dx factor; in this case it is the -2. *The derivative is incomplete, and therefore incorrect, without this factor.*

Example B

Find the derivative of $y = (1 - 3x^2)^4$.

In this example, $n = 4$ and $u = 1 - 3x^2$. Hence

$$\frac{dy}{dx} = 4(1 - 3x^2)^3(-6x) = -24x(1 - 3x^2)^3$$

▶ (We must not forget the $-6x$.)

$$4 - 12x^2$$

Example C

Find the derivative of

$$y = \frac{(3x - 1)^3}{1 - x}$$

QUOATION RULE FIRST

$x - 1$ IS DEVT.

BoTT × DERR TOP
– TOP × DEVT BoTTOM
DIVIDE BY BOTTOM SR.
THEN FACTOR

Here we must use the quotient rule in combination with the power rule. We find the derivative of the numerator by using the power rule.

$$\frac{dy}{dx} = \frac{(1 - x)3(3x - 1)^2(3) - (3x - 1)^3(-1)}{(1 - x)^2}$$

$$= \frac{(3x - 1)^2(9 - 9x + 3x - 1)}{(1 - x)^2} = \frac{(3x - 1)^2(8 - 6x)}{(1 - x)^2}$$

$$= \frac{2(3x - 1)^2(4 - 3x)}{(1 - x)^2}$$

It is normally better to have the derivative in a factored, simplified form, since this is the only form from which useful information may readily be obtained. In this form we can determine where the derivative is undefined (denominator equal to zero) or where the slope is zero (numerator equal to zero); we can also make other required analyses of the derivative. Thus, all derivatives should be expressed in simplest algebraic form.

So far, we have derived the formulas for the derivatives of powers of x and for functions of x for integral powers. We shall now establish that these formulas are also valid for any rational number used as an exponent. If $y = u^{p/q}$, and if each side is raised to the qth power, we have $y^q = u^p$. Applying the power rule to each side of this equation, we have

$$qy^{q-1}\left(\frac{dy}{dx}\right) = pu^{p-1}\left(\frac{du}{dx}\right)$$

Solving for dy/dx, we have

$$\frac{dy}{dx} = \frac{pu^{p-1}(du/dx)}{qy^{q-1}} = \frac{p}{q}\frac{u^{p-1}}{(u^{p/q})^{q-1}}\frac{du}{dx} = \frac{p}{q}\frac{u^{p-1}}{u^{p-p/q}}\frac{du}{dx}$$

$$= \frac{p}{q}u^{p-1-p+(p/q)}\frac{du}{dx}$$

Thus,

$$\frac{du^{p/q}}{dx} = \frac{p}{q}u^{(p/q)-1}\frac{du}{dx} \qquad\qquad (2\text{--}30)$$

We see that in finding the derivative we multiply the function by the rational exponent and subtract 1 from it to find the exponent of the function in the derivative. This, of course, is the same rule derived for integral exponents. Since the only restriction on the values of p and q is that they are integers ($q \neq 0$), *the power rule can be used for all rational exponents, positive and negative.*

Example D

We can now find the derivative of $y = \sqrt{x^2 + 1}$.

By use of Eq. (2–30) or Eq. (2–29), and writing the square root as the fractional exponent $\frac{1}{2}$, we can easily derive the result:

$$y = (x^2 + 1)^{1/2}$$

$$\frac{dy}{dx} = \frac{1}{2}(x^2 + 1)^{-1/2}(2x) = \frac{x}{(x^2 + 1)^{1/2}}$$

To avoid introducing apparently significant factors into the numerator, we do not usually rationalize such fractions.

Having shown that we may use fractional exponents to find derivatives of roots of functions of x, we may also use them to find derivatives of roots of x itself. Consider the following example.

Example E

For reference, Eq. (2–23) is

$$\frac{dx^n}{dx} = nx^{n-1}$$

Find the derivative of $y = 6\sqrt[3]{x}$.

We can write this function as $y = 6x^{1/3}$. In finding the derivative we may use Eq. (2–29), where we have $u = x$. This means that $du/dx = dx/dx = 1$. Therefore, using Eq. (2–29) is equivalent to using Eq. (2–23) for a power of x (not for a power of a function of x). Thus,

$$y = 6x^{1/3}$$

$$\frac{dy}{dx} = 6\left(\frac{1}{3}\right)x^{-2/3} = \frac{2}{x^{2/3}} = \frac{2}{\sqrt[3]{x^2}}$$

Example F

Find the derivative of

$$y = \frac{1}{(1-4x)^2}$$

We can find the derivative of this function by means of the quotient rule. However, by writing the function as

$$y = (1-4x)^{-2}$$

we can actually find the derivative more readily by use of the power rule. Therefore, the derivative is

$$\frac{dy}{dx} = (-2)(1-4x)^{-3}(-4) = \frac{8}{(1-4x)^3}$$

▶ (Remember, subtracting 1 from -2 gives -3.)

We can now see the value of fractional exponents in calculus. They are useful in algebraic operations, but are almost essential in calculus. Without fractional exponents, it would be necessary to develop more formulas to find derivatives of radicals. To find the derivative of an algebraic function, we need only those equations which we have already developed. Often it is necessary to combine these, as we saw in Example C. Actually, most derivatives are combinations. The problem in finding the derivative is recognizing the form of the function with which you are dealing. When you have recognized the form, completing the problem is only a matter of mechanics and algebra. You should now see the importance of being able to handle algebraic operations with facility.

Example G

Evaluate the derivative of

$$y = \frac{x}{\sqrt{1 - 4x}}$$

for $x = -2$.

Here we have a quotient, and in order to find the derivative of this quotient, we must also use the power rule (and a derivative of a polynomial form). With sufficient practice in taking derivatives, we can recognize the rule to use almost automatically. Thus, we find the derivative:

$$\frac{dy}{dx} = \frac{(1 - 4x)^{1/2}(1) - x(\frac{1}{2})(1 - 4x)^{-1/2}(-4)}{1 - 4x}$$

$$= \frac{(1 - 4x)^{1/2} + \dfrac{2x}{(1 - 4x)^{1/2}}}{1 - 4x}$$

$$= \frac{\dfrac{(1 - 4x)^{1/2}(1 - 4x)^{1/2} + 2x}{(1 - 4x)^{1/2}}}{1 - 4x} = \frac{(1 - 4x) + 2x}{(1 - 4x)^{1/2}(1 - 4x)}$$

$$= \frac{1 - 2x}{(1 - 4x)^{3/2}}$$

Now, evaluating the derivative for $x = -2$, we have

$$\left.\frac{dy}{dx}\right|_{x=-2} = \frac{1 - 2(-2)}{[1 - 4(-2)]^{3/2}} = \frac{1 + 4}{(1 + 8)^{3/2}} = \frac{5}{9^{3/2}}$$

$$= \frac{5}{27}$$

Exercises 2–8

In Exercises 1 through 24, find the derivative of each of the given functions.

1. $y = \sqrt{x}$

2. $y = \sqrt[4]{x}$

3. $y = \dfrac{1}{x^2}$

4. $y = \dfrac{2}{x^4}$

5. $y = \dfrac{3}{\sqrt[3]{x}}$

6. $y = \dfrac{1}{\sqrt{x}}$

7. $y = x\sqrt{x} - \dfrac{1}{x}$

8. $y = 2x^{-3} - x^{-2}$

9. $y = (x^2 + 1)^5$

10. $y = (1 - 2x)^4$

11. $y = 2(7 - 4x^3)^8$

12. $y = 3(8x^2 - 1)^6$

13. $y = (2x^3 - 3)^{1/3}$

14. $y = (1 - 6x)^{3/2}$

15. $y = \dfrac{1}{(1 - x^2)^4}$

16. $y = \dfrac{1}{\sqrt{1 - x}}$

17. $y = 4(2x^4 - 5)^{3/4}$

18. $y = 5(3x^7 - 4)^{2/3}$

19. $y = \sqrt[4]{1 - 8x^2}$

20. $y = \sqrt[3]{4x^6 + 2}$

21. $y = x\sqrt{x - 1}$

22. $y = x^2(1 - 3x)^5$ 23. $y = \dfrac{\sqrt{4x + 3}}{8x + 1}$ 24. $y = \dfrac{x\sqrt{x - 1}}{2x + 1}$

In Exercises 25 through 28, evaluate the derivatives of the given functions at the given values of x.

25. $y = \sqrt{3x + 4}$, $x = 7$ 26. $y = (4 - x^2)^{-1}$, $x = -1$

27. $y = \dfrac{\sqrt{x}}{1 - x}$, $x = 4$ 28. $y = x^2\sqrt[3]{3x + 2}$, $x = 2$

In Exercises 29 through 44, solve the given problems by finding the appropriate derivatives.

29. Find the derivative of $y = 1/x^3$ as (a) a quotient and (b) a negative power of x, and show that the results are the same.

30. Find the derivative of $y = \dfrac{2}{4x + 3}$ as (a) a quotient and (b) as a negative power of $4x + 3$, and show that the results are the same.

31. Find any values of x for which the derivative of $y = \dfrac{x^2}{\sqrt{x^2 + 1}}$ is zero.

32. Find any values of x for which the derivative of $y = \dfrac{x}{\sqrt{4x - 1}}$ is zero.

33. Find the slope of a line tangent to the parabola $y^2 = 4x$ at the point $(1, 2)$.

34. Find the slope of a tangent to the circle $x^2 + y^2 = 25$ at the point $(4, 3)$.

35. Find the velocity at $t = 4$, given that the displacement in terms of the time for a particular object is $s = (t^3 - t)^{4/3}$.

36. Find the velocity at $t = 11$ given that the displacement as a function of the time for a certain object is $s = \dfrac{1}{\sqrt{2t + 3}}$.

37. When the volume of a gas changes very rapidly, an approximate relation is that the pressure varies inversely as the $\frac{3}{2}$ power of the volume. If P is 300 kPa when $V = 100 \text{ cm}^3$, find the derivative of P with respect to V. Evaluate this derivative for $V = 100 \text{ cm}^3$.

38. The period of a pendulum is directly proportional to the square root of its length. Find the derivative of the period with respect to the length for a pendulum 2 ft long which has a period of $\pi/2$ s.

39. The total solar radiation H, in watts per square meter, on a certain surface during an average clear day is given by $H = \dfrac{4000}{\sqrt{t^6 + 100}}$, where t $(-6 < t < 6)$ is the number of hours from noon. Find the rate at which H is changing with time at 4 P.M.

40. Under certain conditions, the velocity of an object moving with simple harmonic motion is given by $v = k\sqrt{A^2 - x^2}$, where x is the x-coordinate of the position of the object, A is the amplitude, and k is a constant. Find the derivative of v with respect to x.

41. Under certain conditions, due to the presence of a charge q, the electric potential V along a line is given by

$$V = kq/\sqrt{x^2 + b^2}$$

where k is a constant and b is the minimum distance from the charge to the line. Find the expression for the rate of change of V with respect to x.

42. The current in a circuit containing a resistance R and an inductance L is found from the expression

$$I = \frac{V}{\sqrt{R^2 + (\omega L)^2}}$$

Find the expression for the rate of change of current with respect to L, assuming that the other quantities remain constant.

43. The magnetic field H due to a magnet of length l at a distance r is given by

$$H = \frac{k}{[r^2 + (l/2)^2]^{3/2}}$$

where k is a constant for a given magnet. Find the expression for the derivative of H with respect to r.

44. Under certain circumstances the velocity of a water wave is given by

$$v = \sqrt{\frac{a}{\lambda} + b\lambda}$$

where λ is the wavelength and a and b are constants. Find the derivative of v with respect to λ.

2–9 Differentiation of Implicit Functions

To this point the functions which we have differentiated have been of the form $y = f(x)$. There are, however, occasions when we need to find the derivative of a function determined by an equation which does not express the dependent variable explicitly in terms of the independent variable.

An equation in which y is not expressed explicitly in terms of x may determine one or more functions. *Any such function, where y is defined implicitly as a function of x, is called an* **implicit function.** Some equations defining implicit functions may be solved to determine the explicit functions, and for others it is not possible to solve for the explicit functions. Also, not all such equations define y as a function of x for real values of x.

Example A

The equation $3x + 4y = 5$ is an equation which defines a function, although it is not in explicit form. In solving for y as $y = -\frac{3}{4}x + \frac{5}{4}$, we have the explicit form of the function.

The equation $y^2 + x = 3$ is an equation which defines two functions, although we do not have the explicit forms. When we solve for y, we obtain the explicit functions $y = \sqrt{3 - x}$ and $y = -\sqrt{3 - x}$.

The equation $y^5 + xy^2 + 3x^2 = 5$ defines y as a function of x, although we cannot actually solve for the algebraic form of the explicit form of the function.

The equation $x^2 + y^2 + 4 = 0$ is not satisfied by any pair of real values of x and y.

Even when it is possible to determine the explicit form of a function given in implicit form, it is not always desirable to do so. In some cases the implicit form is more convenient than the explicit form.

The derivative of an implicit function may be found directly without having to solve for the explicit function. Thus, *to find dy/dx when y is defined as an implicit function of x, we differentiate each term of the equation with respect to x, regarding y as a function of x.* We then solve for *dy/dx*, which will usually be in terms of *x* and *y*.

Example B

For reference, Eq. (2–29) is

$$\frac{du^n}{dx} = nu^{n-1} \frac{du}{dx}.$$

Find dy/dx if $y^2 + 2x^2 = 5$.

Here we find the derivative of each term, and then solve for dy/dx. Thus,

$$\frac{d(y^2)}{dx} + \frac{d(2x^2)}{dx} = \frac{d(5)}{dx}$$

$$2y\frac{dy}{dx} + 4x = 0$$

$$\frac{dy}{dx} = -\frac{2x}{y}$$

▶ The factor dy/dx arises from the derivative of the first term as a result of using the derivative of a power of a function of x (Eq. 2–29). The factor dy/dx corresponds to the du/dx of the formula. In the second term, no factor of dy/dx appears, since there are no y factors in the term.

Example C

Find dy/dx if $3y^4 + xy^2 + 2x^3 - 6 = 0$.

In finding the derivative, we note that the second term is a product, and we must use the product rule for derivatives on it. Thus, we have

$$\frac{d(3y^4)}{dx} + \frac{d(xy^2)}{dx} + \frac{d(2x^3)}{dx} - \frac{d(6)}{dx} = \frac{d(0)}{dx}$$

▶ $$12y^3\frac{dy}{dx} + \left[x\left(2y\frac{dy}{dx}\right) + y^2(1)\right] + 6x^2 - 0 = 0$$

$$12y^3\frac{dy}{dx} + 2xy\frac{dy}{dx} + y^2 + 6x^2 = 0$$

$$(12y^3 + 2xy)\frac{dy}{dx} = -y^2 - 6x^2$$

$$\frac{dy}{dx} = \frac{-y^2 - 6x^2}{12y^3 + 2xy}$$

Example D

Find dy/dx if $2x^3y + (y^2 + x)^3 = x^4$.

In this case we use the product rule on the first term and the power rule on the second term.

$$\frac{d(2x^3y)}{dx} + \frac{d(y^2 + x)^3}{dx} = \frac{d(x^4)}{dx}$$

$$2x^3\left(\frac{dy}{dx}\right) + y(6x^2) + 3(y^2 + x)^2\left(2y\frac{dy}{dx} + 1\right) = 4x^3$$

$$2x^3\frac{dy}{dx} + 6x^2y + 3(y^2 + x)^2\left(2y\frac{dy}{dx}\right) + 3(y^2 + x)^2 = 4x^3$$

$$[2x^3 + 6y(y^2 + x)^2]\frac{dy}{dx} = 4x^3 - 6x^2y - 3(y^2 + x)^2$$

$$\frac{dy}{dx} = \frac{4x^3 - 6x^2y - 3(y^2 + x)^2}{2x^3 + 6y(y^2 + x)^2}$$

Example E

Find the slope of a line tangent to the curve of $2y^3 + xy + 1 = 0$ at the point $(-3, 1)$.

Here we must find dy/dx and evaluate it for $x = -3$ and $y = 1$.

$$\frac{d(2y^3)}{dx} + \frac{d(xy)}{dx} + \frac{d(1)}{dx} = \frac{d(0)}{dx}$$

$$6y^2\frac{dy}{dx} + x\frac{dy}{dx} + y + 0 = 0$$

$$\frac{dy}{dx} = \frac{-y}{6y^2 + x}$$

$$\frac{dy}{dx}\bigg|_{(-3,1)} = \frac{-1}{6(1^2) - 3} = \frac{-1}{6 - 3}$$

$$= -\frac{1}{3}$$

Thus, the slope is $-\frac{1}{3}$.

Exercises 2–9

In Exercises 1 through 20, find dy/dx by differentiating implicitly. When applicable, express the result in terms of x and y.

1. $3x + 2y = 5$

2. $6x - 3y = 4$

3. $4y - 3x^2 = x$

4. $x^5 - 5y = 6 - x$

5. $x^2 - y^2 - 9 = 0$

6. $x^2 + 2y^2 - 11 = 0$

7. $y^5 = x^2 - 1$

8. $y^4 = 3x^3 - x$

9. $y^2 + y = x^2 - 4$

10. $2y^3 - y = 7 - x^4$

11. $y + 3xy - 4 = 0$

12. $8y - xy - 7 = 0$

13. $xy^3 + 3y + x^2 = 9$

14. $y^2x - x^2y + 3x = 4$

15. $3x^2y^2 + y = 3x - 1$

16. $2x - x^3y^2 = y - x^2 - 1$

17. $(2y - x)^4 + x^2 = y + 3$

18. $(y^2 + 2)^3 = x^4y + 11$

19. $2(x^2 + 1)^3 + (y^2 + 1)^2 = 17$

20. $(2x + 1)(1 - 3y) + y^2 = 13$

In Exercises 21 through 24, evaluate the derivatives of the given functions at the given points.

21. $3x^3y^2 - 2y^3 = -4$, $(1, 2)$

22. $2y + 5 - x^2 - y^3 = 0$, $(2, -1)$

23. $5y^4 + 7 = x^4 - 3y$, $(3, -2)$

24. $(xy - y^2)^3 = 5y^2 + 22$, $(4, 1)$

In Exercises 25 through 28, solve the given problems by use of implicit differentiation.

25. Find the slope of a line tangent to the curve of $xy + y^2 + 2 = 0$ at the point $(-3, 1)$.

26. Find the velocity of an object moving such that its displacement s and the time t are related by the equation $s^3 - t^2 = 7$ for $s = 2$ and $t = 1$.

27. For a right triangle in which the hypotenuse is always a units in length, the legs x and y are related by the equation $x^2 + y^2 = a^2$. Find the expression for dy/dx.

28. The polar moment of inertia I of a rectangular area is given by $I = \frac{1}{12}(b^3h + bh^3)$, where b and h are the base and height, respectively, of the rectangle. If I is constant, find the expression for db/dh.

2-10 Higher Derivatives

In the previous sections of this chapter we developed certain important formulas for finding the derivative of a function. The derivative itself is a function, and we may therefore take its derivative. In technology and in mathematics there are many applications which require the use of the derivative of a derivative. Therefore, in this section we will develop the concept and necessary notation, as well as show some of the applications.

The derivative of a function is called the **first derivative** *of the function. The derivative of the first derivative is the* **second derivative.** Since the second derivative is a function, we may find its derivative, which is called the third derivative. We may continue to find the fourth derivative, fifth derivative, and so on (provided each derivative is defined). *The second derivative, third derivative, and so on, are known as* **higher derivatives.**

The notations used for higher derivatives follow closely those used for the first derivative. As shown in Section 2-4, the notations for the first derivative are y', D_xy, $f'(x)$, and dy/dx. The notations for the second derivative are y'', D_x^2y, $f''(x)$, and d^2y/dx^2. Similar notations are used for other higher derivatives.

Example A

Find the higher derivatives of $y = 5x^3 - 2x$.

First we find the first derivative as

$$\frac{dy}{dx} = 15x^2 - 2 \quad \text{or} \quad y' = 15x^2 - 2$$

Next, we obtain the second derivative by finding the derivative of the first derivative:

$$\frac{d^2y}{dx^2} = 30x \quad \text{or} \quad y'' = 30x$$

Continuing to find the successive derivatives, we have

$$\frac{d^3y}{dx^3} = 30 \quad \text{or} \quad y''' = 30$$

$$\frac{d^4y}{dx^4} = 0 \quad \text{or} \quad y^{(4)} = 0$$

Since the third derivative is a constant, the fourth derivative and all successive derivatives will be zero. This can be shown as $d^n y/dx^n = 0$ for $n \geq 4$.

Example B

Find the higher derivatives of $f(x) = x(x^2 - 1)^2$.

Using the product rule, Eq. (2–26), to find the first derivative, we have

For reference, Eq. (2–26) is

$$\frac{d(uv)}{dx} = u\frac{dv}{dx} + v\frac{du}{dx}.$$

$$f'(x) = x(2)(x^2 - 1)(2x) + (x^2 - 1)^2(1)$$
$$= (x^2 - 1)(4x^2 + x^2 - 1) = (x^2 - 1)(5x^2 - 1)$$
$$= 5x^4 - 6x^2 + 1$$

Continuing to find the higher derivatives, we have

$$f''(x) = 20x^3 - 12x$$
$$f'''(x) = 60x^2 - 12$$
$$f^{(4)}(x) = 120x$$
$$f^{(5)}(x) = 120$$
$$f^{(n)}(x) = 0 \quad \text{for} \quad n \geq 6$$

Example C

Evaluate the second derivative of $y = \dfrac{2}{1-x}$ for $x = -2$.

We first write the function as $y = 2(1-x)^{-1}$, and then find the first and second derivatives:

$$y = 2(1-x)^{-1}$$

$$\frac{dy}{dx} = 2(-1)(1-x)^{-2}(-1) = 2(1-x)^{-2}$$

$$\frac{d^2y}{dx^2} = 2(-2)(1-x)^{-3}(-1) = 4(1-x)^{-3} = \frac{4}{(1-x)^3}$$

Evaluating the second derivative for $x = -2$, we have

$$\left.\frac{d^2y}{dx^2}\right|_{x=-2} = \frac{4}{(1+2)^3} = \frac{4}{27}$$

We note that the function and all of its derivatives are not defined for $x = 1$, since at $x = 1$ we would have division by zero.

Example D

Find y'' for the implicit function defined by $2x^2 + 3y^2 = 6$.

Differentiating with respect to x, we have

$$2(2x) + 3(2yy') = 0$$

$$4x + 6yy' = 0 \quad \text{or} \quad 2x + 3yy' = 0 \qquad (1)$$

Before differentiating again we see that $3yy'$ is a product, and we note that the derivative of y' is y''. Thus, differentiating again, we have

$$2 + 3yy'' + 3y'(y') = 0$$

$$2 + 3yy'' + 3(y')^2 = 0 \qquad (2)$$

Now solving Eq. (1) for y' and substituting this into Eq. (2), we have

$$y' = -\frac{2x}{3y} \quad \text{\textit{y prime}}$$

$$2 + 3yy'' + 3\left(-\frac{2x}{3y}\right)^2 = 0$$

$$2 + 3yy'' + \frac{4x^2}{3y^2} = 0$$

get rid of fractions

$$6y^2 + 9y^3y'' + 4x^2 = 0$$

$$y'' = \frac{-4x^2 - 6y^2}{9y^3} = \frac{-2(2x^2 + 3y^2)}{9y^3}$$

Since $2x^2 + 3y^2 = 6$, we have

$$y'' = \frac{-2(6)}{9y^3} = -\frac{4}{3y^3}$$

As we mentioned earlier, higher derivatives are used often in certain applications. This is particularly true of the second derivative. We will use the first and second derivative in the next chapter in sketching curves, and these and other higher derivatives will be used when we discuss infinite series in Chapter 13. An important technical application of the second derivative is shown in the example which follows, and others are shown in the exercises.

In Section 2–5 we briefly discussed the instantaneous velocity of an object, and in the exercises we mentioned acceleration. From that discussion we recall that the instantaneous velocity is the time rate of change of the displacement, and *the **instantaneous acceleration** is the time rate of change of the instantaneous velocity.* Therefore, we see that the *acceleration is found from the second derivative of the displacement with respect to time.* Consider the following example.

Example E

For the first 12 s after launch, the height s (in meters) of a certain rocket is given by $s = 10\sqrt{t^4 + 25} - 50$. Determine the vertical acceleration of the rocket when $t = 10.0$ s.

Since the velocity is found from the first derivative and the acceleration is found from the second derivative, we must find the second derivative and then evaluate it for $t = 10.0$ s.

$$s = 10\sqrt{t^4 + 25} - 50$$

$$v = \frac{ds}{dt} = 10\left(\frac{1}{2}\right)(t^4 + 25)^{-1/2}(4t^3) = \frac{20t^3}{(t^4 + 25)^{1/2}}$$

$$a = \frac{dv}{dt} = \frac{d^2s}{dt^2} = \frac{(t^4 + 25)^{1/2}(60t^2) - 20t^3(\frac{1}{2})(t^4 + 25)^{-1/2}(4t^3)}{t^4 + 25}$$

$$= \frac{(t^4 + 25)(60t^2) - 40t^6}{(t^4 + 25)^{3/2}} = \frac{20t^6 + 1500t^2}{(t^4 + 25)^{3/2}}$$

$$= \frac{20t^2(t^4 + 75)}{(t^4 + 25)^{3/2}}$$

Finding the value of the acceleration when $t = 10.0$ s, we have

$$a\big|_{t=10.0} = \frac{20(10.0^2)(10.0^4 + 75)}{(10.0^4 + 25)^{3/2}} = 20.1 \text{ m/s}^2$$

Exercises 2–10

In Exercises 1 through 8, find all the higher derivatives of the given functions.

1. $y = x^3 + x^2$ 2. $f(x) = 3x - x^4$ 3. $f(x) = x^3 - 6x^4$

4. $y = 2x^5 + 5x^2$ 5. $y = (1 - 2x)^4$ 6. $f(x) = (3x + 2)^3$

7. $f(x) = x(2x + 1)^3$ 8. $y = x(x - 1)^3$

In Exercises 9 through 28, find the second derivative of each of the given functions.

9. $y = 2x^7 - x^6 - 3x$ 10. $y = 6x - 2x^5$ 11. $y = 2x + \sqrt{x}$

12. $y = x^2 - \dfrac{1}{\sqrt{x}}$ 13. $f(x) = \sqrt[4]{8x - 3}$ 14. $f(x) = \sqrt[3]{6x + 5}$

15. $f(x) = \dfrac{4}{\sqrt{1 - 2x}}$ 16. $f(x) = \dfrac{5}{\sqrt{3 - 4x}}$ 17. $y = 2(2 - 5x)^4$

18. $y = (4x + 1)^6$ 19. $y = (3x^2 - 1)^5$ 20. $y = 3(2x^3 + 3)^4$

21. $f(x) = \dfrac{2x}{1 - x}$ 22. $f(x) = \dfrac{1 - x}{1 + x}$ 23. $y = \dfrac{x^2}{x + 1}$

24. $y = \dfrac{x}{\sqrt{1 - x^2}}$ 25. $x^2 - y^2 = 9$ 26. $xy + y^2 = 4$

27. $x^2 - xy = 1 - y^2$ 28. $xy = y^2 + 1$

In Exercises 29 through 32, evaluate the second derivative for the given value of x.

29. $f(x) = \sqrt{x^2 + 9}$, $x = 4$ 30. $f(x) = x - \dfrac{2}{x^3}$, $x = -1$

31. $y = 3x^{2/3} - \dfrac{2}{x}$, $x = -8$ 32. $y = 3(1 + 2x)^4$, $x = \dfrac{1}{2}$

In Exercises 33 through 38, solve the given problems by finding the appropriate derivatives.

33. A bullet is fired vertically upward. Its distance s, in feet, above the ground is given by $s = 2250t - 16.1t^2$, where t is measured in seconds. Find the acceleration of the bullet.

34. In testing the brakes on a new model automobile it was found that the distance s, in feet, which it traveled under specified conditions after the brakes were applied was given by $s = 57.6t - 1.20t^3$. What were the velocity and acceleration of the automobile for $t = 4.00$ s?

35. The deflection y, in meters, of a 5.00-m beam is given by $y = 0.0001 (x^5 - 25x^2)$, where x is the distance, in meters, from one end. Find the value of d^2y/dx^2 (the rate at which the slope of the beam changes) where $x = 3.00$ m.

36. The number n of grams of a compound formed during a certain chemical reaction is given by $n = \dfrac{2t}{t + 1}$, where t is the time in minutes. Evaluate d^2n/dt^2 (the rate of increase of the amount of the compound being formed) when $t = 4.00$ min.

37. The voltage V, in volts, induced in an inductor in an electric circuit is given by $V = L(d^2q/dt^2)$, where L is the inductance, in henrys (H). Find the expression for the voltage induced in a 1.60-H inductor if $q = \sqrt{2t + 1} - 1$.

38. How fast is the rate of change of solar radiation changing on the surface in Exercise 23 of Section 2–5 at 3 P.M.?

2–11 Review Exercises for Chapter 2

In Exercises 1 through 12, evaluate the given limits.

1. $\lim_{x \to 4} (8 - 3x)$

2. $\lim_{x \to 3} (2x^2 - 10)$

3. $\lim_{x \to -3} \dfrac{2x + 5}{x - 1}$

4. $\lim_{x \to 1} \dfrac{x^2 - 1}{x + 1}$

5. $\lim_{x \to 2} \dfrac{4x - 8}{x^2 - 4}$

6. $\lim_{x \to 5} \dfrac{x^2 - 25}{3x - 15}$

7. $\lim_{x \to 2} \dfrac{x^2 + 3x - 10}{x^2 - x - 2}$

8. $\lim_{x \to 3} \dfrac{x^2 - x - 6}{x - 3}$

9. $\lim_{x \to \infty} \dfrac{2 + \dfrac{1}{x + 4}}{3 - \dfrac{1}{x^2}}$

10. $\lim_{x \to \infty} \left(7 - \dfrac{1}{x + 1} \right)$

11. $\lim_{x \to \infty} \dfrac{x - 2x^3}{1 + x^3}$

12. $\lim_{x \to \infty} \dfrac{2x + 5}{3x^3 - 2x}$

In Exercises 13 through 20, use the delta-process to find the derivative of each of the given functions.

13. $y = 7 + 5x$

14. $y = 6x - 2$

15. $y = 6 - 2x^2$

16. $y = 2x^2 - x^3$

17. $y = \dfrac{2}{x^2}$

18. $y = \dfrac{1}{1 - 4x}$

19. $y = \sqrt{x + 5}$

20. $y = \dfrac{1}{\sqrt{x}}$

In Exercises 21 through 36, find the derivative of each of the given functions.

21. $y = 2x^7 - 3x^2 + 5$

22. $y = 8x^5 - x - 1$

23. $y = 4\sqrt{x} - \dfrac{1}{x}$

24. $y = \dfrac{3}{x^2} - 8\sqrt[4]{x}$

25. $y = \dfrac{x}{1 - x}$

26. $y = \dfrac{2x - 1}{x^2 + 1}$

27. $y = (2 - 3x)^4$

28. $y = (2x^2 - 3)^6$

29. $y = \dfrac{3}{(5 - 2x^2)^{3/4}}$

30. $y = \dfrac{7}{(3x - 1)^3}$

31. $y = x^2\sqrt{1 - 6x}$

32. $y = (x - 1)^3(x^2 - 2)^2$

33. $y = \dfrac{\sqrt{4x + 3}}{2x}$

34. $y = \dfrac{x}{\sqrt{x^2 + 1}}$

35. $(2x - 3y)^3 = x^2 - y$

36. $x^2y^2 = x^2 + y^2$

In Exercises 37 through 40, evaluate the derivatives of the given functions for the given values of x.

37. $y = \dfrac{4}{x} + 2\sqrt[3]{x}$, $x = 8$

38. $y = (3x - 5)^4$, $x = -2$

39. $y = 2x\sqrt{4x + 1}$, $x = 6$

40. $y = \dfrac{\sqrt{2x^2 + 1}}{3x}$, $x = 2$

In Exercises 41 through 44, find the second derivative of each of the given functions.

41. $y = 3x^4 - \dfrac{1}{x}$

42. $y = \sqrt{1 - 8x}$

43. $y = \dfrac{1 - 3x}{1 + 4x}$

44. $y = 2x(6x + 5)^4$

In Exercises 45 through 66, solve the given problems by finding the appropriate derivative or limit.

45. The combined capacitance for two capacitors connected in series is given by

$$C_T = \frac{C_1 C_2}{C_1 + C_2}$$

Find $\lim\limits_{C_1 \to C_2} C_T$ and $\lim\limits_{C_1 \to 0} C_T$.

46. The illuminance I is inversely proportional to the square of the distance x from the source. Find $\lim\limits_{x \to 0} I$.

47. Find the slope of a line tangent to the curve of $y = 7x^4 - x^3$ at $(-1, 8)$.

48. Find the slope of a line tangent to the curve of $y = \sqrt[3]{3 - 8x}$ at $(-3, 3)$.

49. The perimeter p of a rectangular field is a function of its width w according to $p = 2(50 + w)$, where p and w are measured in meters. Find the expression for the instantaneous rate of change of p with respect to w.

50. The centripetal acceleration a of an object moving in a circular path of radius r with a velocity v is given by $a = v^2/r$. Find the expression for the instantaneous rate of change of the acceleration with respect to the velocity for an object moving in a circular path of radius 2 ft. The acceleration is in feet per square second and the velocity is in feet per second.

51. Find the velocity of an object after 3 s if its displacement s (in meters) is given by

$$s = \frac{t}{3t + 1}.$$

52. The distance s (in feet) traveled by a subway train after the brakes are applied is given by $s = 40t - 5t^2$. How far does it travel, after the brakes are applied, in coming to a stop?

53. The voltage induced in an inductor L is given by

$$E = L\frac{dI}{dt}$$

where I is the current in the circuit and t is the time. Find the voltage induced in a 0.4-H inductor if the current I (in amperes) is related to the time (in seconds) by $I = 0.1t + 5.0$.

54. The specific heat c_p of a certain gas as a function of the temperature T is given by $c_p = 6.50 + 0.01T + 0.000002T^2$. Find the expression for the rate of change of the specific heat with respect to temperature.

55. The electric field E at a distance r from a point charge is $E = k/r^2$, where k is a constant. Find an expression for the rate of change of the electric field with respect to r.

56. The distance d between ion layers of a crystalline solid such as table salt is given by $d = \sqrt[3]{\dfrac{M}{2N\rho}}$, where M is the molecular weight, N is called Avogadro's number, and ρ is the density. Find the rate of change of d with respect to ρ.

57. The volume of a certain right circular cylinder is 100 in.^{3}. Find the expression for the instantaneous rate of change of the total surface area of the cylinder with respect to the radius of the base.

58. A 5-Ω resistor and a variable resistor of resistance R are placed in parallel. The expression for the resulting resistance R_T is given by $R_T = \dfrac{5R}{5 + R}$. Determine the instantaneous rate of change of R_T with respect to R, and then evaluate this expression for $R = 8\ \Omega$.

59. The velocity of an object moving with constant acceleration can be found from the equation $v = \sqrt{v_0^2 + 2as}$, where v_0 is the initial velocity, a is the acceleration, and s is the distance traveled. Find dv/ds.

60. The rate of change of angular velocity is angular acceleration. If a wire is moving through a magnetic field in a circular path such that its angular velocity is given by

$$\omega = \frac{2t^2}{1 + t}$$

where t is the time (in seconds), find the expression for the angular acceleration (in radians per second squared). What is the value of the angular acceleration when $t = 2$ s?

61. Under certain conditions, the efficiency of an internal combustion engine is given by

$$\text{eff (in percent)} = 100\left(1 - \frac{1}{(V_1/V_2)^{0.4}}\right)$$

where V_1 and V_2 are the maximum and minimum volumes of air in a cylinder, respectively. Assuming that V_2 is kept constant, find the expression for the rate of change of efficiency with respect to V_1.

62. Water is being drained from a pond such that the volume of water in the pond (in cubic meters) after t hours is given by $V = 5000(60 - t)^2$. Find the rate at which the pond is being drained after 4 h.

63. A rectangle is inscribed in the first quadrant under the parabola $y = 4 - x^2$ (see Fig. 2–23). Express the area of the rectangle in terms of x. Then find the expression for dA/dx.

64. An airplane flies over an observer at a velocity of 400 mi/h at an altitude of 2640 ft (see Fig. 2–24). If the plane flies horizontally in a straight line, find the rate at which the distance from the observer to the plane is changing 0.6 min after the plane passes over the observer.

65. A computer analysis showed that a specialized piece of machinery has a value, in dollars, given by $V = 1{,}500{,}000/(2t + 10)$, where t is the number of years after purchase. Calculate the value of dV/dt and d^2V/dt^2 for $t = 5$ years. What is the meaning of these values?

66. The *radius of curvature* of $y = f(x)$ at the point (x, y) on a curve is given by

$$R = \frac{[1 + (y')^2]^{3/2}}{|y''|}$$

A certain roadway follows the parabola $y = 1.2x - x^2$ for $0 < x < 1.2$, where x is measured in miles. Find R for $x = 0.2$ mi and $x = 0.6$ mi. See Fig. 2–25.

Figure 2–23

Figure 2–24

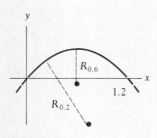

Figure 2–25

CHAPTER 3

Applications of the Derivative

In Chapter 2 we developed the meaning of the derivative of a function and then went on to find several formulas by which we can differentiate functions. In establishing the concept of the derivative as an instantaneous rate of change, numerous applications of the derivative were indicated in the examples and exercises. It was not necessary to develop any of these applications in any detail, since finding the derivative was the primary concern. There are, however, certain types of problems in geometry and technology in which the derivative plays a key role in the solution, although other concepts are also involved. In this chapter we shall consider some of these basic types of applications of the derivative.

3–1 Tangents and Normals

The first application of the derivative which we consider involves finding the equations of lines tangent and lines normal (perpendicular) to a given curve.

To find the equation of a line tangent to a curve at a given point, we first find the derivative of the function. The derivative is then evaluated at the point, and this gives us the slope of a line tangent to the curve at the point. Then, by using the point-slope form of the equation of a straight line, we find the equation of the tangent line. The following examples illustrate the method.

Example A

Find the equation of the line tangent to the parabola $y = x^2 - 1$ at the point $(-2, 3)$.

The derivative of this function is

$$\frac{dy}{dx} = 2x$$

The value of this derivative for the value $x = -2$ (the y-value of 3 is not used since the derivative does not directly contain y) is

$$\frac{dy}{dx} = -4$$

which means that the slope of the tangent line at $(-2, 3)$ is -4. Thus, by using the point-slope form of the equation of the straight line we obtain the desired equation. Thus, we have

$$y - 3 = -4(x + 2)$$
$$y - 3 = -4x - 8$$

or

$$y = -4x - 5$$

The parabola and the tangent line $y = -4x - 5$ are shown in Fig. 3–1.

Figure 3–1

Example B

Find the equation of the line tangent to the ellipse $4x^2 + 9y^2 = 40$ at the point $(1, 2)$.

The easiest method of finding the derivative of this equation is to treat it as an implicit function. In this way we have

$$8x + 18yy' = 0 \quad \text{or} \quad y' = -\frac{4x}{9y}$$

Evaluating this derivative at the point $(1, 2)$, we have

$$y' = -\frac{4}{18} = -\frac{2}{9}$$

Thus, the slope of the desired tangent line is $-\frac{2}{9}$. Using the point-slope form of the equation of the straight line, we have

$$y - 2 = -\frac{2}{9}(x - 1)$$
$$9y - 18 = -2x + 2$$
$$2x + 9y - 20 = 0$$

The ellipse and the tangent line $2x + 9y - 20 = 0$ are shown in Fig. 3–2.

It might be noted that the derivative could also have been found by solving for y. In this way we would have to differentiate $y = \frac{2}{3}\sqrt{10 - x^2}$.

Figure 3–2

If we wish to obtain the equation of a line normal (perpendicular to a tangent) to a curve, we recall that the slopes of perpendicular lines are negative reciprocals. Thus, the derivative is found and evaluated at the specified point. Since this gives the slope of a tangent line, we take the negative reciprocal of this number to find the slope of the normal line. Then, by using the point-slope form of the equation of a straight line we find the equation of the normal. The following examples illustrate the method.

Example C

Figure 3–3

Find the equation of the line normal to the hyperbola $y = 2/x$ at the point (2, 1).

The derivative of this function is $dy/dx = -2/x^2$, which evaluated at $x = 2$ gives $dy/dx = -\frac{1}{2}$. Therefore, the slope of a line normal to the curve at the point (2, 1) is 2. The equation of the normal is then

$$y - 1 = 2(x - 2)$$

or

$$y = 2x - 3$$

The hyperbola and the normal line are shown in Fig. 3–3.

Example D

Figure 3–4

Find the y-intercept of the line normal to the curve $y = 3x - x^3$ at (3, −18).

The derivative of this function is $dy/dx = 3 - 3x^2$, which evaluated at (3, −18) is

$$\left. \frac{dy}{dx} \right|_{x=3} = 3 - 3(9) = -24$$

Therefore, the slope of a normal line is $\frac{1}{24}$. The equation of the normal line is

$$y + 18 = \frac{1}{24}(x - 3)$$

or

$$24y - x + 435 = 0$$

To find the y-intercept of this line we set $x = 0$. In this way we find the y-intercept to be $-\frac{145}{8}$. The curve, normal line, and intercept are indicated in Fig. 3–4.

Many of the applications of tangents and normals are geometric. However, there are certain applications in technology, and one of these is shown in the following example. Others are shown in the exercises.

Example E

Figure 3–5

In Fig. 3–5 the cross section of a parabolic solar reflector is shown, along with an incident ray of light and the reflected ray. The angle of incidence i is equal to the angle of reflection r where both angles are measured with respect to the normal to the surface. If the incident ray strikes at the point where the slope of the normal is -1, and the equation of the parabola is $4y = x^2$, what is the equation of the normal line?

If the slope of the normal line is -1, then the slope of a tangent line is $-(\frac{1}{-1}) = 1$. Therefore, we know that the value of the derivative at the point of reflection is 1. This allows us to find the coordinates of the point.

$$4y = x^2$$
$$4\frac{dy}{dx} = 2x$$
$$\frac{dy}{dx} = \frac{1}{2}x$$
$$1 = \frac{1}{2}x$$
$$x = 2$$

This means that the x-coordinate of the point of reflection is 2. We can find the y-coordinate by substituting $x = 2$ into the equation of the parabola. Thus, the point is $(2, 1)$. Since the slope is -1, the equation is

$$y - 1 = (-1)(x - 2)$$

or

$$y = -x + 3$$

Exercises 3–1

In Exercises 1 through 4, find the equations of the lines tangent to the indicated curves at the given points. In each, sketch the curve and the tangent line.

1. $y = x^2 + 2$ at $(2, 6)$ 2. $y = x^3 - 2x$ at $(2, 4)$

3. $y = \dfrac{1}{x^2 + 1}$ at $(1, \frac{1}{2})$ 4. $x^2 + y^2 = 25$ at $(3, 4)$

In Exercises 5 through 8, find the equations of the lines normal to the indicated curves at the given points. In each, sketch the curve and the normal line.

5. $y = 6x - x^2$ at $(1, 5)$ 6. $y = 2x^3 + 10$ at $(-2, -6)$

7. $y = \dfrac{1}{(x^2 + 1)^2}$ at $(1, \frac{1}{4})$ 8. $x^2 - y^2 = 8$ at $(3, 1)$

In Exercises 9 through 12, find the equations of the tangent lines and the normal lines to the indicated curves.

9. $y = \dfrac{1}{\sqrt{x^2 + 1}}$ where $x = \sqrt{3}$ 10. $y = \dfrac{4}{(5 - 2x)^2}$ where $x = 2$

11. The parabola with vertex at $(0, 3)$ and focus at $(0, 0)$, where $x = -1$.

12. The ellipse with focus at $(4, 0)$, vertex at $(5, 0)$, and center at $(0, 0)$, where $x = 2$.

In Exercises 13 through 16, find the equations of the lines tangent or normal to the given curves and with the given slopes.

13. $y = x^2 - 2x$, tangent line with slope 2.

14. $y = \sqrt{2x - 9}$, tangent line with slope 1.

15. $y = (2x - 1)^3$, normal line with slope $-\frac{1}{24}$, $x > 0$.

16. $y = \frac{1}{2}x^4 + 1$, normal line with slope 4.

In Exercises 17 through 24, solve the given problems by finding the equation of the appropriate tangent or normal line.

17. Find the x-intercept of the line tangent to the parabola $y = 4x^2 - 8x$ at $(-1, 12)$.

18. Find the y-intercept of the line normal to the curve $y = x^{3/4}$ where $x = 16$.

19. A certain suspension cable with supports on the same level is closely approximated as being parabolic in shape. If the supports are 200 ft apart and the sag at the center is 30 ft, what is the equation of the line along which the tension acts (tangentially) at the right support? (Choose the origin of the coordinate system at the lowest point of the cable.)

20. In an electric field the lines of force are perpendicular to the curves of equal electric potential. A certain curve of equal electric potential is given by $y = 2x^{2/3}$. What is the equation of the line along which the force acts on an electron positioned on this curve where $x = 8$?

21. An object is moving in a horizontal circle of radius 2 ft at the rate of 2 rad/s. If the object is on the end of a string, and the string breaks after $\pi/3$ s, causing the object to travel along a line tangent to the circle, what is the equation of the path of the object after the string breaks? Sketch the path of the object. (Choose the origin of the coordinate system at the center of the circle and assume the object started on the positive x-axis moving counterclockwise.)

43.60°

$y^2 = 8x$

Figure 3–6

22. A radio wave reflects from a reflecting surface in the same way as a light wave (see Example E). A certain horizontal radio wave reflects off of a parabolic reflector such that the reflected wave is 43.60° below the horizontal, as shown in Fig 3–6. If the equation of the parabola is $y^2 = 8x$, what is the equation of the normal line through the point of reflection?

Figure 3–7

23. In designing a flexible tubing system, the supports for the tubing must be perpendicular to the tubing. If a section of the tubing follows the curve $y = \dfrac{4}{x^2 + 1}$ (units in decimeters and $-2 < x < 2$), along which lines must the supports be directed if they are located at $x = -1$, $x = 0$, and $x = 1$? See Fig. 3–7.

24. On a particular drawing, a pulley wheel can be described by the equation $x^2 + y^2 = 100$, where distances are measured in centimeters. The pulley belt is directed along the lines $y = -10$, and $4y - 3x - 50 = 0$ when first and last making contact with the wheel. What are the first and last points on the wheel where the belt makes contact?

3–2 Newton's Method for Solving Equations

As we know, finding the roots of an equation $f(x) = 0$ is very important in mathematics and in many types of applications. In algebra and trigonometry some equations can be solved by the use of basic formulas and methods. For example, if $f(x)$ is a linear or quadratic polynomial, we can use basic methods, which might include the quadratic formula. Methods of solving other types of equations, such as polynomial equations with rational roots and certain trigonometric equations, are also normally covered in algebra and trigonometry. However, for a great many algebraic and nonalgebraic equations there is no method for finding the roots exactly.

In this section we show a method, known as **Newton's method,** which uses the derivative to locate approximately, but very accurately, the real roots of many kinds of equations. It can be used with polynomial equations of any degree and with nonalgebraic equations.

Newton's method is an example of an **iterative method.** In using such a method, we start with a reasonable guess for a root of the equation. By using the method we obtain a new number, which is a better approximation. This, in turn, gives a still better approximation. Continuing in this way we obtain an approximate answer with the required degree of accuracy. A calculator can be used with Newton's method to find the roots of a great many equations to a very good degree of accuracy. Iterative methods in general are easily programmable for use on a computer.

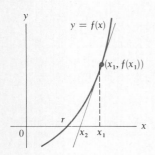

Figure 3–8

Let us consider a section of a curve $y = f(x)$ which (a) crosses the x-axis, (b) always has either a positive slope or a negative slope, and (c) has a slope which either becomes greater or becomes less as x increases. See Fig. 3–8. The curve in the figure crosses the x-axis at $x = r$, which means that $x = r$ is a root of the equation $f(x) = 0$. If x_1 is sufficiently close to r, a line tangent to the curve at $(x_1, f(x_1))$ will cross the x-axis at a point $(x_2, 0)$, which is closer to r than is x_1.

We know that the slope of the tangent line is the value of the derivative at x_1, or $m_{tan} = f'(x_1)$. Therefore, the equation of the tangent line is

$$y - f(x_1) = f'(x_1)(x - x_1)$$

For the point $(x_2, 0)$ on this line, we have

$$-f(x_1) = f'(x_1)(x_2 - x_1)$$

Solving for x_2, we have the formula

$$x_2 = x_1 - \frac{f(x_1)}{f'(x_1)} \tag{3–1}$$

Here x_2 is our first approximation to the root. We can then replace x_1 in Eq. (3–1) by x_2 and find a closer approximation, x_3. This process can be repeated as many times as are needed to find the root to a required accuracy. We can see that this method lends itself well to the use of a calculator or a computer for finding the root.

Example A

Find the root of $x^2 - 3x + 1 = 0$ which lies between $x = 0$ and $x = 1$.

Here $f(x) = x^2 - 3x + 1$. Therefore, $f(0) = 1$ and $f(1) = -1$, which indicates that the root may be near the middle of the interval. Since x_1 must be within the interval, we choose $x_1 = 0.5$.

The derivative is

$$f'(x) = 2x - 3$$

Therefore, $f(0.5) = -0.25$ and $f'(0.5) = -2$, which gives us

$$x_2 = 0.5 - \frac{-0.25}{-2} = 0.375$$

Next, $f(0.375) = 0.015625$ and $f'(0.375) = -2.25$. This gives us

$$x_3 = 0.375 - \frac{0.015625}{-2.25} = 0.3819444$$

Since this is a quadratic equation, we can check this result by use of the quadratic formula. Using this formula we find the root is $x = 0.3819660$. Our result using Newton's method is good to three decimal places. Additional accuracy may be obtained by using the method again as many times as needed.

Example B

Find the root of $x^4 - 5x^3 + 6x^2 - 5x + 5 = 0$ between 1 and 2 to five decimal places.

Here

$$f(x) = x^4 - 5x^3 + 6x^2 - 5x + 5$$

and

$$f'(x) = 4x^3 - 15x^2 + 12x - 5$$

Since $f(1) = 2$ and $f(2) = -5$ it appears that the root is possibly closer to 1 than to 2. Thus, we let $x_1 = 1.3$. Setting up a table, we have the following values.

n	x_n	$f(x_n)$	$f'(x_n)$	$x_n - \dfrac{f(x_n)}{f'(x_n)}$
1	1.3	0.5111	-5.962	1.3857263
2	1.3857263	-0.0245128	-6.5311513	1.3819731
3	1.3819731	-0.0000461	-6.5066242	1.3819660

Since $x_4 = x_3 = 1.38197$ to five decimal places, this is the value of the required root. (Actually, by finding x_5 we see that x_4 is accurate to the value shown.)

If your calculator can store more than one value in memory, the above results can be more easily obtained by storing at least x_n, $f(x_n)$, and $f'(x_n)$ in memory.

Example C

Figure 3–9

Solve the equation $x^2 - 1 = \sqrt{4x - 1}$.

We can see approximately where the root is by sketching the graphs of $y_1 = x^2 - 1$ and $y_2 = \sqrt{4x - 1}$. We see from Fig. 3–9 that they intersect between $x = 1$ and $x = 2$. Therefore, we choose $x_1 = 1.5$. With

$$f(x) = x^2 - 1 - \sqrt{4x - 1}$$

and

$$f'(x) = 2x - \frac{2}{\sqrt{4x - 1}}$$

we now find the values in the following table.

n	x_n	$f(x_n)$	$f'(x_n)$	$x_n - \dfrac{f(x_n)}{f'(x_n)}$
1	1.5	-0.98606798	2.1055728	1.9683134
2	1.9683134	0.25256859	3.1737598	1.8887332
3	1.8887332	0.00705269	2.9962957	1.8863794
4	1.8863794	0.00000620	2.9910265	1.8863773

Since $x_5 = x_4 = 1.88638$ to five decimal places, this is the required solution. (Here, rounded off values of x_n are shown, although additional digits were carried and used.)

Exercises 3–2

In Exercises 1 through 4, find the indicated roots of the given quadratic equations by finding x_3 from Newton's method. Then compare this root with that obtained by using the quadratic formula.

1. $x^2 - 2x - 5 = 0$ (between 3 and 4)
2. $2x^2 - x - 2 = 0$ (between 1 and 2)
3. $3x^2 - 5x - 1 = 0$ (between -1 and 0)
4. $x^2 + 4x + 2 = 0$ (between -4 and -3)

In Exercises 5 through 16, find the indicated roots of the given equations to at least four decimal places.

5. $x^3 - 6x^2 + 10x - 4 = 0$ (between 0 and 1)
6. $x^3 - 3x^2 - 2x + 3 = 0$ (between 0 and 1)
7. $x^3 + 5x^2 + x - 1 = 0$ (the positive root)
8. $2x^3 + 2x^2 - 11x + 3 = 0$ (the larger positive root)
9. $x^4 - x^3 - 3x^2 - x - 4 = 0$ (between 2 and 3)
10. $2x^4 - 2x^3 - 5x^2 - x - 3 = 0$ (between 2 and 3)
11. $x^4 - 2x^3 - 8x - 16 = 0$ (the negative root)
12. $3x^4 - 3x^3 - 11x^2 - x - 4 = 0$ (the negative root)

See Appendix C for a computer program for Newton's method.

13. $2x^2 = \sqrt{2x + 1}$ (the positive real solution)

14. $x^3 = \sqrt{x + 1}$ (the real solution)

15. $x = \dfrac{1}{\sqrt{x + 2}}$ (the real solution)

16. $x^{3/2} = \dfrac{1}{2x + 1}$ (the real solution)

In Exercises 17 through 20, determine the required values.

17. Find all the real roots of the equation $x^3 - 2x^2 - 5x + 4 = 0$.

18. Find all the real roots of the equation $x^3 - 2x^2 - 2x - 7 = 0$.

19. A metal storage tank which holds 500.0 m³ is spherical, with an outside diameter of 10.0000 m. What is the thickness (to four decimal places) of the metal of which the tank is made? (Set up and solve the equation involving the thickness.)

20. The sides of a rectangular box are 3, 5, and 6 ft. If each side is increased by the same amount, the volume is doubled. By how much should each side be increased to accomplish this?

3–3 Curvilinear Motion

A great many phenomena in technology involve the time rate of change of certain quantities. We encountered one of the most fundamental and most important of these when we discussed *velocity, the time rate of change of displacement,* in Section 2–5. In this section we further develop the concepts of velocity and acceleration, which were also introduced in Chapter 2. Other time-rate-of-change problems are discussed in the next section.

When velocity was introduced in Section 2–5, the discussion was limited to rectilinear motion, or motion along a straight line. A more general discussion of velocity is necessary when we discuss the motion of an object in a plane. There are many important applications of motion in the plane, a principal one among these being the motion of a projectile.

An important concept necessary in developing this topic is that of a **vector.** A good elementary definition of *a vector is a quantity which has direction and magnitude.* For example, velocity, acceleration, force, and momentum are vector quantities. Each is specified by giving both its magnitude and direction. For instance, we cannot determine the effect of a force unless we know the magnitude of the force as well as the direction in which it is acting. *To represent a vector, we draw a line segment with length proportional to its magnitude, and in a direction which shows the direction of the vector.*

Figure 3–10

Figure 3–11

The addition of two vectors **OA** *and* **AB**, *directed from O to A and from A to B, respectively, gives the* **resultant** *vector* **OB**, *as shown in Fig. 3–10. The* **initial point** *is O and the* **terminal point** *is B.* This is equivalent to making the two vectors being added the sides of a parallelogram, with the diagonal being the resultant (see Fig. 3–11). (Note: A vector quantity is represented by a letter printed in boldface type. The same letter in italic (lightface) type represents the magnitude only.) In general, *a resultant is a single vector which can replace any number of other vectors and still produce the same physical effect.*

Vectors may be subtracted by reversing the direction of the vector being subtracted and then proceeding as in adding vectors.

Not only do we need to add vectors, but it is also frequently useful to consider a given vector as the sum of two vectors. *These two vectors when added together give the original vector; they are called* **components** *of the original vector.*

In practice there are certain components of a vector which are of particular importance. If a vector is so placed that its initial point is at the origin and its direction is indicated by an angle with respect to the positive x-axis (*standard position*), we may find its components which are directed along the x-axis and along the y-axis. These are known as its x-component and its y-component. *Finding these component vectors is called* **resolving** *the vector into its components.* (It is assumed here that you have a basic knowledge of angles and the trigonometric functions. If a review is needed, see Section 6–1.)

Example A

Figure 3–12

Resolve a vector 10.0 units long and directed at an angle of 120.0° into its x- and y-components (see Fig. 3–12).

Placing the initial point of the vector at the origin and putting the angle in standard position, we see that the vector directed along the x-axis, $\mathbf{V}_x$, is related to the vector $\mathbf{V}$ of magnitude V by

$$V_x = V \cos 120.0° = -V \cos 60.0°$$

(The minus sign indicates that the x-component is directed in the negative direction, that is, to the left.) Since the vector directed along the y-axis, $\mathbf{V}_y$, could also be placed along the vertical dashed line, it is related to the vector $\mathbf{V}$ by

$$V_y = V \sin 120.0° = V \sin 60.0°$$

Thus, the vectors $\mathbf{V}_x$ and $\mathbf{V}_y$ have the magnitudes

$$V_x = -10.0(0.5000) = -5.00, \qquad V_y = 10.0(0.8660) = 8.66$$

Therefore, we have resolved the given vector into two components, one directed along the negative x-axis of magnitude 5.00 and the other directed along the positive y-axis of magnitude 8.66.

Adding vectors by diagrams gives only approximate results. It is possible to find accurate numerical results for the sum of two vectors by the use of components. The method is to resolve (break into components) all vectors being added into their x- and y-components. These can then be added algebraically, since they are all directed along either the x-axis or the y-axis. The result will give us the component of the resultant in the x-direction and the component of the resultant in the y-direction. By addition we may then find the value of the resultant vector.

Example B

See Appendix C for a computer program for adding vectors.

Find the resultant **R** of the two vectors shown in Fig. 3–13(a), **A** of magnitude 8.075 and direction 57.26° and **B** of magnitude 5.437 and direction 322.15°.

$$A_x = 8.075 \cos 57.26° = 4.367$$
$$A_y = 8.075 \sin 57.26° = 6.792$$
$$B_x = 5.437 \cos 322.15° = 4.293$$
$$B_y = 5.437 \sin 322.15° = -3.336$$
$$R_x = A_x + B_x = 4.367 + 4.293 = 8.660$$
$$R_y = A_y + B_y = 6.792 - 3.336 = 3.456$$
$$R = \sqrt{R_x^2 + R_y^2} = \sqrt{(8.660)^2 + (3.456)^2} = 9.324$$
$$\tan \theta = \frac{R_y}{R_x} = \frac{3.456}{8.660} = 0.3991, \qquad \theta = 21.76°$$

The resultant vector is 9.324 units long and is directed at an angle of 21.76°, as shown in Fig. 3–13(c).

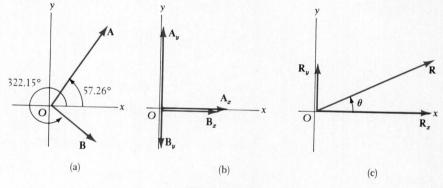

(a)   (b)   (c)

Figure 3–13

Some general formulas can be derived from the previous examples. For a given vector **A**, directed at an angle θ, of magnitude A, and with components A_x and A_y, we have the following relations:

$$A_x = A \cos \theta, \qquad A_y = A \sin \theta \tag{3–2}$$
$$A = \sqrt{A_x^2 + A_y^2} \tag{3–3}$$
$$\tan \theta = \frac{A_y}{A_x} \tag{3–4}$$

Although vectors can be used to represent many physical quantities, we shall restrict our attention to their use in describing the velocity and acceleration of an object moving in a plane along a specified path. Such motion is called **curvilinear motion.**

In describing an object undergoing curvilinear motion, it is very common to express the x and y coordinates of its position separately as functions of time. Equations given in this form, that is, *x and y both given in terms of a third variable (in this case t), are said to be in* **parametric form.** The third variable, t, is called the **parameter.**

To find the velocity of an object whose coordinates are given in parametric form, we find its x-component of velocity v_x by determining dx/dt and its y-component of velocity v_y by determining dy/dt. These are then evaluated, and the resultant velocity is found from $v = \sqrt{v_x^2 + v_y^2}$. The direction in which the object is moving is found from $\tan \theta = v_y/v_x$. The following examples illustrate the method.

Example C

If the horizontal distance x that an object has moved is given by $x = 3t^2$, and the vertical distance y is given by $y = 1 - t^2$, find the resultant velocity when $t = 2$.

To find the resultant velocity, we must find v and θ, by first finding v_x and v_y as indicated above. After the derivatives are found, they are then evaluated for $t = 2$. Thus,

$$v_x = \frac{dx}{dt} = 6t \qquad v_x|_{t=2} = 12$$

$$v_y = \frac{dy}{dt} = -2t \qquad v_y|_{t=2} = -4$$

$$v = \sqrt{12^2 + (-4)^2} = 12.6$$

$$\tan \theta = \frac{-4}{12} \qquad \theta = -18.4°$$

Figure 3–14

The path and velocity vectors are shown in Fig. 3–14.

Example D

Find the velocity and direction of motion when $t = 2$ of an object moving such that its x- and y-coordinates of position are given by $x = 1 + 2t$ and $y = t^2 - 3t$.

$$v_x = \frac{dx}{dt} = 2 \qquad v_x|_{t=2} = 2$$

$$v_y = \frac{dy}{dt} = 2t - 3 \qquad v_y|_{t=2} = 1$$

$$v|_{t=2} = \sqrt{2^2 + 1^2} = 2.24$$

$$\tan \theta = \frac{1}{2} \qquad \theta = 26.6°$$

Figure 3–15

These quantities are shown in Fig. 3–15.

In these examples we note that *we first find the necessary derivatives, and then we evaluate them*. This procedure should always be followed. When a derivative is to be found, it is incorrect to take the derivative of the expression which is the evaluated function.

In Chapter 2 we stated that acceleration is the time rate of change of velocity. Therefore if the velocity, or its components, of an object is known as a function of time, the acceleration can be found by taking the derivative of the velocity with respect to time. If the displacement is known, the acceleration is found by finding the second derivative with respect to time. The determination of the acceleration of an object is illustrated in the following example.

Example E

Find the magnitude and direction of the acceleration when $t = 2$ for an object moving such that its x- and y-coordinates of position are given by $x = t^3$ and $y = 1 - t^2$:

$$v_x = \frac{dx}{dt} = 3t^2 \qquad a_x = \frac{dv_x}{dt} = \frac{d^2x}{dt^2} = 6t \qquad a_x|_{t=2} = 12$$

$$v_y = \frac{dy}{dt} = -2t \qquad a_y = \frac{dv_y}{dt} = \frac{d^2y}{dt^2} = -2 \qquad a_y|_{t=2} = -2$$

$$a|_{t=2} = \sqrt{12^2 + (-2)^2} = 12.2$$

$$\tan \theta = \frac{a_y}{a_x} = -\frac{2}{12}$$

$$\theta = -9.5°$$

The quadrant in which θ lies is determined from the fact that a_y is negative and a_x is positive. Thus, θ must be a fourth-quadrant angle (see Fig. 3–16). We see from this example that the magnitude and direction of acceleration are found from its components just as with velocity.

Figure 3–16

For reference Eq. (2–29) is $\frac{du^n}{dx} = nu^{n-1}\frac{du}{dx}$.

If the curvilinear path which an object follows is given as y as a function of x, the velocity (and acceleration) is found by taking derivatives of each term of the equation with respect to time. In finding the derivatives we must be very careful to use the power rule for finding derivatives, Eq. (2–29), so that the general factor du/dx is not neglected. The following examples illustrate the method.

Example F

Figure 3–17

A particle moves along the parabola $y = \frac{1}{3}x^2$ so that its vertical velocity v_y is always 8. Find the magnitude of the resultant velocity when the particle is at the point $(2, \frac{4}{3})$.

Since both y and x change with time, both can be considered functions of time. Thus, we can take derivatives of $y = \frac{1}{3}x^2$ with respect to time. When we do this we obtain the following result:

$$\blacktriangleright \qquad \frac{dy}{dt} = \frac{2}{3}x\,\frac{dx}{dt} \qquad v_y = \frac{2}{3}xv_x$$

At the point $(2, \frac{4}{3})$ we have $8 = \frac{2}{3}(2)v_x$. Thus, $v_x = 6$ at this point (see Fig. 3–17). Therefore, the resultant velocity is

$$v = \sqrt{8^2 + 6^2} = 10$$

Example G

Figure 3–18

A helicopter is flying at 18.0 m/s and at an altitude of 120 m when a steel ball is released from it. The ball maintains a horizontal velocity and follows a path given by $y = 120 - 0.0151x^2$, as shown in Fig. 3–18. Find the magnitude and direction of the velocity and of the acceleration of the ball 3.00 s after release.

We know that $v_x = dx/dt = 18.0$ m/s. Taking derivatives of $y = 120 - 0.0151x^2$ with respect to t, we have

$$\frac{dy}{dt} = -0.0302x\,\frac{dx}{dt}$$

or

$$v_y = -0.0302xv_x$$

After 3.00 s,

$$x = (3.00)(18.0) = 54.0 \text{ m}$$

This means that

$$v_y|_{t=3.00} = -0.0302(54.0)(18.0) = -29.35 \text{ m/s}$$

or

$$v = \sqrt{18.0^2 + (-29.35)^2} = 34.4 \text{ m/s}$$

Also,

$$\tan\theta = \frac{-29.35}{18.0}, \quad \theta = -58.5°$$

Therefore, the velocity is 34.4 m/s and is directed at an angle of 58.5° below the horizontal.

To find the acceleration, we return to the equation $v_y = -0.0302xv_x$. Since v_x is constant, we can substitute 18.0 for v_x to get

$$v_y = -0.5436x$$

Again taking derivatives with respect to time, we have

$$\frac{dv_y}{dt} = -0.5436\frac{dx}{dt}$$

or

$$a_y = -0.5436v_x$$

Since $v_x = 18.0$ m/s,

$$a_y = -9.78 \text{ m/s}^2$$

We know that v_x is constant, which means that $a_x = 0$. Therefore, the acceleration is 9.78 m/s² and is directed vertically downward.

Exercises 3–3

In Exercises 1 through 4, given that the x- and y-coordinates of a moving particle are given by the indicated parametric equations, find the magnitude and direction of the velocity for the specific value of t. Plot the curves and show the appropriate components of the velocity.

1. $x = 3t$, $y = 1 - t$, $t = 4$
2. $x = 2t^3$, $y = 4t^2$, $t = 1$
3. $x = \dfrac{10}{2t + 3}$, $y = \dfrac{1}{t^2}$, $t = 2$
4. $x = \sqrt{1 + 2t}$, $y = t - t^2$, $t = 4$

In Exercises 5 through 8, use the parametric equations and values of t of Exercises 1 through 4 to find the magnitude and direction of the acceleration in each case.

In Exercises 9 through 20, find the indicated velocities and accelerations.

9. A particle moves along the curve $y = 5x^3$ with a constant x component of velocity equal to 20. Find the magnitude and direction of the velocity at the point (1, 5).

10. A particle moves along the curve $y = 2\sqrt{x}$. What is the ratio of the x-component of velocity to the y-component at (9, 6)?

11. A particle moves along the upper half of the parabola $y^2 = 2x + 1$ such that $v_x = \sqrt{2x + 3}$. Find v_y when $x = 4$.

12. A particle moves counterclockwise along the ellipse $x^2 + 9y^2 = 9$ such that $v_y = 2y$. Find v_x at the point $(2, \frac{1}{3}\sqrt{5})$.

13. A projectile moves according to the parametric equations $x = 120t$ and $y = 160t - 16t^2$, where distances are in feet and time is in seconds. Find the magnitude and direction of the velocity when $t = 3.0$ s and when $t = 6.0$ s.

14. A projectile moves according to the parametric equations $x = 30t$ and $y = -4.9t^2$, where distances are in meters and time is in seconds. Find the magnitude and direction of each of the velocity and acceleration when $t = 3.0$ s.

15. Find the acceleration at the specified times for the projectile in Exercise 13.

16. Find the magnitude and direction of the acceleration of the particle in Exercise 9.

17. A spacecraft moves such that it follows the equation $x = 10(\sqrt{1 + t^4} - 1)$, $y = 40t^{3/2}$ for the first hundred seconds after launch. Here, x and y are measured in meters and t is measured in seconds. Find the magnitude and direction of the velocity of the spacecraft 10 s and 100 s after launch.

18. An electron moves in an electric field according to the parametric equations

$$x = \frac{20}{\sqrt{1 + t^2}} \quad \text{and} \quad y = \frac{20t}{\sqrt{1 + t^2}}$$

where distances are in meters and time in seconds. Find the magnitude and direction of the velocity when $t = 1$ s.

19. A rocket follows a path given by $y = x - \frac{1}{90}x^3$ (distances in miles). If the horizontal velocity is given by $v_x = x$, find the magnitude and direction of the velocity when the rocket hits the ground (assume level terrain) if time is in minutes.

20. A meteor traveling toward the earth has a velocity inversely proportional to the square root of the distance from the earth's center. Show that its acceleration is inversely proportional to the square of the distance from the center of the earth.

3–4 Related Rates

Any two variables which vary with respect to time and between which a relation is known to exist can have the time rate of change of one expressed in terms of the time rate of change of the other. We do this by taking the derivative with respect to time of the expression which relates the variables as we did in Examples F and G of Section 3–3. Since the rates of change are related, this type of problem is referred to as a **related-rate** problem. The following examples illustrate the basic method of solution.

Example A

The voltage of a certain thermocouple as a function of temperature is given by $E = 2.800T + 0.006T^2$. If the temperature is increasing at the rate 1.00°C/min, how fast is the voltage increasing when $T = 100$°C?

Since we are asked to find the time rate of change of voltage, we first take derivatives with respect to time. This gives us

$$\frac{dE}{dt} = 2.800\frac{dT}{dt} + 0.012T\frac{dT}{dt}$$

▶ again being careful to include the factor dT/dt. From the given information we know that $dT/dt = 1.00$°C/min and that we wish to know dE/dt when $T = 100$°C. Thus,

$$\left.\frac{dE}{dt}\right|_{T=100} = 2.800(1.00) + 0.012(100)(1.00) = 4.00 \text{ V/min}$$

The derivative must be taken before values are substituted. In this problem we are finding the time rate of change of the voltage for a specified value of T. For other values of T, dE/dt would have different values.

Example B

The distance q that an image is from a certain lens in terms of p, the distance the object is from the lens, is given by

$$q = \frac{10p}{p - 10}$$

If the object distance is increasing at the rate of 0.200 cm/s, how fast is the image distance changing when $p = 15.0$ cm?

Taking derivatives with respect to time, we have

$$\frac{dq}{dt} = \frac{(p - 10)\left(10\dfrac{dp}{dt}\right) - 10p\left(\dfrac{dp}{dt}\right)}{(p - 10)^2}$$

Now substituting $p = 15.0$ and $dp/dt = 0.200$, we have

$$\frac{dq}{dt}\bigg|_{p=15} = \frac{5(10)(0.200) - 10(15.0)(0.2)}{25.0} = -0.800 \text{ cm/s}$$

Thus, the image distance is decreasing (the significance of the minus sign) at the rate of 0.800 cm/s when $p = 15.0$ cm.

In many related-rate problems the function is not given, but must be set up according to the statement of the problem. The following examples illustrate this type of problem.

Example C

A spherical balloon is being blown up at the constant rate of 2.00 ft³/min. Find the rate at which the radius is increasing when it is 3.00 ft.

We are asked to find the relation between the rate of change of the volume of a sphere with respect to time and the corresponding rate of change of the radius. Thus, we are to take derivatives of the expression for the volume of a sphere with respect to time.

$$V = \frac{4}{3}\pi r^3 \qquad \frac{dV}{dt} = 4\pi r^2\left(\frac{dr}{dt}\right)$$

From the information given, we know the value of dV/dt to be 2.00. We want to find dr/dt when $r = 3.00$. Substituting these values, we have

$$2.00 = 4\pi(3.00)^2\left(\frac{dr}{dt}\right)$$

or

$$\frac{dr}{dt}\bigg|_{r=3} = \frac{1}{18.0\pi} = 0.0177 \text{ ft/min}$$

The power supply on a certain satellite of the earth has an output power P which is inversely proportional to $t + 1000$, where t is the number of days after launch. If the power output is initially 110 W, find the rate at which the power is decreasing after 1000 days (d).

From the first statement we know that

$$P = \frac{k}{t + 1000}$$

where k is the constant of proportionality. Also, we know that $P = 110$ W when $t = 0$. Thus,

$$110 = \frac{k}{1000} \quad \text{or} \quad k = 1.10 \times 10^5 \text{ W} \cdot \text{d}$$

This means that

$$P = \frac{1.10 \times 10^5}{t + 1000}$$

Now taking derivatives with respect to time, we have

$$\frac{dP}{dt} = (1.10 \times 10^5)(-1)(t + 1000)^{-2}(1) = \frac{-1.10 \times 10^5}{(t + 1000)^2}$$

When $t = 1000$, we have

$$\left. \frac{dP}{dt} \right|_{t=1000} = \frac{-1.10 \times 10^5}{2000^2} = -0.0275 \text{ W/d}$$

Thus, the power is decreasing at the rate of 0.0275 W/d after 1000 days.

Example E

Figure 3–19

Two ships, A and B, leave a port at noon. Ship A travels north at 6.00 mi/h, and ship B travels east at 8.00 mi/h. How fast are they separating at 2 P.M.?

We are asked to find how fast the length of a line drawn directly from A to B is increasing at a particular time. First a relation must be found between the distances traveled. The Pythagorean theorem provides the necessary relation (see Fig. 3–19). If y is the distance traveled by A and x is the distance traveled by B, we can find the distance between them, z, from $z^2 = x^2 + y^2$. Taking derivatives of this expression with respect to time, we find that

Derivatives are taken
with respect to time.

$$2z\frac{dz}{dt} = 2x\frac{dx}{dt} + 2y\frac{dy}{dt}$$

From the information given, we know that $dx/dt = 8.00$ mi/h, $dy/dt = 6.00$ mi/h, and for the specified time, $x = 16.0$ mi, $y = 12.0$ mi, and $z = 20.0$ mi. Thus, we have

$$\left. \frac{dz}{dt} \right|_{z=20} = \frac{(16.0)(8.00) + (12.0)(6.00)}{20.0} = 10.0 \text{ mi/h}$$

Exercises 3–4

In the following exercises, solve the given problems in related rates.

1. The electric resistance of a certain resistor as a function of temperature is given by $R = 4.000 + 0.003T^2$, where R is measured in ohms and T in degrees Celsius. If the temperature is increasing at the rate of 0.100°C/s, find how fast the resistance changes when $T = 150$°C.

2. The kinetic energy of an object is given by KE $= \frac{1}{2}mv^2$, where m is the mass of the object and v is its velocity. If a 20.0-kg object is accelerating (remember that the time rate of change of velocity is acceleration) at 5.00 m/s^2, how fast is the kinetic energy (in joules) changing when the velocity is 30.0 m/s?

3. A firm found that its profit was p dollars for the production of x tons per week of a product according to the function $p = 30\sqrt{10 - x^2} - 50$. Determine the rate of change of profit if production is increasing such that $dx/dt = 0.200$ tons/week^2, when $x = 4.00$ tons/week.

4. A refinery is capable of producing two grades of oil. If x is the number of gallons of the lower-grade oil produced per day, and y is the number of gallons of the higher-grade oil produced per day, where $y = \dfrac{100(20 - x)}{50 - x}$, find how fast y is changing if $dx/dt = 2.00$ gal/d^2, when $x = 8.00$ gal/d.

5. The length of a pendulum is slowly decreasing at the rate of 0.100 in./s. What is the rate of change of the period of the pendulum when the length is 16.0 in., if the relation between the period and length is $T = \pi\sqrt{L/96}$?

6. When air expands so that there is no change in heat (an adiabatic change) the relationship between the pressure and volume is $pv^{1.4} = k$, where k is a constant. At a certain instant, the pressure is 300 kPa and the volume is 10.0 cm^3. The volume is increasing at the rate of 2.00 cm^3/s. What is the time rate of change of pressure at this instant?

7. The relation between the voltage V that produces a current I in a wire of radius r is $V = 0.03I/r^2$. If the current increases at the rate of 0.020 A/s in a wire of 0.040 in. radius, find the rate at which the voltage is increasing.

8. Two resistances in parallel vary with time. At a certain instant R_1 is 8.00 Ω and is increasing at the rate of 0.500 Ω/s. At the same instant, R_2 is 6.00 Ω and is increasing at the rate of 0.600 Ω/s. What is the rate of change of the resistance R of the combination if

$$R = \frac{R_1R_2}{R_1 + R_2}$$

9. The side of a square is increasing at the rate of 5.00 ft/min. How fast is the area changing when the side is 10.0 ft long?

10. A circular plate is contracting by being cooled. Find the rate of decrease in area if the radius decreases at the rate of 0.100 ft/h, when the radius is 4.00 ft.

11. The radius of a certain steam engine cylinder is 3.00 in. What is the speed of the piston if steam is entering the cylinder at 2.00 ft^3/s? Assume no compression of the steam at this moment.

12. An ice cube is melting so that an edge decreases in length by 0.100 cm/min. How fast is the volume changing when an edge is 4.50 cm?

13. One statement of Boyle's law is that the pressure of a gas varies inversely as the volume for constant temperature. If a certain gas occupies 600 cm³ when the pressure is 200 kPa and the volume is increasing at the rate of 20.0 cm³/min, how fast is the pressure changing when the volume is 800 cm³?

14. The intensity of heat varies directly as the strength of the source and inversely as the square of the distance from the source. If an object approaches a heated object of strength of 8.00 units at the rate of 50.0 cm/s, how fast is the intensity changing when it is 100 cm from the source?

15. A spherical metal object is ejected from an Earth satellite and reenters the atmosphere. It heats up so that the radius increases at the rate of 5.00 mm/s. What is the rate of change of volume when the radius is 200 mm?

16. A pile of sand in the shape of a cone has a radius which always equals the altitude. If 100 ft³ of sand are poured onto the pile each minute, how fast is the radius increasing when the pile is 10.0 ft high?

17. Two cars leave the same point, traveling on straight roads which are at right angles. If the cars are traveling at 80.0 km/h and 100 km/h, how fast is the distance between them changing after 3.00 min (0.0500 h)?

18. A ladder is slipping down along a vertical wall. If the ladder is 10.0 ft long, and the top of it is slipping at the constant rate of 10.0 ft/s, how fast is the bottom of the ladder moving along the ground when the bottom is 6.00 ft from the wall?

19. A rope that is attached to a boat is being pulled in at a rate of 10.0 ft/s. If the water is 20.0 ft below the level at which the rope is being drawn in, how fast is the boat approaching the wharf when 36.0 ft of rope are yet to be pulled in? See Fig. 3–20.

20. A man 6.00 ft tall approaches a street light 15.0 ft above the ground at the rate of 5.00 ft/s. How fast is the end of the man's shadow moving when he is 10.0 ft from the base of the light?

21. A supersonic jet leaves an airfield traveling due east at 1600 mi/h. A second jet leaves the same airfield at the same time and travels at 1800 mi/h along a line north of east such that it remains due north of the first jet. After a half-hour, how fast are the jets separating?

22. A roller mechanism, as shown in Fig. 3–21, moves such that the right roller is always in contact with the bottom surface and the left roller is always in contact with the left surface. If the right roller is moving to the right at 1.50 cm/s when x = 10.0 cm, how fast is the left roller moving?

10.0 ft/s

20.0 ft

Figure 3–20

25.0 cm

y

25.0 cm

120°

x

Figure 3–21

3–5 Using Derivatives in Curve Sketching

Derivatives can be used effectively in the sketching of curves. An analysis of the first two derivatives can provide useful information as to the graph of a function. This information, possibly along with two or three key points on the curve, is often sufficient to obtain a good, although approximate, graph of the function. Graphs where this information must be supplemented with other analyses are the subject of the next section.

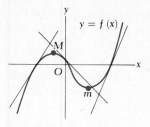

Figure 3-22

Considering the function $f(x)$ as shown in Fig. 3-22, we see that as x increases (from left to right) the y-values also increase until the point M is reached. From M to m the values of y decrease. To the right of m the values of y again increase. We also note that any tangent line to the left of M or to the right of m will have a positive slope. Any tangent line between M and m will have a negative slope. Since the derivative of a function determines the slope of a tangent line, we can conclude that, *as x increases, y increases if the derivative is positive, and decreases if the derivative is negative*. This can be stated as

$$f(x) \text{ increases if } f'(x) > 0$$

and

$$f(x) \text{ decreases if } f'(x) < 0$$

▶ *It is always assumed that x is increasing. Also, we assume in our present analysis that f(x) and its derivatives are continuous over the indicated interval.*

Example A

Determine those values of x for which the function $f(x) = x^3 - 3x^2$ is increasing and those values for which it is decreasing.

We find the solution to this problem by determining the values of x for which the derivative is positive and those for which it is negative. Finding the derivative, we have

$$f'(x) = 3x^2 - 6x = 3x(x - 2)$$

This now becomes a problem of solving an inequality. To find the values of x for which $f(x)$ is increasing we must solve the inequality

$$3x(x - 2) > 0$$

We now recall that the solution of an inequality consists of *all* values of x which may satisfy it. Normally, this consists of certain intervals of values of x. *These intervals are found by first setting the left side of the inequality equal to zero, thus obtaining the* **critical values** *of x*. The function on the left will have the same *sign* for all values of x less than the leftmost critical value. The sign of the function will also be the same within any given interval between critical values and to the right of the rightmost critical value. Those intervals which give the proper sign will satisfy the inequality. In this case the critical values are $x = 0$ and $x = 2$. Therefore, we have the following analysis:

If $x < 0$, $3x(x - 2) > 0$ or $f'(x) > 0$.
If $0 < x < 2$, $3x(x - 2) < 0$ or $f'(x) < 0$.
If $x > 2$, $3x(x - 2) > 0$ or $f'(x) > 0$.

Figure 3-23

Therefore, the solution of the above inequality is $x < 0$ or $x > 2$, which means that for these values $f(x)$ is increasing. We can also see that for $0 < x < 2$, $f'(x) < 0$, which means that $f(x)$ is decreasing for these values of x. The solution is now complete. The graph of $f(x) = x^3 - 3x^2$ is shown in Fig. 3-23.

The points M and m in Fig. 3–22 are called a **relative maximum point** and a **relative minimum point,** respectively. *This means that M has a greater y-value than any other point near it, and that m has the least y-value of any point near it.* This does not necessarily mean that M has the greatest y-value of any point on the curve, or that m has the least y-value of any point on the curve. However, the points M and m are the greatest or least values of y for that part of the curve (that is why we use the word "relative"). Examination of Fig. 3–22 verifies this point. *The characteristic of both M and m is that the derivative is zero at each point.* (We see that this is so since a tangent line would have a slope of zero at each.) *This is how relative maximum and relative minimum points are located. The derivative is found and then set equal to zero. The solutions of the resulting equation give the x-coordinates of the maximum and minimum points.*

It remains now to determine whether a given value of x, for which the derivative is zero, is the coordinate of a maximum or a minimum point (or neither, which is also possible). From the discussion of increasing and decreasing values for y, we can see that *the derivative changes sign from plus to minus when passing through a relative maximum point, and from minus to plus when passing through a relative minimum point.* Thus, we find maximum and minimum points by determining those values of x for which the derivative is zero and by properly analyzing the sign change of the derivative. If the sign of the derivative does not change, it is neither a maximum nor a minimum point. This is known as the **first-derivative test for maxima and minima.**

In Fig. 3–24 a diagram for the first derivative test is shown. The test for a relative maximum is shown in Fig. 3–24(a), and that for a relative minimum is shown in Fig. 3–24(b). For the curves shown in Fig. 3–24, $f(x)$ and $f'(x)$ are continuous throughout the interval shown. (Although $f(x)$ must be continuous, $f'(x)$ may be discontinuous at the maximum point or minimum point, and the sign changes of the first derivative test remain valid.)

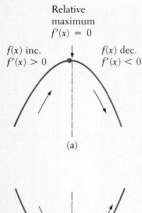

Relative
maximum
$f'(x) = 0$

$f(x)$ inc. $f(x)$ dec.
$f'(x) > 0$ $f'(x) < 0$

(a)

$f'(x) < 0$ $f'(x) > 0$
$f(x)$ dec. $f(x)$ inc.

Relative
minimum
$f'(x) = 0$

(b)

Figure 3–24

Example B

Find any maximum and minimum points on the graph of the function

$$y = 3x^5 - 5x^3$$

Finding the derivative and setting it equal to zero, we have

$$y' = 15x^4 - 15x^2 = 15x^2(x^2 - 1)$$

Therefore,

$$15x^2(x - 1)(x + 1) = 0 \quad \text{for} \quad x = 0, \quad x = 1 \quad \text{and} \quad x = -1$$

Thus, the sign of the derivative is the same for all points to the left of $x = -1$. For these values $y' > 0$ (thus y is increasing). For values of x between -1 and 0, $y' < 0$. For values of x between 0 and 1, $y' < 0$. For values of x greater than 1, $y' > 0$. Thus, the curve has a maximum at $(-1, 2)$ and a minimum at $(1, -2)$. The point $(0, 0)$ is neither a maximum nor a minimum, since the sign of the derivative did not change for this value of x. The graph of $y = 3x^5 - 5x^3$ is shown in Fig. 3–25.

Neither
max. nor
min.

Max.
$(-1, 2)$

$(0, 0)$

$(1, -2)$
Min.

Figure 3–25

Figure 3-26

Points of inflection *I*

Figure 3-27

Looking again at the slope of tangents drawn to a curve [see Fig. 3-26(a)], we see that the slope goes from large to small positive values and then to negative values for this curve. At the point *I* the values of the slope again start to increase so that values of the slope again become positive after the point *m*. *The conclusion is that the slope decreases as a point approaches* I, *and that the slope increases as a point moves to the right of* I. The curve in Fig. 3-26(b) is that of the derivative, and therefore indicates the values of the slope of $f(x)$. If the slope changes, this means that we are dealing with the rate of change of slope, or the rate of change of the derivative. This function is the second derivative.

The curve in Fig. 3-26(c) is that of the second derivative. This means that we wish to be able to find an expression for the derivative of a derivative. *Where the second derivative of a function is negative, the slope is decreasing, or the curve is* **concave down** *(opens down). Where the second derivative is positive, the slope is increasing, or the curve is* **concave up** *(opens up).* This may be summarized as follows:

If $f''(x) > 0$, the curve is concave up.

If $f''(x) < 0$, the curve is concave down.

We can also now use this information in the determination of maximum and minimum points. By the nature of the definition of maximum and minimum points and of concavity, it is apparent that *a curve is concave down at a maximum point and concave up at a minimum point.* We can see these properties when we make a close analysis of the curve in Fig. 3-26. Therefore, at $x = a$,

if $f'(a) = 0$ and $f''(a) < 0$,

then $f(x)$ has a relative maximum at $x = a$, or

if $f'(a) = 0$ and $f''(a) > 0$,

then $f(x)$ has a relative minimum at $x = a$.

These statements comprise what is known as the **second-derivative test for maxima and minima.** This test is often easier to use than the first-derivative test. However, it can happen that $y'' = 0$ at a maximum or minimum point, and in such cases it is necessary that we use the first-derivative test.

In using the second-derivative test we should note carefully that $f''(x)$ is *negative* at a *maximum* point and *positive* at a *minimum* point. This is contrary to a natural inclination to think of "maximum" and "positive" together, or "minimum" and "negative" together.

The points at which the curve changes from concave up to concave down, or from concave down to concave up, are known as **points of inflection.** Thus point *I* is a point of inflection. Inflection points are found by determining those values of *x* for which the second derivative changes sign. This is analogous to finding maximum and minimum points by the first-derivative test. In Fig. 3-27 various types of points of inflection are illustrated.

Example C

Figure 3–28

Determine the concavity and find any points of inflection of the function $y = x^3 - 3x$.

This requires inspection and analysis of the second derivative. So we write

$$y' = 3x^2 - 3, \qquad y'' = 6x$$

The second derivative is positive where the function is concave up, and this occurs if $x > 0$. The curve is concave down for $x < 0$, since y'' is negative. Thus, $(0, 0)$ is a point of inflection, since the concavity changes there. The graph of $y = x^3 - 3x$ is shown in Fig. 3–28.

At this point we shall summarize the information regarding the derivatives of a function.

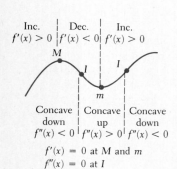

$f'(x) = 0$ at M and m
$f''(x) = 0$ at I

Figure 3–29

$f(x)$ increases if $f'(x) > 0$; $f(x)$ decreases if $f'(x) < 0$.

$f'(x) = 0$ at a maximum point if $f'(x)$ changes from $+$ to $-$, or if $f''(x) < 0$.

$f'(x) = 0$ at a minimum point if $f'(x)$ changes from $-$ to $+$, or if $f''(x) > 0$.

$f''(x) > 0$ where $f(x)$ is concave up; $f''(x) < 0$ where $f(x)$ is concave down.

$f''(x) = 0$ at a point of inflection if $f''(x)$ changes from $+$ to $-$ or $-$ to $+$.

(See Figure 3-29.)

The following examples illustrate how the above information is put together to obtain the graph of a function.

Example D

Sketch the graph of $y = 6x - x^2$.

Finding the first two derivatives, we have

$$y' = 6 - 2x = 2(3 - x) \quad \text{and} \quad y'' = -2.$$

We now note that $y' = 0$ for $x = 3$. For $x < 3$ we see that $y' > 0$, which means that y is increasing over this interval. Also, for $x > 3$, we note that $y' < 0$, which means that y is decreasing over this interval.

Since y' changes from positive on the left of $x = 3$, to negative on the right of $x = 3$, the curve has a maximum point where $x = 3$. Since $y = 9$ for $x = 3$, this maximum point is $(3, 9)$.

Since $y'' = -2$, this means that its value remains constant for all values of x. Therefore, there are no points of inflection and the curve is concave down for all values of x. This also shows that the point $(3, 9)$ is a maximum point.

Summarizing, we know that y is increasing for $x < 3$, y is decreasing for $x > 3$, there is a maximum point at $(3, 9)$, and that the curve is always concave down. Using this information we sketch the curve shown in Fig. 3–30.

Figure 3–30

Example E

Sketch the graph of $y = 2x^3 + 3x^2 - 12x$.

Finding the first two derivatives, we have

$$y' = 6x^2 + 6x - 12 = 6(x + 2)(x - 1)$$

and

$$y'' = 12x + 6 = 6(2x + 1)$$

We note that $y' = 0$ when $x = -2$ and $x = 1$. Using these values in the second derivative, we find that y'' is negative (-18) for $x = -2$ and y'' is positive $(+18)$ when $x = 1$. When $x = -2$, $y = 20$; and when $x = 1$, $y = -7$. Therefore, $(-2, 20)$ is a maximum point and $(1, -7)$ is a minimum point.

Next we see that $y' > 0$ if $x < -2$ or $x > 1$. Also, $y' < 0$ for the interval $-2 < x < 1$. Therefore, y is increasing if $x < -2$ or $x > 1$, and y is decreasing if $-2 < x < 1$.

Now we note that $y'' = 0$ when $x = -\frac{1}{2}$, and that $y'' < 0$ when $x < -\frac{1}{2}$, and $y'' > 0$ when $x > -\frac{1}{2}$. When $x = -\frac{1}{2}$, $y = \frac{13}{2}$. Therefore, there is a point of inflection at $(-\frac{1}{2}, \frac{13}{2})$, the curve is concave down if $x < -\frac{1}{2}$, and the curve is concave up if $x > -\frac{1}{2}$.

Finally, by locating the points $(-2, 20)$, $(-\frac{1}{2}, \frac{13}{2})$, and $(1, -7)$, we draw the curve *up* to $(-2, 20)$ and then *down* to $(-\frac{1}{2}, \frac{13}{2})$, with the curve *concave down.* Continuing *down,* but *concave up,* we draw the curve to $(1, -7)$, at which point we start *up* and continue up. This gives us an excellent graph indicating the shape of the curve. If more precision is required, additional points may be used (see Fig. 3–31).

Figure 3–31

Example F

Sketch the graph of $y = x^5 - 5x^4$.

The first two derivatives are

$$y' = 5x^4 - 20x^3 = 5x^3(x - 4) \quad \text{and} \quad y'' = 20x^3 - 60x^2 = 20x^2(x - 3)$$

We now see that $y' = 0$ when $x = 0$ and $x = 4$. When $x = 0$, $y'' = 0$ also, which means we cannot use the second-derivative test for maximum and minimum points for this value of x. When $x = 4$, y'' is positive $(+320)$, which means $(4, -256)$ is a minimum point. Next we note that $y' > 0$ if $x < 0$ or $x > 4$ and $y' < 0$ if $0 < x < 4$. Thus, by the first-derivative test, there is a maximum at $(0, 0)$. Also, we see that y is increasing if $x < 0$ or $x > 4$ and decreasing if $0 < x < 4$. The second derivative indicates that there is a point of inflection at $(3, -162)$. It also indicates that the function is concave down if $x < 3$ $(x \neq 0)$ and concave up if $x > 3$. There is no point of inflection at $(0, 0)$ since the second derivative does not change sign when $x = 0$. From this information we sketch the curve in Fig. 3–32.

Figure 3–32

Exercises 3–5

In Exercises 1 through 4, determine those values of x for which the given functions are increasing and those values of x for which they are decreasing.

1. $y = x^2 - 2x$ 2. $y = 4 - x^2$
3. $y = 12x - x^3$ 4. $y = x^4 - 6x^2$

See Appendix C for a computer program to find the maximum value of a function.

In Exercises 5 through 8, find any maximum or minimum points of the given functions. (These are the same functions as in Exercises 1 through 4.)

5. $y = x^2 - 2x$ 6. $y = 4 - x^2$
7. $y = 12x - x^3$ 8. $y = x^4 - 6x^2$

In Exercises 9 through 12, determine the values of x for which the curve of the given function is concave up, those values for which it is concave down, and any points of inflection. (These are the same functions as in Exercises 1 through 4.)

9. $y = x^2 - 2x$ 10. $y = 4 - x^2$
11. $y = 12x - x^3$ 12. $y = x^4 - 6x^2$

In Exercises 13 through 16, use the information from Exercises 1 through 12 to sketch the graphs of the given functions.

13. $y = x^2 - 2x$ 14. $y = 4 - x^2$
15. $y = 12x - x^3$ 16. $y = x^4 - 6x^2$

In Exercises 17 through 24, sketch the graphs of the given functions by determining the appropriate information and points from the first and second derivatives.

17. $y = 12x - 2x^2$ 18. $y = 3x^2 - 1$
19. $y = 2x^3 + 6x^2$ 20. $y = x^3 - 9x^2 + 15x + 1$
21. $y = x^5 - 5x$ 22. $y = x^4 + 8x + 2$
23. $y = 4x^3 - 3x^4$ 24. $y = x^5 - 20x^2$

In Exercises 25 through 30, sketch the indicated curves by the methods of this section.

25. A certain projectile follows a path given by $y = x - 0.00025x^2$, where distances are measured in meters. Sketch the graph of the path of the projectile.

26. The power, in watts, in a certain electric circuit is given by $P = 4i - 0.5i^2$, where i is the current in amperes. Sketch the graph of P vs. i.

27. For the force exerted by the cam on the lever in Exercise 41 of Section 2–6, sketch the graph of F vs. x. (*Hint:* Use methods for equations of higher degree to analyze the first derivative.)

28. The deflection y of a beam at a horizontal distance x from one end is given by $y = k(x^3 - 60x^2)$. Sketch the curve representing the beam if it is 10 ft long and $k = 1/5000$.

29. A rectangular box is made from a piece of cardboard 8 in. by 12 in., by cutting a square from each corner and bending up the sides. Express the volume of the box as a function of the side of the square which is cut out, and then sketch the curve of the resulting equation.

30. A tool box whose end is square is to be made from 600 in.^{2} of sheet metal. Express the volume of the box as a function of the side of the square of the end. Sketch the curve of the resulting function.

31. Sketch a continuous curve having the following characteristics.

$$f(1) = 0, \qquad f'(x) > 0 \quad \text{for all } x, \qquad f''(x) < 0 \quad \text{for all } x$$

32. Sketch a continuous curve having the following characteristics,

$$f(0) = -1$$
$$f'(0) > 0 \quad \text{for} \quad x < 0 \quad \text{and} \quad x > 2, \qquad f'(x) < 0 \quad \text{for} \quad 0 < x < 2$$
$$f''(x) < 0 \quad \text{for } x < 1, \qquad f''(x) > 0 \quad \text{for} \quad x > 1$$

3–6 More on Curve Sketching

We are now in a position to combine the information from the derivative with information obtainable from the function itself to sketch the graph of a function. The information we can obtain from the function includes topics introduced previously. We can obtain intercepts, symmetry, the behavior of the curve as x becomes large, and the vertical asymptotes of the curve. Also, continuity is important in sketching certain functions. The examples which follow demonstrate how these concepts are used along with the information from the derivatives to sketch the graph of a function. We will find that some of these considerations are of much more value than others in graphing any particular curve.

Example A

Sketch the graph of

$$y = \frac{2}{x^2 + 1}$$

Intercepts: If $x = 0$, $y = 2$, which means that $(0, 2)$ is an intercept. We see that if $y = 0$, there is no corresponding value of x, since $2/(x^2 + 1)$ is a fraction greater than zero for all x. This also indicates that all points on the curve are above the x-axis.

Symmetry: The curve is symmetric to the y-axis, since

$$y = \frac{2}{(-x)^2 + 1} \quad \text{is the same as} \quad y = \frac{2}{x^2 + 1}$$

The curve is not symmetric to the x-axis, since

$$-y = \frac{2}{x^2 + 1} \quad \text{is not the same as} \quad y = \frac{2}{x^2 + 1}$$

The curve is not symmetric to the origin, since

$$-y = \frac{2}{(-x)^2 + 1} \quad \text{is not the same as} \quad y = \frac{2}{x^2 + 1}$$

(*continued on next page*)

The value in knowing the symmetry is that we should find those portions of the curve on either side of the y-axis to be reflections of each other. It is possible to use this fact directly or to use it as a check.

Behavior as x becomes large: We note that as $x \to \infty$, $y \to 0$ since $2/(x^2 + 1)$ is always a fraction greater than zero, but which becomes smaller as x becomes larger. Therefore, we see that $y = 0$ is an asymptote. Either from the symmetry or the function, we also see that $y \to 0$ as $x \to -\infty$.

Vertical asymptotes: From the discussion of the hyperbola we recall that an asymptote is a line which a curve approaches. We have already noted that $y = 0$ is an asymptote for this curve. This asymptote, the x-axis, is a horizontal line. *Vertical asymptotes, if any exist, are found by determining those values of x for which the denominator of any term is zero.* Such a value of x makes y undefined. Since $x^2 + 1$ cannot be zero, this curve has no vertical asymptotes. The next example illustrates a curve which has a vertical asymptote.

Derivatives: Since

$$y = \frac{2}{x^2 + 1} = 2(x^2 + 1)^{-1}$$

then

$$y' = -2(x^2 + 1)^{-2}(2x) = \frac{-4x}{(x^2 + 1)^2}$$

Since $(x^2 + 1)^2$ is positive for all values of x, the sign of y' is determined by the numerator. Thus, we note that $y' = 0$ for $x = 0$ and that $y' > 0$ for $x < 0$ and $y' < 0$ for $x > 0$. The curve, therefore, is increasing for $x < 0$, decreasing for $x > 0$, and has a maximum point at $(0, 2)$. Now,

$$y'' = \frac{(x^2 + 1)^2(-4) + 4x(2)(x^2 + 1)(2x)}{(x^2 + 1)^4} = \frac{-4(x^2 + 1) + 16x^2}{(x^2 + 1)^3}$$

$$= \frac{12x^2 - 4}{(x^2 + 1)^3} = \frac{4(3x^2 - 1)}{(x^2 + 1)^3}$$

We note that y'' is negative for $x = 0$, which confirms that $(0, 2)$ is a maximum point. Also, points of inflection are found for the values of x satisfying $3x^2 - 1 = 0$. Thus, $(-\frac{1}{3}\sqrt{3}, \frac{3}{2})$ and $(\frac{1}{3}\sqrt{3}, \frac{3}{2})$ are points of inflection. The curve is concave up if $x < -\frac{1}{3}\sqrt{3}$, or $x > \frac{1}{3}\sqrt{3}$, and concave down if $-\frac{1}{3}\sqrt{3} < x < \frac{1}{3}\sqrt{3}$.

Putting this information together we sketch the curve shown in Fig. 3–33. It might be noted that this curve could have been sketched primarily by use of the fact that $y \to 0$ as $x \to +\infty$ and as $x \to -\infty$, and the fact that a maximum point exists at $(0, 2)$. However, the other parts of the analysis, such as symmetry and concavity, serve as excellent checks and also make the curve more accurate.

Figure 3–33

Figure 3–34

Example B   Sketch the graph of $y = x + \dfrac{4}{x}$.

Intercepts: If we set $x = 0$, y is undefined. This means that the curve is not continuous at $x = 0$ and there are no y-intercepts. If we set $y = 0$, $x + 4/x = (x^2 + 4)/x$ cannot be zero since $x^2 + 4$ cannot be zero. Therefore, there are no intercepts. This may seem to be of little value, but we must realize *this curve does not cross either axis*. This, in itself, will be of value when we sketch the graph in Fig. 3–34.

Symmetry: In testing for symmetry, we find that it is not symmetric to either axis. However, this curve does possess symmetry to the origin. This is determined by the fact that when $-x$ replaces x and at the same time $-y$ replaces y, the equation does not change.

Behavior as x becomes large: As $x \to +\infty$, and as $x \to -\infty$, $y \to x$ since $4/x \to 0$. Thus, $y = x$ is an asymptote of the curve.

Vertical asymptotes: As we noted in Example A, vertical asymptotes exist for values of x for which y is undefined. In this equation, $x = 0$ makes the second term on the right undefined and therefore y is undefined. In fact as $x \to 0$ from the positive side, $y \to +\infty$, and as $x \to 0$ from the negative side, $y \to -\infty$. This is derived from the sign of $4/x$ in each case.

Derivatives: Finding the first derivative we have

$$y' = 1 - \frac{4}{x^2} = \frac{x^2 - 4}{x^2}$$

The x^2 in the denominator indicates that the sign of the first derivative is the same as its numerator. The numerator is zero if $x = -2$ or $x = 2$. If $x < -2$ or $x > 2$, then $y' > 0$; and if $-2 < x < 2$, $x \neq 0$, $y' < 0$. Thus, y is increasing if $x < -2$ or $x > 2$, and also y is decreasing if $-2 < x < 2$, except at $x = 0$ (y is undefined). Also, $(-2, -4)$ is a maximum point and $(2, 4)$ is a minimum point. The second derivative is $y'' = 8/x^3$. This cannot be zero, but it is negative if $x < 0$ and positive if $x > 0$. Thus, the curve is concave down if $x < 0$ and concave up if $x > 0$. Using this information we have the graph shown in Fig. 3–34.

Example C

Sketch the graph of

$$y = \frac{1}{\sqrt{1 - x^2}}$$

Intercepts: If $x = 0$, $y = 1$. If $y = 0$, $1/\sqrt{1 - x^2}$ would have to be zero, but it cannot since it is a fraction with 1 as the numerator for all values of x. Thus, $(0, 1)$ is an intercept.

Symmetry: The curve is symmetric to the y-axis.

Behavior as x becomes large: The values of x cannot be considered beyond 1 or -1, for any value of $x < -1$ or $x > 1$ gives imaginary values for y. Thus, the curve does not exist for values of $x < -1$ or $x > 1$.

Vertical asymptotes: If $x = 1$ or $x = -1$, y is undefined. In each case as $x \to 1$ and as $x \to -1$, $y \to +\infty$.

Derivatives:

$$y' = -\tfrac{1}{2}(1 - x^2)^{-3/2}(-2x) = \frac{x}{(1 - x^2)^{3/2}}$$

We see that $y' = 0$ if $x = 0$. If $-1 < x < 0$, $y' < 0$ and also if $0 < x < 1$, $y' > 0$. Thus the curve is decreasing if $-1 < x < 0$ and increasing if $0 < x < 1$. There is a minimum point at $(0, 1)$.

$$y'' = \frac{(1 - x^2)^{3/2} - x(\tfrac{3}{2})(1 - x^2)^{1/2}(-2x)}{(1 - x^2)^3} = \frac{(1 - x^2) + 3x^2}{(1 - x^2)^{5/2}}$$

$$= \frac{2x^2 + 1}{(1 - x^2)^{5/2}}$$

The second derivative cannot be zero since $2x^2 + 1$ is positive for all values of x. The second derivative is also positive for all permissible values of x, which means the curve is concave up for these values.

Using this information we sketch the graph in Fig. 3–35.

Figure 3–35

Exercises 3–6

In the following exercises, use the method of the examples of this section to sketch the indicated curves.

1. $y = \dfrac{4}{x^2}$ 2. $y = \dfrac{2}{x^3}$ 3. $y = \dfrac{2}{x + 1}$ 4. $y = \dfrac{x}{x - 2}$

5. $y = x^2 + \dfrac{2}{x}$ 6. $y = x + \dfrac{4}{x^2}$ 7. $y = x - \dfrac{1}{x}$ 8. $y = 3x + \dfrac{1}{x^3}$

9. $y = \dfrac{x^2}{x + 1}$ 10. $y = \dfrac{9x}{x^2 + 9}$ 11. $y = \dfrac{1}{x^2 - 1}$ 12. $y = x\sqrt{1 - x^2}$

Wall

20,000 m² x

Figure 3–36

13. Sketch R_T as a function of R for the function relating the combined resistance of the circuit given in Exercise 58 of Section 2–11.

14. A fence is to be constructed to enclose a rectangular area of 20,000 m². A previously constructed wall is to be used for one side. Sketch the length ℓ of fence to be built as a function of the side x of the fence parallel to the wall. See Fig. 3–36.

15. A cylindrical oil drum is to be made such that it will contain 20 kL. Sketch the area of sheet metal required for construction as a function of the radius of the drum.

16. Sketch P as a function of r for the function relating power and resistance of the circuit given in Exercise 39 of Section 2–7. Assume values of $E = 6$ V and $R = 1\ \Omega$.

3–7 Applied Maximum and Minimum Problems

Problems from various applied situations frequently occur which require finding a maximum or minimum value of some function. If the function is known, the methods we have already discussed can be used directly. This is discussed in the following example.

Example A

An automobile manufacturer, in testing a new engine on one of its models, found that the efficiency e of the engine as a function of the speed s of the car was given by $e = 0.768s - 0.00004s^3$. Here, e is given in percent and s is given in km/h. What is the maximum efficiency of the engine?

Since we wish to find a maximum value, we find the derivative of e with respect to s.

$$\frac{de}{ds} = 0.768 - 0.00012s^2$$

We then set the derivative equal to zero in order to find the value of s for which a maximum may occur.

$$0.768 - 0.00012s^2 = 0$$
$$0.00012s^2 = 0.768$$
$$s^2 = 6400 \qquad s = 80.0 \text{ km/h}$$

We know that s must be positive to have meaning in this problem. Therefore, the apparent solution of $s = -80$ is discarded. The second derivative is

$$\frac{d^2e}{ds^2} = -0.00024s$$

which is negative for any positive value of s. Thus, a maximum occurs for $s = 80.0$. Substituting $s = 80.0$ in the function for e, we obtain

$$e = 0.768(80.0) - 0.00004(80.0^3)$$
$$= 61.44 - 20.48 = 40.96$$

Therefore, to the nearest tenth, the maximum efficiency is 41.0%, and this occurs for $s = 80.0$ km/h.

In many problems for which a maximum or minimum value is to be determined, the function is itself not given explicitly. It is necessary to find the function from the statement of the problem. The principal difficulty which arises in these problems is finding this function. Once we determine this function, we take a derivative with respect to a variable in which it is expressed, and then set the derivative equal to zero. Sufficient information must be given so that a unique solution is possible. Often this information is available, but we must analyze the wording of the problem carefully to find it. Also, many times it is more convenient to set up more than one function, based on the given information, and to find a derivative of each with respect to the same variable. We can then eliminate by substitution any undesired variables or derivatives which appear in each. The following examples illustrate these methods.

Example B

Find the number which exceeds its square by the greatest amount.

We must set up a function which expresses the difference between a number and its square. Let the number be x, and let the difference be D. Then $D = x - x^2$. Taking a derivative, we have

$$\frac{dD}{dx} = 1 - 2x$$

Setting the derivative equal to zero, we have $0 = 1 - 2x$. Solving for x, we obtain the result $x = \frac{1}{2}$. The second derivative gives $d^2D/dx^2 = -2$, which states that the second derivative is always negative. This means that whenever the first derivative is zero, it represents a maximum. In many problems it is not necessary to test for maximum or minimum, since the nature of the problem will indicate which must be the case. For example, in this problem we know that numbers greater than 1 do not exceed their squares at all. The same is true for all negative numbers. Thus, the answer must lie between 0 and 1.

Example C

Find the maximum possible area of a rectangle whose perimeter is 16 units.

When we reflect on this problem, we see that there are limitless possibilities for rectangles of a perimeter of 16 units and differing areas (see Fig. 3–37). For example, if the sides are 7 and 1, the area is 7. If the sides are 6 and 2, the area is 12. Therefore, we set up a function for the area of a rectangle in terms of its sides x and y:

$$A = xy$$

We know, however, one other important fact: that the perimeter is 16. Thus, $2x + 2y = 16$. Solving for y, we have $y = 8 - x$. Substituting this equation into the expression for the area, we have

$$A = x(8 - x) = 8x - x^2$$

Taking a derivative of this function and setting it equal to zero, we have

Figure 3–37

$$\frac{dA}{dx} = 8 - 2x$$

$$8 - 2x = 0$$

$$x = 4$$

By noting values of the derivative near 4 or by finding the second derivative, we can show that we have a maximum for $x = 4$. Thus, $x = 4$ and $y = 4$ give the maximum area of 16 square units.

Example D

Find the point on the parabola $y = x^2$ which is nearest the point $(6, 3)$.

In this example we must set up a function for this distance between a general point (x, y) on the parabola and the point $(6, 3)$. This relation is

$$D = \sqrt{(x - 6)^2 + (y - 3)^2}$$

However, to make it easier for us to take derivatives, we shall square both sides of this expression. If a function is a minimum, then so is its square. We shall also use the fact that the point (x, y) is on $y = x^2$ by replacing y by x^2. Thus, we have

$$D^2 = (x - 6)^2 + (x^2 - 3)^2 = x^2 - 12x + 36 + x^4 - 6x^2 + 9$$

$$= x^4 - 5x^2 - 12x + 45$$

$$\frac{dD^2}{dx} = 4x^3 - 10x - 12$$

$$0 = 2x^3 - 5x - 6$$

Figure 3–38

Using synthetic division or some similar method, we find that the solution to this equation is $x = 2$. Thus, the required point on the parabola is $(2, 4)$. (See Fig. 3–38.) We can show that we have a minimum by analyzing the first derivative, the second derivative, or by noting that points at much greater distances exist (therefore, it cannot be a maximum).

Example E

A company determines that it can sell 1000 units of a product per month if the price is $5 for each unit. It also estimates that for each 1¢ reduction in unit price, 10 more units can be sold. Under these conditions, what is the maximum possible income and what price per unit gives this income?

If we let $x =$ the number of units over 1000 sold, the total number of units sold is $1000 + x$. The price for each unit is $5 less 1¢ ($0.01) for each block of ten units over 1000 which are sold. Thus, the price for each unit is

$$5 - 0.01\left(\frac{x}{10}\right) \quad \text{or} \quad 5 - 0.001x \text{ dollars}$$

The income I is the number of units sold times the price of each unit. Therefore, we have

$$I = (1000 + x)(5 - 0.001x)$$

(*continued on next page*)

6. The electric resistance (in ohms) of a particular type of resistor as a function of the temperature (in degrees Celsius) is given by

$$R = \sqrt{0.001T^4 - 4T + 100}$$

What is the minimum resistance of this type of resistor?

7. The product of two positive numbers is 64. Find the numbers if their sum is a minimum.

8. The sum of two numbers is 40. Find the numbers such that their product is a maximum.

9. A rectangular field bounded on one side by a river is to be fenced in. If no fencing is required along the river, what is the maximum area which can be enclosed if 2000 ft of fencing are used?

10. A rectangular microprocessor chip is designed to have an area of 25 mm². What are to be its dimensions if its perimeter is to be a minimum?

11. A metal plate is to be in the shape of an isosceles triangle with each of the equal sides being 12.0 cm. What should be the length of the base if the plate is to have maximum area?

12. An architect is designing a rectangular building in which the front wall costs twice as much per linear meter as the other three walls. The building is to cover 1350 m². What dimensions must it have such that the cost of the walls is a minimum?

13. Find the dimensions of the largest rectangle which can be inscribed in a quarter-circle of radius 4.

14. What is the area of the largest rectangle which can be inscribed in the first quadrant under the parabola $y = 4 - x^2$? See Fig. 3–40.

15. If the hypotenuse of a right triangle is 10, what must the sides be so that the area is a maximum?

16. A box with a square base is open at the top. If 108 ft² of material are used, what is the maximum volume possible for the box?

17. What is the maximum slope of the curve $y = 6x^2 - x^3$?

18. What is the minimum slope of the curve $y = x^5 - 10x^2$?

19. A person is in a boat 4 km from the nearest point P on a straight shoreline. The person wishes to go to point A, which is on the shore, 5 km from P. If the person can row at 3 km/h and walk at 5 km/h, how far along the shore from P towards A should the boat land in order that the person can reach A in the least time? See Fig. 3–41.

20. A poster is to have margins of 4 in. at the top and bottom and margins of 2 in. at each side. The printed matter of the poster is to cover 50 in.². Find the dimensions of the poster if the total area is to be a minimum.

21. An open box is to be made from a square piece of cardboard whose sides are 8 in. long by cutting equal squares from the corners and bending up the sides. Determine the side x of the square which is to be cut out so that the volume of the box may be a maximum. See Fig. 3–42.

22. Find the relation between the radius and the height of a right circular cylinder for which the total surface area is to be a minimum for a given volume.

23. A certain area is a rectangle with a semicircle at each end. Find the open side (not between the rectangle and the semicircle) of the rectangle if the perimeter is 8 units and the area of the rectangle is to be a maximum. See Fig. 3–43.

Figure 3–40

Figure 3–41

Figure 3–42

Figure 3–43

24. A company finds that there is a net profit of $10 for each of the first 1000 units produced each week. For each unit over 1000 produced, there is 2 cents less profit per unit. How many units should be produced each week to net the greatest profit?

25. The deflection y of a beam of length L at a horizontal distance x from one end is given by $y = k(2x^4 - 5Lx^3 + 3L^2x^2)$, where k is a constant. For what value of x does the maximum deflection occur?

26. The electric power (in watts) produced by a certain source is given by

$$P = \frac{144r}{(r + 0.6)^2}$$

$16 = w^2 + d^2$

where r is the resistance in the circuit. For what value of r is the power a maximum?

27. The strength of a rectangular beam is proportional to the product of its width and the square of its depth. Find the dimensions of the strongest beam that can be cut from a circular log 16 in. in diameter. See Fig. 3–44.

Figure 3–44

28. A specially made cylindrical container is made of stainless steel sides and bottom and a silver top. If silver is ten times as expensive as stainless steel, what are the most economical dimensions of the container if it is to hold 10π cm³?

3–8 Review Exercises for Chapter 3

In Exercises 1 through 6, find the equations of the tangent and normal lines.

1. Find the equation of the line tangent to the parabola $y = 3x - x^2$ at the point $(-1, -4)$.

2. Find the equation of the line tangent to the curve $y = x^2 - \dfrac{2}{x}$ at the point $(2, 3)$.

3. Find the equation of the line normal to $y = \dfrac{x}{4x - 1}$ at the point $(1, \frac{1}{3})$.

4. Find the equation of the line normal to $y = \dfrac{1}{\sqrt{x - 2}}$ at the point $(6, \frac{1}{2})$.

5. Find the equation of the line tangent to the curve $y = \sqrt{x^2 + 3}$ and which has a slope of $\frac{1}{2}$.

6. Find the equation of the line normal to the curve $y = \dfrac{1}{2x + 1}$ and which has a slope of $\frac{1}{2}$ if $x \geq 0$.

In Exercises 7 through 12, determine the required velocities and accelerations.

7. Given that the x- and y-coordinates of a moving particle are given as a function of time by the parametric equations $x = t^4 - t$, $y = \dfrac{1}{t}$, find the magnitude and direction of the velocity when $t = 2$.

8. If the x- and y-coordinates of a moving object as functions of time are given by $x = 0.1t^2 + 1$, $y = \sqrt{4t + 1}$, find the magnitude and direction of the velocity when $t = 6$.

9. An object moves along the curve $y = 0.5x^2 + x$ such that $v_x = 0.5\sqrt{x}$. Find v_y at $(2, 4)$.

10. A particle moves along the curve of $y = \dfrac{1}{x + 2}$ with a constant velocity in the x-direction of 4 cm/s. Find v_y at $(2, \frac{1}{4})$.

11. Find the magnitude and direction of the acceleration for the particle in Exercise 7.

12. Find the magnitude and direction of the acceleration of the particle in Exercise 10.

In Exercises 13 through 16, find the indicated roots of the given equations to at least four decimal places by the use of Newton's method.

13. $x^3 - 3x^2 - x + 2 = 0$ (between 0 and 1)

14. $x^3 + 3x^2 - 6x - 2 = 0$ (between 1 and 2)

15. $3x^3 - x^2 - 8x - 2 = 0$ (between 1 and 2)

16. $x^4 + 3x^3 + 6x + 4 = 0$ (between -1 and 0)

In Exercises 17 through 24, sketch the graph of the given functions by information obtained from the function as well as information obtained from the derivatives.

17. $y = 4x^2 + 16x$ 18. $y = 2x^2 + x - 1$

19. $y = \cdot 27x - x^3$ 20. $y = x^3 + 2x^2 + x + 1$

21. $y = x^4 - 32x$ 22. $y = x^4 + 4x^3 - 16x$

23. $y = \dfrac{x}{x + 1}$ 24. $y = x^3 + \dfrac{3}{x}$

In Exercises 25 through 44, solve the given problems.

25. The parabolas $y = x^2 + 2$ and $y = 4x - x^2$ are tangent to each other. Find the equation of the line tangent to them at the point of tangency.

26. Find the equation of the line tangent to $y = x^4 - 8x$ and parallel to $y + 4x + 3 = 0$.

27. The deflection y of a beam at a horizontal distance x from one end is given by $y = k(x^4 - 30x^3 + 1000x)$, where k is a constant. Observing the equation and using Newton's method, find the values of x where the deflection is zero, if the beam is 10.000 m long.

28. If the edge of a cube is increased by 1.0000 mm, its volume is doubled. Find the edge of the cube by use of Newton's method.

29. A projectile moves according to the equations $x = 1200t$ and $y = -490t^2$, where distances are measured in centimeters and time is measured in seconds. Find the magnitude and direction of velocity after 3.00 s.

30. An electron is moving along the path $y = 1/x$ at the constant velocity of 100 m/s. Find the velocity in the x-direction when the electron is at the point $(2, \frac{1}{2})$.

31. In Fig. 3–45 the tension T supports the 40.0-N weight. The relation between the tension T and the deflection d is $d = \dfrac{1000}{\sqrt{T^2 - 400}}$. If the tension is increasing at 2.00 N/s when $T = 28.0$ N, how fast is the deflection, in centimeters, changing?

32. The altitude h, in feet, of a rocket as a function of time after launching is given by $h = 1600t - 16t^2$. What is the maximum altitude which the rocket attains?

Figure 3–45

33. Sketch a continuous curve having the following characteristics.

$$f(0) = 2 \qquad\qquad f'(x) < 0 \text{ for } x < 0$$
$$f''(x) > 0 \quad \text{for all } x \qquad f'(x) > 0 \text{ for } x > 0$$

34. Sketch a continuous curve having the following characteristics.

$$f(0) = 1 \qquad\qquad f'(0) = 0$$
$$f''(x) < 0 \quad \text{for } x < 0 \qquad f'(x) > 0 \quad \text{for } |x| > 0$$
$$f''(x) > 0 \quad \text{for } x > 0$$

35. Side a of a triangle is increasing at the rate of 2.50 in./min, and side b is increasing at the rate of 3.70 in./min. If the angle between sides a and b is always 60°, find the rate of change of side c when $a = 5.60$ in. and $b = 8.00$ in.

36. The current I through a circuit with a resistance R and a battery whose voltage is E and whose internal resistance is r is given by $I = E/(R + r)$. If R changes at the rate of 0.250 Ω/min. how fast is the current changing when $R = 6.25$ Ω, if $E = 3.10$ V and $r = 0.230$ Ω?

Figure 3–46

37. A machine part is to be in the shape of a circular sector of radius r and central angle θ. Find r and θ if the area is one unit and the perimeter is a minimum. See Fig. 3–46.

38. A beam of rectangular cross section is to be cut from a log 2 ft in diameter. The stiffness varies as the width and the cube of the depth. What dimensions will give the beam maximum stiffness? See Fig. 3–47.

Figure 3–47

39. Sketch a graph of the van der Waals equation (see Exercise 38 of Section 2–7), assuming the following values: $R = T = a = 1$ and $b = 0$. For many gases the value of a is much greater than that for b. Even though the values of $R = T = 1$ are not realistic, the *shape* of the curve will be correct for the assumed value of $b = 0$.

40. The weekly profits of a corporation for a particular year are given by $p = x^4 - 40x^3$, where x is the week of the year. Sketch the graph of p versus x, assuming it to be a continuous function over the appropriate interval.

41. Two ships leave the same port at the same time, one traveling south at the rate of 12 km/h and the other traveling east at the rate of 5 km/h. At what rate is the distance between them changing 2 h after they leave the port?

42. The base of a conical machine part is being milled such that the height is decreasing at the rate of 0.05 cm/min. If the part originally had a radius of 1.0 cm and height of 3.0 cm, how fast is the volume changing when the height is 2.8 cm?

43. An alpha particle moves through a magnetic field along the parabolic path $y = x^2 - 4$. Determine the closest that the particle comes to the origin.

44. A Norman window has the form of a rectangle surmounted by a semicircle. Find the dimensions (radius of circular part and height of rectangular part) of the window that will admit the most light if the perimeter of the window is 12 ft. See Fig. 3–48.

Figure 3–48

CHAPTER 4

Integration

In the study of physical and technical applications, it is common to find information about the rate of change of a variable. With such information we have to reverse the process of differentiation in order to arrive at the functional relationship between the variables. This leads us to consider the problem of finding the function when we know its derivative, a process known as *integration*.

As we study integration in this chapter and the next, we shall see that it can also be applied to finding areas and volumes, as well as to physical concepts such as work and pressure. Although these appear to be quite different, we will show that they have a very similar mathematical formulation.

4–1 Differentials

Although the primary concern of this chapter is the development of the inverse process of differentiation, it is necessary to first introduce certain concepts and notation in this section before proceeding to the inverse process. The material presented here is necessary to properly relate differentiation and the methods we shall develop.

We *shall now define the* **differential** *of a function* $y = f(x)$ *as*

$$dy = f'(x)\,dx \tag{4–1}$$

In Eq. (4–1), the quantity *dy is the differential of y, and dx is the differential of x. The differential dx is defined as equal to* Δx, *the increment in x.* We define it purposely in this way, so that $f'(x) = dy/dx$. In this way we can interpret the derivative as the ratio of the differential of y to the differential of x. That is, now the derivative can be considered as a fraction. Although we had previously used the notation dy/dx for the derivative, we had not interpreted it as a fraction.

Example A

Find the differential of $y = 3x^5 - x$.

In this example, $f(x) = 3x^5 - x$, which in turn means that $f'(x) = 15x^4 - 1$. Thus,

$$dy = (15x^4 - 1)\, dx$$

Example B

Find the differential of $y = (2x^3 - 1)^4$.

$$dy = 4(2x^3 - 1)^3(6x^2)\, dx = 24x^2(2x^3 - 1)^3\, dx$$

Example C

Find the differential of $y = \dfrac{4x}{x^2 + 4}$.

$$dy = \frac{(x^2 + 4)(4) - (4x)(2x)}{(x^2 + 4)^2}\, dx = \frac{4x^2 + 16 - 8x^2}{(x^2 + 4)^2}\, dx$$

$$= \frac{-4x^2 + 16}{(x^2 + 4)^2}\, dx = \frac{-4(x^2 - 4)}{(x^2 + 4)^2}\, dx$$

We shall find that the definition of the differential will clarify the connection between the various interpretations of the inverse process of differentiation. Although the principal purpose in introducing it here is to be able to use the notation, there are useful direct applications of the differential.

The applications of the differential are based upon the fact that the differential of y, dy, closely approximates the increment in y, Δy, if the differential of x, dx, is small. To understand this statement, let us look at Fig. 4–1. Recalling the meaning of Δx and Δy, we see that the points $P(x, y)$ and $Q(x + \Delta x, y + \Delta y)$ lie on the curve of $f(x)$. However, $f'(x) = dy/dx$ at P, which means if we draw a tangent line at P, its slope may be indicated by dy/dx. By choosing $\Delta x = dx$, we can see the difference between Δy and dy. It can be seen that as dx becomes smaller, Δy more nearly equals dy.

For given changes in x, it is necessary to use the delta-process to find the exact change, Δy, in y. However, *for small values of Δx, dy can be used to approximate Δy closely.* Generally dy is much more easily determined than is Δy.

Figure 4–1

Example D

Calculate Δy and dy for $y = x^3 - 2x$ for $x = 3$ and $\Delta x = 0.1$.

By the delta-process we find

$$\Delta y = 3x^2 \, \Delta x + 3x(\Delta x)^2 + (\Delta x)^3 - 2 \, \Delta x$$

Using the given values we find

$$\Delta y = 3(9)(0.1) + 3(3)(0.01) + (0.001) - 2(0.1) = 2.591$$

The differential of y is

$$dy = (3x^2 - 2) \, dx$$

Since $dx = \Delta x$, we have

$$dy = [3(9) - 2](0.1) = 2.5$$

Thus, $\Delta y = 2.591$ and $dy = 2.5$. We see that in this case dy is very nearly equal to Δy.

Example E

If finding Δy by the delta-process is complicated or lengthy, we can use a calculator to find the difference between Δy and dy. For the function

$$y = f(x) = \sqrt{8x + 3}$$

the differential is

$$dy = \frac{1}{2}(8x + 3)^{-1/2}(8) \, dx = \frac{4 \, dx}{\sqrt{8x + 3}}$$

For $x = 2$ and $\Delta x = 0.003$, we have

$$f(x) = \sqrt{8(2) + 3} = \sqrt{19} = 4.3588989$$

and

$$f(x + \Delta x) = \sqrt{8(2.003) + 3} = \sqrt{19.024} = 4.3616511$$

This means that

$$\Delta y = f(x + \Delta x) - f(x)$$
$$= 4.3616511 - 4.3588989 = 0.0027522$$

Now calculating the value of dy, we have

$$dy = \frac{4(0.003)}{\sqrt{8(2) + 3}} = 0.0027530$$

We can again see that the values of dy and Δy are very nearly equal.

The fact that dy can be used to approximate Δy is useful in determining the error in a result, if the data are in error, or the equivalent problem of finding the change in a result if a change is made in the data. Even though such changes could be found by use of a calculator, the differential can be used to set up a general expression for the change in a particular function. The following example illustrates the method.

Example F

The edge of a cube was determined to be 9.00 in. From this value the volume was found. Later it was discovered that the value of the edge was 0.20 in. too small. Approximately how much was the volume in error?

The volume V of a cube, in terms of an edge, e, is $V = e^3$. Since we wish to find the change in V for a given change in e, this means we want the value of dV for $de = 0.20$ in. Thus,

$$dV = 3e^2\, de$$
$$dV = 3(9.00)^2(0.20) = 48.6 \text{ in.}^3$$

We see that in this case the volume was in error by about 49 in.^{3}. As long as de is small compared to e, we can calculate an error or change in the volume of a cube by calculating the value of $3e^2\, de$.

Often when considering the error of a given value or result, *the actual numerical value of the error, the* **absolute error,** is not as important as its size is in relation to the size of the quantity itself. *The ratio of the absolute error to the size of the quantity itself is known as the* **relative error.** The relative error is commonly expressed as a percentage.

Example G

Referring to Example F, we see that the absolute error in the edge was 0.20 in. The relative error in the edge was $0.20/9.00 = 0.022 = 2.2\%$. The absolute error in the volume was 49 in.^{3}, whereas the relative error in the volume was $49/729 = 0.067 = 6.7\%$.

Exercises 4–1

In Exercises 1 through 12, find the differential of each of the given functions.

1. $y = x^5 + x$

2. $y = 3x^2 + 6$

3. $y = \dfrac{2}{x^5} + 3$

4. $y = 2x - \dfrac{1}{x}$

5. $y = (x^2 - 1)^4$

6. $y = (4 + 3x)^{1/3}$

7. $y = \dfrac{2}{3x^2 + 1}$

8. $y = \dfrac{1}{\sqrt{1 - x^3}}$

9. $y = x^2(1 - x)^3$

10. $y = x\sqrt{1 - 4x}$

11. $y = \dfrac{x}{x + 1}$

12. $y = \dfrac{3x + 1}{\sqrt{2x - 1}}$

In Exercises 13 through 16, find the values of Δy and dy for the given values of x and Δx.

13. $y = 7x^2 + 4x$, $x = 4$, $\Delta x = 0.2$

14. $y = 2x^2 - 3x + 1$, $x = 5$, $\Delta x = 0.15$

15. $y = 2x^3 - 4x$, $x = 2.5$, $\Delta x = 0.5$

16. $y = x^4$, $x = 3.2$, $\Delta x = 0.08$

In Exercises 17 through 20, determine the value of dy for the given values of x and Δx. Compare with values of $f(x + \Delta x) - f(x)$ found by use of a calculator.

17. $y = (1 - 3x)^5$, $x = 1$, $\Delta x = 0.01$

18. $y = (x^2 + 2x)^3$, $x = 7$, $\Delta x = 0.02$

19. $y = x\sqrt{1 + 4x}$, $x = 12$, $\Delta x = 0.06$

20. $y = \dfrac{x}{\sqrt{6x - 1}}$, $x = 3.5$, $\Delta x = 0.025$

In Exercises 21 through 28, solve the given problems by finding the appropriate differential.

0.50 in.

16.0 in.

A

16.0 in.

dA

0.50 in.

Figure 4–2

21. If the side of a square is measured to be 16.0 in., and an error of 0.50 in. is found in the measurement, by approximately how much is the original calculation of area in error? See Fig. 4–2.

22. A circular solar cell is to be made with a radius of 12.00 cm. If the maximum allowable error in the radius is 0.05 cm, what is the maximum possible relative error in the area of the cell?

23. The voltage of a certain thermocouple as a function of temperature is given by $E = 6.2T + 0.0002T^3$. What is the approximate change in voltage if the temperature changes from 100°C to 101°C?

24. The velocity of an object rolling down a certain inclined plane is given by $v = \sqrt{100 + 16h}$, where h is the distance traveled along the plane by the object. What is the increase in velocity (in feet per second) of an object in moving from 20.0 to 20.5 ft along the plane? What is the relative change in the velocity?

25. What is the volume of metal used to make a right circular cylindrical container of radius 3.00 in. and height 4.00 in., if the thickness of sheet metal it was made from is 0.02 in.?

26. A precisely measured 10.0-Ω resistor (any error in its value will be considered negligible) is put in parallel with a variable resistor of resistance R. The combined resistance of the two resistors is $R_T = \dfrac{10R}{10 + R}$. What is the relative error of the combined resistance, if R is measured at 40.0 Ω, with a possible error of 1.5 Ω?

27. Show that an error of 2% in the measurement of the side of a square results in an error of approximately 4% in the calculation of the area.

28. Show that the relative error in the calculation of the volume of a sphere is approximately three times the relative error in the measurement of the radius.

4–2 Antiderivatives

Since many kinds of problems in many areas, including science and technology, can be solved by reversing the process of finding a derivative or a differential, we shall introduce the basic technique of this procedure in this section. *This reverse process is known as* **antidifferentiation.** In the next section we shall formalize the process, but it is only the basic idea that is the topic of this section. Many of the applications are found in Chapter 5. The following example illustrates the method.

Example A

Find a function for which the derivative is $8x^3$. That is, find an antiderivative of $8x^3$.

We know that, when we set out to find the derivative of a polynomial, we reduce the power of x by 1. Also we multiply the coefficient by the power of x. Thus, the power of the function must have been 4, since the power in the derivative is 3. A factor of 4 of the derivative must also be divided out in the process of finding the function. If we write the derivative as $2(4x^3)$, we recognize $4x^3$ as the derivative of x^4. Therefore, the desired function is $2x^4$, which means that an antiderivative of $8x^3$ is $2x^4$. We can verify our result by finding the derivative.

Example B

Find an antiderivative of $x^2 + 2x$.

As for the x^2, we know that the power of x required in an antiderivative is 3. Also, to make the coefficient correct, we must multiply by $\frac{1}{3}$. The $2x$ should be recognized as the derivative of x^2. Therefore, we have the antiderivative as $\frac{1}{3}x^3 + x^2$.

In Examples A and B, we note that we could add any constant to the antiderivative given as the result, and still have a correct antiderivative. This is due to the fact that the derivative of a constant is zero. This is considered further in the following section. For the examples and exercises in this section, we will not include any such constants in the results.

We note that when we find the antiderivative of a given function, we obtain another function. Thus, we can define the antiderivative of the function $f(x)$ to be the function $F(x)$ such that $F'(x) = f(x)$.

Example C

Find the antiderivative of the function $f(x) = \sqrt{x} - \dfrac{2}{x^3}$.

Since we wish to find the antiderivative of $f(x)$, we know that $f(x)$ is the derivative of the required function.

Considering the term $\sqrt{x}$, we first write it as $x^{1/2}$. To have x to the $\frac{1}{2}$ power in the derivative, we must have x to the $\frac{3}{2}$ power in the antiderivative. Knowing that the derivative of $x^{3/2}$ is $\frac{3}{2}x^{1/2}$, we write $x^{1/2}$ as $\frac{2}{3}(\frac{3}{2}x^{1/2})$. Thus, the first term of the antiderivative is $\frac{2}{3}x^{3/2}$.

As for the term $-2/x^3$, we write it as $-2x^{-3}$. This we recognize as the derivative of x^{-2}, or $1/x^2$.

This means that the antiderivative of the function

$$f(x) = \sqrt{x} - \frac{2}{x^3}$$

is the function

$$F(x) = \frac{2}{3}x^{3/2} + \frac{1}{x^2}.$$

A great many functions of which we must find the antiderivative are not polynomials or simple powers of x. It is these functions which may cause more difficulty in the general process of antidifferentiation. The reader is advised to pay special attention to the following examples, for they illustrate a type of problem which *will* be found to be very important.

Example D

Find an antiderivative of the function $f(x) = 3(x^3 - 1)^2(3x^2)$.

Noting that we have a power of $x^3 - 1$ in the derivative, it is reasonable that the antiderivative may include a power of $x^3 - 1$. Since, in the derivative, $x^3 - 1$ is raised to the power 2, the antiderivative would then have $x^3 - 1$ raised to the power 3. Noting that the derivative of $(x^3 - 1)^3$ is $3(x^3 - 1)^2(3x^2)$, the desired antiderivative is

$$F(x) = (x^3 - 1)^3$$

▶ We note that the factor of $3x^2$ does not appear in the antiderivative, for it was included from the process of finding a derivative. Therefore, it must be present for $(x^3 - 1)^3$ to be the proper antiderivative, but it must also be excluded in the process of antidifferentiation.

Example E

Find an antiderivative of the function $f(x) = (2x + 1)^{1/2}$.

Here we note a power of $2x + 1$ in the derivative, which implies that the antiderivative has a power of $2x + 1$. Since in finding a derivative, 1 is subtracted from the power of $2x + 1$, we should add 1 in finding the antiderivative. Thus, we should have $(2x + 1)^{3/2}$ as part of the antiderivative. Finding a derivative of $(2x + 1)^{3/2}$, we obtain $\frac{3}{2}(2x + 1)^{1/2}(2) = 3(2x + 1)^{1/2}$. This differs from the given derivative by the factor of 3. If we write $(2x + 1)^{1/2} = \frac{1}{3}[3(2x + 1)^{1/2}]$, we have the required antiderivative as

$$F(x) = \tfrac{1}{3}(2x + 1)^{3/2}$$

Checking, the derivative of $\frac{1}{3}(2x + 1)^{3/2}$ is $\frac{1}{3}(\frac{3}{2})(2x + 1)^{1/2}(2) = (2x + 1)^{1/2}$.

Exercises 4-2

In the following exercises, find antiderivatives of the given functions.

1. $f(x) = 3x^2$
2. $f(x) = 5x^4$
3. $f(x) = 6x^5$
4. $f(x) = 10x^9$
5. $f(x) = 6x^3 + 1$
6. $f(x) = 12x^5 + 2x$
7. $f(x) = 2x^2 - x$
8. $f(x) = x^2 - 5$
9. $f(x) = \frac{5}{2}x^{3/2}$
10. $f(x) = \frac{4}{3}x^{1/3}$
11. $f(x) = 2\sqrt{x} + 3$
12. $f(x) = 3\sqrt[3]{x} - 4$
13. $f(x) = -\dfrac{1}{x^2}$
14. $f(x) = -\dfrac{7}{x^6}$
15. $f(x) = -\dfrac{6}{x^4}$
16. $f(x) = -\dfrac{8}{x^5}$
17. $f(x) = 2x^4 + 1$
18. $f(x) = 3x^3 - 5x^2 + 3$
19. $f(x) = 6x + \dfrac{1}{x^4}$
20. $f(x) = \dfrac{1}{2\sqrt{x}} + 4$
21. $f(x) = x^2 + 2 + x^{-2}$
22. $f(x) = x + 1 + x^{-3}$
23. $f(x) = 6(2x + 1)^5(2)$
24. $f(x) = 3(x^2 + 1)^2(2x)$
25. $f(x) = 4(x^2 - 1)^3(2x)$
26. $f(x) = 5(2x^4 + 1)^4(8x^3)$
27. $f(x) = x^3(2x^4 + 1)^4$
28. $f(x) = x(1 - x^2)^7$
29. $f(x) = \frac{3}{2}(6x + 1)^{1/2}(6)$
30. $f(x) = \frac{5}{4}(1 - x)^{1/4}(-1)$
31. $f(x) = (3x + 1)^{1/3}$
32. $f(x) = (4x + 3)^{1/2}$

4-3 The Indefinite Integral

In the previous section, in developing the basic technique of finding an antiderivative, we noted that the results given are not unique. That is, we could have added any constant to the answers and the result would still have been correct. Again, this is the case since the derivative of a constant is zero.

Example A

The derivatives of x^3, $x^3 + 4$, $x^3 - 7$, and $x^3 + 4\pi$ are all $3x^2$. This means that any of the functions listed, as well as others, would be a proper answer to the problem of finding an antiderivative of $3x^2$.

From Section 4–1, we know that the differential of a function $F(x)$ can be written as $d[F(x)] = F'(x)\,dx$. Therefore, since finding a differential of a function is very closely related to finding the derivative, so is the antiderivative very closely related to the process of finding the function for which the differential is known.

The notation used for finding the general form of the antiderivative, the **indefinite integral,** *is written in terms of the differential.* Thus, *the indefinite integral of a function f(x), for which dF(x)/dx = f(x), or dF(x) = f(x) dx, is defined as*

$$\int f(x)\,dx = F(x) + C \qquad (4\text{--}2)$$

Here f(x) is called the **integrand,** *F(x) + C is the indefinite integral, and C is an arbitrary constant, called the* **constant of integration.** It represents any of the constants which may be attached to an antiderivative to have a proper result. We must have additional information beyond a knowledge of the differential to assign a specific value to C. *The symbol ∫ is the* **integral sign,** *and indicates that the inverse of the differential is to be found. Determining the indefinite integral is called* **integration,** which we can see is essentially the same as finding an antiderivative.

Example B

In performing the integration

$$\int 5x^4\,dx = x^5 + C$$

we might think that the inclusion of this constant C would affect the derivative of the function x^5. However, the only effect of the C is to raise or lower the curve. The slope of $x^5 + 2$, $x^5 - 2$, or any function of the form $x^5 + C$ is the same for any given value of x. As Fig. 4–3 shows, tangents drawn to the curves are all parallel for the same value of x.

Figure 4–3

At this point we shall derive some basic formulas for integration. Since

$$\frac{d(cu)}{dx} = c\frac{du}{dx}$$

where u is a function of x and c is a constant, we can write

$$\int c\ du = c\int du = cu + C \qquad\qquad (4\text{--}3)$$

Also, since the derivative of a sum of functions equals the sum of the derivatives, we write

$$\int (du + dv) = u + v + C \qquad\qquad (4\text{--}4)$$

To find the derivative of a power of a function, we multiply by the power and subtract 1 from it. *To find the integral, we reverse this by adding 1 to the power of f(x) in f(x) dx, and dividing by this new power. The power formula for integration is therefore*

$$\int u^n\ du = \frac{u^{n+1}}{n+1} + C \qquad n \neq -1 \qquad\qquad (4\text{--}5)$$

When we use the formulas above, we must be able to recognize not only the proper form but also the component parts of the given formula. Unless you can do this and unless you have a good knowledge of differentiation, you will have trouble in applying Eq. (4–5). Most of the difficulty, if it exists, arises from improper identification of du. The following examples illustrate the use of the above formulas.

Example C

Integrate: $\int 6x\ dx$.

First we must identify u, n, du, and any multiplying constants. By noting that 6 is a multiplying constant, we identify x as u, which in turn means that dx must be du and $n = 1$. Hence,

Do not forget the constant of integration.

$$\int 6x\ dx = 6\int x^1 dx = 6\left(\frac{x^2}{2}\right) + C = 3x^2 + C$$

We see that our result checks since the differential of $3x^2 + C$ is $6x\ dx$.

Example D

Integrate: $\int (5x^3 - 6x^2 + 1)\, dx$.

In solving this problem, we have to use a combination of all three of the above formulas. Thus, we have

$$\int (5x^3 - 6x^2 + 1)\, dx = \int 5x^3\, dx + \int (-6x^2)\, dx + \int dx$$

$$= 5 \int x^3\, dx - 6 \int x^2\, dx + \int dx$$

In the first of these integrals, $u = x$, $n = 3$, and $du = dx$. In the second, $u = x$, $n = 2$, and $du = dx$. The third is a direct application of Eq. (4–3), with $c = 1$ and $du = dx$. Thus, we have

$$5 \int x^3\, dx - 6 \int x^2\, dx + \int dx = 5\left(\frac{x^4}{4}\right) - 6\left(\frac{x^3}{3}\right) + x + C$$

$$= \frac{5}{4}x^4 - 2x^3 + x + C$$

Example E

Integrate: $\int \left(\sqrt{x} - \dfrac{1}{x^3}\right) dx$.

We write

$$\int \left(\sqrt{x} - \frac{1}{x^3}\right) dx = \int x^{1/2}\, dx - \int x^{-3}\, dx$$

$$= \frac{1}{3/2}x^{3/2} - \frac{1}{-2}x^{-2} + C$$

$$= \frac{2}{3}x^{3/2} + \frac{1}{2}x^{-2} + C = \frac{2}{3}x^{3/2} + \frac{1}{2x^2} + C$$

Example F

Integrate: $\int (x^2 + 1)^3 (2x\, dx)$.

We first note that $n = 3$, for this is the power involved in the function being integrated. If $n = 3$, then $x^2 + 1$ must be u. If $u = x^2 + 1$, then $du = 2x\, dx$. Thus the integral is in proper form for integration *as it stands*. Using the power formula, we have

$$\int (x^2 + 1)^3 (2x\, dx) = \frac{(x^2 + 1)^4}{4} + C$$

It is in this type of problem that recognition of the proper form of a differential is very important. Only a good knowledge of differential forms, along with sufficient experience, makes ready recognition of this type possible. *It cannot be overemphasized that the entire quantity (2x dx) must be equated to du if we are to integrate properly.* Normally u and n are recognized first, and then the proper form of du is derived from u.

Showing the use of u directly, we can write the integration as

$$\int (x^2 + 1)^3 (2x\, dx) = \int u^3\, du = \tfrac{1}{4}u^4 + C = \frac{(x^2 + 1)^4}{4} + C$$

Example G

Integrate: $\int x^2 \sqrt{x^3 + 2}\ dx$.

We first note that $n = \frac{1}{2}$ and u is then $x^3 + 2$. Since $u = x^3 + 2$, $du = 3x^2\ dx$. In order to integrate properly, we must group the quantity $3x^2\ dx$ as du, with other factors isolated from this quantity. Since there is no 3 under the integral sign, we introduce one. In order not to change the numerical value, we also introduce $\frac{1}{3}$, normally before the integral sign. In this way we take full advantage of the fact that a constant (and only a constant) factor may be moved across the integral sign. The next form we should write is

$$\int x^2 \sqrt{x^3 + 2}\ dx = \frac{1}{3}\int 3x^2 \sqrt{x^3 + 2}\ dx = \frac{1}{3}\int \sqrt{x^3 + 2}(3x^2\ dx)$$

In this last form we indicate the proper grouping to result in the correct form of Eq. (4–5). Thus,

$$\int x^2 \sqrt{x^3 + 2}\ dx = \frac{1}{3}\cdot\frac{2}{3}(x^3 + 2)^{3/2} + C = \frac{2}{9}(x^3 + 2)^{3/2} + C$$

The $1/\frac{3}{2}$ was written as $\frac{2}{3}$, since this form is generally more convenient when we are working with fractions.

Showing the use of u directly, we can write the integral as

$$\int x^2 \sqrt{x^3 + 2}\ dx = \int (x^3 + 2)^{1/2}(x^2\ dx)$$

$$= \int u^{1/2}\left(\frac{1}{3}\ du\right) = \frac{1}{3}\int u^{1/2}\ du$$

$$= \frac{1}{3}\cdot\frac{2}{3}u^{3/2} + C = \frac{2}{9}u^{3/2} + C$$

$$= \frac{2}{9}(x^3 + 2)^{3/2} + C$$

We mentioned earlier that, in addition to knowing the differential, we need more information in order to find the constant of integration. Such information usually consists of a set of values which the function is known to satisfy. That is, a point through which the curve of the function passes would provide the necessary information.

Example H

Find y in terms of x, given that $dy/dx = 3x - 1$, and the curve passes through (1, 4).

When we write the equation as $dy = (3x - 1)\ dx$ and then indicate the integration, $\int dy = \int (3x - 1)\ dx$, the result of this integration is $y = \frac{3}{2}x^2 - x + C$. We know that the required curve passes through (1, 4), which means that the coordinates of this point must satisfy the equation. Thus, $4 = \frac{3}{2} - 1 + C$, or $C = \frac{7}{2}$. The complete solution is

$$y = \frac{3}{2}x^2 - x + \frac{7}{2} \qquad \text{or} \qquad 2y = 3x^2 - 2x + 7$$

Example I

At each point of a certain curve, the slope is given by $x\sqrt{2 - x^2}$. Given that the curve passes through $(1, 2)$, find its equation.

We write

$$\frac{dy}{dx} = x\sqrt{2 - x^2}, \qquad y = \int x\sqrt{2 - x^2}\, dx$$

To integrate this last expression, we recognize that $n = \frac{1}{2}$, $u = 2 - x^2$, and that therefore $du = -2x\, dx$. Hence, we need a -2 with the x and dx to form the proper du, which in turn means we place a $-\frac{1}{2}$ before the integral sign.

$$y = \int x\sqrt{2 - x^2}\, dx = -\frac{1}{2}\int (-2x)\sqrt{2 - x^2}\, dx$$

$$= -\frac{1}{2}\int (2 - x^2)^{1/2}(-2x\, dx) = -\frac{1}{2}\cdot\frac{2}{3}(2 - x^2)^{3/2} + C$$

$$= -\frac{1}{3}(2 - x^2)^{3/2} + C$$

Using the coordinates of the point $(1, 2)$, we have

$$2 = -\frac{1}{3}(2 - 1)^{3/2} + C \quad \text{or} \quad C = \frac{7}{3}$$

and

$$3y = 7 - (2 - x^2)^{3/2}$$

In this section we have discussed the integration of certain basic types of functions. There are many other methods used to integrate other functions, and some of these methods are discussed in Chapters 8 and 9. Also, there are many functions which cannot be integrated.

Exercises 4–3

In Exercises 1 through 32, integrate each of the given expressions.

1. $\displaystyle\int 2x\, dx$ 2. $\displaystyle\int 5x^4\, dx$ 3. $\displaystyle\int x^7\, dx$ 4. $\displaystyle\int x^5\, dx$

5. $\displaystyle\int x^{3/2}\, dx$ 6. $\displaystyle\int \sqrt[3]{x}\, dx$ 7. $\displaystyle\int x^{-4}\, dx$ 8. $\displaystyle\int \frac{1}{\sqrt{x}}\, dx$

9. $\displaystyle\int (x^2 - x^5)\, dx$ 10. $\displaystyle\int (1 - 3x)\, dx$ 11. $\displaystyle\int (9x^2 + x + 3)\, dx$

12. $\displaystyle\int (4x^2 - 2x - 5)\, dx$ 13. $\displaystyle\int \left(\frac{1}{x^3} + \frac{1}{2}\right) dx$ 14. $\displaystyle\int \left(4x - \frac{2}{x^3}\right) dx$

15. $\displaystyle\int \sqrt{x}\,(x^2 - x)\, dx$ 16. $\displaystyle\int (x\sqrt{x} - 5x^2)\, dx$ 17. $\displaystyle\int (2x^{-2/3} + 3^{-2})\, dx$

18. $\int (x^{1/3} + x^{1/5} + x^{-1/7})\, dx$ 19. $\int (1 + 2x)^2\, dx$

20. $\int (1 - x)^2\, dx$ 21. $\int (x^2 - 1)^5 (2x\, dx)$ 22. $\int (x^3 - 2)^6 (3x^2\, dx)$

23. $\int (x^4 + 3)^4 (4x^3\, dx)$ 24. $\int (1 - 2x)^{1/3} (-2\, dx)$ 25. $\int (x^5 + 4)^7 x^4\, dx$

26. $\int x^2 (1 - x^3)^{4/3}\, dx$ 27. $\int \sqrt{8x + 1}\, dx$ 28. $\int \sqrt[3]{4 - 3x}\, dx$

29. $\int \dfrac{x\, dx}{\sqrt{6x^2 + 1}}$ 30. $\int \dfrac{x^2\, dx}{\sqrt{2x^3 + 1}}$

31. $\int \dfrac{x - 1}{\sqrt{x^2 - 2x}}\, dx$ 32. $\int (x^2 - x)(x^3 - \tfrac{3}{2}x^2)^8\, dx$

In Exercises 33 through 36, find y in terms of x.

33. $\dfrac{dy}{dx} = 6x^2$,   curve passes through $(0, 2)$.

34. $\dfrac{dy}{dx} = x + 1$,   curve passes through $(-1, 4)$.

35. $\dfrac{dy}{dx} = x^2(1 - x^3)^5$,   curve passes through $(1, 5)$.

36. $\dfrac{dy}{dx} = 2x^3(x^4 - 6)^4$,   curve passes through $(2, 10)$.

In Exercises 37 through 40, find the required equations.

37. Find the equation of the curve whose slope is $-x\sqrt{1 - 4x^2}$ and which passes through $(0, 7)$.

38. Find the equation of the curve whose slope is $\sqrt{6x - 3}$ and which passes through $(2, -1)$.

39. Find the equation of the curve for which the second derivative is 6. The curve passes through $(1, 2)$ with a slope of 8.

40. Find the equation of the curve for which the second derivative is $12x^2$. The curve passes through $(1, 6)$ with a slope of 4.

4–4 The Area Under a Curve

Another basic problem which can be solved by integration is that of finding the area under a curve. In geometry, there are methods and formulas for finding the area of regular figures. By means of the calculus we will find it possible to find the area between curves for which we know the equations. First let us look at an example which illustrates the basic idea behind the method.

(a)

(b)

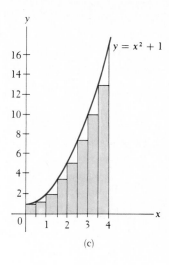

(c)

Figure 4–4

Example A

Approximate the area in the first quadrant and to the left of the line $x = 4$ under the parabola $y = x^2 + 1$. First make this approximation by inscribing two rectangles of equal width in the area and finding the sum of the areas of these rectangles. Then improve the approximation by repeating the process using eight rectangles.

The area to be approximated is shown in Fig. 4–4(a). The area with two rectangles inscribed under the curve is shown in Fig. 4–4(b). The first approximation, admittedly small, of the area can be found by adding the areas of the two rectangles. Both rectangles have a width of 2. The left rectangle is 1 unit high and the right rectangle is 5 units high. Thus, the area of the two rectangles is

$$A = 2(1 + 5) = 12$$

A much better approximation is found by inscribing the eight rectangles as shown in Fig. 4–4(c). Each of these rectangles has a width of $\frac{1}{2}$. The leftmost rectangle has a height of 1. The next has a height of $\frac{5}{4}$, which is determined by finding y for $x = \frac{1}{2}$. The next rectangle has a height of 2, which is found by evaluating y for $x = 1$. Finding the heights of all the rectangles and multiplying their sum by $\frac{1}{2}$ gives the area of the eight rectangles as

$$A = \frac{1}{2}\left(1 + \frac{5}{4} + 2 + \frac{13}{4} + 5 + \frac{29}{4} + 10 + \frac{53}{4}\right) = \frac{43}{2} = 21.5$$

Table 4–1

Number of rectangles n	Total area of rectangles
8	21.5
100	25.0144
1,000	25.301344
10,000	25.330134

An even better approximation could be obtained by inscribing more rectangles under the curve. The greater the number of rectangles, the more nearly the sum of their areas equals the area under the curve. See Table 4–1. By a method involving integration developed later in this section, we determine the *exact* area to be $\frac{76}{3} = 25\frac{1}{3}$.

We can now develop the basic method involved in finding an area under a curve. We consider the sum of the areas of rectangles inscribed under the curve as the number of rectangles is assumed to increase without bound. The reason for this last condition is that, as we saw in Example A, as the number of rectangles increases, the approximation of the area is better. Following is a specific example of this technique.

Example B

Figure 4–5

Find the area under the straight line $y = 2x$, above the x-axis, and to the left of the line $x = 4$.

We might mention at the beginning that this area can easily be found, since the figure is a triangle. However, the method we shall use here is the important concept at the moment. We first subdivide the interval from $x = 0$ to $x = 4$ into n inscribed rectangles of width Δx. Then the extremities of the intervals are labeled $a, x_1, x_2, \ldots, b \ (=x_n)$, as shown in Fig. 4–5, where

$$x_1 = \Delta x$$
$$x_2 = 2\,\Delta x$$
$$\vdots$$
$$x_{n-1} = (n-1)\,\Delta x$$
$$b = n\,\Delta x$$

We then find the area for each of the rectangles. These areas are (for the respective rectangles):

First $f(a)\,\Delta x$, where $f(a) = f(0) = 2(0) = 0$ is the height.

Second $f(x_1)\,\Delta x$, where $f(x_1) = 2(\Delta x) = 2\,\Delta x$ is the height.

Third $f(x_2)\,\Delta x$, where $f(x_2) = 2(2\,\Delta x) = 4\,\Delta x$ is the height.

Fourth $f(x_3)\,\Delta x$, where $f(x_3) = 2(3\,\Delta x) = 6\,\Delta x$ is the height.

$$\vdots$$

Last $f(x_{n-1})\,\Delta x$, where $f[(n-1)\,\Delta x] = 2(n-1)\,\Delta x$ is the height.

These areas are summed up as follows:

$$A_n = f(a)\,\Delta x + f(x_1)\,\Delta x + f(x_2)\,\Delta x + \cdots + f(x_{n-1})\,\Delta x$$
$$= 0 + 2\,\Delta x(\Delta x) + 4\,\Delta x(\Delta x) + \cdots + 2[(n-1)\,\Delta x]\,\Delta x$$
$$= 2(\Delta x)^2[1 + 2 + 3 + \cdots + (n-1)]$$

Now $b = n\,\Delta x$, or $4 = n\,\Delta x$, or $\Delta x = 4/n$. Thus,

$$A_n = 2\left(\frac{4}{n}\right)^2[1 + 2 + 3 + \cdots + (n-1)]$$

The sum $1 + 2 + 3 + \cdots + n - 1$ is the indicated sum of an arithmetic progression with the first term equal to 1 and the nth term equal to $n - 1$. This

(*continued on next page*)

sum is

$$s = \frac{n-1}{2}(1 + n - 1) = \frac{n(n-1)}{2} = \frac{n^2 - n}{2}$$

Now the expression for the sum of the areas can be set forth as

$$A_n = \frac{32}{n^2}\left(\frac{n^2 - n}{2}\right) = 16\left(1 - \frac{1}{n}\right)$$

This expression is an approximation of the actual area under consideration. The larger n becomes, the better the approximation. If we let $n \to \infty$ (which is equivalent to letting $\Delta x \to 0$), the limit of this sum will equal the area in question. Thus,

$$A = \lim_{n \to \infty} 16\left(1 - \frac{1}{n}\right) = 16$$

(This checks with the geometric result.) The area under the curve can be considered as the limit of the sum of the inscribed rectangles as the number of rectangles approaches infinity.

Figure 4–6

The method indicated in Example B illustrates the interpretation of finding an area as a summation process, although it should not be considered as a proof. However, we shall find that integration proves to be a much more useful method for finding an area. Let us now see how integration can be used directly.

Let ΔA represent the area *BCEG* under the curve, as indicated in Fig. 4–6. We see that the following inequality is true for the indicated areas:

$$A_{BCDG} < \Delta A < A_{BCEF}$$

If the point G is now designated as (x, y) and E as $(x + \Delta x, y + \Delta y)$, we have $y\,\Delta x < \Delta A < (y + \Delta y)\,\Delta x$. Dividing through by Δx, we have

$$y < \frac{\Delta A}{\Delta x} < y + \Delta y$$

Now we take the limit as $\Delta x \to 0$ (Δy then also approaches 0). This results in

$$\frac{dA}{dx} = y \tag{4–6}$$

This is true since the left member of the inequality is y and the right member approaches y. Also remember that

$$\lim_{\Delta x \to 0} \frac{\Delta A}{\Delta x} = \frac{dA}{dx}$$

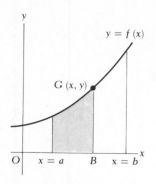

Figure 4–7

We shall now use Eq. (4–6) to show the method of finding the complete area under a curve. We now let $x = a$ be the left boundary of the desired area, and $x = b$ be the right boundary (Fig. 4–7). The area under the curve to the right of $x = a$ and bounded on the right by the line GB is now designated as A_{ax}. From Eq. (4–6) we have

$$dA_{ax} = [y\,dx]_a^x \quad \text{or} \quad A_{ax} = \left[\int y\,dx\right]_a^x$$

where $[\quad]_a^x$ is the notation used to indicate the boundaries of the area. Thus,

$$A_{ax} = \left[\int f(x)\,dx\right]_a^x = [F(x) + C]_a^x \tag{4–7}$$

But we know that if $x = a$, then $A_{aa} = 0$. Thus, $0 = F(a) + C$, or $C = -F(a)$. Therefore,

$$A_{ax} = \left[\int f(x)\,dx\right]_a^x = F(x) - F(a) \tag{4–8}$$

Now, to find the area under the curve that reaches from a to b, we write

$$A_{ab} = F(b) - F(a) \tag{4–9}$$

Thus the area under the curve that reaches from a to b is given by

$$A_{ab} = \left[\int f(x)\,dx\right]_a^b = F(b) - F(a) \tag{4–10}$$

Figure 4–8

This shows that *the area under the curve may be found by integrating the function f(x) to find the function F(x), which is then evaluated at each boundary value. The area is the difference between these values of F(x).* See Fig. 4–8.

In Example B, we found an area under a curve by finding the limit of the sum of the inscribed rectangles as the number of rectangles approaches infinity. Equation (4–10) expresses the area under a curve in terms of integration. We can now see that we have obtained the area by summation and also expressed it in terms of integration. *Thus, we conclude that summations can be evaluated by integration.* Also, we have grasped the connection between the problem of finding the slope of a tangent to a curve (differentiation) and the problem of finding an area (integration). We would not normally suspect that these two problems would have solutions which lead to reverse processes. We have also seen that the definition of integration has much more application than originally anticipated.

Example C

Figure 4-4(a)

Find the area under the curve $y = x^2 + 1$ between the y-axis and the line $x = 4$. This is the area of Example A shown in Fig. 4–4(a), which is given again here for reference.

In Eq. (4–10), we note that $f(x) = x^2 + 1$. This means that

$$\int (x^2 + 1)\, dx = \frac{1}{3}x^3 + x + C$$

Therefore, with $F(x) = \frac{1}{3}x^3 + x$, the area is given by

$$A_{0,4} = F(4) - F(0)$$

$$= \left[\frac{1}{3}(4^3) + 4\right] - \left[\frac{1}{3}(0^3) + 0\right] = \frac{1}{3}(64) + 4 = \frac{76}{3}$$

We note that this is about 4 square units greater than the value obtained by the approximation in Example A. This result means that the exact area is $25\frac{1}{3}$, as stated at the end of Example A.

Example D

Figure 4–9

Find the area under the curve $y = x^3$ between the lines $x = 1$ and $x = 2$.

In Eq. (4–10), $f(x) = x^3$. Therefore

$$\int x^3\, dx = \frac{1}{4}x^4 + C$$

$$A_{1,2} = F(2) - F(1) = \left[\frac{1}{4}(2^4)\right] - \left[\frac{1}{4}(1^4)\right]$$

$$= 4 - \frac{1}{4} = \frac{15}{4}$$

Therefore, the required area is $\frac{15}{4}$. Again, this is the exact area, not an approximation. See Fig. 4–9.

Example E

Figure 4–10

Find the area under the curve $y = 4 - x^2$ which lies in the first quadrant.

By solving the equation $4 - x^2 = 0$, we determine that the area to be found extends from $x = 0$ to $x = 2$ (see Fig. 4–10). Thus,

$$\int (4 - x^2)\, dx = 4x - \frac{x^3}{3} + C$$

$$A_{0,2} = \left(8 - \frac{8}{3}\right) - (0 - 0) = \frac{16}{3}$$

We can see from these examples that we do not have to include the constant of integration when we are finding areas. Any constant which might be added to $F(x)$ would cancel out when $F(a)$ is subtracted from $F(b)$.

Exercises 4–4

See Appendix C for a computer program for finding the area by summing the areas of n rectangles

In each of Exercises 1 through 8, find the approximate area under the given curves by dividing the indicated intervals into n subintervals and adding up the areas of the inscribed rectangles. There are two values of n for each, and therefore two approximations for each area. The height of each rectangle may be found by evaluating the function for the proper value of x. See Example A.

1. $y = 3x$, between $x = 0$ and $x = 3$, for (a) $n = 3$ ($\Delta x = 1$), (b) $n = 10$ ($\Delta x = 0.3$).

2. $y = 2x$, between $x = 0$ and $x = 2$, for (a) $n = 4$ ($\Delta x = 0.5$), (b) $n = 10$ ($\Delta x = 0.2$).

3. $y = x^2$, between $x = 0$ and $x = 2$, for (a) $n = 5$ ($\Delta x = 0.4$), (b) $n = 10$ ($\Delta x = 0.2$).

4. $y = x^2 + 2$, between $x = 0$ and $x = 3$, for (a) $n = 3$ ($\Delta x = 1$), (b) $n = 10$ ($\Delta x = 0.3$).

5. $y = 4x - x^2$, between $x = 1$ and $x = 4$, for (a) $n = 6$, (b) $n = 10$.

6. $y = 1 - x^2$, between $x = 0.5$ and $x = 1$, for (a) $n = 5$, (b) $n = 10$.

7. $y = \dfrac{1}{x^2}$, between $x = 1$ and $x = 5$, for (a) $n = 4$, (b) $n = 8$.

8. $y = \sqrt{x}$, between $x = 1$ and $x = 4$, for (a) $n = 3$, (b) $n = 12$.

In Exercises 9 through 16, find the exact area under the given curves between the indicated values of x. The functions are the same as those for which approximate areas were found in Exercises 1 through 8.

9. $y = 3x$, between $x = 0$ and $x = 3$.

10. $y = 2x$, between $x = 0$ and $x = 2$.

11. $y = x^2$, between $x = 0$ and $x = 2$.

12. $y = x^2 + 2$, between $x = 0$ and $x = 3$.

13. $y = 4x - x^2$, between $x = 1$ and $x = 4$.

14. $y = 1 - x^2$, between $x = 0.5$ and $x = 1$.

15. $y = \dfrac{1}{x^2}$, between $x = 1$ and $x = 5$.

16. $y = \sqrt{x}$, between $x = 1$ and $x = 4$.

4–5 The Definite Integral

Using reasoning similar to that in the preceding section, *we define the* **definite integral** *of a function* $f(x)$ *as*

$$\int_a^b f(x)\ dx = F(b) - F(a) \qquad\qquad (4\text{–}11)$$

where $F'(x) = f(x)$. We call this a *definite integral* because the final result of integrating and evaluating is a *number*. (*Indefinite* integrals are *functions*.) *The numbers a and b are called the* **lower limit** *and the* **upper limit,** *respectively. We can see that the value of a definite integral is found by evaluating the function (found by integration) at the upper limit and subtracting the value of this function at the lower limit.*

We know, from the analysis in the preceding section, that this definite integral can be interpreted as a summation process, where the size of the subdivision approaches a limit of zero. This fact explains the choice of the $\int$ symbol for integration: It is an elongated S, representing the sum. It is this interpretation of integration which we shall apply to many kinds of problems.

Example A

Evaluate the integral $\int_0^2 x^4\ dx$.
We write

$$\int_0^2 x^4\ dx = \frac{x^5}{5}\Big|_0^2 = \frac{2^5}{5} - 0 = \frac{32}{5}$$

Note that a vertical line—with the limits written at the top and the bottom—is the way the value is indicated after integration, but before evaluation.

Example B

Evaluate $\int_1^3 (x^{-2} - 1)\ dx$.
We set this forth as

Substitute upper limit   Subtract   Substitute lower limit

$$\int_1^3 (x^{-2} - 1)\ dx = -\frac{1}{x} - x\Big|_1^3 = \left(-\frac{1}{3} - 3\right) - (-1 - 1)$$

$$= -\frac{10}{3} + 2 = -\frac{4}{3}$$

Example C

Evaluate $\int_0^1 5x(x^2+1)^5 \, dx$.

For purposes of integration, $n=5$, $u=x^2+1$, and $du=2x\,dx$. Hence,

$$\int_0^1 5x(x^2+1)^5 \, dx = \frac{5}{2}\int_0^1 (x^2+1)^5 (2x\,dx)$$

$$= \frac{5}{2}\cdot\frac{1}{6}(x^2+1)^6\Big|_0^1$$

$$= \frac{5}{12}(2^6 - 1^6)$$

$$= \frac{5(63)}{12} = \frac{105}{4}$$

Example D

Evaluate $\int_{-1}^3 x(1-3x^2)^{1/3} \, dx$.

For purposes of integration, $n=\frac{1}{3}$, $u=1-3x^2$, and $du=-6x\,dx$. Thus

$$\int_{-1}^3 x(1-3x^2)^{1/3} \, dx = -\frac{1}{6}\cdot\frac{3}{4}(1-3x^2)^{4/3}\Big|_{-1}^3$$

$$= -\frac{1}{8}(-26)^{4/3} + \frac{1}{8}(-2)^{4/3}$$

$$= \frac{1}{8}(2\sqrt[3]{2} - 26\sqrt[3]{26})$$

$$= \frac{1}{4}(\sqrt[3]{2} - 13\sqrt[3]{26})$$

The result shown is the exact value. The approximate value to three decimal places (from a calculator) is -9.313.

Example E

Evaluate $\int_{0.1}^{2.7} \frac{dx}{\sqrt{4x+1}}$.

In order to integrate this we have $n=-\frac{1}{2}$, $u=4x+1$, and $du=4\,dx$. Therefore,

$$\int_{0.1}^{2.7} \frac{dx}{\sqrt{4x+1}} = \int_{0.1}^{2.7}(4x+1)^{-1/2}\,dx = \frac{1}{4}\int_{0.1}^{2.7}(4x+1)^{-1/2}(4\,dx)$$

$$= \frac{1}{4}\cdot\frac{1}{\frac{1}{2}}(4x+1)^{1/2}\Big|_{0.1}^{2.7} = \frac{1}{2}(4x+1)^{1/2}\Big|_{0.1}^{2.7}$$

$$= \frac{1}{2}(\sqrt{11.8} - \sqrt{1.4}) = 1.126$$

Example F   Evaluate

$$\int_0^4 \frac{x + 1}{(x^2 + 2x + 2)^3} \, dx$$

For purposes of integration

$$n = -3, \, u = x^2 + 2x + 2 \quad \text{and} \quad du = (2x + 2) \, dx$$

Therefore,

$$\int_0^4 (x^2 + 2x + 2)^{-3}(x + 1) \, dx = \frac{1}{2} \int_0^4 (x^2 + 2x + 2)^{-3}[2(x + 1) \, dx]$$

$$= \frac{1}{2} \cdot \frac{1}{-2}(x^2 + 2x + 2)^{-2} \Big|_0^4$$

$$= -\frac{1}{4}(16 + 8 + 2)^{-2} + \frac{1}{4}(0 + 0 + 2)^{-2}$$

$$= \frac{1}{4}\left(-\frac{1}{26^2} + \frac{1}{2^2}\right) = \frac{1}{4}\left(\frac{1}{4} - \frac{1}{676}\right)$$

$$= \frac{1}{4}\left(\frac{168}{676}\right) = \frac{21}{338}$$

It should be pointed out that the definition of the definite integral is valid regardless of the source of $f(x)$. That is, we may apply the definite integral whenever we want to sum a function in a manner similar to that which we use to find an area.

Exercises 4–5

In the following exercises, evaluate the given definite integrals.

1. $\int_0^1 2x \, dx$

2. $\int_0^2 3x^2 \, dx$

3. $\int_1^4 x^{5/2} \, dx$

4. $\int_4^9 (x^{3/2} - 1) \, dx$

5. $\int_3^6 \left(\frac{1}{\sqrt{x}} + 2\right) dx$

6. $\int_{1.2}^{1.6} \left(5 + \frac{6}{x^4}\right) dx$

7. $\int_{-1}^1 (1 - x)^{1/3} \, dx$

8. $\int_1^5 \sqrt{2x - 1} \, dx$

9. $\int_0^3 (x^4 - x^3 + x^2) \, dx$

10. $\int_1^2 (3x^5 - 2x^3) \, dx$

11. $\int_{0.5}^{2.2} (\sqrt[3]{x} - 2) \, dx$

12. $\int_{2.7}^{5.3} \left(\frac{1}{x\sqrt{x}} + 4\right) dx$

13. $\int_0^4 (1 - \sqrt{x})^2 \, dx$

14. $\int_1^4 \frac{x^2 + 1}{\sqrt{x}} \, dx$

15. $\int_1^2 2x(4 - x^2)^3 \, dx$

16. $\int_0^1 x(3x^2 - 1)^3 \, dx$

17. $\int_0^4 \frac{x \, dx}{\sqrt{x^2 + 9}}$

18. $\int_0^3 x^2(x^3 + 2)^{3/2} \, dx$

19. $\int_{2.75}^{3.25} \dfrac{dx}{\sqrt[3]{6x+1}}$ 20. $\int_{2}^{6} \dfrac{2\,dx}{\sqrt{4x+1}}$ 21. $\int_{1}^{3} \dfrac{2x\,dx}{(2x^2+1)^3}$

22. $\int_{12.6}^{17.2} \dfrac{3\,dx}{(6x-1)^2}$ 23. $\int_{3}^{7} 3\sqrt{4x-3}\,dx$ 24. $\int_{-5}^{1} \sqrt{6-2x}\,dx$

25. $\int_{0}^{2} 2x(9-2x^2)^2\,dx$ 26. $\int_{-1}^{0} x^3(1-2x^4)^3\,dx$

27. $\int_{0}^{1} (x^2+3)(x^3+9x+6)\,dx$ 28. $\int_{2}^{3} \dfrac{x^2+1}{(x^3+3x)^2}\,dx$

4–6 Numerical Integration: The Trapezoidal Rule

For data and functions which cannot be directly integrated by available methods, it is possible to develop numerical methods of integration. These numerical methods are of greater importance today since they are readily adaptable for use on a calculator or computer. There are a great many such numerical techniques for approximating the value of an integral. In this section we shall develop one of these, the trapezoidal rule. In the following section, another numerical method is discussed.

We know from Sections 4–4 and 4–5 that we can interpret a definite integral as the area under a curve. We shall therefore show how we can approximate the value of the integral by approximating the appropriate area by a set of inscribed trapezoids. The basic idea here is very similar to that used when rectangles were inscribed under a curve. However, the use of trapezoids reduces the error and provides a better approximation.

The area to be found is subdivided into n intervals of equal width. Perpendicular lines are then dropped from the curve (or points, if only a given set of numbers is available). If the points on the curve are joined by straight-line segments, the area of successive parts under the curve is then approximated by finding the area of each of the trapezoids formed. However, if these points are not too far apart, the approximation will be very good (see Fig. 4–11). From

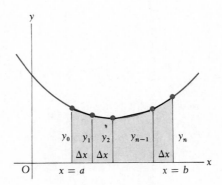

Figure 4–11

geometry we recall that the area of a trapezoid equals one-half the product of the sum of the bases times the altitude. For these trapezoids the bases are the y-coordinates and the altitudes are Δx. When we thus indicate the sum of these trapezoidal areas, we have

$$A_T = \frac{1}{2}(y_0 + y_1)\,\Delta x + \frac{1}{2}(y_1 + y_2)\,\Delta x + \frac{1}{2}(y_2 + y_3)\,\Delta x + \cdots$$
$$+ \frac{1}{2}(y_{n-2} + y_{n-1})\,\Delta x + \frac{1}{2}(y_{n-1} + y_n)\,\Delta x$$

We note, when this addition is performed, that the result is

$$A_T = \left(\frac{1}{2}y_0 + y_1 + y_2 + \cdots + y_{n-1} + \frac{1}{2}y_n\right)\Delta x \qquad \textbf{(4–12)}$$

The y-values to be used are either derived from the function as $y_1 = f(x_1)$, or are the y-coordinates of a set of data.

Since A_T approximates the area under the curve, it also approximates the value of the definite integral, or

$$\int_a^b f(x)\,dx \approx \left(\frac{1}{2}y_0 + y_1 + y_2 + \cdots + y_{n-1} + \frac{1}{2}y_n\right)\Delta x \qquad \textbf{(4–13)}$$

Equation (4–13) is known as the **trapezoidal rule.**

Example A

Approximate the value of $\int_1^3 \frac{1}{x}\,dx$ by the trapezoidal rule. Let $n = 4$.

We are to approximate the area under $y = 1/x$ from $x = 1$ to $x = 3$ by dividing the area into 4 trapezoids. This area is found by applying Eq. (4–12), which is the approximate value of the integral, as shown in Eq. (4–13). Figure 4–12 shows the graph. In this example, $f(x) = 1/x$, and

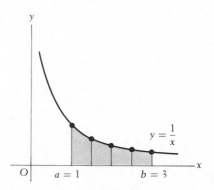

Figure 4–12

$$\Delta x = \frac{3-1}{4} = \frac{1}{2}, \qquad y_0 = f(a) = f(1) = 1$$

$$y_1 = f\left(\frac{3}{2}\right) = \frac{2}{3}, \qquad y_2 = f(2) = \frac{1}{2}$$

$$y_3 = f\left(\frac{5}{2}\right) = \frac{2}{5} \qquad y_n = y_4 = f(b) = f(3) = \frac{1}{3}$$

And so we write

$$A_T = \left[\frac{1}{2}(1) + \frac{2}{3} + \frac{1}{2} + \frac{2}{5} + \frac{1}{2}\left(\frac{1}{3}\right)\right]\frac{1}{2}$$

$$= \left(\frac{15 + 20 + 15 + 12 + 5}{30}\right)\frac{1}{2} = \frac{67}{30}\left(\frac{1}{2}\right) = \frac{67}{60}$$

Therefore,

$$\int_1^3 \frac{1}{x}\, dx \approx \frac{67}{60}$$

Table 4–2

Number of trapezoids n	Total area of trapezoids
4	1.1166667
100	1.0986419
1,000	1.0986126
10,000	1.0986123

Note that we cannot perform this integration directly by methods developed up to this point. As we increase the number of trapezoids, the value becomes more accurate. See Table 4–2. The actual value to 7 decimal places is 1.0986123.

Example B

Approximate the value of $\int_0^1 \sqrt{x^2 + 1}\, dx$ by the trapezoidal rule. Let $n = 5$. Figure 4–13 shows the graph. In this example,

$$\Delta x = \frac{1-0}{5} = 0.2$$

Figure 4–13

Therefore,

$y_0 = f(0) = 1$ $y_1 = f(0.2) = \sqrt{1.04} = 1.0198039$

$y_2 = f(0.4) = \sqrt{1.16} = 1.0770330$ $y_3 = f(0.6) = \sqrt{1.36} = 1.1661904$

$y_4 = f(0.8) = \sqrt{1.64} = 1.2806248$ $y_5 = f(1) = \sqrt{2.00} = 1.4142136$

Hence we have

$$A_T = \left[\frac{1}{2}(1) + 1.0198039 + 1.0770330 + 1.1661904\right.$$

$$\left. + 1.2806248 + \frac{1}{2}(1.4142136)\right](0.2)$$

$$= 1.150 \quad \text{(rounded off)}$$

This means that

$$\int_0^1 \sqrt{x^2 + 1}\, dx \approx 1.150$$

(The actual value is 1.148 to three decimal places.) The above values of y_0, y_1, and so on do not have to be tabulated when using a calculator. The entire calculation can be done directly on a calculator.

Example C

Figure 4–14

Approximate the value $\int_2^3 x\sqrt{x+1}\ dx$ by use of the trapezoidal rule. Use $n = 10$.

In Fig. 4–14 the graph of the function and the area used in the trapezoidal rule are shown. From the given values we have $\Delta x = \frac{3-2}{10} = 0.1$. Therefore,

$$y_0 = f(2) = 2\sqrt{3} = 3.4641016$$
$$y_1 = f(2.1) = 2.1\sqrt{3.1} = 3.6974315$$
$$y_2 = f(2.2) = 2.2\sqrt{3.2} = 3.9354796$$
$$y_3 = f(2.3) = 2.3\sqrt{3.3} = 4.1781575$$
$$y_4 = f(2.4) = 2.4\sqrt{3.4} = 4.4253813$$
$$y_5 = f(2.5) = 2.5\sqrt{3.5} = 4.6770717$$
$$y_6 = f(2.6) = 2.6\sqrt{3.6} = 4.9331532$$
$$y_7 = f(2.7) = 2.7\sqrt{3.7} = 5.1935537$$
$$y_8 = f(2.8) = 2.8\sqrt{3.8} = 5.4582048$$
$$y_9 = f(2.9) = 2.9\sqrt{3.9} = 5.7270411$$
$$y_{10} = f(3) = 3\sqrt{4} = 6.0000000$$

$$A_T = \left[\frac{1}{2}(3.4641016) + 3.6974315 + \cdots + 5.7270411\right.$$

$$\left. + \frac{1}{2}(6.0000000)\right](0.1)$$

$$= 4.6958$$

Therefore, $\int_2^3 x\sqrt{x+1}\ dx \approx 4.6958$. (The actual value, to five significant digits, is 4.6954.) Again, the entire calculation can be done directly on a calculator.

Example D

Figure 4–15

The following points were found empirically.

x	0	1	2	3	4	5
y	5.68	6.75	7.32	7.35	6.88	6.24

Approximate the value of the integral of the function defined by these points between $x = 0$ and $x = 5$ by the trapezoidal rule.

In order to find A_T, we use the values of y_0, y_1, etc. directly from the table. We also note that $\Delta x = 1$. The graph is shown in Fig. 4–15. Therefore, we have

$$A_T = \left[\frac{1}{2}(5.68) + 6.75 + 7.32 + 7.35 + 6.88 + \frac{1}{2}(6.24)\right] = 34.26$$

Although we do not know the algebraic form of the function, we can state that

$$\int_0^5 f(x)\ dx \approx 34.26$$

See Appendix C for a computer program for finding areas by use of the trapezoidal rule.

Exercises 4-6

In Exercises 1 through 4, (a) approximate the value of each of the given integrals by use of the trapezoidal rule, using the given value of n, and (b) check by direct integration.

1. $\int_0^2 2x^2\, dx$, $n = 4$

2. $\int_0^1 (1 - x^2)\, dx$, $n = 3$

3. $\int_1^4 (1 + \sqrt{x})\, dx$, $n = 6$

4. $\int_3^8 \sqrt{1 + x}\, dx$, $n = 5$

In Exercises 5 through 12, approximate each of the given integrals by use of the trapezoidal rule, using the given value of n.

5. $\int_2^3 \frac{1}{2x}\, dx$, $n = 2$

6. $\int_2^4 \frac{dx}{x + 3}$, $n = 4$

7. $\int_0^2 \sqrt{4 - x^2}\, dx$, $n = 4$

8. $\int_0^2 \sqrt{x^3 + 1}\, dx$, $n = 4$

9. $\int_1^5 \frac{1}{x^2 + x}\, dx$, $n = 10$

10. $\int_2^4 \frac{1}{x^2 + 1}\, dx$, $n = 10$

11. $\int_0^4 2^x\, dx$, $n = 12$

12. $\int_0^{1.5} 10^x\, dx$, $n = 15$

In Exercises 13 through 16 approximate the value of the integrals defined by the given sets of points.

13.
x	3	4	5	6	7
y	1.08	1.65	3.23	3.67	2.97

14.
x	2	4	6	8	10	12	14
y	0.67	2.34	4.56	3.67	3.56	4.78	6.87

15.
x	1.4	1.7	2.0	2.3	2.6	2.9	3.2
y	0.18	7.87	18.23	23.53	24.62	20.93	20.76

16.
x	0.4	0.8	1.2	1.6	2.0	2.4
y	1.24	3.54	5.52	4.40	3.63	3.88

4-7 Simpson's Rule

The numerical method of integration developed in this section is also readily programmable for use on a computer, or easily useable such that the necessary calculations can be done on a calculator. It is obtained by interpreting the definite integral as the area under a curve, as we did in developing the trapezoidal rule, and by approximating the curve by a set of parabolic arcs. The use of parabolic arcs, rather than chords as with the trapezoidal rule, usually gives a better approximation.

Figure 4–16

Since we will be using parabolic arcs, we first derive a formula for the area under a parabolic arc. The curve in Fig. 4–16 represents the parabola $y = ax^2 + bx + c$. The indicated points on the curve are $(-h, y_0)$, $(0, y_1)$ and (h, y_2). The area under the parabola is given by

$$A = \int_{-h}^{h} y \, dx = \int_{-h}^{h} (ax^2 + bx + c) \, dx = \frac{ax^3}{3} + \frac{bx^2}{2} + cx \Big|_{-h}^{h}$$

$$= \frac{2}{3} ah^3 + 2ch$$

or

$$A = \frac{h}{3}(2ah^2 + 6c) \qquad (4\text{--}14)$$

The coordinates of the three points also satisfy the equation $y = ax^2 + bx + c$. This means that

$$y_0 = ah^2 - bh + c$$
$$y_1 = c$$
$$y_2 = ah^2 + bh + c$$

By finding the sum of $y_0 + 4y_1 + y_2$, we have

$$y_0 + 4y_1 + y_2 = 2ah^2 + 6c \qquad (4\text{--}15)$$

Substituting Eq. (4–15) into Eq. (4–14), we have

$$A = \frac{h}{3}(y_0 + 4y_1 + y_2) \qquad (4\text{--}16)$$

We note that the area depends only on the distance h and the three y-coordinates.

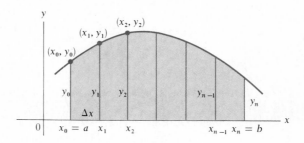

Figure 4–17

Now let us consider the area under the curve in Fig. 4–17. If a parabolic arc is passed through the points (x_0, y_0), (x_1, y_1), and (x_2, y_2) we may use Eq. (4–16) to approximate the area under the curve between x_0 and x_2. We also note that the distance h used in finding Eq. (4–16) is the difference in the x-coordinates, or

$h = \Delta x$. Therefore, the area under the curve between x_0 and x_2 is

$$A_1 = \frac{\Delta x}{3}(y_0 + 4y_1 + y_2)$$

Similarly, if a parabolic arc is passed through the three points starting with (x_2, y_2), the area between x_2 and x_4 is

$$A_2 = \frac{\Delta x}{3}(y_2 + 4y_3 + y_4)$$

The sum of these areas is

$$A_1 + A_2 = \frac{\Delta x}{3}(y_0 + 4y_1 + 2y_2 + 4y_3 + y_4) \qquad (4\text{--}17)$$

We can continue this procedure until the approximate value of the entire area has been found. We must note, however, that the number of intervals n of width Δx must be even. Therefore, generalizing on Eq. (4–17) and recalling again that the value of the definite integral is the area under the curve, we have the following result

$$\int_a^b f(x)\ dx = \frac{\Delta x}{3}(y_0 + 4y_1 + 2y_2 + 4y_3 + 2y_4 + \cdots + 4y_{n-1} + y_n) \qquad (4\text{--}18)$$

Equation (4–18) is known as **Simpson's rule.**

Example A

Approximate the value of $\int_0^1 \frac{dx}{x + 1}$ by Simpson's rule. Let $n = 2$.

We are to approximate the integral by use of Eq. (4–18). We therefore note that $f(x) = 1/(x + 1)$. Also, $x_0 = a = 0$, $x_1 = 0.5$, and $x_2 = b = 1$. This is due to the fact that $n = 2$ and $\Delta x = 0.5$ since the total interval is 1 unit (from $x = 0$ to $x = 1$). Therefore,

$$y_0 = \frac{1}{0 + 1} = 1.0000, \quad y_1 = \frac{1}{0.5 + 1} = 0.6667, \quad y_2 = \frac{1}{1 + 1} = 0.5000$$

Substituting, we have

$$\int_0^1 \frac{dx}{x + 1} = \frac{0.5}{3}[1.0000 + 4(0.6667) + 0.5000]$$
$$= 0.694$$

To three decimal places, the actual value of the integral is 0.693. The method of integrating this function will be considered in a later chapter.

Example B

Approximate the value of $\int_2^3 x\sqrt{x + 1}\ dx$ by Simpson's rule. Let $n = 10$.

Since the necessary values for this function are shown in Example C of Section 4–6, we shall simply tabulate them here. ($\Delta x = 0.1$.)

$$y_0 = 3.4641016 \qquad y_1 = 3.6974315 \qquad y_2 = 3.9354796$$
$$y_3 = 4.1781575 \qquad y_4 = 4.4253813 \qquad y_5 = 4.6770717$$
$$y_6 = 4.9331532 \qquad y_7 = 5.1935537 \qquad y_8 = 5.4582048$$
$$y_9 = 5.7270411 \qquad y_{10} = 6.0000000$$

Therefore, we evaluate the integral as follows.

$$
\begin{aligned}
\int_2^3 x\sqrt{x + 1}\ dx = \frac{0.1}{3}[&3.4641016 + 4(3.6974315) + 2(3.9354796) \\
&+ 4(4.1781575) + 2(4.4253813) + 4(4.6770717) \\
&+ 2(4.9331532) + 4(5.1935537) + 2(5.4582048) \\
&+ 4(5.7270411) + 6.000000] \\
= \frac{0.1}{3}&(140.86156) = 4.6953854
\end{aligned}
$$

This result agrees with the actual value to the eight significant digits shown. The value we obtained with the trapezoidal rule was 4.6958.

Exercises 4–7

In Exercises 1 through 4, (a) approximate the value of each of the given integrals by use of Simpson's rule, using the given value of n, and (b) check by direct integration.

1. $\int_0^2 (1 + x^3)\ dx,\ n = 2$ 2. $\int_0^8 x^{1/3}\ dx,\ n = 2$

3. $\int_1^4 (2x + \sqrt{x})\ dx,\ n = 6$ 4. $\int_0^2 x\sqrt{x^2 + 1}\ dx,\ n = 4$

In Exercises 5 through 10, approximate each of the given integrals by use of Simpson's rule, using the given values of n. These are the same as Exercises 5 through 10 of Section 4–6.

5. $\int_2^3 \frac{1}{2x}\ dx,\ n = 2$ 6. $\int_2^4 \frac{dx}{x + 3},\ n = 4$ 7. $\int_0^2 \sqrt{4 - x^2}\ dx,\ n = 4$

8. $\int_0^2 \sqrt{x^3 + 1}\ dx,\ n = 4$ 9. $\int_1^5 \frac{1}{x^2 + x}\ dx,\ n = 10$ 10. $\int_2^4 \frac{1}{x^2 + 1}\ dx,\ n = 10$

In Exercises 11 and 12, approximate the value of the integrals defined by the given set of points. These are the same as Exercises 13 and 15 of Section 4–6.

11.

x	3	4	5	6	7
y	1.08	1.65	3.23	3.67	2.97

12.

x	1.4	1.7	2.0	2.3	2.6	2.9	3.2
y	0.18	7.87	18.23	23.53	24.62	20.93	20.76

4–8 Review Exercises for Chapter 4

In Exercises 1 through 24, evaluate the given integrals.

1. $\displaystyle \int (4x^3 - x)\, dx$

2. $\displaystyle \int (5 - 3x^2)\, dx$

3. $\displaystyle \int 2(x - x^{3/2})\, dx$

4. $\displaystyle \int x(x - 3x^4)\, dx$

5. $\displaystyle \int_1^4 \left(\sqrt{x} + \frac{1}{\sqrt{x}}\right) dx$

6. $\displaystyle \int_1^2 \left(x + \frac{1}{x^2}\right) dx$

7. $\displaystyle \int_0^2 x(4 - x)\, dx$

8. $\displaystyle \int_0^1 x(x^3 + 1)\, dx$

9. $\displaystyle \int \left(3 + \frac{2}{x^3}\right) dx$

10. $\displaystyle \int \left(3\sqrt{x} + \frac{1}{2\sqrt{x}} - \frac{1}{4}\right) dx$

11. $\displaystyle \int_{-1.5}^{0.5} \frac{dx}{\sqrt{1 - x}}$

12. $\displaystyle \int_{0.35}^{0.85} x(\sqrt{1 - x^2} + 1)\, dx$

13. $\displaystyle \int \frac{dx}{(2 - 5x)^2}$

14. $\displaystyle \int x(1 - 2x^2)^5\, dx$

15. $\displaystyle \int (7 - 2x)^{3/4}\, dx$

16. $\displaystyle \int (3x - 1)^{1/3}\, dx$

17. $\displaystyle \int_0^2 \frac{3x\, dx}{\sqrt[3]{1 + 2x^2}}$

18. $\displaystyle \int_1^6 \frac{2\, dx}{(3x - 2)^{3/4}}$

19. $\displaystyle \int x^2(1 - 2x^3)^4\, dx$

20. $\displaystyle \int 3x^3(1 - 5x^4)^{1/3}\, dx$

21. $\displaystyle \int \frac{(2 - 3x^2)\, dx}{(2x - x^3)^2}$

22. $\displaystyle \int \frac{x^2 - 3}{\sqrt{6 + 9x - x^3}}\, dx$

23. $\displaystyle \int_1^3 (x^2 + x + 2)(2x^3 + 3x^2 + 12x)\, dx$

24. $\displaystyle \int_0^2 (4x + 18x^2)(x^2 + 3x^3)^2\, dx$

In Exercises 25 through 28, find the differential of each of the given functions.

25. $\displaystyle y = \frac{1}{(x^2 - 1)^3}$

26. $\displaystyle y = \frac{1}{(2x - 1)^2}$

27. $y = x\sqrt[3]{1 - 3x}$

28. $\displaystyle y = \frac{3 + x}{4 - x^2}$

In Exercises 29 and 30, evaluate $\Delta y - dy$ for the given functions and values.

29. $y = x^3$, $x = 2$, $\Delta x = 0.1$

30. $y = 6x^2 - x$, $x = 3$, $\Delta x = 0.2$

In Exercises 31 and 32, find the required equations.

31. Find the equation of the curve which passes through $(-1, 3)$ for which the slope is given by $3 - x^2$.

32. Find the equation of the curve which passes through $(1, -2)$ for which the slope is $x(x^2 + 1)^2$.

In Exercises 33 and 34, perform the indicated integrations as directed.

33. Perform the integration $\int (1 - 2x)\, dx$ (a) term by term, labeling the constant of integration as C_1, and then (b) by letting $u = 1 - 2x$, using the general power rule, and labeling the constant of integration as C_2. Compare C_1 with C_2.

34. Following the methods (a) and (b) in Exercise 33, perform the integration $\int (3x + 2)\, dx$. In (b) let $u = 3x + 2$. Compare the constants of integration C_1 and C_2.

In Exercises 35 and 36, use Eq. (4–10) to find the indicated areas.

35. The area under $y = 6x - 1$ between $x = 1$ and $x = 3$

36. The first quadrant area under $y = 8x - x^4$

In Exercises 37 and 38, solve the given problems by the trapezoidal rule.

37. Approximate $\displaystyle \int_1^3 \frac{dx}{2x - 1}$ with $n = 4$.

38. Approximate the value of the integral defined by the following set of points.

x	6.0	9.0	12	15	18	21
y	2.0	1.2	0.2	1.0	6.0	12

In Exercises 39 and 40, solve the given problems by Simpson's rule.

39. Approximate $\displaystyle \int_1^3 \frac{dx}{2x - 1}$ with $n = 4$ (see Exercise 37).

40. Approximate the value of the integral of the function defined by the following points.

x	1.0	1.4	1.8	2.2	2.6	3.0	3.4
y	1.45	1.89	2.66	3.50	3.22	3.04	2.44

In Exercises 41 and 42, use the function $y = \dfrac{x}{x^2 + 2}$ and approximate the area under the curve in the first quadrant to the left of the line $x = 5$ by the indicated method.

41. Inscribe five rectangles and find the sum of the areas of the rectangles.

42. Use the trapezoidal rule with $n = 5$.

In Exercises 43 and 44, use the function $y = x\sqrt{x^3 + 1}$ and approximate the area under the curve between $x = 1$ and $x = 3$ by the indicated method.

43. Use Simpson's rule with $n = 10$.

44. Inscribe ten rectangles, and find the sum of the areas of the rectangles.

In Exercises 45 through 48, solve the given problems by finding the appropriate differentials.

45. A ball bearing 4.00 mm in radius is coated with a special metal of thickness 0.02 mm. Approximately what volume of the special metal is used?

46. The electric resistance of a certain resistor as a function of temperature is given by $R = 12.00 + 0.06T + 0.0001T^2$. Approximately by how much does the resistance (in ohms) change as T changes from 100°C to 102°C?

47. The impedance Z of an electric circuit as a function of the resistance R and the reactance X is given by $Z = \sqrt{R^2 + X^2}$. Derive an expression for the relative error in impedance for an error in R and given value of X.

48. Show that the relative error of the nth root of a given measurement equals approximately $1/n$ of the relative error of the measurement.

CHAPTER 5

Applications of Integration

The applications of integration in engineering and technology are numerous. In the first section of this chapter, we shall present some of the applications of the indefinite integral. In the remaining sections, we shall present applications of the definite integral in geometry and certain technical areas.

5–1 Applications of the Indefinite Integral

In the examples of this section we shall show two basic applications of the indefinite integral. Some other applications are shown in the exercises.

The first of these applications deals with velocity and acceleration. The concepts of velocity as a first derivative and acceleration as a second derivative were introduced in Chapters 2 and 3. Here we shall apply integration to the problem of finding the distance as a function of time, when we know the relationship between acceleration and time, as well as certain specific values of distance and velocity. These latter values are necessary for determining the values of the constants of integration which are introduced. Recalling now that the acceleration a of an object is given by $a = dv/dt$, we can find the expression for the velocity in terms of a, t, and the constant of integration. We write

$$dv = a \, dt$$

or

$$v = \int a \, dt \qquad (5-1)$$

If the acceleration is constant, we have

$$v = at + C_1 \tag{5-2}$$

Of course, Eq. (5–1) can be used in general to find the velocity as a function of time so long as we know the acceleration as a function of time. However, since the case of constant acceleration is often encountered, Eq. (5–2) is often encountered. If the velocity is known for some specified time, the constant C_1 may be evaluated.

Example A

Find the expression for the velocity if $a = 12t$, given that $v = 8$ when $t = 1$.
  Using Eq. (5–1), we have

$$v = \int (12t)\, dt = 6t^2 + C_1$$

Substituting the known values, we obtain

$$8 = 6 + C_1 \quad \text{or} \quad C_1 = 2$$

Thus, $v = 6t^2 + 2$.

Example B

For an object falling under the influence of gravity, the acceleration due to gravity is essentially constant. Its value is -32 ft/s². (The minus sign is chosen so that all quantities directed up are positive, and all quantities directed down are negative.) Find the expression for the velocity of an object under the influence of gravity if $v = v_0$ when $t = 0$.
  We write

$$v = \int (-32)\, dt = -32t + C_1, \qquad v_0 = 0 + C_1, \qquad v = v_0 - 32t$$

The velocity v_0 is called the *initial velocity*. If the object is given an initial upward velocity of 100 ft/s, $v_0 = 100$ ft/s. If the object is dropped, $v_0 = 0$. If the object is given an initial downward velocity of 100 ft/s, $v_0 = -100$ ft/s.

Once we obtain the expression for velocity, we can integrate to find the expression for displacement in terms of the time. Since $v = ds/dt$, we can write $ds = v\, dt$, or

$$s = \int v\, dt \tag{5-3}$$

Example C

Find the expression for displacement in terms of time, if $a = 6t^2$, $v = 0$ when $t = 2$, and $s = 4$ when $t = 0$.

$$v = \int 6t^2 \, dt = 2t^3 + C_1; \qquad 0 = 2(2^3) + C_1, \qquad C_1 = -16$$

$$v = 2t^3 - 16$$

$$s = \int (2t^3 - 16) \, dt = \frac{1}{2}t^4 - 16t + C_2; \qquad 4 = 0 - 0 + C_2, \qquad C_2 = 4$$

$$s = \frac{1}{2}t^4 - 16t + 4$$

Example D

Find the expression for the distance above the ground of an object, given a vertical velocity of v_0 from the ground.

From Example B, we know that $v = v_0 - 32t$. In this problem we know that $s = 0$ when $t = 0$ (given velocity v_0 *from the ground*) if distances are measured from ground level. Therefore,

$$s = \int (v_0 - 32t) \, dt = v_0 t - 16t^2 + C_2; \qquad 0 = 0 - 0 + C_2, \qquad C_2 = 0$$

$$s = v_0 t - 16t^2$$

Example E

40 ft/s

Time of flight = 5s

200 ft

An object is thrown vertically from the top of a building 200 ft high, and hits the ground 5 s later. What initial velocity was the object given?

Measuring vertical distances from the ground, we know that $s = 200$ ft when $t = 0$. Also, we know that $v = v_0 - 32t$. Thus,

$$s = \int (v_0 - 32t) \, dt = v_0 t - 16t^2 + C$$

$$200 = v_0(0) - 16(0) + C, \qquad C = 200$$

$$s = v_0 t - 16t^2 + 200$$

We also know that $s = 0$ when $t = 5$ s. Thus,

$$0 = v_0(5) - 16(5^2) + 200$$

$$5v_0 = 200$$

$$v_0 = 40 \text{ ft/s}$$

Figure 5–1

This means that the initial velocity was 40 ft/s upward. See Fig. 5–1.

The second basic application of the indefinite integral which we shall discuss comes from the field of electricity. By definition, *the current i in an electric circuit equals the time rate of change of the charge q (in coulombs) which passes a given point in the circuit, or*

$$i = \frac{dq}{dt} \tag{5-4}$$

Rewriting this expression in differential notation as $dq = i\,dt$ and integrating both sides of the equation, we have

$$q = \int i\,dt \tag{5-5}$$

Now, the voltage V_C across a capacitor C is given by $V_C = q/C$. By combining equations, the voltage V_C is given by

$$V_C = \frac{1}{C} \int i\,dt \tag{5-6}$$

Here, V_C is measured in volts, C in farads, i in amperes, and t in seconds.

Example F

The current in a certain electric circuit as a function of time is given by $i = 6t^2 + 4$. Find an expression for the amount of charge which passes a point in the circuit as a function of time. Assuming that $q = 0$ when $t = 0$, determine the total charge which passes the point in 2 s.

Since $q = \int i\,dt$, we have

$$q = \int (6t^2 + 4)\,dt$$

$$= 2t^3 + 4t + C$$

This last expression is the desired expression giving charge as a function of time. We note that when $t = 0$, then $q = C$, which means that the constant of integration represents the initial charge, or the charge which passed a given point before we started timing. Using q_0 to represent this charge, we have

$$q = 2t^3 + 4t + q_0$$

Now, returning to the second part of the problem, we see that $q_0 = 0$. Therefore, evaluating q for $t = 2\,s$, we have

$$q = 2(8) + 4(2) = 24 \text{ C}$$

(Here, the symbol C represents coulombs, and is not the C for capacitance of Eq. (5-6) or the constant of integration.) This is the charge which passes any specified point in the circuit in 2 s.

Example G

The voltage across a 5.0-μF capacitor is zero. What is the voltage after 20 ms if a current of 75 mA charges the capacitor?

From Eq. (5–6) we have

$$V_C = \frac{1}{C} \int i \, dt$$

 In substituting, we must use the proper power of 10 which corresponds to each prefix which is used. Since 5.0 μF = 5.0 $\times$ 10^{-6} F and 75 mA = 7.5 $\times$ 10^{-2} A, we have

$$V_C = \frac{1}{5.0 \times 10^{-6}} \int 7.5 \times 10^{-2} \, dt$$

$$= 1.5 \times 10^4 \int dt$$

$$= 1.5 \times 10^4 t + C_1$$

From the given information we know that $V_C = 0$ when $t = 0$. Thus,

$$0 = 1.5 \times 10^4(0) + C_1 \quad \text{or} \quad C_1 = 0$$

This means that

$$V_C = 1.5 \times 10^4 t$$

Evaluating this expression for $t = 20 \times 10^{-3}$ s, we have

$$V_C = 1.5 \times 10^4(20 \times 10^{-3})$$

$$= 30 \times 10 = 300 \text{ V}$$

Example H

A certain capacitor is measured to have a voltage of 100 V across it. At this instant a current as a function of time given by $i = 0.06\sqrt{t}$ is sent through the circuit. After 0.25 s, the voltage across the capacitor is measured to be 140 V. What is the capacitance of the capacitor?

From Eq. (5–6) we have

$$V_C = \frac{1}{C} \int i \, dt$$

Substituting $i = 0.06\sqrt{t}$, we find that

$$V_C = \frac{1}{C} \int (0.06\sqrt{t} \, dt) = \frac{0.06}{C} \int t^{1/2} \, dt$$

$$= \frac{0.04}{C} t^{3/2} + C_1$$

From the given information we know that $V_C = 100$ V when $t = 0$. Thus,

$$100 = \frac{0.04}{C}(0) + C_1 \quad \text{or} \quad C_1 = 100 \text{ V}$$

This means that

$$V_C = \frac{0.04}{C} t^{3/2} + 100$$

We also know that $V_C = 140$ V when $t = 0.25$ s. Therefore,

$$140 = \frac{0.04}{C} (0.25)^{3/2} + 100$$

$$40 = \frac{0.04}{C} (0.125)$$

or

$$C = 1.25 \times 10^{-4} \text{ F} = 125 \ \mu\text{F}$$

Exercises 5-1

1. What is the velocity after 3 s of an object which is dropped and falls under the influence of gravity?

2. The acceleration of an object rolling down an inclined plane is 6.0 ft/s². If it is given an upward velocity along the plane of 20 ft/s, what is its velocity after 4 s?

3. The acceleration of an object is given by $6t$. What is the relation between the velocity and time for this object if $v = 8$ when $t = 1$?

4. An object starts from rest and has an acceleration given by $a = \sqrt{t + 1}$. Find the velocity (in meters per second) after 8 s.

5. An object moves in a straight line with a constant velocity of 80 cm/s. How far is it from its starting point after 7 s?

6. The velocity of an object is given by $v = 2t + 1$. Find the relation between s and t if $s = 0$ when $t = 0$.

7. A rocket is fired vertically upward. When it reaches an altitude of 16,500 m, the engines cut off and the rocket is moving upward at 450 m/s. What will be its altitude 3.00 s later ($a = -9.80$ m/s²)?

8. An object is given an upward initial velocity of 60 ft/s from the ground. What is its distance above the ground after 2.5 s?

9. Standing on a cliff, a woman throws a ball with a vertical velocity of 64 ft/s upward. If the ball hits the ground at the base of the cliff 7 s after it is released, how far is it from the top of the cliff to the ground below?

10. A man wishes to throw an object to a height 100 ft above him. What initial vertical velocity must he give the object?

11. The acceleration of an object is given by $a = 24t$. Find the displacement in meters after 1.5 s if $s = 0$ and $v = 0$ when $t = 0$.

12. The acceleration of a certain particle is given by $a = t^4$. Find the displacement in centimeters of the particle after 2 s if $v = 6$ cm/s and $s = 0$ when $t = 0$.

13. The current in a certain integrated electric circuit is 0.500 mA. How many coulombs of charge pass a given point in 2.00 s?

14. The current in a certain circuit is a function of the time and is given by $i = 2t + 3$. How many coulombs pass a given point in the circuit in the first 4 s?

15. The current in a certain wire changes with time according to the relation $i = \sqrt[3]{1 + 3t}$. How many coulombs of charge pass a specified point in the first 3 s?

16. The current i, in microamperes, in a certain microprocessor circuit is given by $i = 8 - t$, where t is the time in microseconds and $0 \le t \le 20\ \mu\text{s}$. If $q_0 = 0$, for what value of t, greater than zero, is $q = 0$? What interpretation can be given to this result?

17. The voltage across a 3.0-μF capacitor is zero. What is the voltage after 10 ms if a current of 0.20 A charges the capacitor?

18. The voltage across a 7.5-μF capacitor is zero. What is the voltage after 5.0 ms if a current of $i = 0.1t$ charges the capacitor?

19. The voltage across a 150-μF capacitor is 50 V. What is the voltage after 1 s if a current $i = 0.012t^{1/5}$ further charges the capacitor?

20. A current $i = t/\sqrt{t^2 + 1}$ is sent into a circuit containing a previously uncharged 4.0-μF capacitor. How long does it take for the capacitor voltage to be 100 V?

21. The angular velocity ω is the time rate of change of the angular displacement θ of a rotating object. If the angular velocity of an object is given by $\omega = 4t$, find an expression for the angular displacement θ if $\theta = 0$ when $t = 0$. See Fig. 5–2.

22. The angular acceleration α is the time rate of change of the angular velocity. If the angular acceleration of a rotating object is given by $\alpha = \sqrt{2t + 1}$, find the expression for the angular displacement if $\omega = 0$ and $\theta = 0$ when $t = 0$. See Fig. 5–2.

23. An inductor in an electric circuit is essentially a coil of wire in which the voltage is affected by a changing current. By definition, the voltage caused by the changing current is given by

$$V_L = L\frac{di}{dt}$$

where L is the inductance and is measured in henrys. If $V_L = 12.0 - 0.2t$ for a 3-H inductor, find the current in the circuit after 20 s if the initial current was zero.

24. If the inner and outer walls of a container are at different temperatures, the rate of change of temperature with respect to the distance from one wall is a function of the distance from the wall. Symbolically this is stated as $dT/dx = f(x)$, where T is the temperature. If x is measured from the outer wall, at 20°C, and $f(x) = 72x^2$, find the temperature at the inner wall if the container walls are 0.5 cm thick.

25. Surrounding an electrically charged particle is an electric field. The rate of change of electric potential with respect to the distance from the particle creating the field equals the negative of the value of the electric field. That is,

$$\frac{dV}{dx} = -E$$

where E is the electric field. If $E = k/x^2$, where k is a constant, find the electric potential at a distance x_1 from the particle, if $V \to 0$ as $x \to \infty$.

$$\omega = \frac{d\theta}{dt} \qquad \alpha = \frac{d\omega}{dt}$$

Figure 5–2

26. The rate of change of the vertical deflection y with respect to the horizontal distance x from one end of a beam is a function of x. For a particular beam, this function is $k(x^5 + 1350x^3 - 7000x^2)$, where k is a constant. Find y as a function of x.

27. Fresh water is flowing into a brine solution, with an equal volume of mixed solution flowing out. The amount of salt in the solution decreases, but more slowly as time increases. Under certain conditions the time rate of change of mass of salt (in grams per minute) is given by $-1/\sqrt{t+1}$. Find the mass of salt as a function of time if 1000 g were originally present. Under these conditions, how long would it take for all the salt to be removed?

28. The rate of change of resistance of a certain electric resistor with respect to the temperature is given by

$$\frac{0.002T}{\sqrt[3]{3T^2 + 1}}$$

Find the resistance as a function of temperature if $R = 0.5$ mΩ when $T = 0°C$.

5-2 Areas by Integration

In Section 4–4 we introduced the method of finding the area under a curve by means of integration. In the same section we also showed that the area under a curve can be found by a summation process performed on the rectangles inscribed under the curve. In this way it was shown that integration can be interpreted as a summation process. *The basic applications of the definite integral use this summation interpretation of the integral.* In this section we shall formulate a general procedure for finding the area for which the bounding curves are known. The method is based on the summing of the areas of inscribed rectangles and using integration for the summation.

The first step in finding any area is to make a sketch of the area to be found. Next a representative **element of area** dA (a typical rectangle) should be drawn. In Fig. 5–3 the width of the element is dx. The length of the element is determined by the y-coordinate (of the vertex of the element) of the point on the curve. Thus, we call the length y. The area of this element is $y\,dx$, which in turn means that $dA = y\,dx$, or

$$A = \int_a^b y\,dx = \int_a^b f(x)\,dx \tag{5-7}$$

This equation states that the elements are to be summed (this is the meaning of the integral sign) from a (the leftmost element) to b (the rightmost element).

Figure 5-3

Example A

Find the area bounded by the curves $y = 2x^2$, $y = 0$, $x = 1$, and $x = 2$.

The desired area is shown in Fig. 5–4. The rectangle shown is the representative element. Its area is $y\,dx$. The elements are to be summed from $x = 1$ to $x = 2$. Thus,

$$A = \int_1^2 y\,dx = \int_1^2 2x^2\,dx = \frac{2}{3}x^3 \Big|_1^2 = \frac{2}{3}(8) - \frac{2}{3}(1) = \frac{14}{3}$$

In Figs. 5–3 and 5–4 the elements are vertical. It is also possible to use horizontal elements, and many problems are simplified by using them. The difference is that the length (longest dimension) is now measured in terms of the x-coordinate of the point on the curve, and the width becomes dy. In the following example, the area is found by both types of elements.

Example B

Find the area in the first quadrant bounded by $y = 9 - x^2$.

The area to be found is shown in Fig. 5–5. Using first the vertical element of length y and width dx, we have

$$A = \int_0^3 y\,dx = \int_0^3 (9 - x^2)\,dx = \left(9x - \frac{x^3}{3}\right)\Big|_0^3 = (27 - 9) - 0 = 18$$

Now using the horizontal element of length x and width dy, we have

$$A = \int_0^9 x\,dy = \int_0^9 \sqrt{9 - y}\,dy = -\int_0^9 (9 - y)^{1/2}(-dy)$$

$$= -\frac{2}{3}(9 - y)^{3/2}\Big|_0^9 = -\frac{2}{3}(9 - 9)^{3/2} + \frac{2}{3}(9 - 0)^{3/2} = \frac{2}{3}(27) = 18$$

 Note that the limits for the vertical elements were 0 and 3, while those for the horizontal elements were 0 and 9. These limits are determined by the direction in which the elements are summed. *By definition, vertical elements are summed from left to right, and horizontal elements are summed from bottom to top.* Doing it this way means that the summation will be in a positive direction.

The choice of vertical or horizontal elements is determined by (1) which one leads to the simplest solution, or (2) the form of the resulting integral. In some problems it makes little difference which is chosen. However, our present methods of integration do not include many types of integrals.

Figure 5–4

Figure 5–5

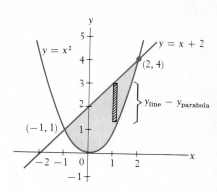

Figure 5–6

It is also possible to find the area between two curves if one is not an axis. In such a case, the length of the element becomes the difference in the y- or x-coordinates, depending on which element is used. The following examples show how we find this type of area.

Example C

Find the area bounded by $y = x^2$ and $y = x + 2$.

This area is indicated in Fig. 5–6. Here we choose vertical elements, since they are all bounded on the top by the line and on the bottom by the parabola. If we choose horizontal elements, the bounding curves are different above the point $(-1, 1)$ (points of intersection are found by solving equations simultaneously) than below this point. Choosing horizontal elements would thus require two separate integrals for solution. Therefore, using vertical elements, we have

$$A = \int_{-1}^{2} (y \text{ of line} - y \text{ of parabola}) \, dx$$

(The difference in y's is taken so that it is positive.) Continuing, we have

$$A = \int_{-1}^{2} (x + 2 - x^2) \, dx = \left(\frac{x^2}{2} + 2x - \frac{x^3}{3} \right) \Big|_{-1}^{2}$$

$$= \left(2 + 4 - \frac{8}{3} \right) - \left(\frac{1}{2} - 2 + \frac{1}{3} \right)$$

$$= \frac{10}{3} + \frac{7}{6} = \frac{27}{6} = \frac{9}{2}$$

Example D

Find the area between $y = x^2$ and $y = x^3$.

Either horizontal or vertical elements may be used equally well (see Fig. 5–7). If we use horizontal elements, we have

$$A = \int_0^1 (y^{1/3} - y^{1/2}) \, dy = \frac{3}{4} y^{4/3} - \frac{2}{3} y^{3/2} \Big|_0^1 = \frac{3}{4} - \frac{2}{3} = \frac{1}{12}$$

▶ It is important that the length of the element be positive. If the difference is taken incorrectly, the result will be negative. *Getting positive lengths can be assured for vertical elements if we subtract y of the lower curve from y of the upper curve. For horizontal elements we should subtract x of the left curve from x of the right curve.* This becomes of definite importance if part of the area is bounded above by a curve, and a different part of the area is bounded below by this curve. In such a case, the area is found by two integrals. The following example illustrates the necessity of this procedure.

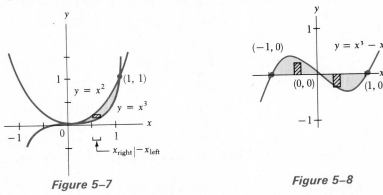

Figure 5–7   Figure 5–8

Example E

Find the area between $y = x^3 - x$ and the x-axis.

We note from Fig. 5–8 that the area to the left of the origin is above the axis and the area to the right is below. If we find the area from

$$A = \int_{-1}^1 (x^3 - x) \, dx = \frac{x^4}{4} - \frac{x^2}{2} \Big|_{-1}^1 = \left(\frac{1}{4} - \frac{1}{2}\right) - \left(\frac{1}{4} - \frac{1}{2}\right) = 0$$

we see that the apparent area is zero. From the figure we know that is not correct.

▶ Noting that the y-values (of the area) are negative to the right of the origin, we set up the integrals

$$A = \int_{-1}^0 (x^3 - x) \, dx + \int_0^1 - (x^3 - x) \, dx$$

$$= \left(\frac{x^4}{4} - \frac{x^2}{2}\right)\Big|_{-1}^0 - \left(\frac{x^4}{4} - \frac{x^2}{2}\right)\Big|_0^1$$

$$= 0 - \left(\frac{1}{4} - \frac{1}{2}\right) - \left(\frac{1}{4} - \frac{1}{2}\right) + 0 = \frac{1}{2}$$

The area under a curve can be applied to various kinds of functions. This is illustrated in the following example and in some of the exercises which follow.

Example F

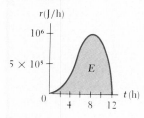

Figure 5-9

Measurements of solar radiation on a particular surface indicated that the rate r, in joules per hour, at which solar energy is received is given by $r = 3600(12t^2 - t^3)$. Since r is a rate, we may write $r = dE/dt$, where E is the energy, in joules, received. This means that $dE = 3600(12t^2 - t^3)\, dt$, and we can find the total energy received at the surface by evaluating the definite integral

$$E = 3600 \int_0^{12} (12t^2 - t^3)\, dt$$

This integral can be interpreted as the area under $f(t) = 3600(12t^2 - t^3)$ from $t = 0$ to $t = 12$, as shown in Fig. 5-9. Evaluating this integral, we have

$$E = 3600 \int_0^{12} (12t^2 - t^3)\, dt = 3600\left(4t^3 - \frac{1}{4}t^4\right)\Big|_0^{12}$$

$$= 3600\left[4(12^3) - \frac{1}{4}(12^4) - 0\right] = 6.22 \times 10^6 \text{ J}$$

Therefore, 6.22 MJ of energy were received in 12 h.

Exercises 5-2

In Exercises 1 through 28, find the areas bounded by the indicated curves.

1. $y = 4x$, $y = 0$, $x = 1$
2. $y = 2x$, $y = 0$, $x = 2$
3. $y = x^2$, $y = 0$, $x = 2$
4. $y = 3x^2$, $y = 0$, $x = 3$
5. $y = 6x$, $x = 0$, $y = 3$
6. $y = 8x$, $x = 0$, $y = 4$
7. $y = x^2$, $x = 0$, $y = 2$
8. $y = x^3$, $x = 0$, $y = 3$
9. $y = 2 - x$, $y = 0$, $x = 1$
10. $y = x^2 - 2x$, $y = 0$
11. $y = x^{-2}$, $y = 0$, $x = 2$, $x = 3$
12. $y = 16 - x^2$, $y = 0$, $x = 1$, $x = 2$
13. $y = \frac{1}{2}x$, $x = 0$, $y = 2$, $y = 4$
14. $y = \frac{1}{3}x$, $x = 0$, $y = 1$, $y = 5$
15. $y = \sqrt{x}$, $x = 0$, $y = 2$, $y = 3$
16. $x = y^2 - y$, $x = 0$
17. $y = 4 - 2x$, $x = 0$, $y = 0$, $y = 3$
18. $y = x$, $y = 2 - x$, $x = 0$
19. $y = x^2$, $y = 2 - x$, $x = 0$ (smaller)
20. $y = x^2$, $y = 2 - x$, $y = 2$ (smallest)
21. $y = x^4$, $y = 16$
22. $y = x^4 - 8x^2 + 16$, $y = 0$
23. $y = \sqrt{x - 1}$, $y = 0$, $x = 2$
24. $y = \sqrt{2x + 1}$, $y = 0$, $x = 4$
25. $y = x^2$, $y = \sqrt{x}$
26. $y = 8 - x^3$, $y = 7x$, $x = 0$
27. $y = x$, $x = -1$, $x = 1$, $y = 0$
28. $y = x^2 + 2x - 8$, $y = x + 4$

Figure 5–10

In Exercises 29 through 38, some applications of areas are shown.

29. Certain physical quantities are often represented as an area under a curve. By definition, power is defined as the time rate of change of performing work. Thus, $p = dw/dt$, or $dw = p\,dt$. Therefore, if $p = 12t - 4t^2$, find the work (in joules) performed in 3 s by finding the area under the curve of p versus t. See Fig. 5–10.

30. When considering the expansion of a gas, the expression for the work done in the expansion is $w = \int_{V_1}^{V_2} p\,dV$, where p is the pressure of the gas and V is its volume. For an adiabatic (no *heat* gained or lost) change, an approximate relation is $p = kV^{-1.4}$. Given that $k = 1$, find a numerical value for the work done if V changes from 1 to 4 units.

31. It is also possible to represent the change in displacement as an area. Since $s = \int v\,dt$, if $v = t^{2/3} + 1$, determine the distance (in feet) that an object moves from $t = 1$ s to $t = 8$ s, by finding the area under the curve of v versus t.

32. The total cost of production can be interpreted as an area. If the cost per unit C', in dollars per unit, of producing x units is given by $\dfrac{100}{(0.01x + 1)^2}$, find the total cost C of producing 100 units by finding the area under the curve of C' versus x.

33. A cam is designed such that one face of it is described as being the area between the curves $y = x^3 - 2x^2 - x + 2$ and $y = x^2 - 1$, with units in centimeters. Show that this description does not uniquely describe the face of the cam. Find the area of the face of the cam, if a complete description requires that $x \le 1$.

34. In designing a lawn area, an architect planned one such area of an estate to be between parabolas $y = 20 - 0.01x^2$ and $y = 0.005x^2 - 15$ and the lines $x = -27$ and $x = 27$, where dimensions are in meters. Find the area of this part of the lawn.

35. An experiment was performed by expanding a gas under the condition of constant temperature. The data found are tabulated below. How much work (in joules) was done in the expansion? Use the trapezoidal rule, with $n = 4$.

p (kilopascals)	102	77.0	60.0	47.0	43.0
V (cubic centimeters)	6.0	8.0	10.0	12.0	14.0

To obtain units of joules we must multiply the numerical value obtained by 10^{-3} to include the powers of 10 indicated by the prefix (1 kPa $= 10^3$ Pa and 1 cm$^3 = (10^{-2}$ m$)^3 = 10^{-6}$ m^3).

36. A certain tract of land is bounded by a bend in a river and a straight fence. If the distances y from the fence to the river are given for various distances x along the fence from one end to the other as in the following table, find the approximate area of the piece of land by use of Simpson's rule, with $n = 8$.

y (feet)	0	15	50	70	95	80	35	20	0
x (feet)	0	10	20	30	40	50	60	70	80

37. A window is designed to be the area between a parabolic section and a straight base, as shown in Fig. 5–11. What is the area of the window?

38. The vertical ends of a fuel storage tank have a parabolic bottom section and a triangular top section, as shown in Fig. 5–12. What volume does the tank hold?

1.60 m

Figure 5–11

Figure 5–12

5-3 Volumes by Integration

Consider an area and its representative element (see Fig. 5–13(a)) to be rotated about the x-axis. When an area is rotated in this manner, it is said to generate a volume, which is also indicated in the figure. We shall now show methods of finding volumes which are generated in this manner.

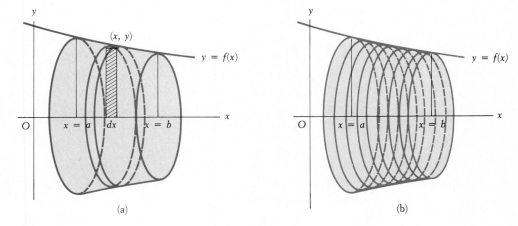

Figure 5–13

As the area rotates about the x-axis, so does its representative element. The element generates a solid for which the volume is known, namely, a thin cylindrical **disk.** We know that the volume of a right circular cylinder is π times the square of the radius times the height (in this case the thickness) of the cylinder. We must now determine the radius and thickness of this disk. Since the element is rotated about the x-axis, the y-coordinate of the point on the curve which touches the element must represent the radius. Also, it can be seen that the thickness is dx (the disk is on its side). This disk, which is the representative **element of volume,** has a volume of $dV = \pi y^2 \, dx$. Summing these elements of volume from left to right, as illustrated in Fig. 5–13(b), we have for the total volume V

$$V = \pi \int_a^b y^2 \, dx = \pi \int_a^b [f(x)]^2 \, dx \qquad (5\text{–}8)$$

Thus, by use of Eq. (5–8), we can find the volume generated by an area bounded by the x-axis, which is rotated about the x-axis.

Example A

Figure 5–14

Find the volume generated by rotating the area bounded by $y = x^2$, $x = 2$, and $y = 0$ about the x-axis.

See Fig. 5–14. By Eq. (5–8), we have

$$V = \pi \int_0^2 (x^2)^2 \, dx$$

$$= \pi \int_0^2 x^4 \, dx$$

$$= \frac{\pi}{5} x^5 \Big|_0^2$$

$$= \frac{32\pi}{5}$$

If an area bounded by the y-axis is rotated about the y-axis, the volume generated is given by

$$V = \pi \int_c^d x^2 \, dy \qquad\qquad (5\text{--}9)$$

In this case the radius of the element of volume is the x-coordinate of the point on the curve, and the thickness of the disk is dy. One should always be careful to identify the radius and thickness properly.

Example B

Find the volume generated by rotating the area bounded by $y = 2x$, $y = 6$, and $x = 0$ about the y-axis.

Figure 5–15 shows the volume to be found. We write

$$V = \pi \int_0^6 x^2 \, dy = \pi \int_0^6 \left(\frac{y}{2}\right)^2 dy = \frac{\pi}{4} \int_0^6 y^2 \, dy = \frac{\pi}{12} y^3 \Big|_0^6 = 18\pi$$

Since this volume is a right circular cone, it is possible to check the result:

$$V = \frac{1}{3}\pi r^2 h = \frac{1}{3}\pi(3^2)(6) = 18\pi$$

Figure 5–15

Figure 5–16

Figure 5–17

Figure 5–18

There is another method of finding a volume of a solid of revolution. If the area in Fig. 5–14 is rotated about the y-axis, the element of area $y\,dx$ generates a different element of volume from that generated when it is rotated about the x-axis. We can see in Fig. 5–16 that this element of volume is a **cylindrical shell.** *The total volume is made up of an infinite number of concentric shells.* When the volumes of these shells are summed, we have the total volume generated. Thus we must now find the approximate volume dV of the representative shell. By finding the circumference of the base and multiplying this by the height, we can obtain an expression for the surface area of the shell. Then, by multiplying this by the thickness of the shell, we obtain its volume. The volume of the representative shell is (see Fig. 5–17)

$$dV = 2\pi(\textbf{radius}) \cdot (\textbf{height}) \cdot (\textbf{thickness}) \qquad (5\text{–}10)$$

Similarly, the volume of a disk is (see Fig. 5–18)

$$dV = \pi(\textbf{radius})^2 \cdot (\textbf{thickness}) \qquad (5\text{–}11)$$

The elements of volume should be remembered in the general forms given in Eqs. (5–10) and (5–11), and not in specific forms such as Eqs. (5–8) and (5–9). If we remember them in this manner, we can readily apply these methods to finding any such volume of a solid of revolution.

Example C

Use the method of cylindrical shells to find the volume generated by rotating the area bounded by $y = 4 - x^2$, $x = 0$, and $y = 0$ about the y-axis.

 From Fig. 5–19, we identify the radius, height, and thickness: $r = x$, $h = y$, $t = dx$ (this determines the limits as $x = 0$ and $x = 2$). And so

$$V = 2\pi \int_0^2 xy\, dx = 2\pi \int_0^2 x(4 - x^2)\, dx = 2\pi \int_0^2 (4x - x^3)\, dx = 8\pi$$

Figure 5–19

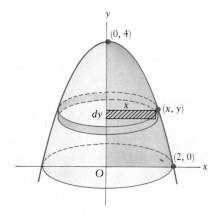

Figure 5–20

Example D

Use the disk method to find the indicated volume of Example C.

From Fig. 5–20, we identify the radius and height: $r = x$, $t = dy$ (which tells us that the limits are from $y = 0$ to $y = 4$). Thus,

$$V = \pi \int_0^4 x^2\, dy = \pi \int_0^4 (4 - y)\, dy = 8\pi$$

Example E

Use shells to find the volume if the area of Example C is rotated about the x-axis.

From Fig. 5–21, we see that $r = y$, $h = x$, $t = dy$ (thus, the limits are $y = 0$ to $y = 4$). Hence

$$V = 2\pi \int_0^4 xy\, dy = 2\pi \int_0^4 \sqrt{4 - y}(y\, dy) = \frac{256\pi}{15}$$

(The method of integrating this function has not yet been discussed. We present the answer here for the reader's information at this time.)

Figure 5-21

Figure 5-22

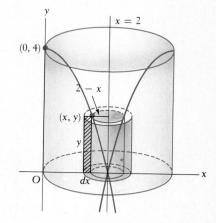

Figure 5-23

Example F

By the use of disks, find the volume indicated in Example E.

From Fig. 5–22, we see that $r = y$, $t = dx$ (the limits are $x = 0$ to $x = 2$). Therefore

$$V = \pi \int_0^2 y^2 \, dx = \pi \int_0^2 (4 - x^2)^2 \, dx$$

$$= \pi \int_0^2 (16 - 8x^2 + x^4) \, dx = \frac{256\pi}{15}$$

Example G

Find the volume generated if the area in Example C is rotated about the line $x = 2$.

Shells are convenient, since the volume of a shell can be expressed as a single integral. From Fig. 5–23, we see that

▶ $r = 2 - x$, $h = y$ and $t = dx$

Thus, we have

$$V = 2\pi \int_0^2 (2 - x) y \, dx = 2\pi \int_0^2 (2 - x)(4 - x^2) \, dx$$

$$= 2\pi \int_0^2 (8 - 2x^2 - 4x + x^3) \, dx = \frac{40\pi}{3}$$

(If the area had been rotated about the line $x = 3$, the only difference in the integral would have been that $r = 3 - x$. Everything else, including the limits, would have remained the same.)

See Appendix C for a
computer program for finding
the volume generated by
rotating an area about the
x-axis.

Exercises 5–3

In Exercises 1 through 12, find the volume generated by the areas bounded by the given curves if they are rotated about the *x*-axis. Use the indicated method in each case.

1. $y = 1 - x$, $x = 0$, $y = 0$ (disks)   2. $y = x$, $y = 0$, $x = 2$ (disks)

3. Area of Exercise 1   (shells)   4. $y = x$, $x = 0$, $y = 2$ (shells)

5. $y = x^2 + 1$, $x = 0$, $x = 3$, $y = 0$ (disks)

6. $y = 2x - x^2$, $y = 0$ (disks)

7. $y = x^3$, $y = 8$, $x = 0$ (shells)   8. $y = x^2$, $y = x$ (shells)

9. $y = 3\sqrt{x}$, $y = 0$, $x = 4$ (disks)

10. $y = 6 - x - x^2$, $x = 0$, $y = 0$ (quadrant I)   (disks)

11. $x = 4y - y^2 - 3$, $x = 0$ (shells)

12. $y = x^4$, $x = 0$, $y = 1$, $y = 2$ (shells)

In Exercises 13 through 24, find the volume generated by the areas bounded by the given curves if they are rotated about the *y*-axis. Use the indicated method in each case.

13. Area of Exercise 1   (disks)   14. Area of Exercise 4   (disks)

15. Area of Exercise 1   (shells)   16. Area of Exercise 2   (shells)

17. $y = 2\sqrt{x}$, $x = 0$, $y = 2$ (disks)   18. $y^2 = x$, $y = 4$, $x = 0$ (disks)

19. Area of Exercise 6   (shells)   20. Area of Exercise 10   (shells)

21. Area of Exercise 11   (disks)   22. $x^2 + 4y^2 = 4$ (quadrant I)   (disks)

23. $y = \sqrt{4 - x^2}$ (quadrant I)   (shells)   24. $y = 8 - x^3$, $x = 0$, $y = 0$ (shells)

In Exercises 25 through 32, find the indicated volumes by integration.

25. Find the volume generated if the area of Exercise 6 is rotated about the line $x = 2$.

26. Find the volume generated if the area of Exercise 8 is rotated about the line $y = 4$.

27. Derive the expression for the volume of a right circular cone obtained by rotating the area bounded by $y = (r/h)x$, $y = 0$ and $x = h$ about the *x*-axis.

28. A conical funnel has an opening at the bottom of radius 0.5 cm. The diameter at the top of the funnel is 20 cm and the depth of the funnel is 19 cm. What is the maximum capacity of the funnel?

29. The capillary tube shown in Fig. 5–24 has circular horizontal cross sections of inner radius 1.1 mm. What is the volume of the liquid in the tube above the level of liquid outside the tube if the top of the liquid in the center vertical cross section is described by the equation $y = x^4 + 1.5$, as shown?

30. The concrete base for a piece of machinery can be described as being the first-quadrant area bounded by $y = 2$ and a section of $y = 8 - x^3$, as shown in Fig. 5–25, being rotated about the *y*-axis. Find the volume, in cubic feet, of the machinery base.

31. The water in a spherical tank 20 m in radius is 15 m deep at the deepest point. How much water is in the tank?

32. All horizontal cross sections of a keg 4 ft tall are circular, and the sides of the keg are parabolic. The diameter at the top and the bottom is 2 ft, and the diameter in the middle is 3 ft. Find the volume that the keg holds.

$y = x^4 + 1.5$

$r = 1.1$ mm

Figure 5–24

$y = 2$

$y = 8 - x^3$

Figure 5–25

5-4 Centroids

In the study of mechanics, a very important property of an object is its center of mass. In this section we shall show the meaning of center of mass and then show how integration is used to determine the center of mass for areas and solids of rotation.

If a mass m is at a distance d from a specified point O, the **moment** *of the mass about O is defined as md.* If several masses $m_1, m_2, \ldots, m_n$ are at distances $d_1, d_2, \ldots d_n$, respectively, from point O, their moment (as a group) about O is defined as $m_1 d_1 + m_2 d_2 + \cdots + m_n d_n$. If all the masses could be concentrated at one point $\bar{d}$ units from O, the moment would be $(m_1 + m_2 + \cdots + m_n)\bar{d}$: this is, however, what is meant by the previous expression. Therefore, we may write

$$m_1 d_1 + m_2 d_2 + \cdots + m_n d_n = (m_1 + m_2 + \cdots + m_n)\bar{d} \qquad (5\text{--}12)$$

In Eq. (5–12), $\bar{d}$ is the distance from O to the **center of mass.** The moment of a mass is a measure of its tendency to rotate about a point. A weight far from the point of balance of a long rod is more likely to make the rod turn than if the same weight were placed near the point of balance. It is easier to open a door if you push near the door knob than if you push near the hinges. This is the type of physical property which the moment of mass measures.

Example A

One of the simplest and most basic illustrations of moments and center of mass is seen in balancing a long rod with masses of different sizes, one on either side of the balance point.

In Fig. 5–26, a mass of 5.0 kg is hung from the rod 0.8 m to the right of point O. We see that this 5.0-kg mass tends to turn the rod clockwise. A mass placed on the opposite side of O will tend to turn the rod counterclockwise. Neglecting the mass of the rod, in order to balance the rod at O, the moments must be equal in magnitude but opposite in sign. Therefore, a 4.0-kg mass would have to be placed 1.0 m to the left.

Thus, with $d_1 = 0.8$ m, $d_2 = -1.0$ m, we see that

$$(5.0 + 4.0)\bar{d} = 5.0(0.8) + 4.0(-1.0) = 4.0 - 4.0 \qquad \bar{d} = 0.0 \text{ m}$$

This means that the center of mass (of the combination of the 5.0-kg mass and the 4.0-kg mass) is at O. Also, we see that we must use *directed* distances in finding moments.

Figure 5–26

Example B

Figure 5–27

On the x-axis a mass of 3 units is placed at $(2, 0)$, another of 6 units at $(5, 0)$, and a third, 7 units, at $(6, 0)$. Find the center of mass of the three objects.

Taking the reference point as the origin, we find $d_1 = 2$, $d_2 = 5$, and $d_3 = 6$. Thus, $m_1 d_1 + m_2 d_2 + m_3 d_3 = (m_1 + m_2 + m_3)\bar{d}$ becomes

$$3(2) + 6(5) + 7(6) = (3 + 6 + 7)\bar{d} \quad \text{or} \quad \bar{d} = 4.88$$

This means that the center of mass of the three objects is at $(4.88, 0)$. Therefore a mass of 16 units placed at this point has the same moment as the three masses as a unit (see Fig. 5–27).

Example C

Figure 5–28

Find the center of mass of the area indicated in Fig. 5–28.

We first note that the center of mass is not *on* either axis. This can be seen from the fact that the major portion of the area is in the first quadrant. *We shall therefore measure the moments with respect to each axis to find the point which is the center of mass. This point is also called the* **centroid** *of the area.*

The easiest method of finding the centroid is to divide the area into rectangles, as indicated by the dashed line in Fig. 5–28, and assume that we may consider the mass of each rectangle to be concentrated at its center. In this way the left rectangle has its center at $(-1, 1)$ and the right rectangle has its center at $(\frac{5}{2}, 2)$. The mass of each rectangle, assumed uniform, is proportional to the area. The area of the left rectangle is 8 units and that of the right rectangle is 12 units. Thus, taking moments with respect to the y-axis, we have

$$8(-1) + 12\left(\frac{5}{2}\right) = (8 + 12)\bar{x}$$

where $\bar{x}$ is the x-coordinate of the centroid. Solving for $\bar{x}$ we have $\bar{x} = \frac{11}{10}$.

Now taking moments with respect to the x-axis we have

$$8(1) + 12(2) = (8 + 12)\bar{y}$$

where $\bar{y}$ is the y-coordinate of the centroid. Thus, $\bar{y} = \frac{8}{5}$. This means that the coordinates of the centroid, the center of mass, are $(\frac{11}{10}, \frac{8}{5})$. This may be interpreted as meaning that an area of this shape would balance on a single support under this point. As an approximate check, we note from the figure that this point appears to be a reasonable balance point for the area.

Figure 5–29

If an area is bounded on the curves of the functions $y_1 = f_1(x)$, $y_2 = f_2(x)$, $x = a$, and $x = b$, as shown in Fig. 5–29, the moment of the element of area about the y-axis is $(k \, dA)x$, where k is the mass per unit area. In this expression, $k \, dA$ is the mass of the element, and x is its distance (moment arm) from the y-axis. The element dA may be written as $(y_1 - y_2) \, dx$, which means that the moment may be written as $kx(y_1 - y_2) \, dx$. If we then sum up the moments of all the elements and express this as an integral (which, of course, means sum), we have $k \int_a^b x(y_1 - y_2) \, dx$. If we consider all the mass of the area to be concentrated at one point $\bar{x}$ units from the y-axis, the moment would be $(kA)\bar{x}$, where kA is the mass of the entire area, and $\bar{x}$ is the distance the center of mass is from the y-axis. By previous discussion these two expressions should be equal. This means $k \int_a^b x(y_1 - y_2) \, dx = kA\bar{x}$. Since k appears on each side of the equation, we divide it out (we are assuming that the mass per unit area is constant). The area A is found by the integral $\int_a^b (y_1 - y_2)\, dx$. Therefore,

$$\bar{x} = \frac{\displaystyle\int_a^b x(y_1 - y_2)\, dx}{\displaystyle\int_a^b (y_1 - y_2)\, dx} \qquad (5\text{–}13)$$

Equation (5–13) gives us the x-coordinate of the centroid of an area if vertical elements are used. It should be noted that the two integrals in Eq. (5–13) must be evaluated separately. We cannot cancel the apparent common factor $y_1 - y_2$, and we cannot combine quantities and perform one integration. The two integrals must be evaluated before cancellations are made.

Following the same reasoning, if an area is bounded by $x_1 = g_1(y)$, $x_2 = g_2(y)$, $y = c$, and $y = d$ (Fig. 5–30), the y-coordinate of the centroid is

Figure 5–30

$$\bar{y} = \frac{\displaystyle\int_c^d y(x_2 - x_1)\, dy}{\displaystyle\int_c^d (x_2 - x_1)\, dy} \qquad (5\text{–}14)$$

In this equation, horizontal elements are used.

In applying Eqs. (5–13) and (5–14), we should keep in mind that the denominators of the right-hand sides give the area, and once we have found this, we may use it for both $\bar{x}$ and $\bar{y}$. Also, we should utilize any symmetry a curve may have.

Example D

Figure 5–31

Find the coordinates of the centroid of the area bounded by $y = x^2$ and the line $y = 4$.

We sketch a graph indicating the area and an element of area (see Fig. 5–31). The curve is a parabola whose axis is the y-axis. Since the area is symmetrical to the y-axis, the centroid must be on this axis. This means that the x-coordinate of the centroid is zero, or $\bar{x} = 0$. To find the y-coordinate of the centroid, we use Eq. (5–14). For this area we have

$$\bar{y} = \frac{\displaystyle\int_0^4 y(2x)\, dy}{\displaystyle\int_0^4 2x\, dy} = \frac{\displaystyle\int_0^4 y(2\sqrt{y})\, dy}{\displaystyle\int_0^4 2\sqrt{y}\, dy} = \frac{2\displaystyle\int_0^4 y^{3/2}\, dy}{2\displaystyle\int_0^4 y^{1/2}\, dy} = \frac{2\left(\frac{2}{5}\right)y^{5/2}\Big|_0^4}{2\left(\frac{2}{3}\right)y^{3/2}\Big|_0^4}$$

$$= \frac{\dfrac{4}{5}(32)}{\dfrac{4}{3}(8)} = \frac{128}{5} \cdot \frac{3}{32} = \frac{12}{5}$$

The coordinates of the centroid are $(0, \frac{12}{5})$. This area would balance if a single pointed support were to be put under this point.

Example E

Find the coordinates of the centroid of an isosceles right triangle with side a.

We must first set up this area in the xy-plane. One choice is to place the triangle with one vertex at the origin and the right angle on the x-axis (see Fig. 5–32). Since each side is a, the hypotenuse passes through the point (a, a). The equation of the hypotenuse is $y = x$. The x-coordinate of the centroid is

$$\bar{x} = \frac{\int_0^a xy\, dx}{\int_0^a y\, dx} = \frac{\int_0^a x(x)\, dx}{\int_0^a x\, dx} = \frac{\int_0^a x^2\, dx}{\frac{1}{2}x^2\Big|_0^a} = \frac{\frac{1}{3}x^3\Big|_0^a}{\frac{a^2}{2}} = \frac{\frac{a^3}{3}}{\frac{a^2}{2}} = \frac{2a}{3}$$

The y-coordinate of the centroid is

$$\bar{y} = \frac{\int_0^a y(a-x)\, dy}{\frac{a^2}{2}} = \frac{\int_0^a y(a-y)\, dy}{\frac{a^2}{2}} = \frac{\int_0^a (ay - y^2)\, dy}{\frac{a^2}{2}}$$

$$= \frac{\frac{ay^2}{2} - \frac{y^3}{3}\Big|_0^a}{\frac{a^2}{2}} = \frac{\frac{a^3}{6}}{\frac{a^2}{2}} = \frac{a}{3}$$

Figure 5–32

Thus, the coordinates of the centroid are $(\frac{2}{3}a, \frac{1}{3}a)$. The results indicate that the center of mass is $\frac{1}{3}a$ units from each of the equal sides.

Another important figure for which we wish to find the centroid is the solid of revolution. If the density of the solid is constant, the coordinates of the centroid will be on the axis of rotation. The problem which remains is to find just where on the axis the centroid is located.

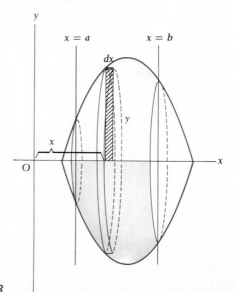

Figure 5–33

If a given area bounded by the x-axis, as shown in Fig. 5–33, is rotated about the x-axis, a vertical element of area generates a disk element of volume. The center of mass of the disk is at its center, and therefore, for purposes of finding moments, we may consider its mass concentrated there. The moment about the y-axis of a typical element is $x(k)(\pi y^2\,dx)$, where x is the moment arm, k is the density, and $\pi y^2\,dx$ is the volume. The sum of the moments of the elements can be expressed as an integral; it equals the volume times the density times the x-coordinate of the centroid of the volume. Since the density k and the factor of π would appear on each side of the equation, they need not be written. Therefore,

$$\bar{x} = \frac{\displaystyle\int_a^b xy^2\,dx}{\displaystyle\int_a^b y^2\,dx} \tag{5-15}$$

is the equation giving the x-coordinate of the centroid of a volume of a solid of revolution about the x-axis.

In the same manner we may find *the y-coordinate of the centroid of a volume of a solid of revolution about the y-axis. It is*

$$\bar{y} = \frac{\displaystyle\int_c^d yx^2\,dy}{\displaystyle\int_c^d x^2\,dy} \tag{5-16}$$

Example F

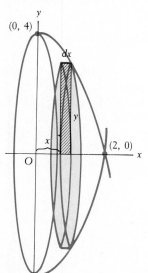

Figure 5-34

Find the coordinates of the centroid of the volume generated by rotating the first-quadrant area under the curve $y = 4 - x^2$ about the x-axis.

Since the curve (see Fig. 5–34) is rotated about the x-axis, the centroid is on the x-axis, which means that $\bar{y} = 0$. We find the x-coordinate by using Eq. (5–15):

$$\bar{x} = \frac{\displaystyle\int_0^2 xy^2\,dx}{\displaystyle\int_0^2 y^2\,dx} = \frac{\displaystyle\int_0^2 x(4-x^2)^2\,dx}{\displaystyle\int_0^2 (4-x^2)^2\,dx} = \frac{\displaystyle\int_0^2 (16x - 8x^3 + x^5)\,dx}{\displaystyle\int_0^2 (16 - 8x^2 + x^4)\,dx}$$

$$= \frac{8x^2 - 2x^4 + \dfrac{1}{6}x^6\Big|_0^2}{16x - \dfrac{8}{3}x^3 + \dfrac{1}{5}x^5\Big|_0^2} = \frac{32 - 32 + \dfrac{64}{6}}{32 - \dfrac{64}{3} + \dfrac{32}{5}} = \frac{5}{8}$$

The coordinates of the centroid are $(\frac{5}{8}, 0)$.

Example G

Find the coordinates of the centroid of the volume generated by rotating the area of Example F about the y-axis.

Since the curve is rotated about the y-axis, $\bar{x} = 0$ (see Fig. 5–35). The y-coordinate is found by Eq. (5–16):

$$\bar{y} = \frac{\displaystyle\int_0^4 yx^2\,dy}{\displaystyle\int_0^4 x^2\,dy} = \frac{\displaystyle\int_0^4 y(4-y)\,dy}{\displaystyle\int_0^4 (4-y)\,dy} = \frac{\displaystyle\int_0^4 (4y-y^2)\,dy}{\displaystyle\int_0^4 (4-y)\,dy} = \frac{2y^2 - \dfrac{1}{3}y^3 \Big|_0^4}{4y - \dfrac{1}{2}y^2 \Big|_0^4}$$

$$= \frac{32 - \dfrac{64}{3}}{16 - 8} = \frac{4}{3}$$

The coordinates of the centroid are $(0, \tfrac{4}{3})$.

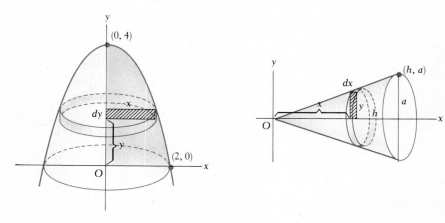

Figure 5–35   Figure 5–36

Example H

Find the centroid of a right circular cone of radius a and altitude h.

To generate a right circular cone, we may rotate a right triangle about one of its legs (Fig. 5–36). Placing the leg of length h along the x-axis, we rotate the right triangle whose hypotenuse is given by $y = (a/h)x$ about the x-axis. The x-coordinate of the centroid is

$$\bar{x} = \frac{\displaystyle\int_0^h xy^2\,dx}{\displaystyle\int_0^h y^2\,dx} = \frac{\displaystyle\int_0^h x\left[\left(\frac{a}{h}\right)x\right]^2 dx}{\displaystyle\int_0^h \left[\left(\frac{a}{h}\right)x\right]^2 dx} = \frac{\left(\dfrac{a^2}{h^2}\right)\left(\dfrac{1}{4}x^4\right)\Big|_0^h}{\left(\dfrac{a^2}{h^2}\right)\left(\dfrac{1}{3}x^3\right)\Big|_0^h} = \frac{3}{4}h$$

Therefore, the centroid is located along the altitude $\tfrac{3}{4}$ of the way from the vertex, or $\tfrac{1}{4}$ of the way from the base.

Exercises 5–4

In Exercises 1 through 4, find the center of mass of the particles of the given masses located at the given points.

1. 5 units at $(1, 0)$, 10 units at $(4, 0)$, 3 units at $(5, 0)$.

2. 2 units at $(2, 0)$, 9 units at $(3, 0)$, 3 units at $(8, 0)$, 1 unit at $(12, 0)$.

3. 4 units at $(-3, 0)$, 2 units at $(0, 0)$, 1 unit at $(2, 0)$, 8 units at $(3, 0)$.

4. 2 units at $(-4, 0)$, 1 unit at $(-3, 0)$, 5 units at $(1, 0)$, 4 units at $(4, 0)$.

In Exercises 5 through 8, find the coordinates of the centroid of the area shown.

5. Fig. 5–37(a)   6. Fig. 5–37(b)

7. Fig. 5–37(c)   8. Fig. 5–37(d)

In Exercises 9 through 24, find the coordinates of the centroids of the indicated figures.

9. Find the coordinates of the centroid of the area bounded by $y = x^2$ and $y = 2$.

10. Find the coordinates of the centroid of a semicircular area.

11. Find the coordinates of the centroid of the area bounded by $y = 4 - x$ and the axes.

12. Find the coordinates of the centroid of the area bounded by $y = x^3$, $x = 2$, and the x-axis.

13. Find the coordinates of the centroid of the area bounded by $y = x^2$ and $y = x^3$.

14. Find the coordinates of the centroid of the area bounded by $y^2 = x$, $y = 2$, $x = 0$.

15. Find the coordinates of the centroid of the volume generated by rotating the area in the first quadrant bounded by $y^2 = 4x$, $y = 0$, and $x = 1$ about the y-axis.

16. Find the coordinates of the centroid of the volume generated by rotating the area bounded by $y = x^2$, $x = 2$, and the x-axis about the x-axis.

17. Find the coordinates of the centroid of the volume generated by rotating the area bounded by $y^2 = 4x$ and $x = 1$ about the x-axis.

18. Find the coordinates of the centroid of the volume generated by rotating the area bounded by $x^2 - y^2 = 9$, $y = 4$, and the x-axis about the y-axis.

19. Find the coordinates of the centroid of the volume generated by rotating the area bounded by $y = 4/x^2$, $x = 1$, $x = 2$, and the x-axis about the x-axis.

20. Find the coordinates of the centroid of the volume generated by rotating the first-quadrant area bounded by the ellipse $4x^2 + 9y^2 = 36$ around the y-axis.

21. Find the location of the centroid of a right triangle with legs a and b.

22. Find the location of the centroid of a hemisphere of radius a.

23. A machine part is in the shape of a frustum of a right circular cone. Find the location of its centroid if the radii of its bases are 2 in. and 6 in. and the height between bases is 5 in.

24. A floodgate is in the shape of an isosceles trapezoid. Find the location of the centroid of the floodgate if the upper base is 20 m, the lower base is 12 m, and the height between bases is 6 m.

(a)

(b)

(c)

(d)

Figure 5–37

5–5 Moments of Inertia

In applying the laws of physics, we encounter an important quantity when we are discussing the rotation of an object: its **moment of inertia.** The moment of inertia of an object rotating about an axis is analogous to the mass of an object moving through space in reference to some point. *In each case the quantity (moment of inertia or mass) is the measure of the tendency of an object to resist a change in motion.*

Suppose that a particle of mass m is rotating about some point; we define its moment of inertia as md^2, where d is the distance from the particle to the point. If a group of particles of masses $m_1, m_2, \ldots, m_n$ are rotating about some axis, the moment of inertia of the group with respect to that axis is

$$I = m_1 d_1^2 + m_2 d_2^2 + \cdots + m_n d_n^2$$

where the d's are the respective distances of the particles from the axis. If all of the masses could be positioned such that they were at the same distance R from the axis of rotation, and so that the total moment of inertia were the same, we would have

$$m_1 d_1^2 + m_2 d_2^2 + \cdots + m_n d_n^2 = (m_1 + m_2 + \cdots + m_n)R^2 \qquad (5\text{–}17)$$

where R is called the **radius of gyration.**

Example A

Find the moment of inertia and radius of gyration of three masses, one of 3 units at $(-2, 0)$, the second of 5 units at $(1, 0)$, and the third of 4 units at $(4, 0)$ with respect to the origin (see Fig. 5–38).

The moment of inertia of the group is

$$I = 3(-2)^2 + 5(1)^2 + 4(4)^2 = 81$$

The radius of gyration is found from $I = (m_1 + m_2 + m_3)R^2$. Thus,

$$81 = (3 + 5 + 4)R^2 \qquad R^2 = \frac{81}{12} \quad \text{or} \quad R = 2.60$$

This means that a mass of 12 units placed at $(2.60, 0)$ [or at $(-2.60, 0)$] would have the same rotational inertia about the origin as the three objects as a unit.

Figure 5–38

Figure 5–39

If an area is bounded by the curves of the functions $y_1 = f_1(x)$, $y_2 = f_2(x)$, and $x = a$ and $x = b$, as shown in Fig. 5–39, the moment of inertia of this area with respect to the y-axis, I_y, is given by the sum of the moments of inertia of the individual elements. The mass of each element is $k(y_1 - y_2)\,dx$ where k is the mass per unit area, and $(y_1 - y_2)\,dx$ is the area of the element. The distance of the element from the y-axis is x. Representing this sum as an integral, we have

$$I_y = k \int_a^b x^2(y_1 - y_2)\,dx \qquad (5\text{–}18)$$

Figure 5-40

To find the radius of gyration of the area with respect to the y-axis, R_y, we would first find the moment of inertia, divide this by the mass of the area, and take the square root of this result.

In the same manner, the moment of inertia of an area, with respect to the x-axis, bounded by $x_1 = g_1(y)$ and $x_2 = g_2(y)$ is given by

$$I_x = k \int_c^d y^2 (x_2 - x_1)\, dy \qquad (5-19)$$

We find the radius of gyration of the area with respect to the x-axis, R_x, in the same manner as we find it with respect to the y-axis (see Fig. 5-40).

Example B

Figure 5-41

Find the moment of inertia and the radius of gyration of the area bounded by $y = 4x^2$, $x = 1$, and the x-axis with respect to the y-axis.

We find the moment of inertia of the area (see Fig. 5-41) by using Eq. (5-18):

$$I_y = k \int_0^1 x^2 y\, dx = k \int_0^1 x^2 (4x^2)\, dx = 4k \int_0^1 x^4\, dx$$

$$= 4k \left(\frac{1}{5} x^5 \right) \Big|_0^1 = \frac{4k}{5}$$

To find the radius of gyration, we first determine the mass of the area:

$$m = k \int_0^1 y\, dx = k \int_0^1 (4x^2)\, dx = 4k \left(\frac{1}{3} x^3 \right) \Big|_0^1 = \frac{4k}{3}$$

$$R_y^2 = \frac{I_y}{m} = \frac{4k}{5} \cdot \frac{3}{4k} = \frac{3}{5} \qquad R_y = \sqrt{\frac{3}{5}} = \frac{\sqrt{15}}{5}$$

Example C

Figure 5-42

Find the moment of inertia of a right triangle with sides a and b with respect to the side b. Assume that $k = 1$.

Placing the triangle as shown in Fig. 5-42, we see that the equation of the hypotenuse is $y = (a/b)x$. The moment of inertia is

$$I_x = \int_0^a y^2 (b - x)\, dy = \int_0^a y^2 \left(b - \frac{b}{a} y \right) dy = b \int_0^a \left(y^2 - \frac{1}{a} y^3 \right) dy$$

$$= b \left(\frac{1}{3} y^3 - \frac{1}{4a} y^4 \right) \Big|_0^a = b \left(\frac{a^3}{3} - \frac{a^3}{4} \right) = \frac{ba^3}{12}$$

Figure 5–43   Figure 5–44

Among the most important moments of inertia are those of solids of revolution. Since all parts of an element of mass should be at the same distance from the axis, the most convenient element of volume to choose is the cylindrical shell (see Fig. 5–43). If the area bounded by the curves $y = f(x)$, $x = a$, $x = b$, and the x-axis is rotated about the y-axis, the moment of inertia of the element of volume is $k(2\pi xy\, dx)(x^2)$, where k is the density, $2\pi xy\, dx$ is the volume of the element, and x^2 is the square of its distance from the y-axis. Expressing the sum of the moments of the elements as an integral, *the moment of inertia of the volume with respect to the y-axis, I_y,* is

$$I_y = 2\pi k \int_a^b yx^3\, dx \qquad\qquad (5\text{--}20)$$

The radius of gyration of the volume with respect to the y-axis, R_y, is found by determining (1) the moment of inertia, (2) the mass of the volume, and (3) the square root of the quotient of the moment of inertia divided by the mass.

The moment of inertia of the volume (see Fig. 5–44) generated by rotating the area bounded by $x = g(y)$, $y = c$, $y = d$, and the y-axis about the x-axis, I_x, is given by

$$I_x = 2\pi k \int_c^d xy^3\, dy \qquad\qquad (5\text{--}21)$$

The radius of gyration of the volume with respect to the x-axis, R_x, is found in the same manner as R_y.

Figure 5-45

Figure 5-46

Example D

Find the moment of inertia and the radius of gyration of the solid (see Fig. 5–45) generated by rotating the area bounded by $y^3 = x$, $y = 2$, and the y-axis about the x-axis.

$$I_x = 2\pi k \int_0^2 xy^3 \, dy = 2\pi k \int_0^2 (y^3)y^3 \, dy = 2\pi k \left(\frac{1}{7}y^7\right)\Big|_0^2 = \frac{256\pi k}{7}$$

$$m = 2\pi k \int_0^2 xy \, dy = 2\pi k \int_0^2 y^3 y \, dy = 2\pi k \left(\frac{1}{5}y^5\right)\Big|_0^2 = \frac{64\pi k}{5}$$

$$R_x^2 = \frac{256\pi k}{7} \times \frac{5}{64\pi k} = \frac{20}{7} \qquad R_x = \sqrt{\frac{20}{7}} = \frac{2}{7}\sqrt{35}$$

Example E

Find the moment of inertia of a disk of radius r with respect to its axis.

To generate a disk (see Fig. 5–46) we shall rotate the area bounded by the axes, $x = r$, and $y = b$, about the y-axis. We then have

$$I_y = 2\pi k \int_0^r x^3 y \, dx = 2\pi k \int_0^r x^3(b) \, dx$$

$$= 2\pi kb \int_0^r x^3 \, dx = 2\pi kb \left(\frac{1}{4}x^4\right)\Big|_0^r = \frac{\pi kbr^4}{2}$$

The mass of the disk is $k(\pi r^2)b$. Rewriting the expression for I_y, we have

$$I_y = \frac{(\pi kbr^2)r^2}{2} = \frac{mr^2}{2}$$

This is the usual way of expressing the moment of inertia of a particular solid. That is, the mass is usually included in the formula for the moment of inertia.

Due to the limited methods of integration available at this point, we cannot integrate the expressions for the moments of inertia of circular areas or of a sphere. These will be introduced in Chapter 9 in the exercises, by which point the proper method of integration will have been developed.

Exercises 5–5

In Exercises 1 through 4, find the moment of inertia and the radius of gyration with respect to the origin of the given masses at the given points.

1. 5 units at $(2, 0)$ and 3 units at $(6, 0)$.

2. 3 units at $(-1, 0)$, 6 units at $(3, 0)$, 2 units at $(4, 0)$.

3. 4 units at $(-4, 0)$, 10 units at $(0, 0)$, 6 units at $(5, 0)$.

4. 6 units at $(-3, 0)$, 5 units at $(-2, 0)$, 9 units at $(1, 0)$, 2 units at $(8, 0)$.

In Exercises 5 through 22, find the indicated moment of inertia or radius of gyration.

5. Find the moment of inertia of the area bounded by $y^2 = x$, $x = 4$, and the x-axis with respect to the x-axis.

6. Find the moment of inertia of the area bounded by $y = 2x$, $x = 1$, $x = 2$, and the x-axis with respect to the y-axis.

7. Find the radius of gyration of the area bounded by $y = x^3$, $x = 2$, and the x-axis with respect to the y-axis.

8. Find the radius of gyration of the first-quadrant area bounded by the curve $y = 1 - x^2$ with respect to the y-axis.

9. Express the moment of inertia of the triangular area in Example C in terms of the mass of the area.

10. Find the moment of inertia of a rectangular area of sides a and b with respect to side a. Express the result in terms of the mass.

11. Find the radius of gyration of the area bounded by $y = x^2$, $x = 2$, and the x-axis with respect to the x-axis.

12. Find the radius of gyration of the area bounded by $y^2 = x^3$, $y = 8$, and the y-axis with respect to the y-axis.

13. Find the radius of gyration of the area of Exercise 12 with respect to the x-axis.

14. Find the radius of gyration of the first-quadrant area bounded by $x = 1$, $y = 2 - x$, and the y-axis with respect to the y-axis.

15. Find the moment of inertia with respect to its axis of the solid generated by rotating the area bounded by $y^2 = x$, $y = 2$, and the y-axis about the x-axis.

16. Find the radius of gyration with respect to its axis of the volume generated by rotating the first-quadrant area under the curve $y = 4 - x^2$ about the y-axis.

17. Find the radius of gyration with respect to its axis of the volume generated by rotating the area bounded by $y = 2x - x^2$ and the x-axis about the y-axis.

18. Find the radius of gyration with respect to its axis of the volume generated by rotating the area bounded by $y = 2x$ and $y = x^2$ about the y-axis.

19. Find the moment of inertia in terms of its mass of a right circular cone of radius r and height h with respect to its axis. See Fig. 5–47.

20. Find the moment of inertia in terms of its mass and with respect to its axis of a circular disk with a hole in the center. See Fig. 5–48.

Figure 5–47

Figure 5–48

21. A rotating drill head is in the shape of a right circular cone. Find the moment of inertia of the drill head with respect to its axis if its radius is 0.6 cm, its height is 0.8 cm, and its mass is 3 g. (See Exercise 19.)

22. Find the moment of inertia of a flywheel with respect to its axis if its inner radius is 4 cm, its outer radius is 6 cm, and its mass is 1.2 kg. (See Exercise 20.)

5–6 Work by a Variable Force

The next application of integration we shall demonstrate is to the physical problem of work done by a variable force. The concept of work is useful for comparing the amounts of mechanical energy required to perform various physical tasks.

Work, *in its physical sense, is the product of a constant force times the distance through which it acts.* When we consider the work done in stretching a spring, the first thing we recognize is that the more the spring is stretched, the greater is the force necessary to stretch it. Thus, the force varies. However, if we are stretching the spring a distance Δx, where we are considering the limit as $\Delta x \to 0$, the force can be considered as approaching a constant over Δx. Adding the products of force$_1$ times Δx_1, force$_2$ times Δx_2, and so forth, we see that the total work is the sum of these products. Thus, the work can be expressed as a definite integral in the form

$$W = \int_a^b f(x)\, dx$$

(5–22)

where $f(x)$ is the force as a function of the distance the spring is stretched. The limits a and b refer to the initial and the final distances the spring is stretched *from its normal length.*

One problem remains: We must find the function $f(x)$. From physics we learn that the force required to stretch a spring is proportional to the amount it is stretched (Hooke's law). If a spring is stretched x units from its normal length, then $f(x) = kx$. From conditions stated for a particular spring, the value of k may be determined. Thus, $W = \int_a^b kx\, dx$ is the formula for finding the total work done in stretching a spring.

Example A

A spring of natural length 12 in. requires a force of 6 lb to stretch it 2 in. (see Fig. 5–49). Find the work done in stretching it 6 in.

From Hooke's law, we have $F = kx$, or $6 = k2$, $k = 3$ lb/in. Thus,

$$W = \int_0^6 3x\, dx$$

$$= \frac{3}{2} x^2 \Big|_0^6$$

$$= 54 \text{ lb-in.}$$

12 in.

2 in.

6

Figure 5–49

Problems involving work by a variable force arise in many fields of technology. An illustration from electricity is found in the motion of an electric charge through an electric field created by another electric charge.

Electric charges are of two types, designated as positive and negative. A basic law is that charges of the same sign repel each other and charges of opposite signs attract each other. *The force between charges is proportional to the product of their charges, and inversely proportional to the square of the distance between them.* Thus for this case

$$f(x) = \frac{kq_1q_2}{x^2} \tag{5-23}$$

when q_1 and q_2 are the charges (in coulombs), x is the distance (in meters), the force is in newtons, and $k = 9 \times 10^9 \, \text{N} \cdot \text{m}^2/\text{C}^2$. For other systems of units, the numerical value of k is different. We can find the work done when electric charges move toward each other or when they separate by use of Eq. (5-23) in Eq. (5-22).

Example B

Find the work done when two α-particles, $q = 0.32$ aC each, move until they are 10 nm apart, if they were originally separated by 1 m.

Applying Eqs. (5-22) and (5-23) we have, with $q = 0.32$ aC $= 0.32 \times 10^{-18}$ C $= 3.2 \times 10^{-19}$ C, and an upper limit of 10 nm $= 10 \times 10^{-9}$ m $= 10^{-8}$ m,

$$W = \int_1^{10^{-8}} \frac{9 \times 10^9 (3.2 \times 10^{-19})^2}{x^2} \, dx$$

$$= 9.2 \times 10^{-28} \int_1^{10^{-8}} \frac{dx}{x^2}$$

$$= 9.2 \times 10^{-28} \left(-\frac{1}{x} \right) \Big|_1^{10^{-8}} = -9.2 \times 10^{-28}(10^8 - 1)$$

$$= -9.2 \times 10^{-20} \, \text{J}$$

Since $10^8 \gg 1$, where $\gg$ means "much greater than," the 1 may be neglected in the calculation. The meaning of the minus sign in the result is that work must be done *on* the system to move the particles together. If free to move, they would tend to separate.

The following two examples illustrate other types of problems in which work by a variable force is involved.

Example C

Figure 5–50

Find the work done in winding up 60 ft of a 100-ft cable which weighs 4 lb/ft.

First we let x denote the length of cable which has been wound up at any time. Then the force which is required to raise the remaining cable equals the weight of the cable which has not yet been wound up. This weight is the product of the unwound cable length $100 - x$ and its weight per unit length 4 lb/ft, or $4(100 - x)$. This force acts through a distance of dx. See Fig. 5–50. Therefore, since 60 ft are to be wound up, the work done is

$$W = \int_0^{60} 4(100 - x)\, dx = \int_0^{60} (400 - 4x)\, dx = 400x - 2x^2 \Big|_0^{60}$$

$$= 16,800 \text{ ft-lb}$$

Example D

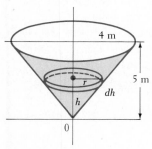

Figure 5–51

A conical tank with a circular cross-section has its axis vertical and its vertex at the bottom. The radius of the top of the tank is 4.00 m and the altitude is 5.00 m. The tank is filled with water. Find the work required to pump all of the water out over the top. The force of gravity on each cubic meter of water is 9800 N (see Fig. 5–51).

Consider the water to be divided into horizontal layers, each of which is dh meters thick. Each of these layers is circular, and therefore its volume is that of a circular disk (just as in finding volumes by use of disk elements). This means that the weight of each disk of water is $9800\pi(\text{radius}^2)(\text{thickness})$. From similar triangles in the figure we see that

$$\frac{r}{h} = \frac{4}{5} \quad \text{or} \quad r = 0.8h.$$

Also, the distance each disk of water must be raised is $5.00 - h$ meters. Therefore, summing the products of weight and distance, we find the work done.

$$W = \int_0^5 \underset{\text{Force}}{\underbrace{(9800\pi r^2\, dh)}}\, \underset{\text{Distance}}{\underbrace{(5 - h)}} = 9800\pi \int_0^5 (0.8h)^2(5 - h)\, dh$$

$$= 6270\pi \int_0^5 (5h^2 - h^3)\, dh$$

$$= 6270\pi \left(\frac{5}{3}h^3 - \frac{1}{4}h^4 \right) \Big|_0^5$$

$$= 6270\pi(52.1) = 1.03 \times 10^6 \text{ N} \cdot \text{m}$$

$$= 1.03 \text{ MJ}$$

Exercises 5–6

1. A spring of natural length 10 in. measures 14 in. when a 12-lb weight is attached to it. How much work is required to stretch it 5 in. from its natural length?

2. How much work is needed to stretch the spring in Exercise 1 from a length of 14 in. to a length of 19 in.?

3. It requires 500 lb to stretch a certain spring 1 ft. How much work is done on the spring if we attach a 300-lb weight to it?

4. If another 300-lb weight is attached to the spring in Exercise 3, how much work is done by the additional weight?

5. Find the work done in separating a 2 μC positive charge from a 4 μC negative charge to a distance of 6 cm from a distance of 2 cm.

6. Find the work done in separating a 0.8-nC negative charge from a 0.6-nC negative charge to a distance of 200 mm from a distance of 100 mm.

7. The gravitational force of attraction between two objects which are x units apart is given by $F = k/x^2$. If two objects are 1 cm apart, find the work done in separating them to a distance of 1 m apart. Express the answer in terms of k.

8. Given two objects which are 10 ft apart, find the work required to separate them further, until they are 100 ft apart. Express the result in terms of k. (See Exercise 7.)

9. Find the work done in winding up a 200-ft cable which weighs 100 lb.

10. In Exercise 9, if an object weighing 50 lb is attached at the end of the cable, find the work done in winding it up.

11. Find the work done in winding up 20 m of a 25-m rope on which the force of gravity is 6.00 N/m.

12. A chain on which the force of gravity is 12.0 N/m is being unwound from a winch. If 20 m have already been unwound, how much work is done by the force of gravity when an additional 30 m are unwound?

13. At lift-off, a rocket weighs 32.5 tons, including the weight of its fuel. It is fired vertically, and, during the first stage of ascent, the fuel is consumed at the rate of 1.25 tons per 1000 ft of ascent. How much work is done in lifting the rocket to an altitude of 12,000 ft?

14. While descending, a 550-N weather balloon enters a zone of freezing rain in which ice forms on it at the rate of 7.50 N per 100 m of descent. Find the work done on the balloon during the first 1000 m of descent through the freezing rain.

15. Find the work done in pumping the water out of the top of a cylindrical tank 3 ft in radius and 10 ft high, if the tank is initially full and water weighs 62.4 lb/ft³.

16. Find the work done in pumping the water from the tank in Exercise 15, if the tank in initially half full.

17. Find the work done in emptying the tank in Example D, if the water is pumped to a level of 1 m above the top of the tank, and the water is initially 4 m deep in the tank.

18. A hemispherical tank of radius 10.0 ft is full of oil. Find the work done in pumping the oil to the top of the tank. The oil weighs 60.0 lb/ft³.

5–7 Force Due to Liquid Pressure

In physics it is shown that at any point within a liquid the pressure is the same in all directions. Also, it is shown that *the pressure is directly proportional to the depth below the surface of the liquid.* If the weight per unit volume of the liquid is w, the pressure at a depth h is

$$p = wh \qquad (5–24)$$

For water, $w = 62.4$ lb/ft^3 or $w = 9800$ N/m^3. *Since pressure is the force per unit area, the force on an area at a given depth is*

$$F = pA \qquad (5–25)$$

Substituting the expression from Eq. (5–24) into Eq. (5–25), we have

$$\underline{F = whA} \qquad (5–26)$$

as the force on an area A at a depth h.

Let us now assume that the plate shown in Fig. 5–52 is submerged vertically in water. At points on the plate at different depths the pressure will be different. Therefore, in using Eq. (5–26) to find the force on the plate, we must choose an area A such that all of its points are at the same depth. We approximate this by choosing the horizontal strip of length l and width dh. The force on this strip is $wh\,(l\,dh)$. By using integration to sum the forces on all such strips from the top of the plate to its bottom, we find *the total force on the plate to be*

Figure 5–52

$$\underline{F = w \int_{a}^{b} lh\,dh} \qquad (5–27)$$

Here l is the length of the element of area, h is the depth of the element of area, w is the weight per unit volume of the liquid, a is the depth of the top and b is the depth of the bottom of the area on which the force is exerted.

Example A

A vertical rectangular floodgate is 5 ft long and 4 ft deep. Find the force on the gate if its upper edge is 3 ft below the surface (see Fig. 5–53).

Each element of area of the floodgate has a length of 5 ft, which means that $l = 5$ ft. Since the top of the gate is 3 ft below the surface, $a = 3$ ft, and since the gate is 4 ft deep, $b = 7$ ft. Using $w = 62.4$ lb/ft^3 we have the force on the gate as

$$F = 62.4 \int_{3}^{7} 5h\,dh = 312 \int_{3}^{7} h\,dh = 156h^2 \Big|_{3}^{7} = 156(49 - 9)$$

$$= 6240 \text{ lb}$$

Figure 5–53

Example B

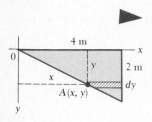

Figure 5–54

The vertical end of a tank of water is in the shape of a right triangle as shown in Fig. 5–54. What is the force on the end of the tank?

In setting up the figure it is convenient to use coordinate axes. It is also convenient to have the y-axis directed downward, since we integrate from the top to the bottom of the area. The equation of the line OA is $y = \frac{1}{2}x$. Thus, we see that the length of an element of area of the end of the tank is $4 - x$, the depth of the element of area is y, the top of the tank is 0 and the bottom is $y = 2$ m. Therefore, the force on the end of the tank is ($w = 9800$ N/m³)

$$F = 9800 \int_0^2 \overset{\overset{\text{Length}}{\downarrow}}{(4 - x)}\,\overset{\overset{\text{Depth}}{\downarrow}}{(y)}(dy) = 9800 \int_0^2 (4 - 2y)(y\,dy)$$

$$= 19{,}600 \int_0^2 (2y - y^2)\,dy = 19{,}600\left(y^2 - \frac{1}{3}y^3\right)\bigg|_0^2 = 26{,}100 \text{ N}$$

Example C

The part of the vertical end of a boat which is under the waterline is a semicircle with a radius of 5 ft, as shown in Fig. 5–55. Find the force exerted on the end of the boat by the water.

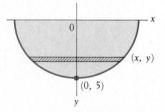

Figure 5–55

Again we select our axes such that the y-axis is directed downward. The equation of the circle is $x^2 + y^2 = 25$. The length of an element of area is $2x$, the depth of an element is y, the top of the water is at $y = 0$, and the bottom is at $y = 5$ ft. Therefore, the force on the end of the boat is

$$F = 62.4 \int_0^5 (2x)(y)\,dy = 62.4 \int_0^5 2y\sqrt{25 - y^2}\,dy$$

$$= -\frac{2}{3}(62.4)(25 - y^2)^{3/2}\bigg|_0^5 = \frac{2}{3}(62.4)(125) = 5200 \text{ lb}$$

Exercises 5–7

1. One end of a spa is a vertical rectangular wall 12 ft wide. What is the force exerted on this wall by the water if it is 2.5 ft deep?

2. Find the force on one end of a square tank 2 m on an edge if the tank is full of water.

3. Find the force on a square floodgate 2 m on an edge if the upper edge is 4 m below the surface of the water.

4. Find the force on a rectangular floodgate 15 ft wide and 6 ft deep if the upper edge is 2 ft below the surface of the water.

5. A rectangular sea aquarium observation window is 10 ft wide and 5 ft high. What is the force on this window if the upper edge is 4 ft below the surface of the water? The density of seawater is 64.0 lb/ft^3.

6. Another observation window in the aquarium is 20 ft below (bottom edge to top edge) the window of Exercise 5. What is the force on this window?

7. The end of a trough in the shape of an isosceles right triangle has a horizontal upper edge of 8 ft and is 8 ft deep. Find the force on the end if the trough is full of water.

8. A right triangular plate of base 2 m and height 1 m is submerged vertically, as shown in Fig. 5–56. Find the force on one side of the plate.

9. Find the force on the plate of Exercise 8 if the upper vertex is 3 m below the surface.

10. The end of a tank is in the shape of a trapezoid, as shown in Fig. 5–57. Find the force on the end of the tank if it is full of gasoline (density 42.0 lb/ft^3).

11. A horizontal tank with vertical circular ends has a radius of 4 ft. It is filled to a depth of 4 ft with oil of density 60.0 lb/ft^3. Find the force on one end of the tank.

12. A horizontal cyclindrical tank has vertical elliptical ends with the minor axis horizontal. The major axis is 6 ft and the minor axis is 4 ft. Find the force on one end if the tank is half filled with fuel oil of density 50.0 lb/ft^3.

13. A small dam is in the shape of the area bounded by $y = x^2$ and $y = 20$. Find the force on the area below $y = 4$ if the surface of the water is at the top of the dam. All distances are in feet.

14. Find the force on the area bounded by $x = 2y - y^2$ and the y-axis if the upper point of the area is at the surface of the water. All distances are in feet.

Surface

1 m   1 m

2 m

Figure 5–56

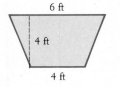

6 ft

4 ft

4 ft

Figure 5–57

5–8 Other Applications

To illustrate the great variety of applications of integration, we shall demonstrate three more types of applied problems in this section. The formulas will be presented without proof, and then examples of their use will be given. The first application is that of finding the length of an arc of a curve.

If $y = f(x)$ is the equation of a given curve, *the **length of arc** s of the curve from $x = a$ to $x = b$ is given by*

$$s = \int_a^b \sqrt{1 + \left(\frac{dy}{dx}\right)^2} \, dx \qquad (5\text{–}28)$$

Example A

Find the length of arc of the curve $y = \frac{2}{3}x^{3/2}$ from $x = 0$ to $x = 3$ (see Fig. 5–58).

First we identify $a = 0$ and $b = 3$. Then we note that we must find the derivative of y with respect to x. Therefore, $dy/dx = x^{1/2}$. Substituting in Eq. (5–28), we find the length of arc.

$$s = \int_0^3 \sqrt{1 + (x^{1/2})^2}\, dx = \int_0^3 \sqrt{1 + x}\, dx = \frac{2}{3}(1 + x)^{3/2}\Big|_0^3$$

$$= \frac{2}{3}[4^{3/2} - 1^{3/2}] = \frac{2}{3}(8 - 1) = \frac{14}{3}$$

This means that the distance between points O and A on the curve, moving along the curve, is $\frac{14}{3}$ units.

The second application of integration in this section is that of finding the area of a surface of revolution. *If the arc of a curve is rotated about the x-axis, it generates a surface of revolution from $x = a$ to $x = b$ for which the area is given by*

$$S = 2\pi \int_a^b y \sqrt{1 + \left(\frac{dy}{dx}\right)^2}\, dx \tag{5–29}$$

Example B

Find the area of the surface generated by rotating about the x-axis the arc of the parabola $y = \sqrt{x}$ from $x = 0$ to $x = 2$ (see Fig. 5–59).

First we note that $a = 0$ and $b = 2$. The derivative of $y = \sqrt{x}$ is $\frac{dy}{dx} = \frac{1}{2x^{1/2}}$. Substituting into Eq. (5–29) we have the area of the surface.

$$S = 2\pi \int_0^2 x^{1/2} \sqrt{1 + \left(\frac{1}{2x^{1/2}}\right)^2}\, dx = 2\pi \int_0^2 x^{1/2} \sqrt{1 + \frac{1}{4x}}\, dx$$

$$= \pi \int_0^2 \sqrt{4x + 1}\, dx = \frac{\pi}{4}\left(\frac{2}{3}\right)(4x + 1)^{3/2}\Big|_0^2 = \frac{\pi}{6}(9^{3/2} - 1^{3/2})$$

$$= \frac{\pi}{6}(27 - 1) = \frac{13\pi}{3}$$

This means that the area of the surface is $\frac{13}{3}\pi$ square units.

The third application of integration in this section is that of the average value of a function. In general *an average is found by summing up the quantities to be*

Figure 5-58

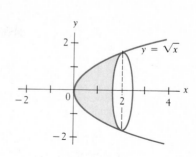

Figure 5-59

averaged, and then dividing by the total number of them. Generalizing on this and using integration for the summation, *the average value of a function y with respect to x from x = a to x = b is given by*

$$y_{av} = \frac{\displaystyle\int_a^b y\, dx}{b - a}$$

(5-30)

The following examples illustrate the applications of the average value of a function.

Example C

The velocity of an object falling under the influence of gravity as a function of time is given by $v = 32t$. What is the average velocity (in feet per second) with respect to time for the first 3 s?

In this case, the average velocity with respect to t is

$$v_{av} = \frac{\displaystyle\int_0^3 v\, dt}{3 - 0} = \frac{\displaystyle\int_0^3 32t\, dt}{3} = \frac{16t^2}{3}\Big|_0^3$$

$$= 48 \text{ ft/s}$$

This result can be interpreted as meaning that an average velocity of 48 ft/s for 3 s would result in the same distance, 144 ft, being traveled as that with the variable velocity. Since $s = \int v\, dt$, the numerator represents the distance traveled.

Example D

The power developed in a certain resistor as a function of current is $P = 6i^2$. What is the average power (in watts) with respect to the current as the current changes from 2 to 5 A?

The average value of P with respect to i is

$$P_{av} = \frac{\displaystyle\int_2^5 P \, di}{5 - 2} = \frac{6 \displaystyle\int_2^5 i^2 \, di}{3} = \frac{2}{3} i^3 \Big|_2^5 = \frac{2}{3}(125 - 8) = 78 \text{ W}$$

In general, it might be noted that the average value of y with respect to x is that value of y, which when multiplied by the length of the interval for x, gives the same area as that under the curve of y as a function of x.

Exercises 5–8

In Exercises 1 through 4, find the length of arc of each of the given curves.

1. $y = \frac{2}{3}x^{3/2}$ from $x = 3$ to $x = 8$
2. $y = \frac{2}{3}(x - 1)^{3/2}$ from $x = 1$ to $x = 4$
3. $y = \frac{2}{3}(x^2 + 1)^{3/2}$ from $x = 0$ to $x = 3$
4. $y = \frac{1}{6}x^3 + \dfrac{1}{2x}$ from $x = 1$ to $x = 3$

In Exercises 5 through 8, find the area generated by rotating the given arc about the x-axis.

5. $y = x$ from $x = 0$ to $x = 2$
6. $y = \frac{4}{3}x$ from $x = 0$ to $x = 3$
7. $y = 2\sqrt{x}$ from $x = 0$ to $x = 3$
8. $y = x^3$ from $x = 1$ to $x = 3$

In Exercises 9 through 12, find the indicated average values.

9. The electric current as a function of time for a certain circuit is given by $i = 4t - t^2$. Find the average value of the current (in amperes), with respect to time, for the first 4 s.

10. Find the average value of the function $y = \sqrt{x}$, with respect to x, from $x = 1$ to $x = 9$.

11. An object moves along the x-axis according to the relation $x = t^3 - 3t$. Find the average value of the velocity with respect to t, as t changes from 1 to 3.

12. Find the average value of the volume of a sphere with respect to the radius.

5–9 Review Exercises for Chapter 5

1. What is the velocity after 5 s of an object which is dropped and falls due to gravity?

2. Find an expression for the distance traveled by an object for which the velocity is given by $v = 3t/\sqrt{t^2 + 1}$, if $s = 0$ when $t = 0$.

3. A weather balloon is rising at the rate of 20 ft/s when a small metal part drops off. If the balloon is 200 ft high at this instant, when will the part hit the ground?

4. The acceleration of an object rolling down an inclined plane is 12 ft/s². If it starts from rest at the top of the plane, find an expression for the distance the object travels from the top as a function of the time.

5. The current in a certain electric resistor is given by $i = 5t^{2/3}$. Find the charge which passes a given point in the circuit during the third second.

6. The current in a certain electric circuit is given by $i = \sqrt{1 + 4t}$, where i is in amperes and t in seconds. Find the charge which passes a given point during the first two seconds.

7. The voltage across a 55.0 nF capacitor is zero. What is the voltage after 20 ms if a current of 10 mA charges the capacitor?

8. What is the capacitance of a capacitor if a current $i = \sqrt[6]{3t + 1}$ charges it such that the voltage across it is 200 V at $t = 0$ and 280 V when $t = 0.7$ s?

9. The distribution of weight on a cable is not uniform. If the slope of the cable at any point is given by $dy/dx = 20 + \frac{1}{40}x^2$, and if the origin of the coordinate system is at the lowest point, find the equation which gives the curve described by the cable.

10. The slope of a certain cable is given by $dy/dx = -0.1/\sqrt{x}$. If the point (0, 0) is on the cable, find the equation which gives the curve described by the cable.

11. Find the area between $y = \sqrt{1 - x}$ and the coordinate axes.

12. Find the area bounded by $y = 3x^2 - x^3$ and the x-axis.

13. Find the area bounded by $y^2 = 2x$ and $y = x - 4$.

14. Find the area bounded by $y = 1/(2x + 1)^2$, $y = 0$, $x = 1$, and $x = 2$.

15. Find the area between $y = x^2$ and $y = x^3 - 2x^2$.

16. Find the area between $y = x^2 + 2$ and $y = 3x^2$.

17. Find the volume generated by rotating the area bounded by $y = 3 + x^2$ and the line $y = 4$ about the x-axis.

18. Find the volume generated by rotating the area of Exercise 12 about the x-axis.

19. Find the volume generated by rotating the area of Exercise 12 about the y-axis.

20. Find the volume generated by rotating the area bounded by $y = x$ and $y = 3x - x^2$ about the y-axis.

21. Find the volume generated by revolving an ellipse about its major axis.

22. A hole of radius 1 cm is bored along the diameter of a sphere of radius 4 cm. Find the volume of the material which is removed from the sphere

23. Find the centroid of the area bounded by $y^2 = x^3$ and $y = 2x$.

24. Find the centroid of the area bounded by $y = 2x - 4$, $x = 1$, and $y = 0$.

25. Find the centroid of the volume generated by rotating the area bounded by $y = \sqrt{x}$, $x = 1$, $x = 4$, and $y = 0$ about the x-axis.

26. Find the centroid of the volume generated by rotating the area bounded by $yx^4 = 1$, $y = 1$, and $y = 4$ about the y-axis.

27. Find the moment of inertia of the area bounded by $y = 3x - x^2$ and $y = x$ with respect to the y-axis.

28. Find the radius of gyration of the first-quadrant area which is bounded by $y = 8 - x^3$ with respect to the y-axis.

29. Find the moment of inertia with respect to its axis of the solid generated by rotating the area bounded by $y = x^{1/2}$, $y = 0$, and $x = 8$ about the x-axis.

30. Find the radius of gyration with respect to its axis of the solid formed by rotating the area bounded by $xy = 1$, $x = 1$, $x = 3$, and $y = \frac{1}{3}$ about the x-axis.

31. If the cost C of producing x units of a product is a function of x, the derivative of C with respect to x is called the marginal cost function. A paint manufacturer determines that the marginal cost function of producing x gallons of paint per day is $dC/dx = 2 + 0.004x$. Find the total cost of producing 300 gal of paint in a day if fixed costs of operation per day are $2500 (i.e., $C = 2500$ when $x = 0$).

32. As a result of the dumping of treated waste water from a manufacturing plant into a stream, the temperature of the stream rises while the plant is in operation. The rate of change of the temperature, T, in degrees Celsius, is measured as a function of time t, in hours from 8 A.M. until 8 A.M. of the following day, to be $dT/dt = 0.0004(t^3 - 36t^2 + 288t)$. If $T = 15.0°C$ at 8 A.M., what is the temperature at 4 P.M.?

33. The deck area of a boat is a parabolic section as shown in Fig. 5–60. What is the area of the deck?

34. The nose cone of a rocket has the shape of a semiellipse rotated about its major axis, as shown in Fig. 5–61. What is the volume of the nose cone?

35. A certain spring is 16 in. long when 12 lb are hung from it. It measures 12 in. when 4 lb are hung from it. How much work is done when the spring is stretched from its natural length by hanging 6 lb on it?

36. If the pressure of a gas is 400 kPa when the volume is 243 cm³, how much work is done in compressing the gas to 32 cm³, if it undergoes an adiabatic change such that $pV^{1.4} = c$?

37. The gravitational force, in pounds, of the earth on a satellite (the weight of the satellite) is given by $F = 10^{11}/x^2$, where x is the vertical distance, in miles, from the center of the earth to the satellite. How much work is done in moving the satellite from the earth's surface to an altitude of 2000 mi? The radius of the earth is 3960 mi.

38. A pail and its contents weigh 80 lb. The pail is attached to the end of a 100-ft rope which weighs 10 lb and is hanging vertically. How much work is done in winding up the rope with the pail attached?

39. Find the work done in pumping water out at the top of a cylindrical tank 4 m in diameter and 3 m high, if the tank is initially full. ($w = 9800$ N/m³.)

40. A chemical holding tank has the shape of a frustum of a right circular cone, as shown in Fig. 5–62. How much work is done in emptying the tank if it is initially filled with a chemical of density 45.0 lb/ft³?

Figure 5–60

Figure 5–61

Figure 5–62

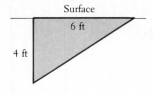

Figure 5–63

41. A square plate is submerged vertically in a tank of water with its upper edge 2 ft below the surface. If the edge of the plate is 4 ft, find the force on one side of the plate.

42. A section of a dam is in the shape of a right triangle. The base of the triangle is 6 ft and is in the surface of the water. If the triangular section goes to a depth of 4 ft, find the force on it. See Fig. 5–63.

43. Find the length of arc of the curve $y = \frac{2}{3}(x^2 - 1)^{3/2}$ from $x = 1$ to $x = 3$.

44. Find the area of the surface generated by rotating $y = \frac{3}{4}x$ from $x = 0$ to $x = 2$ about the x-axis.

45. The electric resistance of a wire is inversely proportional to the square of its radius. If a certain wire has a resistance of $0.3 \ \Omega$ when its radius is 2.0 mm, find the average value of the resistance with respect to the radius if the radius changes from 2.0 mm to 2.1 mm.

46. The velocity of an object as a function of the distance fallen after it is dropped is $v = 8\sqrt{s}$. Find the average value of v with respect to s during the third second, if $s = 16t^2$, and s is measured in feet.

47. The production rate (in pounds per hour) of a certain chemical plant is given by the following table. Using the trapezoidal rule, determine the approximate daily production.

Time	12 M.	4 A.M.	8 A.M.	12 N.	4 P.M.	8 P.M.	12 M.
Production rate	2000	2000	7000	12,000	12,000	8000	2000

48. The current as a function of time was measured in a particular circuit, with the results tabulated below. What charge (in coulombs) was transmitted across a given point in the first 6 s? Use Simpson's rule.

i (amperes)	0.0	1.5	4.0	7.2	7.5	4.3	1.6
t (seconds)	0.0	1.0	2.0	3.0	4.0	5.0	6.0

Derivatives of the Trigonometric and Inverse Trigonometric Functions

Many important applications in technology are periodic in nature. That is, these phenomena have properties that repeat after definite periods of time. Applications of this type arise in alternating-current theory and harmonic motion analysis, as well as in many other fields. Such phenomena may be represented by the use of the trigonometric functions, especially the sine and cosine functions.

In this chapter we review the trigonometric functions and the inverse trigonometric functions, which are among the most important transcendental (nonalgebraic) functions. The primary purpose of the chapter is to find the formulas for the derivatives of these functions and to show some of the important applications of these derivatives.

6–1 The Trigonometric Functions

In this section and the next we present a summary of the trigonometric functions and their properties for purposes of review and reference. The material is essential to the development of later topics in this chapter and later chapters.

Figure 6–1

An **angle** *in* **standard position** *is one with its vertex at the origin and its initial side the positive x-axis. If the angle is measured counterclockwise it is* **positive,** *and if it is measured clockwise it is* **negative** *(see Fig. 6–1).*

There are two widely used measurements of angles. These are the **degree** and the **radian.** The degree is used primarily in triangle solution, and the radian is used in other types of applications and for more theoretical purposes.

The **degree** *is defined as* $\frac{1}{360}$ *of a complete revolution.* One method of subdividing a degree is to use decimal parts of a degree. Most calculators use degrees in decimal form. Another traditional way is to divide each degree into 60 equal parts called **minutes** and each minute into 60 equal parts called **seconds.** The symbols °, ′,″ are used to designate degrees, minutes, and seconds, respectively. Degrees and minutes are commonly used in tables of values of the trigonometric functions.

Although it is assumed that most of your calculational work in using this text will be done on a calculator, it may sometimes be necessary to change an angle expressed in terms of minutes to the same angle expressed in decimal parts, or conversely. The method of making such changes is illustrated in the following example.

Example A

To change the angle 43°24′ to decimal form, we use the fact that $1' = (\frac{1}{60})°$. Therefore, $24' = (\frac{24}{60})° = 0.4°$. This means that

$$43°24' = 43.4°$$

To change 154.36° to an angle expressed to the nearest minute, we use the fact that 1° = 60′. Therefore,

$$0.36° = 0.36(60') = 21.6'$$

To the nearest minute, we have 154.36° = 154°22′.

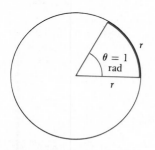

Figure 6–2

A **radian** *is the measure of an angle with its vertex at the center of a circle and with an intercepted arc on the circle equal in length to the radius of the circle.* See Fig. 6–2.

Since the radius may be laid off 2π times along the circumference, it follows that there are 2π radians in one complete rotation. Also, there are 360° in one complete rotation. Therefore, 360° is *equivalent* to 2π radians. It then follows that the relation between degrees and radians is 2π rad = 360°, or

$$\pi \text{ rad} = 180°$$

(6–1)

Example B

$$1° = \frac{\pi}{180} \text{ rad} = 0.01745 \text{ rad}, \qquad 1 \text{ rad} = \frac{180°}{\pi} = 57.3°$$

$$\frac{\pi}{3} \text{ rad} = \frac{180°}{3} = 60°, \qquad \frac{\pi}{10} \text{ rad} = \frac{180°}{10} = 18°$$

We see from Eq. (6–1) that (1) *to convert an angle measured in degrees to the same angle measured in radians, we multiply the number of degrees by* $\pi/180$, and (2) *to convert an angle measured in radians to the same angle measured in degrees, we multiply the number of radians by* $180/\pi$. Also, we should note that due to these conversions it is common to express angles in radians in terms of π.

Example C

$$18.0° = \left(\frac{\pi}{180}\right)(18.0) = \frac{\pi}{10.0} = \frac{3.14}{10.0} = 0.314 \text{ rad}$$

$$2.00 \text{ rad} = \left(\frac{180°}{\pi}\right)(2.00) = \frac{360°}{3.14} = 114.6°$$

$$30° = \left(\frac{\pi}{180}\right)(30) = \frac{\pi}{6} \text{ rad}, \qquad \frac{3\pi}{4} \text{ rad} = \left(\frac{180°}{\pi}\right)\left(\frac{3\pi}{4}\right) = 135°$$

We wish now to make a very important point. Since π is a special way of writing the number (slightly greater than 3) that is the ratio of the circumference of a circle to its diameter, it is the ratio of one distance to another. Thus radians really have no units and *radian measure amounts to measuring angles in terms of real numbers*. It is this property of radians that makes them useful in many situations. Therefore, when radians are being used, it is customary that no units are indicated for the angle. *When no units are indicated, the radian is understood to be the unit of angle measurement.*

Example D

$$60.0° = \left(\frac{\pi}{180}\right)(60.0) = \frac{\pi}{3.00} = 1.05$$

$$2.50 = \left(\frac{180°}{\pi}\right)(2.50) = \frac{450°}{3.14} = 143°$$

Since no units are indicated for 1.05 or 2.50 in this example, they are known to be radian measure.

The six **trigonometric functions** *of an angle* θ *in standard position are sine* θ, *cosine* θ, *tangent* θ, *cotangent* θ, *secant* θ, *and cosecant* θ. They are defined in terms of the coordinates (x, y) of a point on its terminal side, at a distance r (considered positive) from the origin. *The definitions and standard*

Figure 6-3

abbreviations of the functions are as follows (see Fig. 6–3):

$$\sin \theta = \frac{y}{r}, \qquad \cos \theta = \frac{x}{r}, \qquad \tan \theta = \frac{y}{x}$$

$$\cot \theta = \frac{x}{y}, \qquad \sec \theta = \frac{r}{x}, \qquad \csc \theta = \frac{r}{y}$$

(6–2)

Example E

Figure 6-4

Find the values of $\sin \theta$, $\cos \theta$, and $\tan \theta$ for angle θ as shown in Fig. 6–4.

We see that the terminal side of θ passes through $(-3, 4)$. Therefore we find r from the Pythagorean theorem as

$$r = \sqrt{(-3)^2 + 4^2} = \sqrt{25} = 5$$

Now using the values $x = -3$, $y = 4$, and $r = 5$, the values of the functions of θ are

$$\sin \theta = \frac{4}{5} \qquad \cos \theta = -\frac{3}{5} \qquad \tan \theta = -\frac{4}{3}$$

The values of the trigonometric functions can be found with a scientific calculator, and it is assumed that such values will usually be found in this way. To find the values of $\sin \theta$, $\cos \theta$, and $\tan \theta$, we use the keys so marked. Also, from the definitions, we see that *csc θ is the reciprocal of sin θ, sec θ is the reciprocal of cos θ, and cot θ is the reciprocal of tan θ.* Thus, the reciprocal key is also used in finding values of these functions.

In order to find the angle for a given value of a function we make use of the **inverse trigonometric functions,** which are discussed in detail in Section 6–5. For calculator purposes of the moment, it is sufficient to recognize the notation that is used. The notation for "the angle whose sine is x" is $\text{Sin}^{-1}x$ or Arcsin x. Similar meanings are given to $\text{Cos}^{-1}x$, Arccos x, $\text{Tan}^{-1}x$, and Arctan x. On a calculator, the $\boxed{\text{SIN}^{-1}}$, $\boxed{\text{ARCSIN}}$, or the sequence of $\boxed{\text{INV}}$ $\boxed{\text{SIN}}$ keys is used to find x, given $\sin x$.

Example F

To find $\sin 37°$ we use the calculator sequence 37 $\boxed{\text{SIN}}$. The display shows the value 0.6018150. Thus $\sin 37° = 0.6018$, with the result rounded off.

To find $\sec 27.82°$, we use the calculator sequence 27.82 $\boxed{\text{COS}}$ $\boxed{1/x}$. The display is 1.1306869. Therefore, $\sec 27.82° = 1.131$ with the result rounded off.

To find θ, given that $\cos \theta = 0.3527$, we use the calculator sequence $.3527$ $\boxed{\text{INV}}$ $\boxed{\text{COS}}$. This now gives us a display of 69.347452, which means that $\cos 69.35° = 0.3527$. Thus, $\theta = 69.35°$ with the result rounded off.

In Example F we assumed that the calculator was in degree mode. We can also use a calculator when the angle is expressed in radians. When the calculator is in radian mode, it uses values in radians directly, following the procedure shown in Example F.

Before the extensive use of calculators, the values of the trigonometric functions were generally obtained from tables. Since it may be necessary to use such a table to obtain or check the value of a trigonometric function, a table is included in the appendices. Also, since there are occasions when it is useful to have exact values, the following table is included for reference.

Table 6–1

Degrees	0°	30°	45°	60°	90°	180°	270°	360°
Radians	0	$\dfrac{\pi}{6}$	$\dfrac{\pi}{4}$	$\dfrac{\pi}{3}$	$\dfrac{\pi}{2}$	π	$\dfrac{3\pi}{2}$	2π
Sine	0	$\dfrac{1}{2}$	$\dfrac{\sqrt{2}}{2}$	$\dfrac{\sqrt{3}}{2}$	1	0	-1	0
Cosine	1	$\dfrac{\sqrt{3}}{2}$	$\dfrac{\sqrt{2}}{2}$	$\dfrac{1}{2}$	0	-1	0	1
Tangent	0	$\dfrac{\sqrt{3}}{3}$	1	$\sqrt{3}$	Undef.	0	Undef.	0

We note that the signs of the functions differ, depending on the quadrant in which the terminal side lies. It is important that we properly analyze an angle and the sign of its function, for even a calculator will not always give the required angle for a given value of a function. Since any angle in standard position is coterminal with some positive angle less than 360°, we need consider only the problem of finding the values of trigonometric functions of positive angles less than 360°. From the definitions, we see that the signs of the functions in the four quadrants are as follows:

All functions of first-quadrant angles are positive. Sin θ and csc θ are positive for second-quadrant angles. Tan θ and cot θ are positive for third-quadrant angles. Cos θ and sec θ are positive for fourth-quadrant angles. All others are negative.

By use of congruent triangles, it can be seen from Figs. 6–5, 6–6, and 6–7 that $\alpha = \theta_1$, and therefore values of functions of α are the same as those of θ_1. Also, values of functions of θ_2, θ_3, and θ_4 are *numerically* the same as those of α for the respective quadrants. Thus,

$$
\begin{aligned}
F(\theta_2) &= \pm F(\alpha), & \alpha &= 180° - \theta_2 & \theta_2 \text{ in second} \\
& & &= \pi - \theta_2, & \text{quadrant} \\
F(\theta_3) &= \pm F(\alpha), & \alpha &= \theta_3 - 180° & \theta_3 \text{ in third} \\
& & &= \theta_3 - \pi, & \text{quadrant} \\
F(\theta_4) &= \pm F(\alpha), & \alpha &= 360° - \theta_4 & \theta_4 \text{ in fourth} \\
& & &= 2\pi - \theta_4, & \text{quadrant}
\end{aligned}
$$

(6-3)

Figure 6–5

Figure 6–6

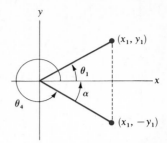

Figure 6–7

Here, F represents any of the trigonometric functions, and α *is the* **reference angle,** *which is the acute angle formed by the terminal side of the angle and the x-axis.* The sign to be used depends on the sign of the function in the given quadrant. By using the reference angle we can express the function of any angle in terms of the same function of an acute angle.

Example G

Illustrations of the use of Eqs. (6–3) are as follows:

$$\sin 160° = +\sin (180° - 160°) = \sin 20° = 0.3420$$
$$\tan 110° = -\tan (180° - 110°) = -\tan 70° = -2.747$$
$$\cos 225° = -\cos (225° - 180°) = -\cos 45° = -0.7071$$
$$\cot 260° = +\cot (260° - 180°) = \cot 80° = 0.1763$$
$$\sec 304° = +\sec (360° - 304°) = \sec 56° = 1.788$$
$$\sin 357° = -\sin (360° - 357°) = -\sin 3° = -0.0523$$

One of the clearest ways to demonstrate the variation in the trigonometric functions is by means of their graphs. Since the graphs of the sine and cosine functions are of primary importance, we shall restrict our attention here to them.

In plotting the trigonometric functions, it is normal to express the angle in radians. In this way both x and y are expressed as numbers.

For the equations $y = a \sin(bx + c)$ and $y = a \cos(bx + c)$, where $b > 0$, we have the following important quantities related to their graphs:

1. *the* **amplitude** *(equal to* $|a|$*)*,

2. *the* **period** $\left(equal\ to\ \dfrac{2\pi}{b}\right)$,

3. *the* **displacement** $\left(equal\ to\ -\dfrac{c}{b}\right)$.

The amplitude is the maximum number of units the curve may be from the x-axis. Any function F has a period P if, for all x, $F(x) = F(x + P)$. The effect of displacement (or phase shift) is to shift the curve of $y = a \sin bx$ or $y = a \cos bx$ to the left if $c > 0$, or to shift it to the right if $c < 0$.

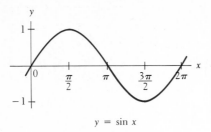

$y = \sin x$

Figure 6–8

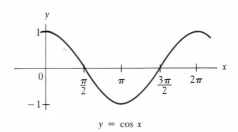

$y = \cos x$

Figure 6–9

Thus, by knowing the basic shape of the sine and cosine curves, which are shown in Figs. 6–8 and 6–9, and using the information obtained from the equation, we may sketch the graphs of sine and cosine functions. The following examples illustrate the method.

Example H

Sketch the graph of $y = 3 \sin(2x + \frac{1}{4}\pi)$.

First we note that $a = 3$, which means that the amplitude of the curve is 3. Next we see that $b = 2$. This means that the period of the curve is $2\pi/2 = \pi$. Therefore, points on the curve will "repeat" every π units along the x-axis. Finally $c = \frac{1}{4}\pi$, which means the displacement is $-(\frac{1}{4}\pi)/2 = -\frac{1}{8}\pi$. This means the curve is displaced $\frac{1}{8}\pi$ units to the left of the curve of $y = 3 \sin 2x$. Therefore, "starting" the curve at $x = -\frac{1}{8}\pi$, we know it starts repeating at $x = \frac{7}{8}\pi$ (π units to the right of $x = -\frac{1}{8}\pi$), again at $\frac{15}{8}\pi$, etc. With this information we can construct the following table, from which the curve can quickly be sketched (see Fig. 6–10).

x	$-\dfrac{\pi}{8}$	$\dfrac{\pi}{8}$	$\dfrac{3\pi}{8}$	$\dfrac{5\pi}{8}$	$\dfrac{7\pi}{8}$	$\dfrac{9\pi}{8}$
y	0	3	0	-3	0	3

Figure 6–10

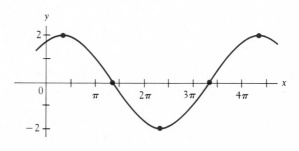

Figure 6–11

Example I

Sketch the graph of the function $y = 2 \cos(\frac{1}{2}x - \frac{1}{6}\pi)$; amplitude = 2, period = $2\pi/\frac{1}{2} = 4\pi$, displacement = $\frac{1}{6}\pi/\frac{1}{2} = \frac{1}{3}\pi$ (to the right).
We construct the table of values:

x	$\frac{\pi}{3}$	$\frac{4\pi}{3}$	$\frac{7\pi}{3}$	$\frac{10\pi}{3}$	$\frac{13\pi}{3}$
y	2	0	-2	0	2

We then sketch the graph in Fig. 6–11. We note that when the coefficient of x is less than 1, the period is extended.

Exercises 6–1

In Exercises 1 and 2, change the given angles to equal angles expressed in decimal form. In Exercises 3 and 4, change the given angles to equal angles expressed to the nearest minute.

1. 15°12′   2. 301°16′   3. 315.8°   4. 24.92°

In Exercises 5 through 8, express the given angle measurements in terms of π.

5. 15°, 150°   6. 12°, 225°   7. 75°, 330°   8. 36°, 315°

In Exercises 9 through 12, the given numbers express angle measurement. Express the measure of each angle in terms of degrees.

9. $\frac{2\pi}{5}, \frac{3\pi}{2}$ 10. $\frac{3\pi}{10}, \frac{5\pi}{6}$ 11. $\frac{\pi}{18}, \frac{7\pi}{4}$ 12. $\frac{7\pi}{15}, \frac{4\pi}{3}$

In Exercises 13 and 14, express the given angles in radian measure. In Exercises 15 and 16, express the given numbers in terms of angle measure in degrees to the nearest 0.1°.

13. 23.0°   14. 178.5°   15. 0.750   16. 16.4

In Exercises 17 through 28, find the values of the indicated trigonometric functions.

17. cos 214°   18. sin 286°   19. csc 137°   20. sec 476°

21. $\sin \frac{\pi}{4}$ 22. $\cos \frac{\pi}{6}$ 23. $\tan \frac{5\pi}{12}$ 24. $\sin \frac{7\pi}{18}$

25. $\cot \frac{5\pi}{6}$ 26. $\tan \frac{4\pi}{3}$ 27. cos 4.596   28. cot 3.27

In Exercises 29 through 32, find the exact values of the indicated trigonometric functions. See Table 6–1.

29. $\sin \frac{3}{4}\pi$ 30. $\tan \frac{5}{6}\pi$ 31. $\cos \frac{5}{3}\pi$ 32. $\csc \frac{5}{2}\pi$

In Exercises 33 through 36, find θ for $0 \leq \theta < 2\pi$.

33. $\sin \theta = 0.3420$ 34. $\cos \theta = 0.4384$

35. $\tan \theta = -1.192$ 36. $\cot \theta = -0.5095$

In Exercises 37 through 44, sketch the graph of each of the given functions.

37. $y = 3 \sin x$ 38. $y = 2 \cos 2x$ 39. $y = -4 \sin 3x$

40. $y = 6 \cos \pi x$ 41. $y = 2 \sin\left(3x - \dfrac{\pi}{2}\right)$ 42. $y = 4 \cos\left(\dfrac{x}{2} + \dfrac{\pi}{8}\right)$

43. $y = \cos\left(\pi x - \dfrac{\pi}{2}\right)$ 44. $y = \sin(\pi x + 1)$

In Exercises 45 through 50, sketch the indicated graphs.

45. The electric current in a certain 60-Hz alternating-current circuit is given by $i = 10 \sin 120\pi t$, where i is the current (in amperes) and t is the time (in seconds). Sketch the graph of i vs. t for $0 \leq t \leq 0.1$ s.

46. A generator produces a voltage given by $v = 200 \cos 50\pi t$, where t is the time (in seconds). Sketch the graph of v vs. t for $0 \leq t \leq 0.1$ s.

47. A wave traveling in a string may be represented by the equation

$$y = A \sin 2\pi\left(\frac{t}{T} - \frac{x}{\lambda}\right)$$

Here A is the amplitude, t is the time the wave has traveled, x is the distance from the origin, T is the time required for the wave to travel one wavelength λ (the Greek lambda). Sketch three cycles of the wave for which $A = 2.00$ cm, $T = 0.100$ s, $\lambda = 20.0$ cm, and $x = 5.00$ cm.

48. The cross section of a particular water wave is

$$y = 0.5 \sin\left(\frac{\pi}{2}x + \frac{\pi}{4}\right)$$

where x and y are measured in feet. Sketch two cycles of y vs. x.

49. If the upper end of a spring is not fixed and is being moved with a sinusoidal motion, the motion of the bob at the end of the spring is affected. Plot the curve of the motion of the upper end of a spring which is being moved by an external force according to the equation $y = 4 \sin 2t - 2 \cos 2t$. This may be done graphically by plotting $y_1 = 4 \sin 2t$ and $y_2 = 2 \cos 2t$ and then finding $y = y_1 - y_2$. This method is referred to as *addition of ordinates*.

50. The current in a certain electric circuit is given by $i = 4 \sin 60\pi t + 2 \cos 120\pi t$. Sketch the graph of i vs. t.

6–2 Basic Trigonometric Relations

Referring to the definitions of the trigonometric functions, Eqs. (6–2), we see that there are a number of relations which exist among the functions. Certain other important relationships may also be established. We shall find that these have very definite application, since certain problems rely on a change of form for solution. The ability to use these basic relations effectively depends to a large extent on being familiar with them and recognizing them in different forms. Therefore, in this section we shall review the important trigonometric relations.

Directly from the definitions we observe that

$$\sin \theta = \frac{1}{\csc \theta}, \qquad \cos \theta = \frac{1}{\sec \theta}, \qquad \tan \theta = \frac{1}{\cot \theta} \tag{6–4}$$

Also, since

$$\tan \theta = \frac{y}{x} = \frac{y/r}{x/r}$$

we have

$$\tan \theta = \frac{\sin \theta}{\cos \theta}, \qquad \cot \theta = \frac{\cos \theta}{\sin \theta} \tag{6–5}$$

Certain important relations can be found by use of the definitions and the Pythagorean relation $x^2 + y^2 = r^2$. By dividing each term of the Pythagorean relation by r^2, we have

$$\sin^2\theta + \cos^2\theta = 1 \tag{6–6}$$

By dividing each term of the Pythagorean relation by x^2, we obtain

$$1 + \tan^2\theta = \sec^2\theta \tag{6–7}$$

By dividing each term of the Pythagorean relation by y^2, we obtain

$$\cot^2\theta + 1 = \csc^2\theta \tag{6–8}$$

Many other relationships which exist among the trigonometric functions may be proven valid by the basic identities (6–4) to (6–8). The following example illustrates the method of proving such trigonometric identities.

Example A

Prove the identity $\sec^2 x + \csc^2 x = \sec^2 x \csc^2 x$.

Here we note the presence of $\sec^2 x$ and $\csc^2 x$ on each side. This suggests the possible use of the square relationships. By replacing the $\sec^2 x$ on the right-hand side by $1 + \tan^2 x$, we can create $\csc^2 x$ plus another term. The left-hand side is the $\csc^2 x$ plus another term, so this procedure should help. Thus, changing only the right side,

$$\sec^2 x + \csc^2 x = \sec^2 x \csc^2 x$$
$$= (1 + \tan^2 x)(\csc^2 x)$$
$$= \csc^2 x + \tan^2 x \csc^2 x$$

Now we note that $\tan x = \sin x / \cos x$ and $\csc x = 1/\sin x$. Thus,

$$\sec^2 x + \csc^2 x = \csc^2 x + \left(\frac{\sin^2 x}{\cos^2 x}\right)\left(\frac{1}{\sin^2 x}\right)$$

$$= \csc^2 x + \frac{1}{\cos^2 x} = \csc^2 x + \sec^2 x$$

We could have used many other variations of this procedure, and they would have been perfectly valid.

Referring to Fig. 6–12 and the definitions of the functions we see, for example, that $\sin \theta = y/r$ and $\sin(-\theta) = -y/r$. Thus, in this way we arrive at the following important relations concerning negative angles:

$$\sin(-\theta) = -\sin \theta, \qquad \cos(-\theta) = \cos \theta,$$
$$\tan(-\theta) = -\tan \theta \tag{6–9}$$

For reference, Eq. (1–1) is
$d = \sqrt{(x_2 - x_1)^2 + (y_2 - y_1)^2}$

By use of the distance formula, Eq. (1–1), and Fig. 6–13, we may establish the **law of cosines.** Thus,

$$c^2 = (b \cos \theta - a)^2 + (b \sin \theta)^2$$
$$= b^2 \cos^2 \theta - 2ab \cos \theta + a^2 + b^2 \sin^2 \theta$$
$$= b^2 (\cos^2 \theta + \sin^2 \theta) + a^2 - 2ab \cos \theta$$

$$c^2 = a^2 + b^2 - 2ab \cos \theta \tag{6–10}$$

Figure 6–12

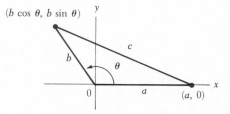

Figure 6–13

We shall now develop a number of important equations, using the distance formula, the law of cosines, and some of the other relations already established. Also, some of those we derive will be used in turn to derive others. Since there are numerous uses of the various equations of the section in the following material, the reader will gain facility in the use of the relationships by following the development.

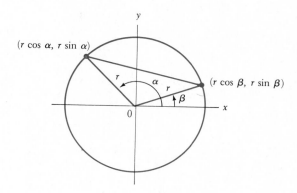

Figure 6–14

From Fig. 6–14 we have

$$(r \cos \alpha - r \cos \beta)^2 + (r \sin \alpha - r \sin \beta)^2$$
$$= r^2 + r^2 - 2(r)(r) \cos(\alpha - \beta)$$
$$r^2 \cos^2\alpha - 2r^2 \cos \alpha \cos \beta + r^2 \cos^2\beta$$
$$+ r^2 \sin^2\alpha - 2r^2 \sin \alpha \sin \beta + r^2 \sin^2\beta = 2r^2 - 2r^2\cos(\alpha - \beta)$$

Dividing through by r^2 and collecting terms we have

$$\cos^2\alpha + \sin^2\alpha - 2(\cos \alpha \cos \beta + \sin \alpha \sin \beta) + \cos^2\beta + \sin^2\beta$$
$$= 2 - 2 \cos(\alpha - \beta)$$

Rearranging terms after using Eq. (6–6) we obtain

$$\cos(\alpha - \beta) = \cos \alpha \cos \beta + \sin \alpha \sin \beta \qquad (6\text{–}11)$$

Equation (6–11) is one of the important **addition formulas,** the remainder of which we shall now develop.

Writing $\cos(\alpha + \beta) = \cos[\alpha - (-\beta)]$, and using Eqs. (6–9) and (6–11), we have

$$\cos(\alpha + \beta) = \cos \alpha \cos \beta - \sin \alpha \sin \beta \qquad (6\text{–}12)$$

In Eq. (6–11), if we let $\alpha = 90°$, the result is

$$\cos(90° - \beta) = \cos 90° \cos \beta + \sin 90° \sin \beta$$

Since $\sin 90° = 1$, $\cos 90° = 0$, we obtain

$$\cos(90° - \beta) = \sin \beta \tag{6–13}$$

Since β is a general representation of an angle, and $\cos \beta$ is a function of β, we may substitute other expressions for angles for β. Therefore, replacing β by $90° - \beta$ in Eq. (6–13) we have

$$\cos[90° - (90° - \beta)] = \sin(90° - \beta)$$

or

$$\sin(90° - \beta) = \cos \beta \tag{6–14}$$

Now, replacing β by $\alpha + \beta$ in Eq. (6–13) we obtain

$$\sin(\alpha + \beta) = \cos[(90° - \alpha) - \beta]$$
$$= \cos(90° - \alpha) \cos \beta + \sin(90° - \alpha) \sin \beta$$

and by using Eqs. (6–13) and (6–14) we arrive at the equation

$$\sin(\alpha + \beta) = \sin \alpha \cos \beta + \cos \alpha \sin \beta \tag{6–15}$$

If we replace β by $-\beta$ in Eq. (6–15) we have

$$\sin(\alpha - \beta) = \sin \alpha \cos \beta - \cos \alpha \sin \beta \tag{6–16}$$

Example B

Show that

$$\sin\left(\frac{\pi}{4} + x\right) \cos\left(\frac{\pi}{4} + x\right) = \frac{1}{2}(\cos^2 x - \sin^2 x)$$

$$\sin\left(\frac{\pi}{4} + x\right) \cos\left(\frac{\pi}{4} + x\right)$$

$$= \left(\sin \frac{\pi}{4} \cos x + \cos \frac{\pi}{4} \sin x\right)\left(\cos \frac{\pi}{4} \cos x - \sin \frac{\pi}{4} \sin x\right)$$

$$= \sin \frac{\pi}{4} \cos \frac{\pi}{4} \cos^2 x - \sin^2 \frac{\pi}{4} \sin x \cos x + \cos^2 \frac{\pi}{4} \sin x \cos x$$

$$\quad - \sin^2 x \sin \frac{\pi}{4} \cos \frac{\pi}{4}$$

$$= \frac{\sqrt{2}}{2} \frac{\sqrt{2}}{2} \cos^2 x - \left(\frac{\sqrt{2}}{2}\right)^2 \sin x \cos x + \left(\frac{\sqrt{2}}{2}\right)^2 \sin x \cos x - \frac{\sqrt{2}}{2} \frac{\sqrt{2}}{2} \sin^2 x$$

$$= \frac{1}{2} \cos^2 x - \frac{1}{2} \sin^2 x = \frac{1}{2}(\cos^2 x - \sin^2 x)$$

Letting $A = \alpha + \beta$ and $B = \alpha - \beta$ leads to $\alpha = \frac{1}{2}(A + B)$ and $\beta = \frac{1}{2}(A - B)$. Using these in Eqs. (6–15) and (6–16) and subtracting, we obtain

$$\sin A - \sin B = 2 \cos \frac{A + B}{2} \sin \frac{A - B}{2} \tag{6–17}$$

We shall find this equation of use in the following section.

If $\alpha = \beta$ in Eq. (6–15) we have

$$\sin 2\alpha = 2 \sin \alpha \cos \alpha \tag{6–18}$$

Also, if $\alpha = \beta$ in Eq. (6–12), we obtain

$$\cos 2\alpha = \cos^2\alpha - \sin^2\alpha \tag{6–19}$$

Using Eq. (6–6), we may rewrite Eq. (6–19) in the form

$$\cos 2\alpha = 2 \cos^2\alpha - 1 = 1 - 2 \sin^2\alpha \tag{6–20}$$

Equations (6–18), (6–19), and (6–20) are the **double-angle formulas.**

By replacing 2α by α in Eq. (6–20) we obtain

$$\sin \frac{\alpha}{2} = \pm \sqrt{\frac{1 - \cos \alpha}{2}} \tag{6–21}$$

and

$$\cos \frac{\alpha}{2} = \pm \sqrt{\frac{1 + \cos \alpha}{2}} \tag{6–22}$$

Equations (6–21) and (6–22) are the **half-angle formulas.**

Example C

Given that $\cos \alpha = \frac{3}{5}$, where $3\pi/2 < \alpha < 2\pi$, find $\sin 2\alpha$ and $\cos \alpha/2$.

Knowing that $\cos \alpha = \frac{3}{5}$ for an angle in the fourth quadrant, we then determine that $\sin \alpha = -\frac{4}{5}$. Thus

$$\sin 2\alpha = 2\left(-\frac{4}{5}\right)\left(\frac{3}{5}\right) = -\frac{24}{25}$$

Also, since α is between $3\pi/2$ and 2π, $\alpha/2$ must be between $3\pi/4$ and π, which means that $\alpha/2$ is in the second quadrant and therefore $\cos \alpha/2$ is negative. Thus,

$$\cos \frac{\alpha}{2} = -\sqrt{\frac{1 + 3/5}{2}} = -0.8944$$

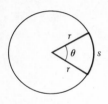

Figure 6–15

Two other useful relations are found from the definition of the radian. From geometry we know that *the length of an arc on a circle is proportional to the central angle,* and that the length of the arc of a complete circle is the circumference. Letting *s* represent the length of arc, we may state that $s = 2\pi r$ for a complete circle, or *the length of arc is*

$$s = \theta r \qquad\qquad (6-23)$$

for a circular arc with central angle θ, measured in radians (see Fig. 6–15).

Another geometric fact is that *areas of sectors of circles are proportional to their central angles.* The area of a complete circle is $A = \pi r^2$. This can be written as $A = \frac{1}{2}(2\pi)r^2$. We now note that the angle for a complete circle is 2π, and therefore *the area of a sector of a circle in terms of the radius and the central angle, in radians, is*

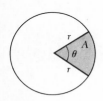

Figure 6–16

$$A = \frac{1}{2}\theta r^2 \qquad\qquad (6-24)$$

See Fig. 6–16.

Example D

(a)

(b)

Figure 6–17

Find the length of a circular arc if the central angle is 218° and the radius of the circle is 5.25 in. Also find the area of the corresponding circular sector.

Expressing θ in radians as $218\left(\dfrac{\pi}{180}\right)$, we have

$$s = (218)\left(\frac{\pi}{180}\right)(5.25) = 20.0 \text{ in.}$$

and

$$A = \frac{1}{2}(218)\left(\frac{\pi}{180}\right)(5.25)^2 = 52.4 \text{ in.}^2$$

See Figs. 6–17(a) and 6–17(b).

Exercises 6-2

In Exercises 1 through 24, prove the given trigonometric identities.

1. $\dfrac{\cot x}{\cos x} = \csc x$

2. $\dfrac{\sin x}{\tan x} = \cos x$

3. $\tan y \csc y = \sec y$

4. $\sin x \cot x = \cos x$

5. $\cos^2 x - \sin^2 x = 1 - 2\sin^2 x$

6. $\cos^2 \theta (1 + \tan^2 \theta) = 1$

7. $\sin x \tan x + \cos x = \sec x$

8. $\tan x + \cot x = \sec x \csc x$

9. $\tan x(\tan x + \cot x) = \sec^2 x$

10. $\tan^2 y \sec^2 y - \tan^4 y = \tan^2 y$

11. $\dfrac{\sec \theta}{\cos \theta} - \dfrac{\tan \theta}{\cot \theta} = 1$

12. $\dfrac{\csc \theta}{\sin \theta} - \dfrac{\cot \theta}{\tan \theta} = 1$

13. $\sin(x + y)\sin(x - y) = \sin^2 x - \sin^2 y$

14. $\cos(x + y)\cos(x - y) = \cos^2 x - \sin^2 y$

15. $\cos(x - y) + \sin(x + y) = (\cos x + \sin x)(\cos y + \sin y)$

16. $\cos^4 x - \sin^4 x = \cos 2x$

17. $2\sin x + \sin 2x = \dfrac{2\sin^3 x}{1 - \cos x}$

18. $2 \csc 2x \tan x = \sec^2 x$

19. $\dfrac{\sin 3x}{\sin x} + \dfrac{\cos 3x}{\cos x} = 4\cos 2x$

20. $\dfrac{\sin 3x}{\sin x} - \dfrac{\cos 3x}{\cos x} = 2$

21. $\dfrac{1 - \cos \alpha}{2\sin(\alpha/2)} = \sin\dfrac{\alpha}{2}$

22. $\tan\dfrac{\alpha}{2} = \pm\sqrt{\dfrac{1 - \cos \alpha}{1 + \cos \alpha}}$

23. $2\sin^2\dfrac{x}{2} + \cos x = 1$

24. $2\cos^2\dfrac{\theta}{2}\sec \theta = \sec \theta + 1$

In Exercises 25 through 34, solve the given problems.

25. The displacements y_1 and y_2 of two waves traveling through the same medium are given by

$$y_1 = A \sin 2\pi\left(\frac{t}{T} - \frac{x}{\lambda}\right) \quad \text{and} \quad y_2 = A \sin 2\pi\left(\frac{t}{T} + \frac{x}{\lambda}\right)$$

Find an expression for the displacement $y_1 + y_2$ of the combination of the waves.

26. In the analysis of the angles of incidence i and reflection r of a light ray subject to certain conditions, the following expression is found:

$$E_2\left(\frac{\tan r}{\tan i} + 1\right) = E_1\left(\frac{\tan r}{\tan i} - 1\right)$$

Show that an equivalent expression is

$$E_2 = E_1\frac{\sin(r - i)}{\sin(r + i)}$$

Figure 6–18

7.535 m

3.755 m   8.250 m

Figure 6–19

27. The equation for the displacement of a certain object at the end of a spring is $y = A \sin 2t + B \cos 2t$. Show that this equation may be written as

$$y = C \sin(2t + \alpha)$$

where $C = \sqrt{A^2 + B^2}$ and $\tan \alpha = B/A$. [*Hint:* Let $A/C = \cos \alpha$ and $B/C = \sin \alpha$.]

28. In developing an expression for the power in an alternating current circuit, the expression $\sin \omega t \sin(\omega t + \phi)$ is found. Show that this expression can be written as $\frac{1}{2}[\cos \phi - \cos(2\omega t + \phi)]$.

29. A pendulum 3.00 ft long oscillates through an angle of 5.0°. Find the distance through which the end of the pendulum swings in going from one extreme position to the other.

30. An ammeter needle is deflected 52.0° by a current of 0.200 A. The needle is 3.00 cm long and a circular scale is used. How long is the scale for a maximum current of 1.00 A?

31. Two streets meet at an angle of 82.0°. What is the length of the piece of curved curbing at the intersection if it is constructed along the arc of a circle 15.0 ft in radius? See Fig. 6–18.

32. What is the floor area of the hallway shown in Fig. 6–19? The outside and inside of the hallway are circular arcs.

33. A horizontal water pipe has a radius of 6.00 cm. If the depth of water in the pipe is 3.00 cm, what percent of the volume of the pipe is filled?

34. A conical tent is made from a circular piece of canvas 15.0 ft in diameter, with a sector of central angle 120° removed. What is the surface area of the tent?

6–3 Derivatives of the Sine and Cosine Functions

In Chapter 2 we introduced the operation of differentiation. The functions we used, however, were algebraic functions in all cases. Now that we have developed the process of finding derivatives and reviewed the properties of the trigonometric functions, we are in a position to develop the formulas necessary for finding the derivatives of the trigonometric functions.

If we find the derivative of the sine function, it is possible to use this derivative as a basis for finding the derivatives of the other trigonometric and inverse trigonometric functions. Therefore, we shall now find the derivative of the sine function, by using the delta-process.

Let $y = \sin u$, where u is a function of x and is expressed in radians. If u changes by Δu, y then changes by Δy. Thus,

$$y + \Delta y = \sin(u + \Delta u)$$
$$\Delta y = \sin(u + \Delta u) - \sin u$$

$$\frac{\Delta y}{\Delta u} = \frac{\sin(u + \Delta u) - \sin u}{\Delta u}$$

Referring now to Eq. (6–17) we have

$$\frac{\Delta y}{\Delta u} = \frac{2 \sin\frac{1}{2}(u + \Delta u - u)\cos\frac{1}{2}(u + \Delta u + u)}{\Delta u}$$

$$= \frac{\sin(\Delta u/2)\cos[u + (\Delta u/2)]}{\Delta u/2}$$

Looking ahead to the next step of letting $\Delta u \to 0$, we see that the numerator and denominator both approach zero. This situation is precisely the same as that in which we were finding the derivatives of the algebraic functions. To find the limit, we must find

$$\lim_{\Delta u \to 0} \frac{\sin(\Delta u/2)}{\Delta u/2}$$

since these are the factors which cause the numerator and the denominator to approach zero.

In finding this limit, we shall let $\theta = \Delta u/2$ for convenience of notation. This means that we are to determine $\lim_{\theta \to 0} \dfrac{\sin \theta}{\theta}$. Of course, it would be convenient to know before proceeding if this limit does actually exist. Therefore, by using a calculator, we can now develop a table of values of $\dfrac{\sin \theta}{\theta}$ as θ becomes very small.

θ (radians)	0.5	0.1	0.05	0.01	0.001
$\dfrac{\sin \theta}{\theta}$	0.9588511	0.9983342	0.9995834	0.9999833	0.9999998

We see from this table that the limit of $\dfrac{\sin \theta}{\theta}$, as θ approaches zero, appears to be 1.

In order to prove that $\lim_{\theta \to 0} \dfrac{\sin \theta}{\theta} = 1$, we use a geometric approach. Considering Fig. 6–20, we can see that the following inequality is true for the indicated areas:

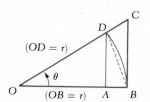

$(OD = r)$

θ

$O \quad (OB = r) \quad A \quad B$

Figure 6–20

Area triangle OBD < area sector OBD < area triangle OBC

$$\frac{1}{2}r(r \sin \theta) < \frac{1}{2}r^2\theta < \frac{1}{2}r(r \tan \theta) \quad \text{or} \quad \sin \theta < \theta < \tan \theta$$

Remembering that we want to find the limit of $(\sin \theta)/\theta$, we next divide through by $\sin \theta$, and then take reciprocals:

$$1 < \frac{\theta}{\sin \theta} < \frac{1}{\cos \theta} \quad \text{or} \quad 1 > \frac{\sin \theta}{\theta} > \cos \theta$$

When we consider the limit as $\theta \to 0$, we see that the left member remains 1 and the right member approaches 1. Thus, $(\sin \theta)/\theta$ must approach 1. This means

$$\lim_{\theta \to 0} \frac{\sin \theta}{\theta} = \lim_{\Delta u \to 0} \frac{\sin(\Delta u/2)}{\Delta u/2} = 1 \qquad (6\text{–}25)$$

Using the result expressed in Eq. (6–25) in the expression for $\Delta y / \Delta u$, we have

$$\lim_{\Delta u \to 0} \frac{\Delta y}{\Delta u} = \lim_{\Delta u \to 0} \left[\cos\left(u + \frac{\Delta u}{2}\right) \frac{\sin(\Delta u / 2)}{\Delta u / 2} \right] = \cos u$$

or

$$\frac{dy}{du} = \cos u \qquad\qquad (6\text{–}26)$$

However, we want the derivative of y with respect to x. This requires the use of the chain rule, Eq. (2–28), which we repeat here for reference.

$$\frac{dy}{dx} = \frac{dy}{du}\frac{du}{dx} \qquad\qquad (6\text{–}27)$$

Combining Eqs. (6–26) and (6–27), we have ($y = \sin u$)

$$\frac{d(\sin u)}{dx} = \cos u \frac{du}{dx} \qquad\qquad (6\text{–}28)$$

Example A

Find the derivative of $y = \sin 2x$.
In this example $u = 2x$. Thus,

$$\frac{dy}{dx} = \frac{d(\sin 2x)}{dx} = \cos 2x \overset{\overset{\textstyle \frac{du}{dx}}{\downarrow}}{\frac{d(2x)}{dx}} = (\cos 2x)(2)$$

$$= 2 \cos 2x$$

Example B

Find the derivative of $y = 2 \sin(x^2)$.
In this example $u = x^2$, which means that $du/dx = 2x$. Hence,

$$\frac{dy}{dx} = 2[\cos(x^2)](2x)$$

$$= 4x \cos(x^2)$$

 It is important here, just as it is in finding the derivatives of powers of all functions, to remember to include the factor du/dx.

Example C

For reference, Eq. (2–29) is
$$\frac{du^n}{dx} = nu^{n-1}\frac{du}{dx}$$

Find the derivative of $y = \sin^2 x$.
This example is a combination of the use of the power rule, Eq. (2–29), and the derivative of the sine function, Eq. (6–28). Since $\sin^2 x$ means $(\sin x)^2$, in using the power rule we have $u = \sin x$. Thus,

$$\frac{dy}{dx} = 2(\sin x)\frac{d \sin x}{dx}$$

$$= 2 \sin x \cos x = \sin 2x$$

Example D

Find the derivative of $y = 2\sin^3(2x^4)$.

In the general power rule, $u = \sin(2x^4)$. In the derivative of the sine function, $u = 2x^4$. Thus, we have

$$\frac{dy}{dx} = \frac{d\,2[\sin(2x^4)]^3}{dx}$$

$$= 2(3)[\sin(2x^4)]^2 \frac{d[\sin(2x^4)]}{dx} \qquad \text{Using } \frac{du^n}{dx} = nu^{n-1}\frac{du}{dx}$$

$$\blacktriangleright \qquad = 6\sin^2(2x^4)\cos(2x^4)\frac{d(2x^4)}{dx} \qquad \text{Using } \frac{d\sin u}{dx} = \cos u\frac{du}{dx}$$

$$= 6\sin^2(2x^4)\cos(2x^4)(8x^3)$$

$$= 48x^3\sin^2(2x^4)\cos(2x^4)$$

To find the derivative of the cosine function, we write it in the form $\cos u = \sin[(\pi/2) - u]$. Thus, if $y = \sin[(\pi/2) - u]$, we have

$$\frac{dy}{dx} = \cos\left(\frac{\pi}{2} - u\right)\frac{d[(\pi/2) - u]}{dx} = \cos\left(\frac{\pi}{2} - u\right)\left(-\frac{du}{dx}\right)$$

$$= -\cos\left(\frac{\pi}{2} - u\right)\frac{du}{dx}$$

Since $\cos[(\pi/2) - u] = \sin u$, we have

$$\frac{d(\cos u)}{dx} = -\sin u\frac{du}{dx} \qquad\qquad\qquad (6\text{–}29)$$

Example E

Find the derivative of $y = \cos^2(2x)$.

Using the power rule and Eq. (6–29), we have

$$\frac{dy}{dx} = 2\cos(2x)(-\sin 2x)(2) = -4\sin 2x\cos 2x$$

$$= -2\sin 4x$$

Example F

Find the derivative of $y = \sin 2x\cos x^2$.

Using the product rule and the derivatives of the sine and cosine functions, we arrive at the following result:

$$\frac{dy}{dx} = \sin 2x(-\sin x^2)(2x) + \cos x^2(\cos 2x)(2)$$

$$= -2x\sin 2x\sin x^2 + 2\cos 2x\cos x^2$$

Example G

Find the slope of a tangent line to the curve of $y = 5 \sin 3x$, where $x = 0.2$.

Here we are to find the derivative of $y = 5 \sin 3x$, and then evaluate the derivative for $x = 0.2$. Therefore, we have the following:

$$y = 5 \sin 3x$$

$$\frac{dy}{dx} = 5(\cos 3x)(3) = 15 \cos 3x$$

$$\left.\frac{dy}{dx}\right|_{x=0.2} = 15 \cos 3(0.2) = 15 \cos 0.6$$

$$= 15(0.8253) = 12.38$$

▶ In evaluating the slope we must remember that $x = 0.2$ means the values are in radians. Therefore, the slope is 12.38.

Exercises 6–3

In Exercises 1 through 32, find the derivatives of the given functions.

1. $y = \sin(x + 2)$ 2. $y = 3 \sin 4x$ 3. $y = 2 \sin(2x - 1)$

4. $y = 5 \sin(3 - x)$ 5. $y = 5 \cos 2x$ 6. $y = \cos(1 - x)$

7. $y = 2 \cos(3x - 1)$ 8. $y = 4 \cos(6x^2 + 5)$ 9. $y = \sin^2 4x$

10. $y = 3 \sin^3(2x + 1)$ 11. $y = 3 \cos^3(5x + 2)$ 12. $y = 4 \cos^2\sqrt{x}$

13. $y = x \sin 3x$ 14. $y = x^2 \sin 2x$ 15. $y = 3x^3 \cos 5x$

16. $y = x \cos 2x^3$ 17. $y = \sin x^2 \cos 2x$ 18. $y = \sin x \cos 4x$

19. $y = 2 \sin 2x \cos x$ 20. $y = 6 \sin 4x \cos 2x^2$ 21. $y = \dfrac{\sin 3x}{x}$

22. $y = \dfrac{2x}{\sin 4x}$ 23. $y = \dfrac{2 \cos x^2}{3x}$ 24. $y = \dfrac{5x}{\cos x}$

25. $y = 2 \sin^2 3x \cos 2x$ 26. $y = \cos^3 4x \sin^2 2x$ 27. $y = \dfrac{\cos^2 3x}{1 + 2 \sin^2 2x}$

28. $y = \dfrac{\sin 5x}{3 - \cos^2 3x}$ 29. $y = \sin^3 x - \cos 2x$ 30. $y = x \sin x + \cos x$

31. $y = x - \dfrac{1}{3} \sin^3 4x$ 32. $y = 2x \sin x + 2 \cos x - x^2 \cos x$

In Exercises 33 through 46, solve the given problems.

33. Verify the values of $(\sin \theta)/\theta$ in the table where $(\sin \theta)/\theta$ is calculated for $\theta = 0.5$, 0.1, 0.05, 0.01, and 0.001.

34. Evaluate $\lim_{\theta \to 0} (\tan \theta)/\theta$. (Use the fact that $\lim_{\theta \to 0} (\sin \theta)/\theta = 1$.)

35. On a calculator find the values of:
    (a) cos 1.0000 and (b) (sin 1.0001 − sin 1.0000)/0.0001.
    Compare the values and give the meaning of each in relation to the derivative of the sine function, where $x = 1$.

36. On a calculator find the values of:
    (a) −sin 1.0000 and (b) (cos 1.0001 − cos 1.0000)/0.0001.
    Compare the values and give the meaning of each in relation to the derivative of the cosine function, where $x = 1$.

37. Show that $\dfrac{d^4 \sin x}{dx^4} = \sin x$.

38. If $y = \cos 2x$, show that $\dfrac{d^2 y}{dx^2} = -4y$.

39. Find the derivative of the members of the identity $\cos 2x = 2 \cos^2 x - 1$ and thereby obtain another trigonometric identity.

40. Find the derivative of each of the members of the identity $2 \sin^2 x = 1 - \cos 2x$ and thereby obtain another trigonometric identity.

41. Evaluate the derivative of $y = 3 \sin 2x$ where $x = \pi/8$.

42. Evaluate the derivative of $y = 6 \cos x^2$ where $x = 0.5$.

43. Find the slope of a line tangent to the curve of $y = x \cos 2x$ where $x = 1.2$.

44. Find the slope of a line tangent to the curve of $y = \dfrac{2 \sin 3x}{x}$ where $x = 0.15$.

45. Find the velocity after 2 s, if the displacement from a reference point of a particle is given by $s = 2 \sin 3t + t^2$, where s is in feet.

46. The current (in amperes) in a certain electric circuit, as a function of time (in seconds), is given by $i = 10 \cos(120\pi t + \pi/6)$. Find the expression for the voltage across a 2-H inductor. (See Exercise 23 of Section 5–1.)

6–4 Derivatives of the Other Trigonometric Functions

We can find the derivatives of the other trigonometric functions by expressing the functions in terms of the sine and cosine. After we perform the differentiation, we use trigonometric relations to put the derivative in a convenient form.

We obtain the derivative of tan u by expressing tan u as sin u/cos u. Therefore, letting $y = \sin u / \cos u$, by employing the quotient rule we have

$$\frac{dy}{dx} = \frac{\cos u [\cos u (du/dx)] - \sin u [-\sin u (du/dx)]}{\cos^2 u}$$

$$= \frac{\cos^2 u + \sin^2 u}{\cos^2 u} \frac{du}{dx} = \frac{1}{\cos^2 u} \frac{du}{dx} = \sec^2 u \frac{du}{dx}$$

$$\frac{d(\tan u)}{dx} = \sec^2 u \frac{du}{dx} \qquad (6\text{–}30)$$

We find the derivative of cot u by letting $y = \cos u / \sin u$, and again using the quotient rule:

$$\frac{dy}{dx} = \frac{\sin u[-\sin u(du/dx)] - \cos u[\cos u(du/dx)]}{\sin^2 u}$$

$$= \frac{-\sin^2 u - \cos^2 u}{\sin^2 u}\frac{du}{dx}$$

$$\frac{d(\cot u)}{dx} = -\csc^2 u \frac{du}{dx} \tag{6–31}$$

To obtain the derivative of sec u, we let $y = 1/\cos u$. Then

$$\frac{dy}{dx} = -(\cos u)^{-2}\left[(-\sin u)\left(\frac{du}{dx}\right)\right] = \frac{1}{\cos u}\frac{\sin u}{\cos u}\frac{du}{dx}$$

$$\frac{d(\sec u)}{dx} = \sec u \tan u \frac{du}{dx} \tag{6–32}$$

We obtain the derivative of csc u by letting $y = 1/\sin u$. And so

$$\frac{dy}{dx} = -(\sin u)^{-2}\left(\cos u \frac{du}{dx}\right) = -\frac{1}{\sin u}\frac{\cos u}{\sin u}\frac{du}{dx}$$

$$\frac{d(\csc u)}{dx} = -\csc u \cot u \frac{du}{dx} \tag{6–33}$$

Example A

Find the derivative of $y = 2 \tan 8x$.

Using Eq. (6–30), we have

$$\frac{dy}{dx} = 2(\sec^2 8x)(\overset{\overset{\textstyle \frac{du}{dx}}{\downarrow}}{8})$$

$$= 16 \sec^2 8x$$

Example B

Find the derivative of $y = \cot^4 2x$.

Using the power rule and Eq. (6–31), we have

$$\frac{dy}{dx} = 4(\cot^3 2x)\frac{d \cot 2x}{dx} \qquad \text{Using } \frac{du^n}{dx} = nu^{n-1}\frac{du}{dx}$$

$$= 4 \cot^3 2x(-\csc^2 2x)\frac{d(2x)}{dx} \qquad \text{Using } \frac{d \cot u}{dx} = -\csc^2 u \frac{du}{dx}$$

$$= -4 \cot^3 2x \csc^2 2x \ (2)$$

$$= -8 \cot^3 2x \csc^2 2x$$

Example C

Find the derivative of $y = 3 \sec^2 4x$.
Using the power rule and Eq. (6–32), we have

$$\frac{dy}{dx} = 3(2)(\sec 4x)(\sec 4x \tan 4x)(4)$$

$$= 24 \sec^2 4x \tan 4x$$

Example D

Find the derivative of $y = x \csc^3 2x$.
Using the power rule, the product rule, and Eq. (6–33), we have

$$\frac{dy}{dx} = x(3 \csc^2 2x)(-\csc 2x \cot 2x)(2) + \csc^3 2x$$

$$= \csc^3 2x(-6x \cot 2x + 1)$$

Example E

Find the derivative of $y = (\tan 2x + \sec 2x)^3$.
Using the power rule and Eqs. (6–30) and (6–32), we have

$$\frac{dy}{dx} = 3(\tan 2x + \sec 2x)^2[\sec^2 2x(2) + \sec 2x \tan 2x(2)]$$

$$= 6 \sec 2x(\tan 2x + \sec 2x)^3$$

Example F

For reference, Eq. (6–28) is
$$\frac{d(\sin u)}{dx} = \cos u \frac{du}{dx}$$

Find the differential of $y = \sin 2x \tan x^2$.
Here we are to find the derivative of the given function and multiply by dx. Therefore, using the product rule along with Eqs. (6–28) and (6–30), we have the following:

$$dy = [(\sin 2x)(\sec^2 x^2)(2x) + (\tan x^2)(\cos 2x)(2)] dx$$

$$= (2x \sin 2x \sec^2 x^2 + 2 \cos 2x \tan x^2) dx$$

Example G

Evaluate the derivative of $y = \dfrac{2x}{1 - \cot 3x}$, where $x = 0.25$.
Finding the derivative, we have

$$\frac{dy}{dx} = \frac{(1 - \cot 3x)(2) - 2x(\csc^2 3x)(3)}{(1 - \cot 3x)^2}$$

$$= \frac{2 - 2 \cot 3x - 6x \csc^2 3x}{(1 - \cot 3x)^2}$$

Now, substituting $x = 0.25$, we have

$$\left.\frac{dy}{dx}\right|_{x=0.25} = \frac{2 - 2 \cot 0.75 - 6(0.25)\csc^2 0.75}{(1 - \cot 0.75)^2} = -626.0$$

In using the calculator we recall that we must have it in radian mode. Also, to evaluate $\cot 0.75$ and $\csc 0.75$, we must use reciprocals of $\tan 0.75$ and $\sin 0.75$, respectively.

Exercises 6–4

In Exercises 1 through 28, find the derivatives of the given functions.

1. $y = \tan 5x$
2. $y = 3 \tan(3x + 2)$
3. $y = \cot(1 - x)^2$
4. $y = 3 \cot 6x$
5. $y = 3 \sec 2x$
6. $y = \sec\sqrt{1 - x}$
7. $y = -3 \csc\sqrt{x}$
8. $y = \csc(1 - 2x)$
9. $y = 5 \tan^2 3x$
10. $y = 2 \tan^2(x^2)$
11. $y = 2 \cot^4 3x$
12. $y = \cot^2(1 - x)$
13. $y = \sqrt{\sec 4x}$
14. $y = \sec^3 x$
15. $y = 3 \csc^4 7x$
16. $y = \csc^2(2x^2)$
17. $y = x^2\tan x$
18. $y = 3x \sec 4x$
19. $y = 4 \cos x \csc x^2$
20. $y = \dfrac{1}{2} \sin 2x \sec x$
21. $y = \dfrac{\csc x}{x}$
22. $y = \dfrac{\cot 4x}{2x}$
23. $y = \dfrac{\cos 4x}{1 + \cot 3x}$
24. $y = \dfrac{\tan^2 3x}{2 + \sin x^2}$
25. $y = \dfrac{1}{3} \tan^3 x - \tan x$
26. $y = \csc 2x - 2 \cot 2x$
27. $y = \tan 2x - \sec 2x$
28. $y = x - \tan x + \sec^2 2x$

In Exercises 29 through 32, find the differentials of the given functions.

29. $y = 4 \tan^2 3x$
30. $y = 5 \sec^3 2x$
31. $y = \tan 4x \sec 4x$
32. $y = 2x \cot 3x$

In Exercises 33 through 42, solve the given problems.

33. On a calculator find the values of (a) $\sec^2 1.0000$ and (b) $(\tan 1.0001 - \tan 1.0000)/0.0001$. Compare the values and give the meaning of each in relation to the derivative of $\tan x$, where $x = 1$.

34. On a calculator find the values of (a) $\sec 1.0000 \tan 1.0000$ and (b) $(\sec 1.0001 - \sec 1.0000)/0.0001$. Compare the values and give the meaning of each in relation to the derivative of $\sec x$, where $x = 1$.

35. Find the derivative of each member of the identity $1 + \tan^2 x = \sec^2 x$ and show that the results are equal.

36. Find the derivative of each member of the identity $1 + \cot^2 x = \csc^2 x$ and show that the results are equal.

37. Find the slope of a line tangent to the curve of $y = 2 \cot 3x$ where $x = \pi/12$.

38. Find the slope of a line normal to the curve of $y = \csc\sqrt{2x + 1}$ where $x = 0.45$.

39. The displacement s of an object as a function of time t is given by $s = 2t^2 + \tan 2t$; find the acceleration a when $t = 0.375$.

40. If the displacement of an object is given by $s = 2t \sec\sqrt{t}$, find the velocity v when $t = 0.04$.

41. An observer of a rocket launch was 1000 ft from the take-off position. He found the angle of elevation of the rocket as a function of time to be $\theta = 3t/(2t + 10)$. Therefore, the height h of the rocket was

$$h = 1000 \tan \frac{3t}{2t + 10}$$

Figure 6–21

Find the time rate of change of height after 5 s. See Fig. 6–21.

42. Show that $y = \cos^3 x \tan x$ satisfies $\cos x \dfrac{dy}{dx} + 3y \sin x - \cos^2 x = 0$.

6–5 The Inverse Trigonometric Functions

There are times when we wish to solve for the angle in trigonometric relations. Such occasions occur when we know the function of an angle and must find the angle. Also, if we wish to find the rate of change of an angle under certain conditions, it might be convenient to have an expression representing the angle, and not simply a function of it. Functions have been defined for this purpose, and as we shall see, they have significant application in calculus.

Therefore, we define the **inverse trigonometric relations.** *The* **inverse sine** *of x is defined by*

$$y = \arcsin x \qquad \text{(the notation } y = \sin^{-1} x \text{ is also used)} \qquad (6\text{–}34)$$

Similar relations exist for the other inverse trigonometric relations. In Eq. (6–34), x is the value of the sine of the angle y, and therefore the most meaningful way of reading it is "*y is the angle whose sine is x.*"

Example A

$y = \arccos x$ would be read as "*y is the angle whose cosine is x.*"

The equation $y = \arctan 2x$ would be read as "*y is the angle whose tangent is 2x.*"

It is important to emphasize that $y = \arcsin x$ and $x = \sin y$ express the same relationship between x and y. The advantage of having both forms is that a trigonometric relation may be expressed in terms of a function of an angle or in terms of the angle itself.

If we consider closely the equation $y = \arcsin x$ and possible values of x, we note that there are an unlimited number of possible values of y for a given value of x. Consider the following example.

Example B

For $y = \arcsin x$, if $x = \frac{1}{2}$, we have $y = \arcsin \frac{1}{2}$. This means that we are to find an angle whose sine is $\frac{1}{2}$. We know that $\sin(\pi/6) = \frac{1}{2}$. Therefore, $y = \pi/6$.

However, we also know that $\sin(5\pi/6) = \frac{1}{2}$. Therefore, $y = 5\pi/6$ is also a proper value. If we consider negative angles, such as $-7\pi/6$, or angles generated by additional rotations, such as $13\pi/6$, we conclude that there are an unlimited number of possible values for y.

To have a properly defined **function** *in mathematics, there must be only one value of the dependent variable for a given value of the independent variable. A* **relation,** *on the other hand, may have more than one such value.* Therefore, we see that $y = \arcsin x$ is not really a function, although it is properly a relation. It is necessary to restrict the values of y in order to define the **inverse trigonometric functions** properly. Therefore, the following values are defined for the given functions.

$$-\frac{\pi}{2} \le \text{Arcsin } x \le \frac{\pi}{2}, \, 0 \le \text{Arccos } x \le \pi, \, -\frac{\pi}{2} < \text{Arctan } x < \frac{\pi}{2}$$

$$0 < \text{Arccot } x < \pi, \, 0 \le \text{Arcsec } x \le \pi, \, -\frac{\pi}{2} \le \text{Arccsc } x \le \frac{\pi}{2}$$

(6–35)

This means that when we are looking for a value of y to correspond to a given value for x, we must use a value of y as defined in Eqs. (6–35). We shall use the initial capital letter to designate the use of the inverse trigonometric *function*.

Example C

$$\text{Arcsin}\left(\frac{1}{2}\right) = \frac{\pi}{6}$$

This is the only value of the function which lies within the defined range. The value $5\pi/6$ is not correct, since it lies outside the defined range of values.

Example D

$$\text{Arccos}\left(-\frac{1}{2}\right) = \frac{2\pi}{3}$$

Other values such as $4\pi/3$ and $-2\pi/3$ are not correct, since they are not within the defined range of values for the function Arccos x.

Example E

$$\text{Arctan}(-1) = -\frac{\pi}{4}$$

 This is the only value within the defined range for the function Arctan x. We must remember that *when x is negative for Arcsin x and Arctan x, the value of y is a fourth-quadrant angle, expressed as a **negative angle**.* This is a direct result of the definition.

Example F

$$\text{Arcsin}\left(-\frac{\sqrt{3}}{2}\right) = -\frac{\pi}{3} \qquad \text{Arccos}(-1) = \pi$$

$$\text{Arctan } 0 = 0 \qquad \text{Arctan}(\sqrt{3}) = \frac{\pi}{3}$$

Example G

By use of a calculator in radian mode, we find the following values.

$$\text{Arcsin } 0.6294 = 0.6808 \qquad \text{Arcsin}(-0.1568) = -0.1574$$

$$\text{Arccos}(-0.8026) = 2.5024 \qquad \text{Arctan}(-1.9268) = -1.0921$$

We note that the calculator gives values which are in the defined ranges of values for each function.

One might logically ask why these values are chosen when there are so many different possibilities. The values are so chosen that, if x is positive, the resulting answer gives an angle in the first quadrant. We must, however, account for the possibility that x might be negative. We could not choose second-quadrant angles for Arcsin x. Since the sine of a second-quadrant angle is also positive, this then would lead to ambiguity. The sine is negative for fourth-quadrant angles, and to have a continuous range of values we must express the fourth-quadrant angles in the form of negative angles. This range is also chosen for Arctan x, for similar reasons. However, Arccos x cannot be chosen in this way, since the cosines of fourth-quadrant angles are also positive. Thus, again to keep a continuous range of values for Arccos x, the second-quadrant angles are chosen for negative values of x.

As for the values for the other functions, we chose values such that if x is positive, the result is also an angle in the first quadrant. As for negative values of x, it rarely makes any difference, since either positive values of x arise, or we can use one of the other functions. Our definitions, however, are those which are generally used.

The graphs of the inverse trigonometric relations can be used to show the fact that many values of y correspond to a given value of x. We can also show the ranges used in defining the inverse trigonometric functions, and that these ranges are specific sections of the curves.

Since $y = \arcsin x$ and $x = \sin y$ are equivalent equations, we can obtain the graph of the inverse sine by sketching the sine curve *along the y-axis*. In Figs. (6–22), (6–23), and (6–24), the graphs of three inverse trigonometric relations are shown, with the heavier portions indicating the graphs of the inverse trigonometric functions. The graphs of the other inverse relations are found in the same way.

Figure 6–22

Figure 6–23

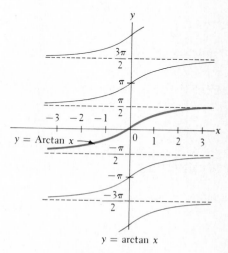

Figure 6–24

If we know the value of x for one of the inverse functions, we can find the trigonometric functions of the angle. If general relations are desired, a representative triangle is very useful. The following examples illustrate these methods.

Example H

Find cos(Arcsin 0.5). (Remember again: The inverse functions yield *angles*.)

We know Arcsin 0.5 is a first-quadrant angle, since 0.5 is positive. Thus we find Arcsin $0.5 = \pi/6$. The problem now becomes one of finding $\cos(\pi/6)$. This is, of course, $\sqrt{3}/2$, or 0.8660.

Example I

$$\sin(\text{Arccot } 1) = \sin \frac{\pi}{4}$$
$$= \frac{\sqrt{2}}{2} = 0.7071$$
$$\tan[\text{Arccos}(-1)] = \tan \pi$$
$$= 0$$

Example J

Figure 6–25

Find sin(Arctan x).

We know that Arctan x is another way of stating ''the angle whose tangent is x.'' Thus, let us draw a right triangle (as in Fig. 6–25) and label one of the acute angles θ, the side opposite θ as x, and the side adjacent to θ as 1. In this way we see that, by definition, $\tan \theta = x/1$, or $\theta = \text{Arctan } x$, which means θ is the desired angle. By the Pythagorean theorem, the hypotenuse of this triangle is $\sqrt{x^2 + 1}$. Now we find that the sin θ, which is the same as sin(Arctan x), is $x/\sqrt{x^2 + 1}$, from the definition of the sine. Thus,

$$\sin(\text{Arctan } x) = \frac{x}{\sqrt{x^2 + 1}}$$

Example K

Find cos(2 Arcsin x).

From Fig. 6–26, we see that $\theta = \text{Arcsin } x$. From the double-angle formulas, we have

$$\cos 2\theta = 1 - 2 \sin^2\theta$$

Thus, since $\sin \theta = x$, we have

$$\cos(2 \text{ Arcsin } x) = 1 - 2x^2$$

Figure 6–26

Example L

For the triangle shown in Fig. 6–27, solve for angle A in terms of the given sides a, b, and c.

From the law of cosines we have

$$a^2 = b^2 + c^2 - 2bc \cos A$$

Solving for angle A, we have

$$2bc \cos A = b^2 + c^2 - a^2$$

$$\cos A = \frac{b^2 + c^2 - a^2}{2bc}$$

$$A = \text{Arccos}\left(\frac{b^2 + c^2 - a^2}{2bc}\right)$$

Figure 6–27

Exercises 6–5

In Exercises 1 through 20, evaluate the given expressions.

1. $\text{Arccos}(\frac{1}{2})$
2. $\text{Arcsin}(1)$
3. $\text{Arcsin } 0$
4. $\text{Arccos } 0$
5. $\text{Arctan}(-\sqrt{3})$
6. $\text{Arcsin}(-\frac{1}{2})$
7. $\text{Arcsec } 2$
8. $\text{Arccot } \sqrt{3}$
9. $\text{Arctan}\left(\frac{\sqrt{3}}{3}\right)$
10. $\text{Arctan } 1$
11. $\text{Arcsin}\left(-\frac{\sqrt{2}}{2}\right)$
12. $\text{Arccos}\left(-\frac{\sqrt{3}}{2}\right)$
13. $\text{Arccsc } \sqrt{2}$
14. $\text{Arccot } 1$
15. $\sin(\text{Arctan } \sqrt{3})$
16. $\tan\left(\text{Arcsin } \frac{\sqrt{2}}{2}\right)$
17. $\cos[\text{Arctan}(-1)]$
18. $\sec[\text{Arccos}(-\frac{1}{2})]$
19. $\cos(2 \text{ Arcsin } 1)$
20. $\sin(2 \text{ Arctan } 2)$

In Exercises 21 through 32, use a calculator to evaluate the given expressions.

21. $\text{Arctan}(-3.7321)$
22. $\text{Arccos}(-0.6561)$
23. $\text{Arcsin}(-0.8326)$
24. $\text{Arctan } 0.2846$
25. $\text{Arccos } 0.1291$
26. $\text{Arcsin } 0.2119$
27. $\text{Arctan } 8.2614$
28. $\text{Arcsin}(-0.8881)$
29. $\tan[\text{Arccos}(-0.6281)]$
30. $\cos[\text{Arctan}(-1.2256)]$
31. $\sin[\text{Arctan}(-0.2297)]$
32. $\tan[\text{Arcsin}(-0.3019)]$

In Exercises 33 through 40, find an algebraic expression for each of the expressions given.

33. $\tan(\text{Arcsin } x)$
34. $\sin(\text{Arccos } x)$
35. $\cos(\text{Arcsec } x)$
36. $\cot(\text{Arccot } x)$
37. $\sec(\text{Arccsc } 3x)$
38. $\tan(\text{Arcsin } 2x)$
39. $\sin(2 \text{ Arcsin } x)$
40. $\cos(2 \text{ Arctan } x)$

In Exercises 41 and 42, prove that the given expressions are equal. This can be done by use of the relation for sin $(\alpha + \beta)$ and by showing that the sine of the sum of angles on the left equals the sine of the angle on the right.

41. $\text{Arcsin } \dfrac{3}{5} + \text{Arcsin } \dfrac{5}{13} = \text{Arcsin } \dfrac{56}{65}$ 42. $\text{Arctan } \dfrac{1}{3} + \text{Arctan } \dfrac{1}{2} = \dfrac{\pi}{4}$

In Exercises 43 and 44, verify the given expressions.

43. $\text{Arcsin } 0.5 + \text{Arccos } 0.5 = \dfrac{\pi}{2}$

44. $\text{Arctan } \sqrt{3} + \text{Arccot } \sqrt{3} = \dfrac{\pi}{2}$

In Exercises 45 and 46, solve for angle A for the triangles in the given figures in terms of the given sides and angles.

45.

46.

Figure 6–28 *Figure 6–29*

In Exercises 47 and 48, derive the given expressions.

47. For the triangle in Fig. 6–30, show that $\alpha = \text{Arctan}(a + \tan \beta)$.

48. Show that the length of the pulley belt indicated in Fig. 6–31 is given by the expression

$$L = 24 + 11\pi + 10 \,\text{Arcsin}\!\left(\dfrac{5}{13}\right)$$

Figure 6–30

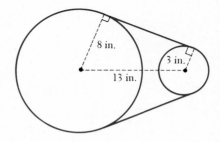

Figure 6–31

In Exercises 49 and 50, derive the required expressions.

49. For the figure shown in Fig. 6–32, express θ as a function of x, using inverse trigonometric functions ($x < 6$).

50. For the building and lines of sight shown in Fig. 6–33, express θ as a function of the distance x from the building, using inverse trigonometric functions.

Figure 6–32

Figure 6–33

6–6 Derivatives of the Inverse Trigonometric Functions

To obtain the derivatives of $y = \text{Arcsin } u$, we first solve for u in the form $u = \sin y$ and then take derivatives with respect to x:

$$\frac{du}{dx} = \cos y \frac{dy}{dx}$$

Solving this equation for dy/dx, we obtain

$$\frac{dy}{dx} = \frac{1}{\cos y}\frac{du}{dx} = \frac{1}{\sqrt{1 - \sin^2 y}}\frac{du}{dx}$$

$$= \frac{1}{\sqrt{1 - u^2}}\frac{du}{dx}$$

We choose the positive square root since $\cos y > 0$ for $-\pi/2 < y < \pi/2$, which is the range of the defined values of Arcsin u. Therefore, we obtain the following result:

$$\frac{d(\text{Arcsin } u)}{dx} = \frac{1}{\sqrt{1 - u^2}}\frac{du}{dx} \qquad\qquad (6\text{–}36)$$

We note here that the derivative of the inverse sine function is an algebraic function.

We find the derivative of the inverse cosine function by letting $y = \text{Arccos } u$, and by following the same procedure as that used in finding the derivative of Arcsin u:

$$u = \cos y, \qquad \frac{du}{dx} = -\sin y \frac{dy}{dx}$$

$$\frac{dy}{dx} = -\frac{1}{\sin y}\frac{du}{dx} = -\frac{1}{\sqrt{1 - \cos^2 y}}\frac{du}{dx}$$

Therefore,

$$\frac{d(\text{Arccos } u)}{dx} = -\frac{1}{\sqrt{1 - u^2}}\frac{du}{dx} \tag{6--37}$$

The positive square root is chosen here since $\sin y > 0$ for $0 < y < \pi$, which is the range of the defined values of Arccos u. We note that the derivative of the inverse cosine is the negative of the derivative of the inverse sine.

By letting $y = \text{Arctan } u$, solving for u, and taking derivatives, we find the derivative of the inverse tangent function:

$$u = \tan y, \qquad \frac{du}{dx} = \sec^2 y\frac{dy}{dx}, \qquad \frac{dy}{dx} = \frac{1}{\sec^2 y}\frac{du}{dx} = \frac{1}{1 + \tan^2 y}\frac{du}{dx}$$

Therefore,

$$\frac{d(\text{Arctan } u)}{dx} = \frac{1}{1 + u^2}\frac{du}{dx} \tag{6--38}$$

We can see that the derivative of the inverse tangent is an algebraic function also.

The inverse sine, inverse cosine, and inverse tangent prove to be of the greatest importance in applications and in further development of mathematics. Therefore, the formulas for the derivatives of the other inverse functions are not presented here, although they are included in the exercises.

Example A

Find the derivative of $y = \text{Arcsin } 4x$.

Using Eq. (6–36), we have

$$\frac{dy}{dx} = \frac{1}{\sqrt{1 - (4x)^2}}\overset{\overset{\frac{du}{dx}}{\downarrow}}{(4)}$$

$$= \frac{4}{\sqrt{1 - 16x^2}}$$

Example B

Find the derivative of $y = \text{Arccos}(1 - 2x)$.
Using Eq. (6–37), we have

$$\frac{dy}{dx} = \frac{-1}{\sqrt{1 - (1 - 2x)^2}}(-2) = \frac{2}{\sqrt{1 - 1 + 4x - 4x^2}}$$

$$= \frac{2}{\sqrt{4x - 4x^2}} = \frac{1}{\sqrt{x - x^2}}$$

Example C

Find the derivative of $y = (x^2 + 1)\text{Arctan } x - x$.
Using the product rule along with Eq. (6–38) on the first term, we have

$$\frac{dy}{dx} = (x^2 + 1)\frac{1}{1 + x^2} + \text{Arctan } x(2x) - 1$$

$$= 2x \text{ Arctan } x$$

Example D

Find the derivative of $y = x \text{ Arctan}^2 2x$.
Using the product rule, power rule, and Eq. (6–38), we have

$$\frac{dy}{dx} = x(2)(\text{Arctan } 2x)\left(\frac{1}{1 + 4x^2}\right)(2) + \text{Arctan}^2 2x$$

$$= \text{Arctan } 2x\left(\frac{4x}{1 + 4x^2} + \text{Arctan } 2x\right)$$

Example E

Find the derivative of $y = x \text{ Arcsin } 2x + \frac{1}{2}\sqrt{1 - 4x^2}$.
We write

$$\frac{dy}{dx} = x\left(\frac{2}{\sqrt{1 - 4x^2}}\right) + \text{Arcsin } 2x + \frac{1}{2}\cdot\frac{1}{2}(1 - 4x^2)^{-1/2}(-8x)$$

$$= \frac{2x}{\sqrt{1 - 4x^2}} + \text{Arcsin } 2x - \frac{2x}{\sqrt{1 - 4x^2}}$$

$$= \text{Arcsin } 2x$$

Example F

Find the slope of a tangent to the curve of $y = \dfrac{\text{Arccos } x}{x^2 + 1}$, where $x = 0.25$.

Here we are to find the derivative and then evaluate it for $x = 0.25$.

$$\frac{dy}{dx} = \frac{(x^2 + 1)\left(-\dfrac{1}{\sqrt{1 - x^2}}\right)(1) - (\text{Arccos } x)(2x)}{(x^2 + 1)^2}$$

$$= \frac{-\dfrac{x^2 + 1}{\sqrt{1 - x^2}} - 2x \text{ Arccos } x}{(x^2 + 1)^2}$$

$$\left.\frac{dy}{dx}\right|_{x=0.25} = \frac{-\dfrac{0.25^2 + 1}{\sqrt{1 - 0.25^2}} - 2(0.25)\text{Arccos } 0.25}{(0.25^2 + 1)^2}$$

$$= -1.5558$$

Exercises 6–6

In Exercises 1 through 28, find the derivatives of the given functions.

1. $y = \text{Arcsin}(x^2)$

2. $y = \text{Arcsin}(1 - x^2)$

3. $y = 2 \text{ Arcsin } 3x^3$

4. $y = \text{Arcsin } \sqrt{1 - x}$

5. $y = \text{Arccos } \frac{1}{2}x$

6. $y = 3 \text{ Arccos } 5x$

7. $y = 2 \text{ Arccos } \sqrt{2 - x}$

8. $y = 3 \text{ Arccos}(x^2 + 1)$

9. $y = \text{Arctan } \sqrt{x}$

10. $y = \text{Arctan}(1 + x)$

11. $y = \text{Arctan}\left(\dfrac{1}{x}\right)$

12. $y = 4 \text{ Arctan } 3x^4$

13. $y = x \text{ Arcsin } x$

14. $y = x^2 \text{Arccos } x$

15. $y = 2x \text{ Arctan } 2x$

16. $y = (x^2 + 1)\text{Arcsin } 4x$

17. $y = \dfrac{3x}{\text{Arcsin } 2x}$

18. $y = \dfrac{\text{Arctan } 2x}{x}$

19. $y = \dfrac{\text{Arcsin } 2x}{\text{Arccos } 2x}$

20. $y = \dfrac{x^2 + 1}{\text{Arctan } x}$

21. $y = 2 \text{ Arccos}^3 4x$

22. $y = 3 \text{ Arcsin}^4 3x$

23. $y = \text{Arcsin}^2 4x$

24. $y = \sqrt{\text{Arcsin }(x - 1)}$

25. $y = 3 \text{ Arctan}^3 x$

26. $y = \dfrac{1}{\text{Arccos } 2x}$

27. $y = \dfrac{1}{1 + 4x^2} - \text{Arctan } 2x$

28. $y = \text{Arcsin } x - \sqrt{1 - x^2}$

In Exercises 29 through 40, solve the given problems.

29. On a calculator find the values of (a) $1/\sqrt{1 - 0.5^2}$ and
(b) $(\text{Arcsin } 0.5001 - \text{Arcsin } 0.5000)/0.0001$. Compare the values and give the meaning of each in relation to the derivative of Arcsin x where $x = 0.5$.

30. On a calculator find the values of (a) $1/(1 + 0.5^2)$ and
(b) $(\text{Arctan } 0.5001 - \text{Arctan } 0.5000)/0.0001$. Compare the values and give the meaning of each in relation to the derivative of Arctan x where $x = 0.5$.

31. Find the differential of the function $y = \text{Arcsin}^3 x$.

32. Find the differential of the function $y = x^3 \text{Arccos } x^2$.

33. Find the second derivative of the function $y = \text{Arctan } 2x$.

34. Show that $\dfrac{d(\text{Arccot } u)}{dx} = -\dfrac{1}{1 + u^2}\dfrac{du}{dx}$.

35. Show that $\dfrac{d(\text{Arcsec } u)}{dx} = \dfrac{1}{\sqrt{u^2(u^2 - 1)}}\dfrac{du}{dx}$.

36. Show that $\dfrac{d(\text{Arccsc } u)}{dx} = -\dfrac{1}{\sqrt{u^2(u^2 - 1)}}\dfrac{du}{dx}$.

37. The angle with the x-axis of the x-component of a vector $\mathbf{R}$ is $\theta = \text{Arccos } (R_x/R)$. If R is constant and R_x changes with time, find an expression for $d\theta/dt$.

38. An equation relating the angle θ at which a rocket is fired, its range R and its initial velocity v_0 is $\theta = \frac{1}{2}\text{Arcsin}(Rg/v_0^2)$, where g is the acceleration due to gravity. If R and g are constant, find an expression for $d\theta/dv_0$.

39. An equation which arises in the theory of solar collectors is $\alpha = \text{Arccos } \dfrac{2f - r}{r}$. Find the expression for $d\alpha/dr$ if f is constant.

40. As a person approaches a building of height h, the angle of elevation of the top of the building is a function of the person's distance from the building. Express the angle of elevation θ in terms of h and the distance x from the building and then find $d\theta/dx$. Assume the person's height is negligible compared to that of the building. See Fig. 6–34.

Figure 6–34

6–7 Applications

With our development of the formulas for the derivatives of the trigonometric and inverse trigonometric functions, it is now possible for us to apply these derivatives in the same manner as we applied the derivatives of algebraic functions. We can now use trigonometric and inverse trigonometric functions to solve tangent and normal, Newton's method, time rate of change, curve tracing, maximum and minimum, and differential application problems. The following examples illustrate the use of these functions in these types of problems.

Example A

Sketch the curve $y = \sin^2 x - \dfrac{x}{2}$ $(0 \le x \le 2\pi)$.

First we write

$$\frac{dy}{dx} = 2 \sin x \cos x - \frac{1}{2} = \sin 2x - \frac{1}{2}, \qquad \frac{d^2y}{dx^2} = 2 \cos 2x$$

The only easily obtained intercept is $(0, 0)$. (See Fig. 6–35.) Maximum and minimum points will occur for $\sin 2x = \frac{1}{2}$. Thus, maximum and minimum points occur if

$$2x = \frac{\pi}{6}, \frac{5\pi}{6}, \frac{13\pi}{6}, \frac{17\pi}{6}$$

Thus, the x-values are

$$x = \frac{\pi}{12}, \frac{5\pi}{12}, \frac{13\pi}{12}, \frac{17\pi}{12}$$

Using the second derivative, we find that d^2y/dx^2 is positive for $x = \pi/12$ and $x = 13\pi/12$, and negative for $x = 5\pi/12$ and $x = 17\pi/12$. Thus, the maximum points are $(5\pi/12, 0.279)$ and $(17\pi/12, -1.29)$. The minimum points are $(\pi/12, -0.064)$ and $(13\pi/12, -1.64)$. Inflection points occur for $\cos 2x = 0$, or

$$2x = \frac{\pi}{2}, \frac{3\pi}{2}, \frac{5\pi}{2}, \frac{7\pi}{2}, \quad \text{or} \quad x = \frac{\pi}{4}, \frac{3\pi}{4}, \frac{5\pi}{4}, \frac{7\pi}{4}$$

Thus, points of inflection are $(\pi/4, 0.11)$, $(3\pi/4, -0.68)$, $(5\pi/4, -1.47)$, and $(7\pi/4, -2.25)$. Using this information, we sketch the curve.

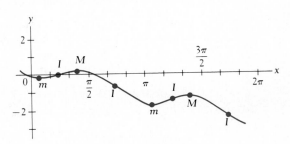

Figure 6–35

Example B

By using Newton's method, solve the equation $2x - 1 = 3 \cos x$.

First we locate the root approximately by sketching $y_1 = 2x - 1$ and $y_2 = 3 \cos x$. From Fig. 6–36 we see that they intersect between $x = 1$ and $x = 2$, nearer to $x = 1$. Therefore, we choose $x_1 = 1.2$. With

$$f(x) = 2x - 1 - 3 \cos x \qquad \text{and} \qquad f'(x) = 2 + 3 \sin x$$

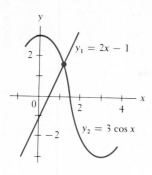

Figure 6-36

we use Eq. (3–1), which is

$$x_2 = x_1 - \frac{f(x_1)}{f'(x_1)}$$

To find x_2, we have

$$f(x_1) = 2(1.2) - 1 - 3 \cos 1.2 = 0.3129267$$

and

$$f'(x_1) = 2 + 3 \sin 1.2 = 4.7961173$$

$$x_2 = 1.2 - \frac{0.3129267}{4.7961173} = 1.1347542$$

Finding the next approximation, we find $x_3 = 1.1342366$, which is accurate to the value shown. Again, when using the calculator it is not necessary to list the values of $f(x_1)$ and $f'(x_1)$, as the complete calculation can be done directly on the calculator.

Example C

Figure 6-37

Find the area of the largest rectangle which can be inscribed in a semicircle of radius 2.

From Fig. 6–37, we see that $A = 2xy$, where $x = 2 \cos \theta$ and $y = 2 \sin \theta$. Thus, $A = 8 \sin \theta \cos \theta = 4 \sin 2\theta$. We shall find the value of θ corresponding to the maximum area. Therefore,

$$\frac{dA}{d\theta} = 8 \cos 2\theta, \qquad 8 \cos 2\theta = 0, \qquad 2\theta = \frac{\pi}{2}, \qquad \theta = \frac{\pi}{4}$$

Thus, the maximum area is $A = 4 \sin 2(\pi/4) = 4$.

Example D

Figure 6-38

A rocket is taking off vertically at a distance of 600 ft from an observer. If, when the angle of elevation is $\pi/4$ it is changing at the rate of 0.5 rad/s, how fast is the rocket ascending?

From Fig. 6–38, we see that

$$\theta = \text{Arctan} \frac{x}{600}$$

Taking derivatives with respect to time, we have

$$\frac{d\theta}{dt} = \frac{1}{1 + x^2/(3.6 \times 10^5)} \frac{dx/dt}{600} = \frac{600 \, dx/dt}{3.6 \times 10^5 + x^2}$$

From the given information, we find that $d\theta/dt = 0.5$ rad/s, $x = 600$ ft when $\theta = \pi/4$. Thus,

$$0.5 = \frac{600 \, dx/dt}{7.2 \times 10^5} \qquad \text{or} \qquad \frac{dx}{dt} = 600 \text{ ft/s}$$

Example E

A particle moves so that its x- and y-coordinates are given by $x = \cos 2t$ and $y = \sin 2t$. Find the magnitude and direction of its velocity when $t = \pi/8$.
    We write

$$v_x = \frac{dx}{dt} = -2\sin 2t \qquad v_y = \frac{dy}{dt} = 2\cos 2t$$

$$v_x|_{t=\pi/8} = -2\sin 2\left(\frac{\pi}{8}\right) = -2\left(\frac{\sqrt{2}}{2}\right) = -\sqrt{2}$$

$$v_y|_{t=\pi/8} = 2\cos 2\left(\frac{\pi}{8}\right) = 2\left(\frac{\sqrt{2}}{2}\right) = \sqrt{2}$$

$$v = \sqrt{v_x^2 + v_y^2} = \sqrt{2+2} = 2 \qquad \tan\theta = \frac{v_y}{v_x} = \frac{\sqrt{2}}{-\sqrt{2}} = -1$$

Since v_x is negative and v_y is positive, $\theta = 135°$. Note that in this example θ is the angle, in standard position, between the horizontal and the resultant velocity.
    By plotting the curve we note that it is a circle (see Fig. 6–39). Thus, the object moves about a circle in a counterclockwise direction. It can be determined in another way that the curve is a circle. If we square each of the expressions defining x and y and add them, we have the equation $x^2 + y^2 = \cos^2 2t + \sin^2 2t = 1$. This is a circle of radius 1.

Figure 6–39

Example F

At a point 180 ft from the base of a building on level ground, the angle of elevation of the top of the building is 30.00°. What would be the error in the height of the building due to an error of 0.25° in this angle?
    From Fig. 6–40 we see that $h = 180\tan\theta$. Finding the differential of h, we have

$$dh = 180\sec^2\theta\, d\theta$$

Since $\theta = 30.00°$, $\sec\theta = 1.1547005$. The possible error in θ is 0.25°, or 0.0043633 rad, which is the value we must use in the calculation. Thus,

$$dh = (180)(1.1547005)^2(0.0043633) = 1.05 \text{ ft}$$

We see that an error of 0.25° in the angle can result in an error of over 1 ft in the calculation of the height.

Figure 6–40

Exercises 6–7

1. Show that the slopes of the sine and cosine curves are negatives of each other at the points of intersection.
2. Show that the tangent curve, when defined, is always increasing.
3. Show that $y = \text{Arctan } x$ is always increasing.
4. Sketch the graph of $y = \sin x + \cos x$ $(0 \le x \le 2\pi)$.

5. Sketch the graph of $y = x - \tan x \, (-\pi/2 < x < \pi/2)$.

6. Sketch the graph of $y = \sin x + \sin 2x \, (0 < x < 2\pi)$.

7. Find the equation of the line tangent to the curve of $y = \sin 2x$ at $x = 5\pi/8$.

8. Find the equation of a line normal to the curve of $y = \text{Arctan}(x/2)$ at $x = 3$.

9. By Newton's method, find the positive root of the equation $x^2 - 4 \sin x = 0$.

10. By Newton's method, find the smallest positive root of the equation $\tan x = 2x$.

11. Find the maximum value of the function $y = 6 \cos x - 8 \sin x$.

12. Find the minimum value of the function $y = \cos 2x + 2 \sin x$.

13. The displacement (in feet) a certain object travels is given as a function of time (in seconds) by $s = \sin t + \cos 2t$. Find its velocity and acceleration when $t = \pi/6$ s.

14. The displacement s (in centimeters) of an object as a function of time is given by $s = t \sin t$. Find the velocity when $t = 0.200$ s.

15. The y-component of a vector of magnitude 20.0 mm is given by $A_y = 20.0 \sin \theta$. If $d\theta/dt = 0.0551$ rad/s, determine the time rate of change of A_y in millimeters per second when $\theta = 0.323$.

16. The *apparent power* P_a of an electric circuit whose power is P and impedance phase angle is θ is given by $P_a = P \sec \theta$. Given P is constant at 12 W, find the time rate of change of P_a if θ is changing at the rate of 0.050 rad/min, when $\theta = 2\pi/9$.

17. Find the magnitude and direction of the velocity of an object which moves so that its x- and y-coordinates are given by $x = \sin 2\pi t$ and $y = \cos 2\pi t$ when $t = \frac{1}{12}$.

18. Find the magnitude and direction of the velocity of an object which moves so that the x- and y-coordinates are given by $x = \cos 2t$ and $y = 2 \sin t$ when $t = \pi/6$.

19. Find the magnitude and direction of the acceleration of the particle in Example E when $t = \pi/8$.

20. Find the magnitude and direction of the acceleration of the object in Exercise 18.

21. A person observes an object dropped from the top of a building 100 ft away. If the building is 200 ft high, how fast is the angle of elevation of the object changing after 1 s? (The distance the object drops is given by $s = 16t^2$.) See Fig. 6–41.

22. An airplane flying horizontally at 500 ft/s passes directly over an observer 4000 ft below. How fast is the angle of elevation changing at this instant?

23. One leg of a right triangle is 6.25 units long. If the acute angle of which this side is the adjacent side is increasing at the rate of 0.120 rad/s, how fast is the hypotenuse changing when it is 10.5 units in length?

24. In a modern hotel, where the elevators are directly observable from the lobby area (and a person can see from the elevators), a person in the lobby observes one of the elevators rising at the rate of 12.0 ft/s. If the person was 50.0 ft from the elevator when it left the lobby, how fast is the angle of elevation of the line of sight to the elevator increasing 10.0 s later?

25. If a block is placed on a plane inclined with the horizontal at an angle θ such that the block just moves down the plane, the coefficient of friction μ is given by $\mu = \tan \theta$. Use differentials to find the change in μ if θ changes from 20° to 21°.

26. The current (in amperes) in a certain electric circuit is given by $i = \sin^2 2t$. Find the approximate change in i between $t = 1.00$ s and $t = 1.05$ s.

200 ft

θ

100 ft

Figure 6–41

Figure 6–42

Figure 6–43

27. Two sides and the included angle of a triangle are measured to be 8.00 cm, 7.50 cm, and 40.0°. What is the error caused in the calculation of the third side by an error of 0.5° in the angle?

28. A firm determined that its total weekly profit from the production of x units is $P = 10,000 \sin 0.1x$, for $0 < x < 30$ and where P is measured in dollars. What is the approximate change in profit when production changes from 20 to 22 units?

29. Use trigonometric functions to find the area of the largest rectangle with a perimeter of 12 in.

30. Find the angle between the equal sides of an isosceles triangle of greatest area for a given perimeter.

31. A wall is 6 ft high and 4 ft from a building. What is the length of the shortest pole that can touch the building and the ground beyond the wall (see Fig. 6–42)? (*Hint:* $y = 6 \csc \theta + 4 \sec \theta$.)

32. A painting 8 ft high is hung such that the lower edge is 2 ft above an oberver's eye level. Assuming that the best view is obtained when the angle subtended by the painting at the eye level is a maximum, how far from the wall should the observer stand? See Fig. 6–43.

6–8 Review Exercises for Chapter 6

In Exercises 1 through 24, find the derivative of each of the given functions.

1. $y = 3 \cos(4x - 1)$ 2. $y = 4 \sec(1 - x^3)$ 3. $y = \tan\sqrt{3 - x}$

4. $y = 5 \sin(1 - 6x)$ 5. $y = 3 \operatorname{Arctan}\left(\dfrac{x}{3}\right)$ 6. $y = 4 \operatorname{Arccos}(2x + 3)$

7. $y = \csc^2(3x + 2)$ 8. $y = \cot^2 5x$ 9. $y = 3 \cos^4 x^2$

10. $y = 2 \sin^3 \sqrt{x}$ 11. $y = \operatorname{Arcsin}(\cos x)$ 12. $y = \sin(\operatorname{Arctan} x)$

13. $y = \sqrt{\csc 4x + \cot 4x}$ 14. $y = (1 + \sin 2x)^4$ 15. $y = x \sec 2x$

16. $y = (x + 1)\cos 2x$ 17. $y = \dfrac{x^2}{\operatorname{Arctan} 2x}$ 18. $y = \dfrac{\operatorname{Arcsin} x}{x}$

19. $y = x \operatorname{Arccos} x - \sqrt{1 - x^2}$

20. $y = x \operatorname{Arcsin}^2 x + 2\sqrt{1 - x^2} \operatorname{Arcsin} x - 2x$ 21. $y^2 \sin 2x + \tan x = 0$

22. $x + \sec^2(xy) = 1$ 23. $x \sin 2y = y \cos 2x$ 24. $\operatorname{Arctan} \dfrac{y}{x} = x^2 y$

In Exercises 25 through 46, solve the given problems.

25. Find the derivative of each of the members of the identity $\sin^2 x + \cos^2 x = 1$ and show that the results are equal.

26. Find the derivative of each of the members of the identity

$$\sin(x + 1) = \sin x \cos 1 + \cos x \sin 1$$

and show that the results are equal.

27. Sketch the graph of the function $y = x - \cos x$.

28. Sketch the graph of the function $y = 4 \sin x + \cos 2x$.

29. Find the equation of the line tangent to the curve $y = 4 \cos^2(x^2)$ at $x = 1$.

30. Find the equation of the line normal to the curve $y = \text{Arctan } x$ at $x = 1$.

31. By Newton's method find the positive root of the equation $\sin x = 1 - x$.

32. By Newton's method find the positive root of the equation $x^2 - \text{Arctan } x = 0$.

33. Power can be defined as the time rate of doing work. If work is being done in an electric circuit according to $W = 10 \cos 2t$, find P as a function of t.

34. A ball on the end of a string is moving in a vertical circle. The distance of its shadow from a certain point is given by

$$x = 4 \cos\left(\frac{\pi}{3}t + \frac{\pi}{6}\right)$$

Find the velocity, in ft/s, of the shadow when $t = 6$ s.

35. Find the magnitude and direction of the velocity of an object which moves so that its x- and y-coordinates are given by $x = \sin 2\pi t$ and $y = \cos 2\pi t$ when $t = \frac{1}{12}$.

36. Find the magnitude and direction of the acceleration of the object in Exercise 35.

37. In the study of polarized light, under certain conditions the intensity of transmitted light is given by $I = I_m \cos^2 \theta$, where I_m is the maximum intensity and θ is the angle of transmission. If $I = 8$ units when $\theta = 60°$, and θ is changing at the rate of 0.03 rad/s, find the rate of change of I when $\theta = 60°$.

38. The force which acts on a pendulum bob to restore it to its equilibrium position is given by $F = -mg \sin \theta$, where m is the mass of the bob, g is the acceleration of gravity, and θ is the angle between the string and the vertical. For a bob of 10 g, find the approximate difference in the force for angles of 3.0° and 3.1° if $g = 980$ cm/s².

39. An object attached to a cord of length l, as shown in Fig. 6–44, moves in a circular path. The angular velocity ω is given by $\omega = \sqrt{g/l \cos \theta}$. By use of differentials find the approximate change in ω if θ changes from 32.50° to 32.75°, given that $g = 9.800$ m/s² and $l = 0.6375$ m.

40. In optics, the index of refraction of a liquid is defined as $n = \sin i / \sin r$, where i is the angle at which the light is incident on the liquid from the air, and r is the angle of the refracted light in the liquid. If the angle of incidence is increasing at the rate of 0.05 rad/min, what is the rate of change of the angle r when $i = 10°$ for benzene, for which $n = 1.50$?

41. A revolving light 300 ft from a straight wall makes 6.00 r/min. Find the velocity of the beam along the wall at the instant it makes an angle of 45.0° with the wall. See Fig. 6–45.

42. A gutter is to be made from a sheet of metal 12 in. wide by turning up strips of width 4 in. along each side to make equal angles θ with the vertical. Sketch a graph of the cross-sectional area A as a function of θ. See Fig. 6–46.

43. A conical filter is made from a circular piece of wire mesh of radius 24.0 cm by cutting out a sector with central angle θ and then taping the cut edges of the remaining piece together (see Fig. 6–47). What is the maximum possible volume which the resulting filter can hold?

Figure 6–44

Figure 6–45

Figure 6–46

Figure 6–47

Figure 6–48

Figure 6–49

44. A force P at an angle θ above the horizontal drags a 50-lb box across a level floor. The coefficient of friction between the floor and box is constant and equals 0.2. The magnitude of the force P is given by

$$P = \frac{(0.2)(50)}{(0.2)\sin \theta + \cos \theta}$$

Find θ such that P is a minimum.

45. The illuminance from a point source of light varies directly as the cosine of the angle of incidence (measured from the perpendicular), and inversely as the square of the distance r from the source. How high above the center of a circle of radius 10 in. should a light be placed so that the illuminance at the circumference will be a maximum? See Fig. 6–48.

46. A metal bracket in the shape of a Y is to be made such that its height is 10 cm and the width across the top is 6 cm. What shape will require the least amount of material? See Fig. 6–49.

CHAPTER 7

Derivatives of the Exponential and Logarithmic Functions

Another important function in many of the applications of technology, especially in mechanics and electricity, is the exponential function. Closely associated with the exponential function is the logarithmic function, which also has numerous applications.

In the first section of this chapter we review the properties of the exponential and logarithmic functions. In the remaining sections we shall develop the formulas for their derivatives and show some of the applications of these derivatives.

7–1 Exponential and Logarithmic Functions

By definition, *the* **exponential function** *is of the form*

$$y = b^x \tag{7–1}$$

where b is the **base,** *a constant, and x is the exponent, the independent variable.* Restricting values of both variables to real numbers leads us to choose the base as a positive number other than 1. We know that 1 raised to any finite power will result in a value of 1. Negative numbers for *b* would result in imaginary values for *y*, if *x* were any fractional exponent with an even integer as its denominator.

Example A

$y = 2^x$ is an exponential function, where we see that x is the exponent of 2. That is, 2 raised to any given power gives us the corresponding value of y. If $x = 2$, $y = 4$. If $x = -4$, $y = \frac{1}{16}$. If $x = \frac{1}{2}$, $y = \sqrt{2} = 1.41$. If $x = \frac{3}{2}$, $y = 2\sqrt{2} = 2.83$. By the definition of the exponential function, any real value of x may be assumed.

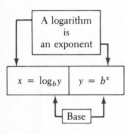

Figure 7–1

In Eq. (7–1), since y is a function of x, it is possible to define x as a function of y. Therefore, in Eq. (7–1) we *define x as the* **logarithm** *of the number y to the base b,* which is written as

$$x = \log_b y \qquad\qquad (7-2)$$

Equations (7–1) and (7–2) state precisely the same relationship, but in different ways. *Equation (7–1) is the* **exponential form,** *and Eq. (7–2) is the* **logarithmic form.** In either case, x is a logarithm, and a logarithm is an exponent. See Fig. 7–1.

Example B

The equation $y = 2^x$ would be written as $x = \log_2 y$ if we put it in logarithmic form. When we choose values of y to find the corresponding values of x from this equation, we ask ourselves, "2 raised to what power gives y?" Hence if $y = 4$, we know that 2^2 is 4, and x would be 2. If $y = 8$, $2^3 = 8$, or $x = 3$.

Example C

$3^2 = 9$ in logarithmic form is $2 = \log_3 9$; $4^{-1} = \frac{1}{4}$ in logarithmic form is $-1 = \log_4 (\frac{1}{4})$. Remember, the exponent may be negative, but the base must be positive.

If $-4 = \log_b (\frac{1}{81})$, and if we write this in exponential form, we may find b. In exponential form it becomes $b^{-4} = \frac{1}{81}$ which means that $b = 3$.

When working with functions, we normally denote the dependent variable as y and the independent variable as x. For this reason, we *define the* **logarithmic function** *to be*

$$y = \log_b x \qquad\qquad (7-3)$$

Equations (7–2) and (7–3) may seem to have different meanings, due to the location of the variables. We must keep in mind that a function is defined by the operation being performed on the independent variable, and not by the letter chosen to represent it. These two equations express the same function, the logarithmic function, since they represent the same operation on the independent variable which appears in each. Equation (7–3) expresses the function with the standard dependent and independent variables.

We note that for the exponential function $y = b^x$ and the logarithmic function $y = \log_b x$, if we change the form of one and then interchange the variables x and y, we then obtain the other function. In general, *such functions are called* **inverse functions.** We note that we may make such changes with the trigonometric functions and inverse trigonometric functions, and therefore they are properly inverse functions.

We shall now show the graphs of the exponential function, Eq. (7–1), and the logarithmic function, Eq. (7–3). In this way we may demonstrate their properties. The following examples are used for this purpose.

Example D

Figure 7–2

Plot the graph of $y = 2^x$.

Assuming values for x and then finding the corresponding values for y, we obtain the following table.

x	-3	-2	-1	0	1	2	3	4
y	$\frac{1}{8}$	$\frac{1}{4}$	$\frac{1}{2}$	1	2	4	8	16

From these values we plot the curve, as shown in Fig. 7–2. We note that the x-axis is an asymptote of the curve.

Example E

Figure 7–3

Plot the graph of $y = \log_3 x$.

We can find the points for this graph more easily if we first put the equation in exponential form: $x = 3^y$. By assuming values for y, we can find the corresponding values for x.

x	$\frac{1}{9}$	$\frac{1}{3}$	1	3	9	27
y	-2	-1	0	1	2	3

Using these values, we construct the graph seen in Fig. 7–3.

Any exponential or logarithmic curve, where $b > 1$, will be similar in shape to those of Examples D and E. From these curves we can draw certain conclusions:

1. *If* $0 < x < 1$, $\log_b x < 0$; *if* $x = 1$, $\log_b 1 = 0$; *if* $x > 1$, $\log_b x > 0$.
2. *If* $x > 1$, x *increases more rapidly than* $\log_b x$.
3. *For all values of* x, $b^x > 0$.
4. *If* $x > 1$, b^x *increases more rapidly than* x.

Since a logarithm is an exponent, it must follow the laws of exponents. Those laws which are of the greatest importance at this time are listed here for reference.

$$b^u \cdot b^v = b^{u+v} \tag{7-4}$$

$$b^u/b^v = b^{u-v} \tag{7-5}$$

$$(b^u)^n = b^{nu} \tag{7-6}$$

We shall now show how these laws for exponents give certain useful properties to logarithms.

If we let $u = \log_b x$ and $v = \log_b y$ and write these equations in exponential form, we have $x = b^u$ and $y = b^v$. Therefore, forming the product of x and y, we obtain

$$xy = b^u b^v = b^{u+v} \quad \text{or} \quad xy = b^{u+v}$$

Writing this last equation in logarithmic form yields

$$u + v = \log_b xy$$

or

$$\log_b x + \log_b y = \log_b xy \tag{7-7}$$

Equation (7–7) states the property that *the logarithm of the product of two numbers is equal to the sum of the logarithms of the numbers.*

Using the same definitions of u and v to form the quotient of x and y, we then have

$$\frac{x}{y} = \frac{b^u}{b^v} = b^{u-v} \quad \text{or} \quad \frac{x}{y} = b^{u-v}$$

Writing this last equation in logarithmic form, we have

$$u - v = \log_b\left(\frac{x}{y}\right)$$

or

$$\log_b x - \log_b y = \log_b\left(\frac{x}{y}\right) \tag{7-8}$$

Equation (7–8) states the property that *the logarithm of the quotient of two numbers is equal to the logarithm of the numerator minus the logarithm of the denominator.*

If we again let $u = \log_b x$ and write this in exponential form, we have $x = b^u$. To find the nth power of x, we write

$$x^n = (b^u)^n = b^{nu}$$

Expressing this equation in logarithmic form yields

$$nu = \log_b (x^n)$$

or

$$n \log_b x = \log_b (x^n) \tag{7–9}$$

This last equation states that *the logarithm of the nth power of a number is equal to n times the logarithm of the number.* The exponent n may be integral or fractional and, therefore, we may use Eq. (7–9) for finding powers and roots of numbers.

Example F

By Eq. (7–7) we can see that

$$\log_2 6 = \log_2(2 \cdot 3) = \log_2 2 + \log_2 3 = 1 + \log_2 3$$

By Eq. (7–8) we have

$$\log_5 100 - \log_5 4 = \log_5\left(\frac{100}{4}\right) = \log_5 25 = 2$$

If we have

$$\log_b y = 2 \log_b x + \log_b a$$

by Eqs. (7–9) and (7–7) we may write

$$\log_b y = \log_b x^2 + \log_b a = \log_b (ax^2)$$

Therefore,

$$y = ax^2$$

There are two bases of logarithms that are, or have been, used widely. Base 10 logarithms were used widely for calculational purposes before the extensive use of calculators. Base e (where e is an irrational number equal approximately to 2.718) logarithms are used widely in certain technical applications and for theoretical purposes. In the next section we shall see why e is chosen for this purpose.

The notation **log** *x will be used for logarithms to the base 10.* Note that no base is indicated.

Since our primary purpose in this chapter is to develop the derivatives of the logarithmic and exponential functions, we shall be using the base e extensively. *The name* **natural logarithm** *is given to a logarithm with the base e, and the notation* **ln** x *is used for a natural logarithm.*

Values of logarithms in both bases may be found on a scientific calculator by use of the $\boxed{\text{LOG}}$ and $\boxed{\text{LN}}$ keys. Tables of values are also available.

Since more than one base is important, it is occasionally useful to be able to change a logarithm in one base to another base. If $u = \log_b x$, $b^u = x$, and taking logarithms of each side to the base a, we have

$$u \log_a b = \log_a x, \qquad u = \frac{\log_a x}{\log_a b}$$

or

$$\log_b x = \frac{\log_a x}{\log_a b} \tag{7-10}$$

Example G

Change $\log 20 = 1.3010$ to a logarithm with base e; that is, find $\ln 20$.

In Eq. (7–10), if we let $b = e$ and $a = 10$, we have

$$\log_e x = \frac{\log_{10} x}{\log_{10} e}$$

or

$$\ln x = \frac{\log x}{\log e}$$

In this example, $x = 20$. Therefore,

$$\ln 20 = \frac{\log 20}{\log e} = \frac{\log 20}{\log 2.718} = \frac{1.3010}{0.4343} = 2.996$$

Exercises 7–1

In Exercises 1 through 12, change those in exponential form to logarithmic form, and those in logarithmic form to exponential form.

1. $4^4 = 256$
2. $3^{-2} = \frac{1}{9}$
3. $\log_8 16 = \frac{4}{3}$
4. $\log_7(\frac{1}{49}) = -2$
5. $8^{1/3} = 2$
6. $81^{3/4} = 27$
7. $\log_{0.5} 16 = -4$
8. $\log_{243} 3 = \frac{1}{5}$
9. $(12)^0 = 1$
10. $(\frac{1}{2})^{-2} = 4$
11. $\log_8(\frac{1}{512}) = -3$
12. $\log_{1/32}(\frac{1}{8}) = \frac{3}{5}$

In Exercises 13 through 16, determine the value of the letter present.

13. $\log_7 y = 3$ 14. $\log_b 4 = \frac{2}{3}$ 15. $\log (10)^{0.2} = x$ 16. $\log_8 y = -\frac{2}{3}$

In Exercises 17 through 20, plot the graphs of the given functions.

17. $y = 3^x$ 18. $y = (\frac{1}{2})^x$ 19. $y = \log_2 x$ 20. $y = 2 \log_4 x$

In Exercises 21 through 28, express each as a single logarithm.

21. $\log_5 9 - \log_5 3$ 22. $\log_2 3 + \log_2 x$

23. $\log_6 x^2 + \log_6 \sqrt{x}$ 24. $\log_7 3^4 - \log_7 9$

25. $2 \log_b 2 + \log_b y$ 26. $3 \log_b 4 - \log_b x$

27. $2 \log_2 x + \log_2 3 - \log_2 5$ 28. $\log_4 7 - 2 \log_4 t + \log_4 a$

In Exercises 29 through 32, evaluate the given logarithms by use of logarithms to the base 10.

29. $\ln 3$ 30. $\ln 0.9$ 31. $\log_7 42$ 32. $\log_{12} 122$

In Exercises 33 through 40, solve the given problems.

33. The atmospheric pressure at a given height h is given by $p = p_0 e^{-kh}$, where p_0 and k are constants. Solve for h.

34. If an amount A is deposited in a bank which pays 8% interest, compounded quarterly, the value of the investment after t years is given by $V = A(1.02)^{4t}$. By use of logarithms to the base 10, solve for t.

35. For an electric circuit containing a battery of voltage E, a resistance R, and an inductance L, the current as a function of time is given by

$$i = \frac{E}{R}(1 - e^{-Rt/L})$$

Solve for t.

36. The vapor pressure P over a liquid may be related to temperature by $\log P = (a/T) + b$, where a and b are constants. Solve for P.

37. Computations can be made by means of natural logarithms, but such computations require adding or subtracting multiples of 2.3026 to locate the decimal point properly. We can see this in a case such as $N = P \cdot 10^k$. Then

$$\ln N = \ln P + k \ln 10 = \ln P + k(2.3026)$$

For example,

$$\ln 2.500 = 0.9163 \quad \text{and} \quad \ln 250.0 = 0.9163 + 2(2.3026) = 5.5215$$

Find the natural logarithms of the following numbers:

    (a) 6.700   (b) 67.00   (c) 3400   (d) 0.3400

Calculate (6.700)(3400) by means of these values.

38. The **hyperbolic sine of u** is defined as

$$\sinh u = \frac{1}{2}(e^u - e^{-u})$$

and the **hyperbolic cosine of u** is defined as

$$\cosh u = \frac{1}{2}(e^u + e^{-u})$$

These functions are called *hyperbolic* functions since, if $x = \cosh u$ and $y = \sinh u$, x and y satisfy the equation of the hyperbola $x^2 - y^2 = 1$. Verify this by substituting the appropriate exponential expressions into the equation of the hyperbola and simplifying.

39. From the definitions of sinh u and cosh u (Exercise 38), find the values of sinh 0.25 and cosh 0.25 and show that $\cosh^2 0.25 - \sinh^2 0.25 = 1$.

40. From the values of sinh 0.25 and cosh 0.25 (see Exercises 38 and 39), show that sinh 0.50 = 2 sinh 0.25 cosh 0.25.

7–2 Derivative of the Logarithmic Function

We shall first find the derivative of the logarithmic function. We can use the derivative of the logarithmic function to find the derivative of the exponential function, which we shall do in the next section. Again we utilize the delta-process.

If we let $y = \log_b u$, where u is a function of x, we have

$$y + \Delta y = \log_b(u + \Delta u)$$

$$\Delta y = \log_b(u + \Delta u) - \log_b u = \log_b\left(\frac{u + \Delta u}{u}\right) = \log_b\left(1 + \frac{\Delta u}{u}\right)$$

$$\frac{\Delta y}{\Delta u} = \frac{\log_b(1 + \Delta u/u)}{\Delta u} = \frac{1}{u} \cdot \frac{u}{\Delta u} \log_b\left(1 + \frac{\Delta u}{u}\right)$$

$$= \frac{1}{u} \log_b\left(1 + \frac{\Delta u}{u}\right)^{u/\Delta u}$$

(We multiply and divide by u for purposes of evaluating the limit, as we shall now show.)

Before we can evaluate the $\lim_{\Delta u \to 0} \Delta y/\Delta u$, it is necessary to determine

$$\lim_{\Delta u \to 0}\left(1 + \frac{\Delta u}{u}\right)^{u/\Delta u}$$

We can see that the exponent becomes unbounded, but the number being raised to this exponent approaches 1. Therefore, we shall investigate this limiting value.

To find an approximate value, let us graph the function $y = (1 + x)^{1/x}$ (for purposes of graphing we let $\Delta u/u = x$). We construct a table of values and then graph the function (see Fig. 7–4).

Figure 7–4

x	-0.5	-0.25	$+0.25$	$+0.50$	$+1.00$
y	4.00	3.16	2.44	2.25	2.00

See Appendix C for a computer program for finding the value for e.

Only these values are shown, since we are interested in the y-value corresponding to $x = 0$. We see from the graph that this value is approximately 2.7. Choosing very small values of x, we may obtain the following values:

x	0.1	0.01	0.001	0.0001
y	2.5937	2.7048	2.7169	2.71815

By methods developed in Chapter 13, it can be shown that this value is about 2.7182818. *The limiting value is the irrational number e*, which is the base of the natural logarithms we introduced in the previous section.

Returning to the derivative of the logarithmic function, we have

$$\lim_{\Delta u \to 0} \frac{\Delta y}{\Delta u} = \lim_{\Delta u \to 0} \left[\frac{1}{u} \log_b \left(1 + \frac{\Delta u}{u} \right)^{u/\Delta u} \right] = \frac{1}{u} \log_b e$$

Therefore,

$$\frac{dy}{du} = \frac{1}{u} \log_b e$$

For reference, Eq. (2–28) is
$$\frac{dy}{dx} = \frac{dy}{du}\frac{du}{dx}$$

Combining this equation with Eq. (2–28), we have

$$\frac{d(\log_b u)}{dx} = \frac{1}{u} \log_b e \frac{du}{dx} \tag{7–11}$$

At this point we see that if we choose e as the basis of a system of logarithms, the above formula becomes

$$\frac{d(\ln u)}{dx} = \frac{1}{u}\frac{du}{dx} \tag{7–12}$$

The choice of e as the base b makes $\log_e e = 1$; thus, this factor does not appear in Eq. (7–12). We can now see why the number e is chosen as the base for a system of logarithms, the natural logarithms. The notation $\ln u$ is the same as that used in Section 7–1 for natural logarithms.

Example A

Find the derivative of $y = \log 4x$.

Using Eq. (7–11), we have

$$\frac{dy}{dx} = \frac{1}{4x} \log e \,(4) \qquad (\log e = 0.4343)$$

$$= \frac{1}{x} \log e$$

Example B

Find the derivative of $y = \ln 3x^4$.
Using Eq. (7–12), we have

$$\frac{dy}{dx} = \frac{1}{3x^4} \overset{\frac{du}{dx}}{(12x^3)}$$

$$= \frac{4}{x}$$

Example C

Find the derivative of $y = \ln \tan x$.
Using Eq. (7–12), along with the derivative of the tangent, we have

$$\frac{dy}{dx} = \frac{1}{\tan x} \sec^2 x = \frac{\cos x}{\sin x} \frac{1}{\cos^2 x}$$

$$= \csc x \sec x$$

   Frequently we can find the derivative of a logarithmic function more simply if we use the properties of logarithms to simplify the expression before the derivative is found. The following examples illustrate this.

Example D

Find the derivative of $y = \ln \dfrac{x - 1}{x + 1}$.

In this example it is easier to write y in the form

$$y = \ln(x - 1) - \ln(x + 1)$$

by using the properties of logarithms. Hence,

$$\frac{dy}{dx} = \frac{1}{x - 1} - \frac{1}{x + 1} = \frac{x + 1 - x + 1}{(x - 1)(x + 1)}$$

$$= \frac{2}{x^2 - 1}$$

Example E

Find the derivative of $y = \ln(1 - 2x)^3$.
First we may rewrite the equation as $y = 3 \ln(1 - 2x)$. Then

$$\frac{dy}{dx} = 3\left(\frac{1}{1 - 2x}\right)(-2) = \frac{-6}{1 - 2x}$$

Be careful to distinguish this function from the function $y = [\ln(1 - 2x)]^3$ [which can also be written as $y = \ln^3(1 - 2x)$]. In this latter function it is the logarithm which is being cubed, whereas in the first function the quantity $(1 - 2x)$ is being cubed. The derivative in the latter case would be

$$\frac{dy}{dx} = 3[\ln(1 - 2x)]^2\left(\frac{1}{1 - 2x}\right)(-2) = -\frac{6[\ln(1 - 2x)]^2}{1 - 2x}$$

Example F

Evaluate the derivative of $y = \ln(\sin 2x)(\sqrt{x^2 + 1})$, where $x = 0.375$.

First, using the properties of logarithms, we rewrite the function as

$$y = \ln \sin 2x + \frac{1}{2} \ln(x^2 + 1)$$

Now, finding the derivative, we have

$$\frac{dy}{dx} = \frac{1}{\sin 2x} (\cos 2x)(2) + \frac{1}{2}\left(\frac{1}{x^2 + 1}\right)(2x) = 2 \cot 2x + \frac{x}{x^2 + 1}$$

Evaluating the derivative for $x = 0.375$ gives us

$$\left.\frac{dy}{dx}\right|_{x=0.375} = 2 \cot 0.750 + \frac{0.375}{0.375^2 + 1} = 2.48$$

Exercises 7-2

In Exercises 1 through 28, find the derivatives of the given functions.

1. $y = \log x^2$
2. $y = \log_2 6x$
3. $y = 2 \log_5 (3x + 1)$
4. $y = 3 \log_7 (x^2 + 1)$
5. $y = \ln(1 - 3x)$
6. $y = 2 \ln(3x^2 - 1)$
7. $y = 2 \ln \tan 2x$
8. $y = \ln \sin x$
9. $y = \ln \sqrt{x}$
10. $y = \ln \sqrt{4x - 3}$
11. $y = \ln(x^2 + 2x)^3$
12. $y = \ln(2x^3 - x)^2$

13. $y = x \ln x$
14. $y = x^2 \ln 2x$
15. $y = \dfrac{3x}{\ln(2x + 1)}$

16. $y = \dfrac{\ln x}{x}$
17. $y = \ln(\ln x)$
18. $y = \ln \cos x^2$

19. $y = \ln \dfrac{x}{1 + x}$
20. $y = \ln(x\sqrt{x + 1})$
21. $y = \sin \ln x$

22. $y = \text{Arctan} \ln 2x$
23. $y = (\ln 2x)^2$
24. $y = (\ln x)^3$

25. $y = \ln(x \tan x)$
26. $y = \ln(x + \sqrt{x^2 - 1})$
27. $y = \ln \dfrac{x^2}{x + 2}$

28. $y = \sqrt{x^2 + 1} - \ln \dfrac{1 + \sqrt{x^2 + 1}}{x}$

In Exercises 29 through 40, solve the given problems.

29. On a calculator find the value of $(\ln 2.0001 - \ln 2.0000)/0.0001$ and compare it with 0.5. Give the meanings of the value found and 0.5 in relation to the derivative of $\ln x$, where $x = 2$.

30. On a calculator find the value of $(\ln 0.5001 - \ln 0.5000)/0.0001$ and compare it with 2. Give the meanings of the value found and 2 in relation to the derivative of $\ln x$, where $x = 0.5$.

31. Verify the values of $(1 + x)^{1/x}$ in the table where $(1 + x)^{1/x}$ is calculated for $x = 0.1$, 0.01, 0.001, and 0.0001.

32. Find the second derivative of the function $y = x^2 \ln x$.

33. Evaluate the derivative of $y = \text{Arctan } 2x + \ln(4x^2 + 1)$, where $x = 0.625$.

34. Evaluate the derivative of $y = \ln \sqrt{\dfrac{2x + 1}{3x + 1}}$, where $x = 2.75$.

35. Find the slope of a line tangent to the curve of $y = \ln \cos x$ at $x = \pi/4$.

36. Find the slope of a line tangent to the curve of $y = x \ln 2x$ at $x = 2$.

37. Find the derivative of $y = x^x$ by first taking logarithms of each side of the equation.

38. Find the derivative of $y = (\sin x)^x$ by first taking logarithms of each side of the equation.

39. If the loudness b (in decibels) of a sound of intensity I is given by $b = 10 \log(I/I_0)$ where I_0 is a constant, find the expression for db/dt in terms of dI/dt.

40. The temperature T from the inner surface of a certain steam pipe to the outer surface is given by

$$T = 100 - 40 \frac{\ln(r/5)}{\ln(6/5)}$$

where r is the distance from the center of the pipe. Determine the rate of change of temperature with respect to r. (This solution is valid only for $5 < r < 6$, since these are the inner and outer radii, respectively.)

7–3 Derivative of the Exponential Function

To obtain the derivative of the exponential function, we let $y = b^u$ and then write the equation in logarithmic form:

$$u = \log_b y, \qquad \frac{du}{dx} = \frac{1}{y} \log_b e \frac{dy}{dx}$$

$$\frac{dy}{dx} = \frac{1}{\log_b e} y \frac{du}{dx} = \frac{1}{\log_b e} b^u \left(\frac{du}{dx}\right)$$

Thus,

$$\frac{d(b^u)}{dx} = \frac{1}{\log_b e} b^u \left(\frac{du}{dx}\right) \tag{7–13}$$

If we let $b = e$, Eq. (7–13) becomes

$$\frac{d(e^u)}{dx} = e^u \left(\frac{du}{dx}\right) \tag{7–14}$$

The simplicity of Eq. (7–14) compared with Eq. (7–13) again shows the advantage of choosing e as the basis of natural logarithms. It is for this reason that e appears so often in applications of calculus.

Example A

Find the derivative of $y = e^x$.

Using Eq. (7–14), we have

$$\frac{dy}{dx} = e^x(1) = e^x$$

We see that the derivative of the function e^x equals itself. This exponential function is widely used in applications of calculus.

We should note carefully that Eq. (2–29) is used with a variable raised to a constant exponent, whereas with Eqs. (7–13) and (7–14) we are finding the derivative of a constant raised to a variable exponent.

$$\underset{\text{Variable}}{\downarrow} \underset{\text{Constant}}{\swarrow} \qquad \underset{\text{Constant}}{\downarrow} \underset{\text{Variable}}{\swarrow}$$

$$\frac{du^n}{dx} = nu^{n-1}\left(\frac{du}{dx}\right) \qquad \frac{db^u}{dx} = \frac{1}{\log_b e} b^u \left(\frac{du}{dx}\right)$$

Example B

Find the derivatives of $y = (4x)^2$ and $y = 2^{4x}$.

Using Eq. (2–29) we have

$$y = (4x)^2 \quad \overset{\frac{du}{dx}}{\underset{\downarrow}{}}$$

$$\frac{dy}{dx} = 2(4x)^1(4)$$

$$= 32x$$

Using Eq. (7–13) we have

$$y = 2^{4x} \quad \overset{\frac{du}{dx}}{\underset{\downarrow}{}}$$

$$\frac{dy}{dx} = \frac{1}{\log_2 e} 2^{4x}(4)$$

$$= \frac{4}{\log_2 e} 2^{4x}$$

We continue now with additional examples of the use of Eq. (7–14).

Example C

Find the derivative of $y = \ln \cos e^{2x}$.

Using Eq. (7–14), along with the derivatives of the logarithmic and cosine functions, we have

$$\frac{dy}{dx} = \frac{1}{\cos e^{2x}} \frac{d \cos e^{2x}}{dx} \qquad \text{Using } \frac{d \ln u}{dx} = \frac{1}{u}\frac{du}{dx}$$

$$= \frac{1}{\cos e^{2x}}(-\sin e^{2x})\frac{de^{2x}}{dx} \qquad \text{Using } \frac{d \cos u}{dx} = -\sin u \frac{du}{dx}$$

$$= -\frac{\sin e^{2x}}{\cos e^{2x}}(e^{2x})(2) \qquad \text{Using } \frac{de^u}{dx} = e^u \frac{du}{dx}$$

$$= -2e^{2x} \tan e^{2x} \qquad \text{Using } \frac{\sin \theta}{\cos \theta} = \tan \theta$$

Example D

Find the derivative of $y = x\,e^{\tan x}$.

Using Eq. (7–14), along with the derivatives of a product and the tangent, we have

$$\frac{dy}{dx} = x(e^{\tan x})\sec^2 x + e^{\tan x}$$

$$= e^{\tan x}(x\sec^2 x + 1)$$

Example E

Find the derivative of $y = (e^{1/x})^2$.

In this example,

$$\frac{dy}{dx} = 2(e^{1/x})(e^{1/x})\left(-\frac{1}{x^2}\right) = \frac{-2(e^{1/x})^2}{x^2}$$

$$= \frac{-2e^{2/x}}{x^2}$$

This problem could have also been solved by first writing the function as $y = e^{2/x}$, which is an equivalent form determined by the laws of exponents. When we use this form, the derivative becomes

$$\frac{dy}{dx} = e^{2/x}\left(-\frac{2}{x^2}\right)$$

$$= \frac{-2e^{2/x}}{x^2}$$

We can see that this change in form of the function simplifies the steps necessary for finding the derivative.

Example F

Find the derivative of $y = (3e^{4x} + 4x^2\ln x)^3$.

Using the general power rule, along with Eq. (7–14), the derivative of a product, and the derivative of a logarithm, we have

$$\frac{dy}{dx} = 3(3e^{4x} + 4x^2\ln x)^2\left[12e^{4x} + 4x^2\left(\frac{1}{x}\right) + (\ln x)(8x)\right]$$

$$= 3(3e^{4x} + 4x^2\ln x)^2(12e^{4x} + 4x + 8x\ln x)$$

Example G

Evaluate the derivative of $y = \dfrac{3e^{2x}}{x^2 + 1}$ for $x = 1.275$.

Finding the derivative, we have

$$\frac{dy}{dx} = \frac{(x^2 + 1)(3e^{2x})(2) - 3e^{2x}(2x)}{(x^2 + 1)^2} = \frac{6e^{2x}(x^2 - x + 1)}{(x^2 + 1)^2}$$

Evaluating this for $x = 1.275$ gives us

$$\frac{dy}{dx} = \frac{6e^{2(1.275)}(1.275^2 - 1.275 + 1)}{(1.275^2 + 1)^2}$$

$$= 15.05$$

Exercises 7-3

In Exercises 1 through 32, find the derivatives of the given functions.

1. $y = 3^{2x}$
2. $y = 3^{1-x}$
3. $y = 4^{6x}$
4. $y = 10^{x^2}$

5. $y = e^{6x}$
6. $y = 3e^{x^2}$
7. $y = e^{\sqrt{x}}$
8. $y = e^{2x^4}$

9. $y = xe^x$
10. $y = x^2e^{2x}$
11. $y = xe^{\sin x}$
12. $y = e^x \sin x$

13. $y = \dfrac{3e^{2x}}{x + 1}$
14. $y = \dfrac{e^x}{x}$ •
15. $y = e^{-3x}\sin 4x$

16. $y = (\cos 2x)(e^{x^2 - 1})$
17. $y = \dfrac{2e^{3x}}{4x + 3}$
18. $y = \dfrac{\ln 2x}{e^{2x} + 2}$

19. $y = \ln(e^{x^2} + 4)$
20. $y = (3e^{2x} + x)^3$
21. $y = (2e^{2x})^3 \sin x^2$

22. $y = (e^{3/x}\cos x)^2$
23. $y = (\ln 2x + e^{2x})^2$
24. $y = (2e^{x^2} + x^2)^3$

25. $y = e^{2x} - e^{-2x}$
26. $y = 4e^{-(2/x)}$
27. $y = e^{2x}\ln x$

28. $y = e^{x^2}\ln \cos x$
29. $y = \ln \sin 2e^{6x}$
30. $y = \tan e^{x+1}$

31. $y = 2 \operatorname{Arcsin} e^{2x}$
32. $y = \operatorname{Arctan} e^{3x}$

In Exercises 33 through 44, solve the given problems.

33. On a calculator find the values of (a) e and (b) $(e^{1.0001} - e^{1.0000})/0.0001$. Compare the values and give the meaning of each in relation to the derivative of e^x where $x = 1$.

34. On a calculator find the values of (a) e^2 and (b) $(e^{2.0001} - e^{2.0000})/0.0001$. Compare the values and give the meaning of each in relation to the derivative of e^x where $x = 2$.

35. Evaluate the derivative of $y = e^{-2x}\cos 2x$ for $x = 0.625$.

36. Evaluate the derivative of $y = \dfrac{e^{-x}}{1 + \ln 4x}$ for $x = 1.842$.

37. Show that $y = xe^{-x}$ satisfies the equation $(dy/dx) + y = e^{-x}$.

38. Show that $y = e^{-x} \sin x$ satisfies the equation

$$\frac{d^2y}{dx^2} + 2\frac{dy}{dx} + 2y = 0$$

39. Show that the slope of the curve $y = e^x$ at $x = a$ equals the ordinate at $x = a$.

40. The Beer-Lambert law of light absorption may be expressed as $I = I_0 e^{-\alpha x}$, where I/I_0 is that fraction of the incident light beam which is transmitted, α is a constant, and x is the distance the light travels through the medium. Find the rate of change of I with respect to x.

41. In a particular electric circuit, the voltage is given by $V = 100(1 - e^{-0.1t})$. Find the expression for dV/dt.

42. The displacement s of an object is given by $s = e^{-0.02t} \cos 2t$. Find the expression for the velocity of the object.

43. By referring to Exercise 38 of Section 7–1, show that

$$\frac{d}{dx} \sinh u = \cosh u \frac{du}{dx}$$

where u is a function of x.

44. By referring to Exercise 38 of Section 7–1, show that

$$\frac{d}{dx} \cosh u = \sinh u \frac{du}{dx}$$

where u is a function of x.

7–4 Applications

The following examples show applications of the logarithmic and exponential functions to curve-tracing, Newton's method, and time-rate-of-change problems. Certain other applications are indicated in the exercises.

Example A

Figure 7–5

Sketch the graph of the function $y = x \ln x$.

The only intercept is $(1, 0)$ (there is no intercept for $x = 0$ since $\log_b 0$ is not defined).

Also,

$$\frac{dy}{dx} = x\left(\frac{1}{x}\right) + \ln x = 1 + \ln x, \qquad \frac{d^2y}{dx^2} = \frac{1}{x}$$

The first derivative is zero if $\ln x = -1$, or $x = e^{-1}$. The second derivative is positive for this value of x. Thus, there is a minimum point at $(1/e, -1/e)$. The only values of x which can be considered are positive values, since $\ln x$ is undefined for negative values. The second derivative indicates that the curve is always concave up. The graph is shown in Fig. 7–5.

Example B

Figure 7-6

Sketch the graph of the function $y = e^{-x}\cos x$ $(0 \le x \le 2\pi)$.

This curve has intercepts for all values for which $\cos x$ is zero. Those values between 0 and 2π for which $\cos x = 0$ are $x = \pi/2$ and $x = 3\pi/2$. Also, $(0, 1)$ is an intercept. We next write

$$\frac{dy}{dx} = -e^{-x}\sin x - e^{-x}\cos x = -e^{-x}(\sin x + \cos x)$$

Setting the first derivative equal to zero, we obtain $\sin x + \cos x = 0$, or $\tan x = -1$. There are maximum or minimum points for $x = 3\pi/4$ and $x = 7\pi/4$ (see Fig. 7-6). We then write

$$\frac{d^2y}{dx^2} = 2e^{-x}\sin x$$

Thus, d^2y/dx^2 is positive when $x = 3\pi/4$ and negative when $x = 7\pi/4$. Hence, $(3\pi/4, -0.067)$ is a minimum and $(7\pi/4, 0.003)$ is a maximum. From the second derivative we find that points of inflection occur for $x = 0$, π, and 2π.

Example C

By Newton's method, find the root of the equation $e^{2x} - 4\cos x = 0$ which lies between 0 and 1.

Here

$$f(x) = e^{2x} - 4\cos x$$

and

$$f'(x) = 2e^{2x} + 4\sin x$$

This means that $f(0) = -3$ and $f(1) = 5.2$. Therefore, we choose $x_1 = 0.5$. Using Eq. (3-1), which is

$$x_2 = x_1 - \frac{f(x_1)}{f'(x_1)}$$

we have the following values.

$$f(x_1) = e^{2(0.5)} - 4\cos 0.5 = -0.7920484$$
$$f'(x_1) = 2e^{2(0.5)} + 4\sin 0.5 = 7.3542658$$
$$x_2 = 0.5 - \frac{-0.7920484}{7.3542658} = 0.6076992$$

Using the method again, we find $x_3 = 0.5979751$, which is correct to three decimal places.

Example D

The population P, in terms of the population P_0 (at $t = 0$) and t, is given by $P = P_0e^{kt}$. Show that the time rate of change of population is proportional to the population present at time t.

By taking a derivative of P with respect to time, we have

$$\frac{dP}{dt} = kP_0e^{kt} = kP$$

Example E

A rocket moving vertically, if the only force acting on it is due to gravity and its mass is decreasing at a constant rate r, has its velocity as a function of time given by

$$v = v_0 - gt - k \ln\left(1 - \frac{rt}{m_0}\right)$$

where v_0 is the initial velocity, m_0 is the initial mass, and k is a constant. Determine the expression for the acceleration.

Since the acceleration is the time rate of change of the velocity, we must find dv/dt. Therefore,

$$\frac{dv}{dt} = -g - k\frac{1}{1 - \dfrac{rt}{m_0}}\left(\frac{-r}{m_0}\right) = -g + \frac{km_0}{m_0 - rt}\left(\frac{r}{m_0}\right)$$

$$= \frac{kr}{m_0 - rt} - g$$

Exercises 7–4

In Exercises 1 through 12, sketch the graphs of the given functions.

1. $y = \ln \cos x$ 2. $y = \dfrac{\ln x}{x}$ 3. $y = xe^{-x}$ 4. $y = xe^{x}$

5. $y = \ln\dfrac{1}{x^2 + 1}$ 6. $y = \ln\dfrac{1}{x}$ 7. $y = e^{-x^2}$ 8. $y = x - e^x$

9. $y = \ln x - x$ 10. $y = e^{-x}\sin x$

11. $y = \frac{1}{2}(e^x - e^{-x})$ (See Exercise 38 of Section 7–1.)

12. $y = \frac{1}{2}(e^x + e^{-x})$ (See Exercise 38 of Section 7–1.)

In Exercises 13 through 28, solve the given problems by finding the appropriate derivatives.

13. Find the equation of the line tangent to the curve $y = x^2 \ln x$ at the point (1, 0).

14. Find the equation of the line tangent to the curve of $y = $ Arctan $2x$, where $x = 1$.

15. Find the equation of the line normal to the curve of $y = 2 \sin \frac{1}{2}x$, where $x = \frac{3}{2}\pi$.

16. Find the equation of the line normal to the curve of $y = e^{2x}/x$ at $x = 1$.

17. By Newton's method, solve the equation $x^2 - 2 + \ln x = 0$. See Fig. 7–7.

18. By Newton's method, solve the equation $e^{-2x} - $ Arctan $x = 0$.

19. The power supply P (in watts) in a satellite is given by $P = 100e^{-0.005t}$, where t is measured in days. Find the time rate of change of power after 100 days.

20. The number N of atoms of radium at any time t is given in terms of the number at $t = 0$, N_0, by $N = N_0 e^{-kt}$. Show that the time rate of change of N is proportional to N.

21. The velocity of an object moving through a resisting medium is found to obey the relation $v = 40(1 - e^{-0.05t})$. Find the acceleration when $t = 10$.

$y = x^2 - 2 + \ln x$

Figure 7–7

22. The charge on a capacitor of capacitance C in a circuit containing the capacitor, a resistance R, and voltage source of voltage E is given by $q = CE(1 - e^{-t/RC})$. Show that this expression for q satisfies the equation

$$R\frac{dq}{dt} + \frac{q}{C} = E$$

23. Assuming that force is proportional to acceleration, show that a particle moving along the x-axis, so that its displacement $x = ae^{kt} + be^{-kt}$, has a force acting on it which is proportional to its displacement.

24. The radius of curvature at any point on a curve is given by

$$R = \frac{[1 + (dy/dx)^2]^{3/2}}{d^2y/dx^2}$$

   Determine the radius of curvature of $y = \ln \sec x$ when $x = \pi/4$.

25. An object on the end of a spring is moving so that its displacement in centimeters from the equilibrium position is given by

$$y = e^{-0.5t}(0.4 \cos 6t - 0.2 \sin 6t)$$

   Find the expression for the velocity of the object. What is the velocity when $t = \pi/12$ s? The motion described by this equation is called *damped harmonic motion*. The general shape of the curve of displacement as a function of time is that of the curve in Example B.

26. In an electronic device, the maximum current density i_m as a function of the temperature T is given by $i_m = AT^2 e^{k/T}$, where A and k are constants. If the temperature is changing with time, find the expression for the time rate of change of i_m.

27. In developing the theory dealing with the friction between a pulley wheel and the pulley belt, the ratio of the tensions in the belt on either side of the wheel is given by

$$\frac{T_2}{T_1} = e^{k/\sin(\theta/2)}$$

   where k is a constant and θ is the angle of the opening of the pulley wheel. Find the expression for a small change in the ratio of tensions for a small change in the angle θ.

28. Find the area of the largest rectangle which can be inscribed in the first quadrant under the curve of $y = e^{-x}$. See Fig. 7–8.

$y = e^{-x}$

Figure 7–8

7–5 Review Exercises for Chapter 7

In Exercises 1 through 24, find the derivative of each of the given functions.

1. $y = 3 \ln(x^2 + 1)$
2. $y = \ln(1 - 5x)^3$
3. $y = (e^{x-3})^2$
4. $y = e^{\sqrt{1-x}}$
5. $y = \ln(\csc x^2)$
6. $y = \ln(3 + \sin x^2)$
7. $y = [\ln(3 + \sin x)]^2$
8. $y = \ln(3 + \sin x)^2$
9. $y = e^{\sin 2x}$
10. $y = 3 e^{\sec 3x}$
11. $y = x^3 e^x$
12. $y = 2x e^{4x}$

13. $y = \ln(x - e^{-x})^2$

14. $y = \ln\sqrt{\sin 2x}$

15. $y = \dfrac{\cos^2 x}{e^{3x} + 1}$

16. $y = \dfrac{\ln\sqrt{3x + 1}}{3x + 1}$

17. $y = x^2 \ln x$

18. $y = (x + 1)\ln(x^3 + 1)$

19. $y = \dfrac{e^{2x}}{x^2 + 1}$

20. $y = \dfrac{\ln x}{2x + 1}$

21. $y = e^{-x}\sec x$

22. $y = e^{3x}\ln x$

23. $x^2 \ln y = y + x$

24. $y = x^2(e^{\cos^2 x})^2$

In Exercises 25 through 48, solve the given problems by finding the appropriate derivatives.

25. Find the equation of the line tangent to the curve $y = \ln\cos x$ where $x = \pi/6$.

26. Find the equation of the line normal to the curve $y = e^{x^2}$ where $x = \frac{1}{2}$.

27. Sketch the graph of $y = \ln(1 + x)$.

28. Sketch the graph of $y = x(\ln x)^2$.

29. By Newton's method, find the negative root of the equation $e^x - x^2 = 0$.

30. By Newton's method, solve the equation $\ln x^2 = \text{Arctan } x$.

31. Fans are exhausting a room such that the number of cubic feet of carbon dioxide in the room after t minutes is given by $n = 3(e^{-0.05t} + 1)$. Find the time rate of change of n with respect to t. Show that the result can be written as $0.05(3 - n)$.

32. A business firm estimates that the value of a machine is given by $V = 25000e^{-0.5t}$, where V is measured in dollars and t in years. What is the time rate of change of the value after 4 years?

33. The voltage in a certain circuit is given by $E = 170e^{-0.015t}$. Find the approximate change in E if t changes from 0.100 s to 0.105 s.

34. An analysis of samples of air for a city showed that the number of parts per million p of sulfur dioxide on a certain day was $p = 0.05\ln(2 + 24t - t^2)$, where t is the hour of the day. Using differentials, find the approximate change in the amount of sulfur dioxide between 10 A.M. and noon.

35. For a circuit containing a capacitance C and a resistance R, the current as a function of time is given by $i = i_0 e^{-t/RC}$. If $i_0 = 10$ A, $R = 2\ \Omega$, and $C = 50\ \mu$F, find the approximate change in the current between 200 μs and 210 μs by use of the appropriate differential.

36. Under certain circumstances, 2% of a substance changes chemically each minute. If there are originally 100 g present, the amount at any time t which is present is $A = 100(0.98)^t$. Using differentials, find the approximate change in the amount present during the fourth second.

37. If inflation makes the dollar worth 5% less each year, then the value of $100 in t years will be $V = 100(0.95)^t$. What is the approximate change in the value during the fourth year?

38. An equation relating the vapor pressure p of a gas and the thermodynamic temperature T is

$$\ln p = \frac{a}{T} + b\ln T + c$$

where a, b, and c are constants. Find the expression for the rate of change of pressure with respect to temperature.

39. In the study of a certain bacterial culture, the time t required for a biological change to occur in x bacteria is found to be

$$t = a \ln \frac{x}{b - x} - c$$

where a, b, and c are constants. Find dt/dx.

40. Under certain conditions, the potential V (in volts) due to a magnet is given by $V = -k \ln\left(1 + \frac{L}{x}\right)$, where L is the length of the magnet and x is the distance from the point where the potential is measured. Find the expression for dV/dx.

41. Find the area of the largest rectangle which can be inscribed under the curve of $y = e^{-x^2}$ in the first quadrant. See Fig. 7–9.

42. A businessman determines that the gross income I he receives by selling x items per week is $I = 100xe^{-x/10}$. How many items should he sell per week to make I a maximum?

43. Under certain conditions the work done by a gas in an expansion from a pressure P_1 to a pressure P_2 is given by $W = k \ln(P_1/P_2)$, where k is a constant depending on the gas. If P_2 varies with time at the rate of 2 kPa/min, find the rate (in joules per minute) at which the work changes when $P_2 = 300$ kPa, if $k = 2400$ J/mol and $P_1 = 200$ kPa.

44. The charge q on a certain capacitor as a function of time is given by

$$q = e^{-0.1t}(0.2 \sin 100t + \cos 100t)$$

The current i in the circuit is the time rate of change of charge. Find the expression for i as a function of t.

45. An object is dropped from a weather balloon. The distance (in feet) it falls, taking air resistance into account, is given by

$$y = 320(t + 10e^{-0.1t} - 10)$$

Find the velocity after 10 s.

46. The velocity of an object moving through a resisting medium is found to obey the relation $v = 40(1 - e^{-0.05t})$. Find the acceleration when $t = 10$.

47. Find the acceleration of an object whose x- and y-components of displacement are given by $x = e^{-t}\sin 2t$ and $y = e^{-t}\cos 2t$ when $t = \pi/4$.

48. The curve formed by a uniform cable hanging from two points under its own weight (e.g., a telephone wire between two poles) is called the *catenary* (see Fig. 7–10). The equation of the curve is

$$y = \frac{H}{w} \cosh \frac{wx}{H}$$

where w and H are constants. Show that the equation of the catenary satisfies the equation

$$\frac{d^2y}{dx^2} = \frac{w}{H} \sqrt{1 + \left(\frac{dy}{dx}\right)^2}$$

(see Exercise 38 of Section 7–1, and Exercises 43 and 44 of Section 7–3).

$y = e^{-x^2}$

Figure 7–9

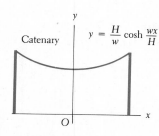

Catenary $\quad y = \frac{H}{w} \cosh \frac{wx}{H}$

Figure 7–10

CHAPTER 8

Integration by Standard Forms

Having developed the derivatives of the basic transcendental functions, we can now expand considerably the functions which we can integrate. In addition to the transcendental functions, many algebraic functions which we were unable to integrate previously will now lend themselves to integration.

In this chapter we shall expand the use of the general power formula for integration for use with integrands which have transcendental functions. We shall then develop additional standard forms for algebraic and transcendental integrands. In the following chapter, other methods of integration will be shown.

8–1 The General Power Formula

The first integration formula we shall discuss is the general power formula. We first used this with basic algebraic functions in Chapter 4.

For reference, we repeat here the general power formula for integration, Eq. (4–5):

$$\int u^n \, du = \frac{u^{n+1}}{n+1} + C \qquad n \neq -1 \tag{8–1}$$

This formula will now be applied to transcendental functions, as well as algebraic functions. When we are applying the general power formula, we must properly recognize the quantities u, n, and du. This requires familiarity with the differential forms presented in Chapters 2, 6, and 7. The following examples illustrate the use of the general power formula.

Example A

Integrate: $\int \sin^3 x \cos x \, dx$.

Since $d(\sin x) = \cos x \, dx$, we note that this integral fits the form of Eq. (8–1) for $u = \sin x$. Thus, with $u = \sin x$, we have $du = \cos x \, dx$, which means that this integral is of the form $\int u^3 \, du$. Therefore, the integration can now be completed.

$$\int \sin^3 x \cos x \, dx = \int \sin^3 x \overset{\overset{\textit{du}}{\swarrow}}{(\cos x \, dx)}$$

$$= \frac{1}{4} \sin^4 x + C \quad \longleftarrow \quad \text{Do not forget the constant of integration.}$$

▶ We note here that the factor $\cos x$ is a necessary part of the du in order to have the proper form of integration, and therefore does not appear in the final result.

Example B

Integrate: $\int \sqrt{1 + \tan x} \, \sec^2 x \, dx$.

Here we note that $d(\tan x) = \sec^2 x \, dx$, which means that the integral fits the form of Eq. (8–1) with $u = 1 + \tan x$, $du = \sec^2 x \, dx$, and $n = \frac{1}{2}$. The integral is of the form $\int u^{1/2} \, du$. Thus.

$$\int \sqrt{1 + \tan x} \, (\sec^2 x \, dx) = \int (1 + \tan x)^{1/2} \overset{\overset{\textit{du}}{\swarrow}}{(\sec^2 x \, dx)}$$

$$= \frac{2}{3} (1 + \tan x)^{3/2} + C$$

Example C

Integrate: $\int \ln x \left(\frac{dx}{x} \right)$.

By noting that $d(\ln x) = \dfrac{dx}{x}$, we have $u = \ln x$, $du = \dfrac{dx}{x}$, and $n = 1$. This means that the integral is of the form $\int u \, du$. Thus,

$$\int \ln x \left(\frac{dx}{x} \right) = \frac{1}{2} (\ln x)^2 + C$$

Example D

Integrate: $\int (2 + 3e^{2t})^3 e^{2t} \, dt$.

In this case $u = 2 + 3e^{2t}$, $du = 6e^{2t} \, dt$, and $n = 3$. This means that we must introduce a 6 to complete the differential so that the integral will fit the proper form. Also, a $\frac{1}{6}$ must then be placed before the integral. This means that the integration is as follows:

$$\int (2 + 3e^{2t})^3 e^{2t} \, dt = \frac{1}{6} \int (2 + 3e^{2t})^3 \overset{\overset{\textit{du}}{\swarrow}}{(6e^{2t} \, dt)}$$

$$= \frac{1}{6} \cdot \frac{1}{4} (2 + 3e^{2t})^4 + C$$

$$= \frac{1}{24} (2 + 3e^{2t})^4 + C$$

Example E

Find the value of

$$\int_0^{0.5} \frac{\text{Arcsin } x}{\sqrt{1-x^2}}\, dx; \qquad n = 1, \qquad u = \text{Arcsin } x, \quad \text{and} \quad du = \frac{dx}{\sqrt{1-x^2}}$$

Therefore,

$$\int_0^{0.5} \frac{\text{Arcsin } x}{\sqrt{1-x^2}}\, dx = \frac{(\text{Arcsin } x)^2}{2}\Big|_0^{0.5} = \frac{(\pi/6)^2}{2} - 0 = \frac{\pi^2}{72}$$

Example F

Figure 8–1

Find the first-quadrant area bounded by $y = \dfrac{e^{2x}}{\sqrt{e^{2x}+1}}$ and $x = 1.5$.

The area to be found is shown in Fig. 8–1. The area of the representative element is $y\, dx$. Therefore, the area is found by evaluating the integral

$$\int_0^{1.5} \frac{e^{2x}\, dx}{\sqrt{e^{2x}+1}}$$

For the purpose of integration, $n = -\frac{1}{2}$, $u = e^{2x} + 1$, and $du = 2e^{2x}\, dx$. Therefore,

$$\int_0^{1.5} (e^{2x} + 1)^{-1/2} e^{2x}\, dx = \frac{1}{2}\int_0^{1.5} (e^{2x} + 1)^{-1/2}(2e^{2x}\, dx)$$

$$= \frac{1}{2}\cdot 2(e^{2x} + 1)^{1/2}\Big|_0^{1.5}$$

$$= (e^{2x} + 1)^{1/2}\Big|_0^{1.5} = \sqrt{e^3 + 1} - \sqrt{2}$$

$$= 3.178$$

This means that the area is 3.178 square units.

Exercises 8–1

In Exercises 1 through 24, integrate each of the given functions.

1. $\displaystyle\int \sin^4 x \cos x\, dx$

2. $\displaystyle\int \cos^5 x(-\sin x\, dx)$

3. $\displaystyle\int \sqrt{\cos x}\, \sin x\, dx$

4. $\displaystyle\int \sin^{1/3} x \cos x\, dx$

5. $\displaystyle\int \tan^2 x \sec^2 x\, dx$

6. $\displaystyle\int \sec^3 x\, (\sec x \tan x)\, dx$

7. $\displaystyle\int_0^{\pi/8} \cos 2x \sin 2x\, dx$

8. $\displaystyle\int_{\pi/6}^{\pi/4} \sqrt{\cot x}\, \csc^2 x\, dx$

9. $\displaystyle\int (\text{Arcsin } x)^3\left(\frac{dx}{\sqrt{1-x^2}}\right)$

10. $\displaystyle\int \frac{(\text{Arccos } 2x)^4\, dx}{\sqrt{1-4x^2}}$

11. $\int \dfrac{\text{Arctan } 5x}{1 + 25x^2} \, dx$

12. $\int \dfrac{\text{Arcsin } 4x \, dx}{\sqrt{1 - 16x^2}}$

13. $\int [\ln(x + 1)]^2 \, \dfrac{dx}{x + 1}$

14. $\int (3 + \ln 2x)^3 \, \dfrac{dx}{x}$

15. $\int_0^{1/2} \dfrac{\ln(2x + 3)}{2x + 3} \, dx$

16. $\int_1^e \dfrac{(1 - 2 \ln x) \, dx}{x}$

17. $\int (4 + e^x)^3 e^x \, dx$

18. $\int \sqrt{1 - e^{-x}} \, (e^{-x} \, dx)$

19. $\int (2e^{2x} - 1)^{1/3} e^{2x} \, dx$

20. $\int \dfrac{(1 + 3e^{-2x})^4}{e^{2x}} \, dx$

21. $\int (1 + \sec^2 x)^4 (\sec^2 x \tan x \, dx)$

22. $\int (e^x + e^{-x})^{1/4} (e^x - e^{-x}) \, dx$

23. $\int_0^{\pi/6} \dfrac{\tan x}{\cos^2 x} \, dx$

24. $\int_{\pi/3}^{\pi/2} \dfrac{\sin \theta \, d\theta}{\sqrt{1 + \cos \theta}}$

In Exercises 25 through 30, solve the given problems by integration.

25. Find the area under the curve $y = \sin x \cos x$ from $x = 0$ to $x = \pi/2$.

26. Find the first quadrant area bounded by $y = \dfrac{\ln(4x + 1)}{4x + 1}$ and $x = 2$. See Fig. 8–2.

27. The general expression for the slope of a given curve is $(\ln x)^2/x$. If the curve passes through $(1, 2)$, find its equation.

28. Find the equation of the curve for which $dy/dx = (1 + \tan 2x)^2 \sec^2 2x$ if the curve passes through $(2, 1)$.

29. In developing the expression for the total pressure P on a wall due to molecules with mass m and velocity v striking the wall, the following equation is found: $P = mnv^2 \int_0^{\pi/2} \sin \theta \cos^2 \theta \, d\theta$. The symbol n represents the number of molecules per unit volume and θ represents the angle between a perpendicular to the wall and the direction of the molecule. Find the final expression for P.

30. The voltage across a certain capacitor is given by $V = 10^6 \int e^t \, dt/\sqrt{3 + e^t}$. Determine the function which gives the voltage in terms of time.

$y = \dfrac{\ln(4x + 1)}{4x + 1}$

$x_1 = 2$

Figure 8–2

8–2 The Basic Logarithmic Form

The general power formula for integration, Eq. (8–1), is valid for all values of n except $n = -1$. If n were set equal to -1, this would cause the result to be undefined. When we obtained the derivative of the logarithmic function, we found that

$$\frac{d(\ln u)}{dx} = \frac{1}{u} \frac{du}{dx}$$

which means the differential of the logarithmic form is $d(\ln u) = du/u$. Reversing the process, we then determine that $\int du/u = \ln u + C$. In other words, when the exponent of the expression being integrated is -1, it is a logarithmic form.

Logarithms are defined only for positive numbers. Thus, $\int du/u = \ln u + C$ is valid if $u > 0$. If $u < 0$, then $-u > 0$. In this case $d(-u) = -du$, or $\int (-du)/(-u) = \ln(-u) + C$. However, $\int du/u = \int (-du)/(-u)$. These results can be combined into the single form

$$\int \frac{du}{u} = \ln |u| + C \qquad\qquad (8\text{--}2)$$

Example A

Integrate: $\int \dfrac{dx}{x+1}$.

Since $d(x + 1) = dx$, this integral fits the form of Eq. (8–2) with $u = x + 1$ and $du = dx$. Therefore, we have

$$\int \frac{dx}{x+1} = \ln|x + 1| + C$$

Example B

Integrate: $\int \dfrac{\cos x}{\sin x}\, dx$.

We note that $d(\sin x) = \cos x\, dx$. This means that this integral fits the form of Eq. (8–2) with $u = \sin x$ and $du = \cos x\, dx$. Thus,

$$\int \frac{\cos x}{\sin x}\, dx = \int \frac{\cos x\, dx}{\sin x} \longleftarrow du$$

$$= \ln|\sin x| + C$$

Example C

Integrate: $\int \dfrac{x\, dx}{4 - x^2}$.

This integral fits the form of Eq. (8–2) with $u = 4 - x^2$ and $du = -2x\, dx$. This means that we must introduce a factor of -2 into the numerator and a factor of $-\frac{1}{2}$ before the integral. Therefore,

$$\int \frac{x\, dx}{4 - x^2} = -\frac{1}{2} \int \frac{-2x\, dx}{4 - x^2}$$

$$= -\frac{1}{2} \ln|4 - x^2| + C$$

▶ We should note that if the quantity $4 - x^2$ were raised to any power other than that in the example, we would have to employ the general power formula for integration. For example,

$$\int \frac{x\, dx}{(4 - x^2)^2} = -\frac{1}{2} \int \frac{-2x\, dx}{(4 - x^2)^2} = -\frac{1}{2}\frac{(4 - x^2)^{-1}}{-1} + C$$

$$= \frac{1}{2(4 - x^2)} + C$$

Example D

Integrate: $\int \dfrac{e^{4x}\,dx}{1 + 3e^{4x}}$; $u = 1 + 3e^{4x}$, $du = 12e^{4x}\,dx$.

We write

$$\int \dfrac{e^{4x}\,dx}{1 + 3e^{4x}} = \dfrac{1}{12}\int \dfrac{12e^{4x}\,dx}{1 + 3e^{4x}} = \dfrac{1}{12}\ln|1 + 3e^{4x}| + C$$

$$= \dfrac{1}{12}\ln(1 + 3e^{4x}) + C \qquad \text{(since } 1 + 3e^{4x} > 0)$$

Example E

Evaluate: $\displaystyle\int_0^{\pi/8} \dfrac{\sec^2 2x}{1 + \tan 2x}\,dx$; $u = 1 + \tan 2x$, $du = 2\sec^2 2x\,dx$.

We write

$$\int_0^{\pi/8} \dfrac{\sec^2 2x}{1 + \tan 2x}\,dx = \dfrac{1}{2}\int_0^{\pi/8} \dfrac{2\sec^2 2x\,dx}{1 + \tan 2x}$$

$$= \dfrac{1}{2}\ln|1 + \tan 2x|\Big|_0^{\pi/8} = \dfrac{1}{2}\ln|2| - \dfrac{1}{2}\ln|1|$$

$$= \dfrac{1}{2}\ln 2 - \dfrac{1}{2}\ln 1 = \dfrac{1}{2}\ln 2 - 0 = \dfrac{1}{2}\ln 2$$

Example F

Figure 8–3

Find the volume generated by rotating the area bounded by $y = \dfrac{3}{\sqrt{4x + 3}}$, $x = 2.5$, and the axes about the x-axis.

Figure 8–3 shows the volume to be found. As shown, the element of volume is a disk. The volume is found as follows.

$$V = \pi \int_0^{2.5} y^2\,dx = \pi \int_0^{2.5} \left(\dfrac{3}{\sqrt{4x + 3}}\right)^2 dx = \pi \int_0^{2.5} \dfrac{9\,dx}{4x + 3}$$

$$= \dfrac{9\pi}{4}\int_0^{2.5} \dfrac{4\,dx}{4x + 3} = \dfrac{9\pi}{4}\ln(4x + 3)\Big|_0^{2.5}$$

$$= \dfrac{9\pi}{4}(\ln 13 - \ln 3) = \dfrac{9\pi}{4}\ln\dfrac{13}{3}$$

$$= 10.36$$

Exercises 8–2

In Exercises 1 through 24, integrate each of the given functions.

1. $\displaystyle\int \dfrac{dx}{1 + 4x}$

2. $\displaystyle\int \dfrac{dx}{1 - 4x}$

3. $\displaystyle\int \dfrac{x\,dx}{4 - 3x^2}$

4. $\displaystyle\int \dfrac{x^2\,dx}{1 - x^3}$

5. $\displaystyle\int \dfrac{\csc^2 2x}{\cot 2x}\,dx$

6. $\displaystyle\int \dfrac{\sin x}{\cos x}\,dx$

7. $\displaystyle\int_0^{\pi/2} \frac{\cos x\, dx}{1 + \sin x}$

8. $\displaystyle\int_0^{\pi/4} \frac{\sec^2 x\, dx}{4 + \tan x}$

9. $\displaystyle\int \frac{e^{-x}}{1 - e^{-x}}\, dx$

10. $\displaystyle\int \frac{e^{3x}}{1 - e^{3x}}\, dx$

11. $\displaystyle\int \frac{1 + e^x}{x + e^x}\, dx$

12. $\displaystyle\int \frac{e^x - e^{-x}}{e^x + e^{-x}}\, dx$

13. $\displaystyle\int \frac{\sec x \tan x\, dx}{1 + 4 \sec x}$

14. $\displaystyle\int \frac{\sin 2x}{1 - \cos^2 x}\, dx$

15. $\displaystyle\int_1^3 \frac{1 + x}{4x + 2x^2}\, dx$

16. $\displaystyle\int_1^2 \frac{4x + 6x^2}{x^2 + x^3}\, dx$

17. $\displaystyle\int \frac{dx}{x \ln x}$

18. $\displaystyle\int \frac{dx}{x(1 + 2 \ln x)}$

19. $\displaystyle\int \frac{2 + \sec^2 x}{2x + \tan x}\, dx$

20. $\displaystyle\int \frac{x + \cos 2x}{x^2 + \sin 2x}\, dx$

21. $\displaystyle\int \frac{dx}{\sqrt{1 - 2x}}$

22. $\displaystyle\int \frac{x\, dx}{(1 + x^2)^2}$

23. $\displaystyle\int_0^{\pi/12} \frac{\sec^2 3x}{4 + \tan 3x}\, dx$

24. $\displaystyle\int_1^2 \frac{x^2 + 1}{x^3 + 3x}\, dx$

In Exercises 25 through 34, solve the given problems by integration.

25. Find the area bounded by $y(x + 1) = 1$, $x = 0$, $y = 0$, and $x = 2$. See Fig. 8–4.

26. Find the area bounded by $xy = 1$, $x = 1$, $x = 2$, and $y = 0$.

27. Find the volume generated by rotating the area bounded by $y = 1/(x^2 + 1)$, $x = 0$, $x = 1$, $y = 0$ about the y-axis. (Use shells.)

28. Find the volume of the solid generated by rotating the area bounded by $y = 2/\sqrt{3x + 1}$, $x = 0$, $x = 3.5$, and $y = 0$ about the x-axis.

29. The general expression for the slope of a curve is $\sin x/(3 + \cos x)$. If the curve passes through $(\pi/3, 2)$, find its equation.

30. Find the average value of the function $xy = 4$ from $x = 1$ to $x = 2$.

31. If the current (in amperes), as a function of time, in a certain electric circuit is given by $i = t/(1 + 2t^2)$, find the total charge to pass a given point in the circuit in the first 2 s.

32. The electric current in a certain inductor is given by

$$i = 5 \int \frac{e^t \cos e^t\, dt}{\sin e^t}$$

Find the function which expresses the current in terms of the time.

33. Under certain conditions an object is subject to a force (in pounds) given by $F = 1/(1 + 3x)$. Find the work done on the object in moving it 5 ft from the origin.

34. Conditions are often such that a force proportional to the velocity tends to retard the motion of an object. Under such conditions, the acceleration of a certain object moving down an inclined plane is given by $20 - v$. This leads to the equation $dv/(20 - v) = dt$. If the object starts from rest, find the expression for the velocity as a function of time.

$y = \dfrac{1}{x + 1}$

$x = 2$

Figure 8–4

8–3 The Exponential Form

In deriving the derivative for the exponential function we obtained the result $de^u/dx = e^u(du/dx)$. This means that the differential of the exponential form is $d(e^u) = e^u \, du$. Reversing this form to find the proper form of the integral for the exponential function, we have

$$\int e^u \, du = e^u + C$$

$(8–3)$

Example A

Integrate: $\int e^{5x} \, dx$.

Since $d(5x) = 5 \, dx$, this integral can be put in the form of Eq. (8–3) with $u = 5x$ and $du = 5 \, dx$. This means that we place a 5 with the dx and a $\frac{1}{5}$ before the integral. Therefore,

$$\int e^{5x} \, dx = \frac{1}{5} \int e^{5x} \overset{du}{\overbrace{(5 \, dx)}}$$

$$= \frac{1}{5} e^{5x} + C$$

Example B

Integrate: $\int xe^{x^2} \, dx$.

Since $d(x^2) = 2x \, dx$, we can write this integral in the form of Eq. (8–3) with $u = x^2$ and $du = 2x \, dx$. Thus,

$$\int xe^{x^2} \, dx = \frac{1}{2} \int e^{x^2} (2x \, dx)$$

$$= \frac{1}{2} e^{x^2} + C$$

Example C

Integrate: $\int \dfrac{dx}{e^{3x}}$.

This integral can be put in proper form by writing it as $\int e^{-3x} \, dx$. In this form $u = -3x$, $du = -3 \, dx$. Thus,

$$\int \frac{dx}{e^{3x}} = \int e^{-3x} \, dx = -\frac{1}{3} \int e^{-3x}(-3 \, dx)$$

$$= -\frac{1}{3} e^{-3x} + C$$

Example D

Integrate: $\displaystyle\int \frac{4e^{3x} - 3e^{x}}{e^{x+1}} \, dx$.

This can be put in the proper form for integration, and then integrated, as follows.

$$\int \frac{4e^{3x} - 3e^{x}}{e^{x+1}} \, dx = \int \frac{4e^{3x}}{e^{x+1}} \, dx - \int \frac{3e^{x}}{e^{x+1}} \, dx$$

$$= 4\int e^{3x-(x+1)} \, dx - 3\int e^{x-(x+1)} \, dx$$

$$= 4\int e^{2x-1} \, dx - 3\int e^{-1} \, dx$$

$$= \frac{4}{2}\int e^{2x-1}(2 \, dx) - \frac{3}{e}\int dx$$

$$= 2e^{2x-1} - \frac{3}{e}x + C$$

Example E

Evaluate: $\int_{0}^{\pi/2} (\sin 2x)(e^{\cos 2x}) \, dx$; $u = \cos 2x$, $du = -2 \sin 2x \, dx$.
We write

$$\int_{0}^{\pi/2} (\sin 2x)(e^{\cos 2x}) \, dx = -\frac{1}{2} \int_{0}^{\pi/2} (e^{\cos 2x})(-2 \sin 2x \, dx)$$

$$= -\frac{1}{2} e^{\cos 2x}\Big|_{0}^{\pi/2} = -\frac{1}{2}\left(\frac{1}{e} - e\right) = 1.175$$

Example F

Find the equation of the curve for which $\dfrac{dy}{dx} = \dfrac{e^{\sqrt{x+1}}}{\sqrt{x+1}}$ if the curve passes through $(0, 1)$.

The solution of this problem requires that we integrate the given function and then evaluate the constant of integration. Hence,

$$dy = \frac{e^{\sqrt{x+1}}}{\sqrt{x+1}} \, dx, \qquad \int dy = \int \frac{e^{\sqrt{x+1}}}{\sqrt{x+1}} \, dx$$

For purposes of integrating the right-hand side,

$$u = \sqrt{x+1} \quad \text{and} \quad du = \frac{1}{2\sqrt{x+1}} \, dx$$

$$y = 2\int e^{\sqrt{x+1}}\left(\frac{1}{2\sqrt{x+1}} \, dx\right) = 2e^{\sqrt{x+1}} + C$$

Letting $x = 0$ and $y = 1$, we have $1 = 2e + C$, or $C = 1 - 2e$. This means that the equation is

$$y = 2e^{\sqrt{x+1}} + 1 - 2e$$

Exercises 8–3

In Exercises 1 through 24, integrate each of the given functions.

1. $\int e^{7x}(7\ dx)$

2. $\int e^{x^4}(4x^3\ dx)$

3. $\int e^{2x+5}\ dx$

4. $\int e^{-4x}\ dx$

5. $\int_0^2 e^{x/2}\ dx$

6. $\int_1^2 3e^{4x}\ dx$

7. $\int x^2 e^{x^3}\ dx$

6. $\int xe^{-x^2}\ dx$

9. $\int \frac{e^{\sqrt{x}}}{\sqrt{x}}\ dx$

10. $\int 4x^3 e^{2x^4}\ dx$

11. $\int (\sec x \tan x)\ e^{2\sec x}\ dx$

12. $\int (\sec^2 x)\ e^{\tan x}\ dx$

13. $\int \frac{(3 - e^x)\ dx}{e^{2x}}$

14. $\int (e^x - e^{-x})^2\ dx$

15. $\int_1^3 3e^{2x}(e^{-2x} - 1)\ dx$

16. $\int_0^{0.5} \frac{3e^{3x+1}}{e^x}\ dx$

17. $\int \frac{2\ dx}{\sqrt{x}\ e^{\sqrt{x}}}$

18. $\int \frac{4\ dx}{\sec x\ e^{\sin x}}$

19. $\int \frac{e^{\text{Arctan } x}}{x^2 + 1}\ dx$

20. $\int \frac{e^{\text{Arcsin } 2x}\ dx}{\sqrt{1 - 4x^2}}$

21. $\int \frac{e^{\cos 3x}\ dx}{\csc 3x}$

22. $\int \frac{e^{2/x}\ dx}{x^2}$

23. $\int_0^{\pi} (\sin 2x)\ e^{\cos^2 x}\ dx$

24. $\int_{-1}^{1} \frac{dx}{e^{2-3x}}$

In Exercises 25 through 32, solve the given problems by integration.

25. Find the area bounded by $y = e^x$, $x = 0$, $y = 0$, and $x = 2$.

26. Prove that the area bounded by $x = a$, $x = b$, $y = 0$, and $y = e^x$ equals the difference of the ordinates at $x = b$ and $x = a$.

27. Find the volume generated by rotating the area bounded by $y = e^{x^2}$, $x = 1$, $y = 0$, and $x = 2$ about the y-axis. See Fig. 8–5.

28. Find the equation of the curve for which $dy/dx = \sqrt{e^{x+3}}$ if the curve passes through $(1, 0)$.

29. Find the average value of the function $y = e^{2x}$ from $x = 0$ to $x = 4$.

30. Find the moment of inertia with respect to the y-axis of the first-quadrant area bounded by $y = e^{x^3}$, $x = 1$, and the axes.

31. If the current in a certain electric circuit is given by $i = 2e^{-30t}$, find the expression for the voltage across a 250 μF capacitor as a function of time. The initial voltage across the capacitor is 60 V.

32. A particle moves such that its velocity is given by $v = e^{\sin 2t} \cos 2t\ dt$. Find the equation for the displacement as a function of the time if $s = 2$ when $t = 0$.

Figure 8–5

$y = e^{x^2}$

8–4 Basic Trigonometric Forms

In this section we shall discuss the integrals of the six trigonometric functions and the trigonometric integrals which arise directly from reversing the formulas for differentiation. Other trigonometric forms will be discussed later.

By directly reversing the differentiation formulas, the following integral formulas are obtained:

$$\int \sin u \ du = -\cos u + C \tag{8-4}$$

$$\int \cos u \ du = \sin u + C \tag{8-5}$$

$$\int \sec^2 u \ du = \tan u + C \tag{8-6}$$

$$\int \csc^2 u \ du = -\cot u + C \tag{8-7}$$

$$\int \sec u \tan u \ du = \sec u + C \tag{8-8}$$

$$\int \csc u \cot u \ du = -\csc u + C \tag{8-9}$$

Example A

Integrate: $\int \sin 3x \ dx$; $u = 3x$, $du = 3 \ dx$.
We have

$$\int \sin 3x \ dx = \frac{1}{3} \int \sin 3x \overset{du}{(3 \ dx)}$$

$$= -\frac{1}{3} \cos 3x + C$$

Example B

Integrate: $\int x \sec^2(x^2) \ dx$; $u = x^2$, $du = 2x \ dx$.
We write

$$\int x \sec^2(x^2) \ dx = \frac{1}{2} \int \sec^2(x^2) \cdot \overset{du}{(2x \ dx)}$$

$$= \frac{1}{2} \tan(x^2) + C$$

Example C

Integrate: $\int \dfrac{\tan 2x}{\cos 2x} \ dx$.

By using the basic identity $\sec \theta = 1/\cos \theta$, we can transform this integral to the form $\int \sec 2x \tan 2x \ dx$. In this form $u = 2x$, $du = 2 \ dx$. Therefore,

$$\int \frac{\tan 2x}{\cos 2x} \ dx = \int \sec 2x \tan 2x \ dx = \frac{1}{2} \int \sec 2x \tan 2x (2 \ dx)$$

$$= \frac{1}{2} \sec 2x + C$$

To find the integrals for the other trigonometric functions, we must change them to a form for which the integral can be determined by methods previously discussed. We can accomplish this by use of the basic trigonometric relations.

The formula for $\int \tan u \, du$ is found by expressing the integral in the form $\int [(\sin u)/(\cos u)] \, du$. We recognize this as being a logarithmic form, where the u of the logarithmic form is $\cos u$ in this integral. The differential of $\cos u$ is $-\sin u \, du$. Therefore, we have

$$\int \tan u \, du = \int \frac{\sin u}{\cos u} \, du = -\int \frac{-\sin u \, du}{\cos u} = -\ln|\cos u| + C$$

The formula for $\int \cot u \, du$ is found by writing it in the form

$$\int \frac{\cos u}{\sin u} \, du$$

In this manner we obtain the following result:

$$\int \cot u \, du = \int \frac{\cos u}{\sin u} \, du = \int \frac{\cos u \, du}{\sin u} = \ln|\sin u| + C$$

The formula for $\int \sec u \, du$ is found by writing it in the form

$$\int \frac{\sec u(\sec u + \tan u)}{\sec u + \tan u} \, du$$

We see that this form is also a logarithmic form, since

$$d(\sec u + \tan u) = (\sec u \tan u + \sec^2 u) \, du$$

The right side of this equation is the expression appearing in the numerator of the integral. Thus,

$$\int \sec u \, du = \int \frac{\sec u(\sec u + \tan u) \, du}{\sec u + \tan u} = \int \frac{\sec u \tan u + \sec^2 u}{\sec u + \tan u} \, du$$
$$= \ln|\sec u + \tan u| + C$$

To obtain the formula for $\int \csc u \, du$, we write it in the form

$$\int \frac{\csc u(\csc u - \cot u) \, du}{\csc u - \cot u}$$

Thus, we have

$$\int \csc u \, du = \int \frac{\csc u(\csc u - \cot u)}{\csc u - \cot u} \, du$$
$$= \int \frac{(-\csc u \cot u + \csc^2 u) \, du}{\csc u - \cot u}$$
$$= \ln|\csc u - \cot u| + C$$

Summarizing these results, we have the following integrals:

$$\int \tan u \, du = -\ln|\cos u| + C \tag{8-10}$$

$$\int \cot u \, du = \ln|\sin u| + C \tag{8-11}$$

$$\int \sec u \, du = \ln|\sec u + \tan u| + C \tag{8-12}$$

$$\int \csc u \, du = \ln|\csc u - \cot u| + C \tag{8-13}$$

Example D

Integrate: $\int \tan 4x \, dx$; $u = 4x$, $du = 4 \, dx$.
We write

$$\int \tan 4x \, dx = \frac{1}{4} \int \tan 4x (4 \, dx) = -\frac{1}{4} \ln|\cos 4x| + C$$

Example E

Integrate: $\int \dfrac{\sec e^{-x} \, dx}{e^x}$; $u = e^{-x}$, $du = -e^{-x} \, dx$.

We have

$$\int \frac{\sec e^{-x} \, dx}{e^x} = -\int (\sec e^{-x})(-e^{-x} \, dx)$$

$$= -\ln|\sec e^{-x} + \tan e^{-x}| + C$$

Example F

Evaluate: $\displaystyle\int_{\pi/6}^{\pi/4} \frac{1 + \cos x}{\sin x} \, dx$.

We write

$$\int_{\pi/6}^{\pi/4} \frac{1 + \cos x}{\sin x} \, dx = \int_{\pi/6}^{\pi/4} \csc x \, dx + \int_{\pi/6}^{\pi/4} \cot x \, dx$$

$$= \ln|\csc x - \cot x|\Big|_{\pi/6}^{\pi/4} + \ln|\sin x|\Big|_{\pi/6}^{\pi/4}$$

$$= \ln|\sqrt{2} - 1| - \ln|2 - \sqrt{3}| + \ln\left|\frac{1}{2}\sqrt{2}\right| - \ln\left|\frac{1}{2}\right|$$

$$= \ln \frac{\left(\frac{1}{2}\sqrt{2}\right)(\sqrt{2} - 1)}{\left(\frac{1}{2}\right)(2 - \sqrt{3})} = \ln \frac{2 - \sqrt{2}}{2 - \sqrt{3}}$$

$$= 0.782$$

Exercises 8-4

In Exercises 1 through 24, integrate each of the given functions.

1. $\int \cos 2x \, dx$

2. $\int \sin(2 - x) \, dx$

3. $\int \sec^2 3x \, dx$

4. $\int \csc 2x \cot 2x \, dx$

5. $\int \sec \frac{1}{2}x \tan \frac{1}{2}x \, dx$

6. $\int e^x \csc^2(e^x) \, dx$

7. $\int x^2 \cot x^3 \, dx$

8. $\int \tan \frac{1}{2}x \, dx$

9. $\int x \sec x^2 \, dx$

10. $\int 2 \csc 3x \, dx$

11. $\int \frac{\sin(1/x)}{x^2} \, dx$

12. $\int \frac{dx}{\sin 4x}$

13. $\int_0^{\pi/6} \frac{dx}{\cos^2 2x}$

14. $\int_0^1 e^x \cos e^x \, dx$

15. $\int \frac{\sec 5x}{\cot 5x} \, dx$

16. $\int \frac{\sin 2x}{\cos^2 x} \, dx$

17. $\int \sqrt{\tan^2 2x + 1} \, dx$

18. $\int (1 + \cot x)^2 \, dx$

19. $\int_0^{\pi/9} \sin 3x(\csc 3x + \sec 3x) \, dx$

20. $\int_{\pi/4}^{\pi/3} (1 + \sec x)^2 \, dx$

21. $\int \frac{1 - \sin x}{1 + \cos x} \, dx$

22. $\int \frac{1 + \sec^2 x}{x + \tan x} \, dx$

23. $\int \frac{1 + \sin 2x}{\tan 2x} \, dx$

24. $\int \frac{1 - \cot^2 x}{\cos^2 x} \, dx$

In Exercises 25 through 32, solve the given problems by integration.

25. Find the area bounded by $y = \tan x$, $x = \pi/4$, and $y = 0$.

26. Find the area under the curve $y = \sin x$ from $x = 0$ to $x = \pi$.

27. Find the volume generated by rotating the area bounded by $y = \sec x$, $x = 0$, $x = \pi/3$, and $y = 0$ about the x-axis.

28. Find the volume generated by rotating the area bounded by $y = \cos x^2$, $x = 0$, $y = 0$, and $x = 1$ about the y-axis.

29. Under certain conditions the expression for the velocity as a function of time for a particle is $v = 4 \cos 3t$. Determine the distance as a function of the time, if $s = 0$ when $t = 0$.

30. If the current in a certain electric circuit is $i = 110 \cos 377t$, find the expression for the voltage across a 500 μF capacitor as a function of time. The initial voltage is zero. Show that the voltage across the capacitor is 90° out of phase with the current.

31. If the area bounded by $y = \sin(x^2)$, $y = 0$, and $x = 1$ is 0.3103, find the x-coordinate of the centroid of this area. See Fig. 8-6.

32. A force is given as a function of the distance from the origin by

$$F = \frac{\sin x}{2 + \cos x}$$

Express the work done by this force as a function of x, if $W = 0$ when $x = 0$.

Figure 8-6

8–5 Other Trigonometric Forms

The basic trigonometric relations developed in Chapter 6 provide the means by which many other integrals involving trigonometric functions may be integrated. By use of the square relations, Eqs. (6–6), (6–7), and (6–8), and the equations for the cosine of the double angle, Eq. (6–20), it is possible to transform integrals involving powers of the trigonometric functions into integrable form. We repeat these equations here for reference:

$$\cos^2 x + \sin^2 x = 1 \tag{8–14}$$
$$1 + \tan^2 x = \sec^2 x \tag{8–15}$$
$$1 + \cot^2 x = \csc^2 x \tag{8–16}$$
$$2\cos^2 x = 1 + \cos 2x \tag{8–17}$$
$$2\sin^2 x = 1 - \cos 2x \tag{8–18}$$

 To integrate a product of powers of the sine and cosine, we use Eq. (8–14) if at least one of the powers is odd. The method is based on transforming the integral so that it is made up of powers of either the sine or cosine and the first power of the other. In this way this first power becomes a factor of *du*.

Example A

Integrate: $\int \sin^3 x \cos^2 x \, dx$.

Since $\sin^3 x = \sin^2 x \sin x = (1 - \cos^2 x)\sin x$, it is possible to write this integral with powers of cos x along with sin x dx. Thus, sin x dx becomes part of the necessary *du* of the integral. Therefore,

$$\int \sin^3 x \cos^2 x \, dx = \int (1 - \cos^2 x)(\sin x)(\cos^2 x)\, dx$$

$$= \int (\cos^2 x - \cos^4 x)(\sin x \, dx)$$

$$= \int \cos^2 x(\sin x \, dx) - \int \cos^4 x(\sin x \, dx)$$

$$= -\int \cos^2 x(-\sin x \, dx) + \int \cos^4 x(-\sin x \, dx)$$

$$= -\frac{1}{3}\cos^3 x + \frac{1}{5}\cos^5 x + C$$

Example B

Integrate: $\int \cos^5 2x \, dx$.

Since $\cos^5 2x = \cos^4 2x \cos 2x = (1 - \sin^2 2x)^2 \cos 2x$, it is possible to write this integral with powers of sin $2x$ along with cos $2x \, dx$. Thus, with the introduction of a factor of 2, $(\cos 2x)(2 \, dx)$ is the necessary du of the integral. Thus,

$$\int \cos^5 2x \, dx = \int (1 - \sin^2 2x)^2 \cos 2x \, dx$$

$$= \int (1 - 2 \sin^2 2x + \sin^4 2x) \cos 2x \, dx$$

$$= \int \cos 2x \, dx - \int 2 \sin^2 2x \cos 2x \, dx + \int \sin^4 2x \cos 2x \, dx$$

$$= \frac{1}{2} \int \cos 2x(2 \, dx) - \int \sin^2 2x(2 \cos 2x \, dx)$$

$$+ \frac{1}{2} \int \sin^4 2x(2 \cos 2x \, dx)$$

$$= \frac{1}{2} \sin 2x - \frac{1}{3} \sin^3 2x + \frac{1}{10} \sin^5 2x + C$$

▶ *In products of powers of the sine and cosine, if the powers to be integrated are even, we use Eqs. (8-17) and (8-18) to transform the integral.* Those most commonly met are $\int \cos^2 u \, du$ and $\int \sin^2 u \, du$.

Example C

Integrate: $\int \sin^2 2x \, dx$.

▶ Using Eq. (8-18) in the form $\sin^2 2x = \frac{1}{2}(1 - \cos 4x)$, this integral can be transformed into a form which can be integrated. (Here we note the x of Eq. (8-18) is treated as $2x$ for this integral.) Therefore, we write

$$\int \sin^2 2x \, dx = \int \left[\frac{1}{2}(1 - \cos 4x) \right] dx = \frac{1}{2} \int dx - \frac{1}{8} \int \cos 4x(4 \, dx)$$

$$= \frac{x}{2} - \frac{1}{8} \sin 4x + C$$

▶ *To integrate even powers of the secant, powers of the tangent, or products of the secant and tangent, we use Eq. (8-15) to transform the integral.* In transforming, the forms we look for are powers of the tangent with $\sec^2 x$, which becomes part of du, or powers of the secant along with sec x tan x, which becomes part of du in this case. Similar transformations are made when we integrate powers of the cotangent and cosecant, with the use of Eq. (8-16).

Example D

Integrate: $\int \tan^5 x \, dx$.

Since $\tan^5 x = \tan^3 x \tan^2 x = \tan^3 x (\sec^2 x - 1)$, we can write this integral with powers of tan x along with $\sec^2 x \, dx$. Thus, $\sec^2 x \, dx$ becomes the necessary du of the integral. It is necessary to replace $\tan^2 x$ with $\sec^2 x - 1$ twice during the integration. Therefore,

$$\int \tan^5 x \, dx = \int \tan^3 x (\sec^2 x - 1) \, dx$$

$$= \int \tan^3 x (\sec^2 x \, dx) - \int \tan^3 x \, dx$$

$$= \frac{1}{4} \tan^4 x - \int \tan x (\sec^2 x - 1) \, dx$$

$$= \frac{1}{4} \tan^4 x - \int \tan x (\sec^2 x \, dx) + \int \tan x \, dx$$

$$= \frac{1}{4} \tan^4 x - \frac{1}{2} \tan^2 x - \ln|\cos x| + C$$

Example E

Integrate: $\int \sec^3 x \tan x \, dx$.

By writing $\sec^3 x \tan x$ as $\sec^2 x (\sec x \tan x)$, we can use the $\sec x \tan x \, dx$ as the du of the integral. Thus,

$$\int \sec^3 x \tan x \, dx = \int (\sec^2 x)(\overset{\overset{\displaystyle du}{\swarrow}}{\sec x \tan x \, dx})$$

$$= \frac{1}{3} \sec^3 x + C$$

Example F

Integrate: $\int \csc^4 2x \, dx$.

By writing $\csc^4 2x = \csc^2 2x \csc^2 2x = \csc^2 2x (1 + \cot^2 2x)$, we can write this integral with powers of cot $2x$ along with $\csc^2 2x \, dx$, which becomes part of the necessary du of the integral. Thus,

$$\int \csc^4 2x \, dx = \int \csc^2 2x (1 + \cot^2 2x) \, dx$$

$$= \frac{1}{2} \int \csc^2 2x (2 \, dx) - \frac{1}{2} \int \cot^2 2x (-2 \csc^2 2x \, dx)$$

$$= -\frac{1}{2} \cot 2x - \frac{1}{6} \cot^3 2x + C$$

Example G

Integrate: $\displaystyle\int_0^{\pi/4} \frac{\tan^3 x}{\sec^3 x}\, dx.$

This integral requires the use of several trigonometric relationships to obtain integrable forms.

$$\int_0^{\pi/4} \frac{\tan^3 x}{\sec^3 x}\, dx = \int_0^{\pi/4} \frac{(\sec^2 x - 1)\tan x}{\sec^3 x}\, dx$$

$$= \int_0^{\pi/4} \frac{\tan x}{\sec x}\, dx - \int_0^{\pi/4} \frac{\tan x\, dx}{\sec^3 x}$$

$$= \int_0^{\pi/4} \sin x\, dx - \int_0^{\pi/4} \frac{\sec x \tan x\, dx}{\sec^4 x}$$

$$= -\cos x + \frac{1}{3}\cos^3 x \Big|_0^{\pi/4}$$

$$= -\frac{\sqrt{2}}{2} + \frac{1}{3}\left(\frac{\sqrt{2}}{2}\right)^3 - \left(-1 + \frac{1}{3}\right) = \frac{8 - 5\sqrt{2}}{12}$$

Example H

The *root-mean-square value of a function* with respect to x is defined by

$$y_{rms} = \sqrt{\frac{1}{T}\int_0^T y^2\, dx} \qquad\qquad\qquad (8–19)$$

Usually the value of T which is of importance is the period of the function. Find the root-mean-square value of the current in an electric circuit for one period, if $i = 3\cos \pi t$.

The period of the current is $2\pi/\pi = 2$ s. Therefore, we must find the square root of the integral

$$\frac{1}{2}\int_0^2 (3\cos \pi t)^2\, dt = \frac{9}{2}\int_0^2 \cos^2 \pi t\, dt$$

Hence, we have

$$\frac{9}{2}\int_0^2 \cos^2 \pi t\, dt = \frac{9}{4}\int_0^2 (1 + \cos 2\pi t)\, dt$$

after substituting $\frac{1}{2}(1 + \cos 2\pi t)$ for $\cos^2 \pi t$. Evaluating, we have

$$\frac{9}{4}\int_0^2 (1 + \cos 2\pi t)\, dt = \frac{9}{4}t \Big|_0^2 + \frac{9}{8\pi}\int_0^2 \cos 2\pi t(2\pi\, dt)$$

$$= \frac{9}{2} + \frac{9}{8\pi}\sin 2\pi t \Big|_0^2 = \frac{9}{2}$$

Thus, the root-mean-square current is

$$i_{rms} = \sqrt{\frac{9}{2}} = \frac{3\sqrt{2}}{2} = 2.12 \text{ A}$$

(continued on next page)

This value of the current, often referred to as the *effective current,* is the value of the direct current which would develop the same quantity of heat in the same time.

Exercises 8–5

In Exercises 1 through 28, integrate each of the given functions.

1. $\displaystyle\int \sin^2 x \cos x \, dx$ 2. $\displaystyle\int \sin x \cos^5 x \, dx$ 3. $\displaystyle\int \sin^3 2x \, dx$

4. $\displaystyle\int \cos^3 x \, dx$ 5. $\displaystyle\int \sin^2 x \cos^3 x \, dx$ 6. $\displaystyle\int \sin^3 x \cos^6 x \, dx$

7. $\displaystyle\int_0^{\pi/4} \sin^5 x \, dx$ 8. $\displaystyle\int_{\pi/3}^{\pi/2} \sqrt{\cos x} \, \sin^3 x \, dx$ 9. $\displaystyle\int \sin^2 x \, dx$

10. $\displaystyle\int \cos^2 2x \, dx$ 11. $\displaystyle\int \cos^2 3x \, dx$ 12. $\displaystyle\int \sin^2 4x \, dx$

13. $\displaystyle\int \tan^3 x \, dx$ 14. $\displaystyle\int \cot^3 x \, dx$ 15. $\displaystyle\int \tan x \sec^4 x \, dx$

16. $\displaystyle\int \cot 4x \csc^4 4x \, dx$ 17. $\displaystyle\int \tan^4 2x \, dx$ 18. $\displaystyle\int \cot^4 x \, dx$

19. $\displaystyle\int \tan^3 3x \sec^3 3x \, dx$ 20. $\displaystyle\int \sqrt{\tan x} \, \sec^4 x \, dx$

21. $\displaystyle\int (\sin x + \cos x)^2 \, dx$ 22. $\displaystyle\int (\tan 2x + \cot 2x)^2 \, dx$

23. $\displaystyle\int \frac{1 - \cot x}{\sin^4 x} \, dx$ 24. $\displaystyle\int \frac{1 + \sin x}{\cos^4 x} \, dx$ 25. $\displaystyle\int_{\pi/6}^{\pi/4} \cot^5 x \, dx$

26. $\displaystyle\int_{\pi/6}^{\pi/3} \frac{dx}{1 + \sin x}$ 27. $\displaystyle\int \sec^6 x \, dx$ 28. $\displaystyle\int \tan^7 x \, dx$

In Exercises 29 through 36, solve the given problems by integration.

29. Find the volume generated by rotating the area bounded by $y = \sin x$ and $y = 0$, from $x = 0$ to $x = \pi$, about the x-axis.

30. Find the volume generated by rotating the area bounded by $y = \tan^3(x^2)$, $y = 0$, and $x = \pi/4$ about the y-axis.

31. In determining the rate of radiation by an accelerated charge, the following integral must be evaluated: $\int_0^\pi \sin^3 \theta \, d\theta$. Find the value of the integral.

32. Find the length of the curve $y = \ln \cos x$ from $x = 0$ to $x = \pi/3$. See Fig. 8–7.

33. Find the root-mean-square value of the current for one period if $i = 2 \sin t$.

34. Find the root-mean-square value of the voltage for one period if the voltage is given by $V = 100 \sin 120\pi t$.

Figure 8–7

35. For a current $i = i_0 \sin \omega t$, show that the root-mean-square value of the current for one period is $i_0/\sqrt{2}$.

36. Show that $\int \sec^2 x \tan x \, dx$ can be integrated in two ways. Explain the difference in the answers.

8–6 Inverse Trigonometric Forms

For reference, Eq. (6–36) is
$$\frac{d(\text{Arcsin } u)}{dx} = \frac{1}{\sqrt{1 - u^2}}\frac{du}{dx}$$

Referring to Eq. (6–36), we can find the differential of $\text{Arcsin}(u/a)$, where a is a constant:

$$d\left(\text{Arcsin }\frac{u}{a}\right) = \frac{1}{\sqrt{1 - (u/a)^2}}\frac{du}{a} = \frac{a}{\sqrt{a^2 - u^2}}\frac{du}{a} = \frac{du}{\sqrt{a^2 - u^2}}$$

Reversing this differentiation formula, we have the important integration formula

$$\int \frac{du}{\sqrt{a^2 - u^2}} = \text{Arcsin }\frac{u}{a} + C \tag{8–20}$$

By finding the differential of $\text{Arctan}(u/a)$ and then reversing the equation, we derive another important integration formula:

$$d\left(\text{Arctan }\frac{u}{a}\right) = \frac{1}{1 + (u/a)^2}\frac{du}{a} = \frac{a^2}{a^2 + u^2}\frac{du}{a} = \frac{a\,du}{a^2 + u^2}$$

Thus,

$$\int \frac{du}{a^2 + u^2} = \frac{1}{a}\text{Arctan }\frac{u}{a} + C \tag{8–21}$$

This shows one of the principal uses of the inverse trigonometric functions: They provide a solution to the integration of important algebraic functions.

Example A

Integrate: $\int \frac{dx}{\sqrt{9 - x^2}}$.

This integral fits the form of Eq. (8–20) with $u = x$, $du = dx$, and $a = 3$. Thus,

$$\int \frac{dx}{\sqrt{9 - x^2}} = \text{Arcsin }\frac{x}{3} + C$$

Example B

Integrate: $\int \dfrac{dx}{4x^2 + 25}$.

This integral fits the form of Eq. (8–21) with $u = 2x$, $du = 2\,dx$, and $a = 5$. Thus, we must insert a 2 in the numerator and a $\frac{1}{2}$ before the integral. Therefore,

$$\int \frac{dx}{4x^2 + 25} = \frac{1}{2} \int \frac{\overset{du}{\overbrace{2\,dx}}}{(2x)^2 + (5)^2}$$

$$= \frac{1}{2} \cdot \frac{1}{5} \operatorname{Arctan} \frac{2x}{5} + C$$

$$= \frac{1}{10} \operatorname{Arctan} \frac{2x}{5} + C$$

Example C

Integrate: $\int_{-1}^{3} \dfrac{dx}{x^2 + 6x + 13}$.

At first glance it does not appear that this integral fits any of the forms presented up to this point. However, by writing the denominator in the form $(x^2 + 6x + 9) + 4 = (x + 3)^2 + 2^2$, we recognize that $u = x + 3$, $du = dx$, and $a = 2$. Thus,

$$\int_{-1}^{3} \frac{dx}{x^2 + 6x + 13} = \int_{-1}^{3} \frac{dx}{(x + 3)^2 + 2^2} = \frac{1}{2} \operatorname{Arctan} \frac{x + 3}{2} \bigg|_{-1}^{3}$$

$$= \frac{1}{2} (\operatorname{Arctan} 3 - \operatorname{Arctan} 1)$$

$$= 0.2318$$

Now we can see the use of completing the square when we are transforming integrals into proper form. We also see the use of inverse trigonometric functions in evaluating integrals.

Example D

Integrate: $\int \dfrac{2x + 5}{x^2 + 9}\,dx$.

By writing this integral as the sum of two integrals, we may integrate each of these separately:

$$\int \frac{2x + 5}{x^2 + 9}\,dx = \int \frac{2x\,dx}{x^2 + 9} + \int \frac{5\,dx}{x^2 + 9}$$

The first of these integrals is seen to be a logarithmic form, and the second the inverse tangent form. For the first, $u = x^2 + 9$, $du = 2x\,dx$. For the second, $u = x$, $du = dx$, $a = 3$. Thus,

$$\int \frac{2x\,dx}{x^2 + 9} + 5\int \frac{dx}{x^2 + 9} = \ln|x^2 + 9| + \frac{5}{3}\operatorname{Arctan}\frac{x}{3} + C$$

 The inverse trigonometric integral forms of this section show very well the importance of proper recognition of the form of the integral. It is important that these forms not be confused with those of the general power rule or the logarithmic form. Consider the following examples.

Example E

The integral $\int \dfrac{dx}{\sqrt{1 - x^2}}$ is of the inverse sine form, with $u = x$, $du = dx$, and $a = 1$. Thus,

$$\int \frac{dx}{\sqrt{1 - x^2}} = \operatorname{Arcsin} x + C$$

The integral $\int \dfrac{x\,dx}{\sqrt{1 - x^2}}$ is not of the inverse sine form due to the factor of x in the numerator. It is integrated by use of the general power rule, with $u = 1 - x^2$, $du = -2x\,dx$, and $n = -\frac{1}{2}$. Thus,

$$\int \frac{x\,dx}{\sqrt{1 - x^2}} = -\sqrt{1 - x^2} + C$$

The integral $\int \dfrac{x\,dx}{1 - x^2}$ is of logarithmic form with $u = 1 - x^2$ and $du = -2x\,dx$. If $1 - x^2$ were raised to any power other than 1 in the denominator, we would use the general power rule. To be of the inverse sine form, we would have the square root of $1 - x^2$ and no factor of x, as in the first illustration. Thus,

$$\int \frac{x\,dx}{1 - x^2} = -\frac{1}{2}\ln|1 - x^2| + C$$

Example F

See Appendix C for a computer program illustrating two of these forms.

The following integrals are of the form indicated.

$$\int \frac{dx}{1 + x^2} \qquad \text{Inverse tangent form}$$

$$\int \frac{x\,dx}{1 + x^2} \qquad \text{Logarithmic form}$$

$$\int \frac{x\,dx}{\sqrt{1 + x^2}} \qquad \text{General power form}$$

$$\int \frac{dx}{1 + x} \qquad \text{Logarithmic form}$$

There are integrals whose forms appear to be similar to those in Examples E and F, but which do not fit the forms which we have discussed. They include

$$\int \frac{dx}{\sqrt{x^2 - 1}}, \quad \int \frac{dx}{\sqrt{1 + x^2}}, \quad \int \frac{dx}{1 - x^2}, \quad \text{and} \quad \int \frac{dx}{x\sqrt{1 + x^2}}$$

We will develop methods to integrate some of these, and all of them can be integrated by the tables discussed in Section 9–6.

Exercises 8–6

In Exercises 1 through 24, integrate each of the given functions.

1. $\int \frac{dx}{\sqrt{4 - x^2}}$ 	2. $\int \frac{dx}{\sqrt{49 - x^2}}$ 	3. $\int \frac{dx}{64 + x^2}$

4. $\int \frac{dx}{4 + x^2}$ 	5. $\int \frac{dx}{\sqrt{1 - 16x^2}}$ 	6. $\int \frac{dx}{\sqrt{9 - 4x^2}}$

7. $\int_0^2 \frac{dx}{1 + 9x^2}$ 	8. $\int_1^3 \frac{dx}{49 + 4x^2}$ 	9. $\int \frac{2\,dx}{\sqrt{4 - 5x^2}}$

10. $\int \frac{4x\,dx}{\sqrt{3 - 2x^2}}$ 	11. $\int \frac{8x\,dx}{9x^2 + 16}$ 	12. $\int \frac{3\,dx}{25 + 16x^2}$

13. $\int_1^2 \frac{dx}{5x^2 + 7}$ 	14. $\int_0^1 \frac{x\,dx}{1 + x^4}$ 	15. $\int \frac{e^x\,dx}{\sqrt{1 - e^{2x}}}$

16. $\int \frac{\sec^2 x\,dx}{\sqrt{1 - \tan^2 x}}$ 	17. $\int \frac{dx}{x^2 + 2x + 2}$ 	18. $\int \frac{2\,dx}{x^2 + 8x + 17}$

19. $\int \frac{4\,dx}{\sqrt{-4x - x^2}}$ 	20. $\int \frac{dx}{\sqrt{2x - x^2}}$ 	21. $\int_{\pi/6}^{\pi/2} \frac{\cos 2x}{1 + \sin^2 2x}\,dx$

22. $\int_{-4}^0 \frac{dx}{x^2 + 4x + 5}$ 	23. $\int \frac{2 - x}{\sqrt{4 - x^2}}\,dx$ 	24. $\int \frac{3 - 2x}{1 + 4x^2}\,dx$

In Exercises 25 through 28, identify the form of each integral as being inverse sine, inverse tangent, logarithmic or general power, as in Examples E and F. Do not integrate.

25. (a) $\int \frac{2\,dx}{4 + 9x^2}$ 	(b) $\int \frac{2\,dx}{4 + 9x}$ 	(c) $\int \frac{2x\,dx}{\sqrt{4 + 9x^2}}$

26. (a) $\int \frac{2x\,dx}{4 - 9x^2}$ 	(b) $\int \frac{2\,dx}{\sqrt{4 - 9x}}$ 	(c) $\int \frac{2x\,dx}{4 + 9x^2}$

27. (a) $\int \frac{2x\,dx}{\sqrt{4 - 9x^2}}$ 	(b) $\int \frac{2\,dx}{\sqrt{4 - 9x^2}}$ 	(c) $\int \frac{2\,dx}{4 - 9x}$

28. (a) $\int \frac{2\,dx}{9x^2 + 4}$ 	(b) $\int \frac{2x\,dx}{\sqrt{9x^2 - 4}}$ 	(c) $\int \frac{2x\,dx}{9x^2 + 4}$

In Exercises 29 through 36, solve the given problems by integration.

$y = \dfrac{1}{\sqrt{4 - x^2}}$

Figure 8–8

29. Find the area bounded by the curve $y(1 + x^2) = 1$, $x = 0$, $y = 0$, and $x = 2$.
30. Find the area bounded by $y\sqrt{4 - x^2} = 1$, $x = 0$, $y = 0$, and $x = 1$. See Fig. 8–8.
31. Find the volume generated by rotating the area bounded by $y = 1/\sqrt{16 + x^2}$, $x = 0$, $y = 0$, and $x = 3$ about the x-axis.

32. The electric current in a certain inductor is given by $i = 8 \int \dfrac{dt}{100 + t^2}$. Determine the function for i in terms of t.

33. In dealing with the theory for simple harmonic motion, it is necessary to solve the following equation:

$$\frac{dx}{\sqrt{A^2 - x^2}} = \sqrt{\frac{k}{m}}\, dt \qquad \text{(with } k, m, \text{ and } A \text{ as constants)}$$

Determine the solution to this equation if $x = x_0$ when $t = 0$.

34. Determine the function of distance in terms of time if the velocity of an object is given by $v = 1/(4 + t^2)$ and $s = 0$ when $t = 0$.

35. Find the moment of inertia with respect to the y-axis for the area bounded by $y = 1/(1 + x^6)$, the x-axis, $x = 1$, and $x = 2$.

36. Find the length of arc along the curve $y = \sqrt{1 - x^2}$ between $x = 0$ and $x = 1$.

8–7 Review Exercises for Chapter 8

In Exercises 1 through 36, integrate each of the given functions.

1. $\displaystyle\int e^{-2x}\, dx$

2. $\displaystyle\int e^{\cos 2x} \sin x \cos x\, dx$

3. $\displaystyle\int \frac{dx}{x(\ln 2x)^2}$

4. $\displaystyle\int x^{1/3}\sqrt{x^{4/3} + 1}\, dx$

5. $\displaystyle\int \frac{\cos x\, dx}{1 + \sin x}$

6. $\displaystyle\int \frac{\sec^2 x\, dx}{2 + \tan x}$

7. $\displaystyle\int \frac{2\, dx}{25 + 49x^2}$

8. $\displaystyle\int \frac{dx}{\sqrt{1 - 4x^2}}$

9. $\displaystyle\int_0^{\pi/2} \cos^3 2x\, dx$

10. $\displaystyle\int_0^{\pi/8} \sec^3 2x \tan 2x\, dx$

11. $\displaystyle\int_0^2 \frac{x\, dx}{4 + x^2}$

12. $\displaystyle\int_1^e \frac{\ln x^2\, dx}{x}$

13. $\displaystyle\int \sec^4 3x \tan 3x\, dx$

14. $\displaystyle\int \frac{\sin^3 x\, dx}{\sqrt{\cos x}}$

15. $\displaystyle\int \tan 3x\, dx$

16. $\displaystyle\int \sec 4x\, dx$

17. $\displaystyle\int \sec^4 3x\, dx$

18. $\displaystyle\int \frac{\sin^2 2x\, dx}{1 + \cos 2x}$

19. $\displaystyle\int_0^{0.5} \frac{2e^{2x} - 3e^x}{e^{2x}}\, dx$

20. $\displaystyle\int \frac{4 - e^{\sqrt{x}}}{\sqrt{x}\, e^{\sqrt{x}}}\, dx$

21. $\displaystyle\int \frac{3x\, dx}{4 + x^4}$

22. $\displaystyle\int_1^3 \frac{x\, dx}{2(3 + x^2)}$

23. $\displaystyle\int \frac{e^x\, dx}{\sqrt{9 - e^{2x}}}$

24. $\displaystyle\int \frac{2 + 4e^{2x}}{x + e^{2x}}\, dx$

25. $\displaystyle\int \frac{e^{2x}\, dx}{\sqrt{e^{2x} + 1}}$

26. $\displaystyle\int \frac{(4 + \ln 2x)^3\, dx}{x}$

27. $\displaystyle\int \sin^2 3x\, dx$

28. $\displaystyle\int \sin^4 x\, dx$

29. $\displaystyle\int \frac{\tan^2 x\, dx}{\tan x - x}$

30. $\displaystyle\int \frac{dx}{\sqrt{5 - (4x + x^2)}}$

31. $\displaystyle\int e^{2x}\cos e^{2x}\, dx$

32. $\displaystyle\int \frac{dx}{x^2 + 6x + 10}$

33. $\displaystyle\int_1^e \frac{(\ln x)^2\, dx}{x}$

34. $\displaystyle\int_1^3 \frac{dx}{x^2 - 2x + 5}$

35. $\displaystyle\int \frac{x^2 - 1}{x + 2}\, dx$

36. $\displaystyle\int [\cos(\ln x)]\frac{dx}{x}$

In Exercises 37 through 50, solve the given problems by integration.

37. Show that $\int e^x(e^x + 1)^2\, dx$ can be integrated in two ways. Explain the difference in the answers.

38. Show that $\int \dfrac{1}{x}(1 + \ln x)\, dx$ can be integrated in two ways. Explain the difference in the answers.

39. The change in the thermodynamic entity of entropy ΔS may be expressed as $\Delta S = \int (c_v/T)\, dT$, where c_v is the heat capacity at constant volume and T is the temperature. For increased accuracy, c_v is often given by the equation $c_v = a + bT + cT^2$, where a, b, and c are constants. Express ΔS as a function of temperature.

40. A second-order chemical reaction leads to the equation

$$dt = \frac{k_1\, dx}{a - x} + \frac{k_2\, dx}{b - x}$$

where k_1 and k_2 are constants, a and b are initial concentrations, t is the time, and x is the decrease in concentration. Solve for t as a function of x.

41. Assuming a resisting force, the velocity (in feet per second) of a falling object in terms of time (in seconds) is given by

$$\frac{dv}{32 - 0.1v} = dt$$

If $v = 0$ when $t = 0$, find v as a function of t.

42. Find the area bounded by $y(1 + x^4) = 2x$, $y = 0$, and $x = 1$. See Fig. 8–9.

43. Find the area bounded by $y = 4e^{2x}$, $x = 1.5$, and the axes.

44. Find the equation of the curve for which $dy/dx = e^x(2 - e^x)^2$, if the curve passes through $(0, 4)$.

45. Find the equation of the curve for which $dy/dx = \sec^4 x$, if the curve passes through the origin.

46. Find the distance from the origin after $\frac{1}{2}$ s of the particle for which the velocity as a function of time is given by $v = 4 \cos 3t$. (That is, integrate the function with limits of 0 and $\frac{1}{2}$, or substitute into the function derived in Exercise 29 of Section 8–4. Compare the results of these two methods.)

47. Find the volume of the piece of tubing in an oil distribution line shown in Fig. 8–10. All cross sections are circular.

48. A metal plate has the shape shown in Fig. 8–11. Find the x-coordinate of the centroid of the plate.

49. Find the length of arc along the curve of $y = \ln \sin x$ from $x = \frac{1}{3}\pi$ to $x = \frac{2}{3}\pi$.

50. The power delivered to an electric circuit is given by $P = ei$, where e and i are the instantaneous voltage and the instantaneous current in the circuit, respectively. The mean power, averaged over a period $2\pi/\omega$, is given by

$$P_{\text{av}} = \frac{\omega}{2\pi}\int_0^{2\pi/\omega} ei\, dt$$

If $e = 20 \cos 2t$ and $i = 3 \sin 2t$, find the average power over a period of $\pi/4$.

Figure 8–9

Figure 8–10

Figure 8–11

CHAPTER 9

Methods of Integration

There are many methods of transforming integrals into forms which can be integrated by use of one of the basic formulas. In Chapter 8 we saw that completing the square and the trigonometric identities can be used for this purpose. In this chapter we shall develop several general methods of integration. At the end of the chapter we will demonstrate the use of a table of integrals.

9–1 Integration by Parts

Since the derivative of a product of functions is found by use of the formula

$$\frac{d(uv)}{dx} = u\frac{dv}{dx} + v\frac{du}{dx}$$

the differential of a product of functions is given by $d(uv) = u\,dv + v\,du$. Integrating both sides of this equation we have $uv = \int u\,dv + \int v\,du$. Solving for $\int u\,dv$, we obtain

$$\int u\,dv = uv - \int v\,du \tag{9–1}$$

Integration by use of Eq. (9–1) is called **integration by parts.**

Example A

Integrate: $\int x \sin x \, dx$.

This integral does not fit any of the previous forms which have been discussed, since neither x nor $\sin x$ can be made a factor of a proper du. However, by choosing $u = x$, and $dv = \sin x \, dx$, integration by parts may be used. Thus,

$$u = x, \qquad dv = \sin x \, dx$$

By finding the differential of u and integrating dv, we find du and v. This gives us

$$du = dx, \qquad v = -\cos x + C_1$$

Now substituting in Eq. (9–1), we have

$$
\begin{array}{ccccccc}
\int & u & dv & = & u & v & - \int & v & du \\
 & \downarrow & \downarrow & & \downarrow & \downarrow & & \downarrow & \downarrow
\end{array}
$$

$$\int (x)(\sin x \, dx) = (x)(-\cos x + C_1) - \int (-\cos x + C_1)(dx)$$

$$= -x \cos x + C_1 x + \int \cos x \, dx - \int C_1 \, dx$$

$$= -x \cos x + C_1 x + \sin x - C_1 x + C$$

$$= -x \cos x + \sin x + C$$

Other choices of u and dv may be made, but they do not prove to be useful. For example, if we choose $u = \sin x$ and $dv = x \, dx$, then $du = \cos x \, dx$ and $v = \frac{1}{2}x^2 + C_2$. This makes $\int v \, du = \int (\frac{1}{2}x^2 + C_2)(\cos x \, dx)$, which is more complex than the integrand of the original problem.

We also note that the constant C_1 which was introduced when we integrated dv does not appear in the final result. This constant will always cancel out and therefore *we will not show any constant of integration when finding v*.

As in Example A, there is often more than one choice as to the part of the integrand which is selected to be u and the part which is selected to be dv. There are no set rules which may be stated for the best choice of u and dv, but there are two guidelines which may be stated: (1) *The quantity u is normally chosen such that du/dx is of simpler form than u.* (2) *The differential dv is normally chosen such that $\int$ dv is easily obtained.* Working examples, and thereby gaining experience in methods of integrating, is the best way to determine when this method should be used and how to employ it.

Example B

Integrate: $\int x \sqrt{1 - x} \, dx$.

We see that this does not fit the general power rule, for $x \, dx$ is not a factor of the differential of $1 - x$. By choosing $u = x$ and $dv = \sqrt{1 - x} \, dx$ we have $du/dx = 1$, and dv can readily be determined. Thus,

$$u = x, \qquad dv = \sqrt{1 - x} \, dx = (1 - x)^{1/2} \, dx$$

$$du = dx, \qquad v = -\frac{2}{3}(1 - x)^{3/2}$$

Substituting in Eq. (9–1), we have

$$\underset{\underset{u}{\downarrow}}{\int} \underset{\underset{dv}{\downarrow}}{} = \underset{\underset{u}{\downarrow}}{u} \quad \underset{\underset{v}{\downarrow}}{} - \int \underset{\underset{v}{\downarrow}}{} \quad \underset{\underset{du}{\downarrow}}{}$$

$$\int x\left[(1 - x)^{1/2} \, dx\right] = x\left[-\frac{2}{3}(1 - x)^{3/2}\right] - \int\left[-\frac{2}{3}(1 - x)^{3/2}\right] dx$$

At this point we see that we can complete the integration. Thus,

$$\int x(1 - x)^{1/2} \, dx = -\frac{2x}{3}(1 - x)^{3/2} + \frac{2}{3}\int (1 - x)^{3/2} \, dx$$

$$= -\frac{2x}{3}(1 - x)^{3/2} + \frac{2}{3}\left(-\frac{2}{5}\right)(1 - x)^{5/2} + C$$

$$= -\frac{2}{3}(1 - x)^{3/2}\left[x + \frac{2}{5}(1 - x)\right] + C$$

$$= -\frac{2}{15}(1 - x)^{3/2}(2 + 3x) + C$$

Example C

Integrate: $\int \sqrt{x} \ln x \, dx$.
  For this integral we have

$$u = \ln x, \qquad dv = x^{1/2} \, dx$$

$$du = \frac{1}{x} \, dx, \qquad v = \frac{2}{3}x^{3/2}$$

This leads to

$$\int \sqrt{x} \ln x \, dx = \frac{2}{3}x^{3/2} \ln x - \frac{2}{3}\int x^{1/2} \, dx$$

$$= \frac{2}{3}x^{3/2} \ln x - \frac{4}{9}x^{3/2} + C$$

Example D

Integrate: $\int \text{Arcsin } x \, dx$.
  We write

$$u = \text{Arcsin } x, \qquad dv = dx$$

$$du = \frac{dx}{\sqrt{1 - x^2}}, \qquad v = x$$

$$\int \text{Arcsin } x \, dx = x \text{ Arcsin } x - \int \frac{x \, dx}{\sqrt{1 - x^2}}$$

$$= x \text{ Arcsin } x + \frac{1}{2}\int \frac{-2x \, dx}{\sqrt{1 - x^2}}$$

$$= x \text{ Arcsin } x + \sqrt{1 - x^2} + C$$

Example E

Integrate: $\int_0^1 xe^{-x}\, dx$.
We write

$$u = x, \qquad dv = e^{-x}\, dx$$
$$du = dx, \qquad v = -e^{-x}$$

$$\int_0^1 xe^{-x}\, dx = -xe^{-x}\Big|_0^1 + \int_0^1 e^{-x}\, dx = -xe^{-x} - e^{-x}\Big|_0^1$$

$$= -e^{-1} - e^{-1} + 1$$

$$= 1 - \frac{2}{e}$$

Example F

Integrate: $\int e^x \sin x\, dx$. Let $u = \sin x$, $dv = e^x\, dx$, $du = \cos x\, dx$, $v = e^x$.
We write

$$\int e^x \sin x\, dx = e^x \sin x - \int e^x \cos x\, dx$$

At first glance it appears that we have made no progress in applying the method of integration by parts. We note, however, that when we integrated $\int e^x \sin x\, dx$, part of the result was a term of $\int e^x \cos x\, dx$. This implies that if $\int e^x \cos x\, dx$ were integrated, a term of $\int e^x \sin x\, dx$ might result. Thus, the method of integration by parts is now applied to the integral $\int e^x \cos x\, dx$:

$$u = \cos x, \qquad dv = e^x\, dx, \qquad du = -\sin x\, dx, \qquad v = e^x$$

And so $\int e^x \cos x\, dx = e^x \cos x + \int e^x \sin x\, dx$. Substituting this expression into the expression for $\int e^x \sin x\, dx$, we obtain

$$\int e^x \sin x\, dx = e^x \sin x - \left(e^x \cos x + \int e^x \sin x\, dx \right)$$

$$= e^x \sin x - e^x \cos x - \int e^x \sin x\, dx$$

$$2\int e^x \sin x\, dx = e^x(\sin x - \cos x) + 2C$$

$$\int e^x \sin x\, dx = \frac{e^x}{2}(\sin x - \cos x) + C$$

Thus, by combining integrals of like form, we obtain the desired result.

Exercises 9–1

In Exercises 1 through 16, integrate each of the given functions.

1. $\int x \cos x\, dx$
2. $\int x \sin 2x\, dx$
3. $\int xe^{2x}\, dx$

4. $\int xe^x\,dx$

5. $\int x\sec^2 x\,dx$

6. $\int_0^{\pi/4} x\sec x\tan x\,dx$

7. $\int \operatorname{Arctan} x\,dx$

8. $\int \ln x\,dx$

9. $\int \dfrac{x\,dx}{\sqrt{1-x}}$

10. $\int x\sqrt{x+1}\,dx$

11. $\int x\ln x\,dx$

12. $\int x^2\ln 4x\,dx$

13. $\int x^2\sin 2x\,dx$

14. $\int x^2 e^{2x}\,dx$

15. $\int_0^{\pi/2} e^x\cos x\,dx$

16. $\int e^{-x}\sin 2x\,dx$

In Exercises 17 through 24, solve the given problems by integration.

17. Find the area bounded by $y = xe^{-x}$, $y = 0$, and $x = 2$. See Fig. 9–1.

18. Find the volume generated by rotating the area bounded by $y = \sin x$ and $y = 0$ (from $x = 0$ to $x = \pi$) about the y-axis.

19. Find the x-coordinate of the centroid of the area bounded by $y = \cos x$ and $y = 0$ for $0 \le x \le \pi/2$.

20. Find the moment of inertia with respect to its axis of the volume generated by rotating the area bounded by $y = e^x$, $x = 1$, and the coordinate axes about the y-axis.

21. Find the root-mean-square value of the function $y = \sqrt{\operatorname{Arcsin} x}$ between $x = 0$ and $x = 1$. (See Example H of Section 8–5.)

22. The general expression for the slope of a curve is $dy/dx = x^3\sqrt{1+x^2}$. Find the equation of the curve if it passes through the origin.

23. The current in a given circuit is given by $i = e^{-2t}\cos t$. Find an expression for the amount of charge which passes a given point in the circuit as a function of the time, if $q_0 = 0$.

24. A particle moves such that its velocity v is given by $v = t\sqrt{t+1}$. Find the expression for the displacement s as a function of time if $s = 0$ when $t = 0$.

y

$y = xe^{-x}$

O

x

$x = 2$

Figure 9–1

9–2 Integration by Substitution

Another method for integrating some expressions is to put them into a standard form by changing the variable of integration by means of an appropriate substitution. There are many different substitutions which can be made, depending on the expression to be integrated. In this section one type of substitution is considered, and in the following section we will develop another.

Certain integrals which contain a root of a linear expression may be integrated by substituting a new variable for the radical. Thus, if an expression of the form $(ax + b)^{p/q}$ appears in an integral which does not fit a standard form, a substitution which may be tried to put it in a standard form is $u = (ax + b)^{1/q}$. Consider the following examples.

Example A

Integrate: $\int x\sqrt{1-x}\,dx$.

Since the differential of $1-x$ is $-dx$ and does not contain a factor of x, we cannot use the power rule. Therefore, let

$$u = \sqrt{1-x}$$

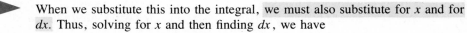

When we substitute this into the integral, we must also substitute for x and for dx. Thus, solving for x and then finding dx, we have

$$u^2 = 1 - x$$
$$x = 1 - u^2$$
$$dx = -2u\,du$$

Now substituting for x, $\sqrt{1-x}$, and dx in the integral, we have

$$\int x\sqrt{1-x}\,dx = \int (1-u^2)(u)(-2u\,du) = -2\int (u^2 - u^4)\,du$$

$$= -\frac{2}{3}u^3 + \frac{2}{5}u^5 + C$$

We have now integrated the expression, but it is in terms of a variable different from that in the original integral. Therefore, we must put it in terms of x.

$$-\frac{2}{3}u^3 + \frac{2}{5}u^5 + C = -\frac{2}{3}(1-x)^{3/2} + \frac{2}{5}(1-x)^{5/2} + C$$

This result may be written in another form which is sometimes more convenient.

$$-\frac{2}{3}(1-x)^{3/2} + \frac{2}{5}(1-x)^{5/2} + C = (1-x)^{3/2}\left[-\frac{2}{3} + \frac{2}{5}(1-x)\right] + C$$

$$= -\frac{2}{15}(1-x)^{3/2}(2 + 3x) + C$$

Thus,

$$\int x\sqrt{1-x}\,dx = -\frac{2}{3}(1-x)^{3/2} + \frac{2}{5}(1-x)^{5/2} + C$$

$$= -2(1-x)^{3/2}\left[\frac{1}{3} - \frac{1}{5}(1-x)\right] + C$$

$$= -\frac{2}{15}(1-x)^{3/2}(2 + 3x) + C$$

See Example B of Section 9–1. We see that some integrals can be evaluated by more than one method.

Example B

Integrate: $\displaystyle\int \frac{2x\,dx}{(x+3)^{2/3}}$

Here we let

$$u = (x+3)^{1/3}$$

Now solving for x and then finding dx, we have

$$u^3 = x + 3$$
$$x = u^3 - 3$$
$$dx = 3u^2\,du$$

$$\int \frac{2x\,dx}{(x+3)^{2/3}} = \int \frac{2(u^3 - 3)(3u^2\,du)}{u^2}$$

$$= 6\int (u^3 - 3)\,du$$

$$= 6\left(\frac{1}{4}u^4 - 3u\right) + C$$

Now substituting back in terms of x, we have

$$6\left(\frac{1}{4}u^4 - 3u\right) + C = 6\left[\frac{1}{4}(x+3)^{4/3} - 3(x+3)^{1/3}\right] + C$$

$$= \frac{6}{4}(x+3)^{1/3}(x + 3 - 12) + C$$

$$= \frac{3}{2}(x+3)^{1/3}(x - 9) + C$$

or

$$\int \frac{2x\,dx}{(x+3)^{2/3}} = \frac{3}{2}(x+3)^{1/3}(x - 9) + C$$

When we use the method of substitution with a definite integral, there are two ways of handling the limits of integration. We can substitute back and use the given limits, or we can change the limits to the corresponding values of the new variable. We will use the latter method, for it is generally easier and eliminates the need to substitute back.

Example C

Evaluate: $\displaystyle\int_0^4 \frac{4x\,dx}{\sqrt{2x+1}}$

We let

$$u = \sqrt{2x+1}$$

which means that

$$u^2 = 2x+1$$

$$x = \frac{1}{2}(u^2-1)$$

$$dx = \frac{1}{2}(2u\,du) = u\,du$$

Also, when $x = 0$ (the lower limit), $u = \sqrt{2(0)+1} = 1$, and when $x = 4$ (the upper limit), $u = \sqrt{2(4)+1} = 3$. Thus,

$$\int_0^4 \frac{4x\,dx}{\sqrt{2x+1}} = \int_1^3 \frac{4(\frac{1}{2})(u^2-1)(u\,du)}{u}$$

$$= 2\int_1^3 (u^2-1)du = \frac{2}{3}u^3 - 2u\Big|_1^3 = \frac{2}{3}(27-1) - 2(3-1)$$

$$= \frac{52}{3} - 4 = \frac{40}{3}$$

Example D

Find the volume generated by rotating the first quadrant area bounded by $y = \sqrt{x} + 4$ and $x = 5$ about the y-axis. See Fig. 9–2.

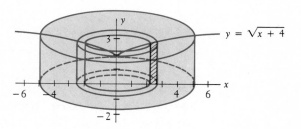

Figure 9–2

From the figure we see that by using shells, the volume is given by

$$V = 2\pi \int_0^5 xy\,dx = 2\pi \int_0^5 x\sqrt{x+4}\,dx$$

Here we let

$$u = \sqrt{x + 4}, \quad \text{or} \quad u^2 = x + 4$$
$$x = u^2 - 4$$
$$dx = 2u\, du$$

For $x = 0$, $u = \sqrt{0 + 4} = 2$, and for $x = 5$, $u = \sqrt{5 + 4} = 3$. Therefore,

$$V = 2\pi \int_2^3 (u^2 - 4)(u)(2u\, du) = 4\pi \int_2^3 (u^4 - 4u^2)\, du$$

$$= 4\pi \left(\frac{1}{5}u^5 - \frac{4}{3}u^3 \right) \Big|_2^3 = 4\pi \left[\frac{1}{5}(243 - 32) - \frac{4}{3}(27 - 8) \right] = 4\pi \left(\frac{211}{5} - \frac{76}{3} \right)$$

$$= \frac{1012\pi}{15} = 212$$

Exercises 9–2

In Exercises 1 through 16, integrate each of the given functions.

1. $\displaystyle\int x\sqrt{x + 1}\, dx$

2. $\displaystyle\int x\sqrt{x + 3}\, dx$

3. $\displaystyle\int x\sqrt{2x + 1}\, dx$

4. $\displaystyle\int x\sqrt{3 - x}\, dx$

5. $\displaystyle\int \frac{x\, dx}{\sqrt{x + 3}}$

6. $\displaystyle\int \frac{x\, dx}{\sqrt{2x + 7}}$

7. $\displaystyle\int \frac{x^2\, dx}{\sqrt{x - 2}}$

8. $\displaystyle\int_3^{11} \frac{3x\, dx}{\sqrt{2x + 3}}$

9. $\displaystyle\int_0^3 x^2\sqrt{1 + x}\, dx$

10. $\displaystyle\int \frac{3x^2\, dx}{\sqrt{1 - x}}$

11. $\displaystyle\int x\sqrt[3]{x - 1}\, dx$

12. $\displaystyle\int x\sqrt[3]{8 - x}\, dx$

13. $\displaystyle\int \frac{x\, dx}{\sqrt[4]{2x + 3}}$

14. $\displaystyle\int x(x - 4)^{2/3}\, dx$

15. $\displaystyle\int_0^7 x(x + 1)^{2/3}\, dx$

16. $\displaystyle\int_0^2 \frac{x^2\, dx}{(4x + 1)^{5/2}}$

In Exercises 17 through 24, solve the given problems by integration.

17. Find the area bounded by $y = x^3\sqrt{x - 1}$, $x = 2$, and $y = 0$. See Fig. 9–3.

18. Find the first-quadrant area bounded by $y = x\sqrt[3]{x + 8}$ and $x = 8$.

Figure 9–3

19. Find the volume generated by rotating the first-quadrant area bounded by $y = x\sqrt[3]{x - 1}$ and $x = 9$ about the y-axis.

20. Find the volume generated by rotating the area of Exercise 19 about the x-axis.

21. The force F, in pounds, required to move a particular lever mechanism is a function of the displacement s, in feet, according to $F = 4s\sqrt{4s + 3}$. Find the work done if the displacement changes from zero to 2.50 ft.

22. The electric current i, in microamperes, in a particular microprocessor circuit varies with the time t, in nanoseconds, according to $i = t^3\sqrt{6t + 1}$ for $0 \le t < 3$ ns. Find the charge which passes through the circuit during the first 2.50 ns.

23. Find the integral in Example A by using the substitution $v = 1 - x$.

24. Find the integral in Example B by using the substitution $v = x + 3$.

9–3 Integration by Trigonometric Substitution

Trigonometric relations not only provide a means of transforming trigonometric integrals, but they can also be used to transform algebraic integrals. Substitutions based on Eqs. (8–14), (8–15), and (8–16) prove to be particularly useful for integrals involving radicals. The method is illustrated in the following examples.

Example A

Integrate: $\displaystyle\int \frac{dx}{x^2\sqrt{1 - x^2}}$.

If we let $x = \sin\theta$, the radical becomes $\sqrt{1 - \sin^2\theta} = \cos\theta$. Therefore, by making this substitution, the integral can be transformed into a trigonometric integral. We must be careful to replace all factors of the integral by proper expressions in terms of θ; $x = \sin\theta$, $dx = \cos\theta\, d\theta$.

And so we obtain

$$\int \frac{dx}{x^2\sqrt{1 - x^2}} = \int \frac{\cos\theta\, d\theta}{\sin^2\theta\sqrt{1 - \sin^2\theta}} = \int \frac{\cos\theta\, d\theta}{\sin^2\theta\,\cos\theta} = \int \csc^2\theta\, d\theta$$

$$= -\cot\theta + C$$

We have now performed the integration, but the answer is expressed in terms of a variable different from the original integral. We must express the result in terms of x. Making a right triangle with an angle θ such that $\sin\theta = x/1$ (see Fig. 9–4), we may express any of the trigonometric functions in terms of x. (This is the method used with inverse trigonometric functions.) Thus,

Figure 9–4

$$\cot\theta = \frac{\sqrt{1 - x^2}}{x}$$

Therefore, the result of the integration becomes

$$\int \frac{dx}{x^2\sqrt{1 - x^2}} = -\cot\theta + C = -\frac{\sqrt{1 - x^2}}{x} + C$$

Example B

Integrate: $\int \dfrac{dx}{\sqrt{x^2 + 4}}$. Let $x = 2 \tan \theta$, $dx = 2 \sec^2 \theta \, d\theta$.

We write

$$\int \frac{dx}{\sqrt{x^2 + 4}} = \int \frac{2 \sec^2 \theta \, d\theta}{\sqrt{4 \tan^2 \theta + 4}}$$

$$= \int \frac{2 \sec^2 \theta \, d\theta}{2\sqrt{\tan^2 \theta + 1}} = \int \frac{2 \sec^2 \theta \, d\theta}{2 \sec \theta}$$

$$= \int \sec \theta \, d\theta = \ln|\sec \theta + \tan \theta| + C$$

$$= \ln \left| \frac{\sqrt{x^2 + 4}}{2} + \frac{x}{2} \right| + C$$

$$= \ln \left| \frac{\sqrt{x^2 + 4} + x}{2} \right| + C$$

Figure 9–5

See Fig. 9–5. This answer is acceptable, but there is another form. By using the properties of logarithms, we have

$$\ln \left| \frac{\sqrt{x^2 + 4} + x}{2} \right| + C = \ln \left| \sqrt{x^2 + 4} + x \right| + (C - \ln 2)$$

$$= \ln \left| \sqrt{x^2 + 4} + x \right| + C'$$

▶ Here C' is another arbitrary constant, which combines all constants of the previous expression. Combining constants in this manner is a common practice in integration problems.

Example C

Integrate: $\int \dfrac{dx}{x \sqrt{x^2 - 9}}$. Let $x = 3 \sec \theta$, $dx = 3 \sec \theta \tan \theta \, d\theta$.

$$\int \frac{dx}{x \sqrt{x^2 - 9}} = \int \frac{3 \sec \theta \tan \theta \, d\theta}{3 \sec \theta \sqrt{9 \sec^2 \theta - 9}}$$

$$= \int \frac{\tan \theta \, d\theta}{3 \sqrt{\sec^2 \theta - 1}} = \int \frac{\tan \theta \, d\theta}{3 \tan \theta}$$

$$= \frac{1}{3} \int d\theta = \frac{1}{3} \theta + C = \frac{1}{3} \text{Arcsec} \frac{x}{3} + C$$

In this solution it is not necessary to refer to a triangle to express the integral in terms of x. This solution is found by solving $x = 3 \sec \theta$ for θ, as indicated.

From the preceding examples we can see that by making the proper substitution in terms of a trigonometric function, we can solve integrals of algebraic functions by means of simpler equivalent trigonometric forms, due to the properties of the square relations of the trigonometric functions. In summary, for the indicated radical form the following trigonometric substitutions are used.

For	$\sqrt{a^2 - x^2}$	use	$x = a \sin \theta$
For	$\sqrt{a^2 + x^2}$	use	$x = a \tan \theta$
For	$\sqrt{x^2 - a^2}$	use	$x = a \sec \theta$

Example D

Evaluate: $\displaystyle\int_1^4 \frac{\sqrt{9 + x^2}}{x}\, dx.$

Since we have the form $\sqrt{a^2 + x^2}$, where $a = 3$, we make the substitution $x = 3 \tan \theta$. This means that $dx = 3 \sec^2 \theta\, d\theta$. Thus,

$$\int \frac{\sqrt{9 + x^2}}{x}\, dx = \int \frac{\sqrt{9 + 9 \tan^2 \theta}}{3 \tan \theta}\, (3 \sec^2 \theta\, d\theta)$$

$$= 3 \int \frac{\sqrt{1 + \tan^2 \theta}}{\tan \theta}\, \sec^2 \theta\, d\theta = 3 \int \frac{\sec^3 \theta}{\tan \theta}\, d\theta$$

$$= 3 \int \frac{1 + \tan^2 \theta}{\tan \theta}\, \sec \theta\, d\theta$$

$$= 3 \left(\int \frac{\sec \theta}{\tan \theta}\, d\theta + \int \tan \theta \sec \theta\, d\theta \right)$$

$$= 3 \left(\int \csc \theta\, d\theta + \int \tan \theta \sec \theta\, d\theta \right)$$

$$= 3[\ln |\csc \theta - \cot \theta| + \sec \theta]$$

$$= 3 \left[\ln \left| \frac{\sqrt{x^2 + 9}}{x} - \frac{3}{x} \right| + \frac{\sqrt{x^2 + 9}}{3} \right] \quad \text{(see Fig. 9–6)}$$

$$= 3 \left[\ln \left| \frac{\sqrt{x^2 + 9} - 3}{x} \right| + \frac{\sqrt{x^2 + 9}}{3} \right]$$

Figure 9–6

Limits have not been included, due to the changes in variables. The actual evaluation may now be completed:

$$\int_1^4 \frac{\sqrt{9 + x^2}}{x}\, dx = 3 \left[\ln \left| \frac{\sqrt{x^2 + 9} - 3}{x} \right| + \frac{\sqrt{x^2 + 9}}{3} \right]_1^4$$

$$= 3 \left[\left(\ln \frac{1}{2} - \ln \frac{\sqrt{10} - 3}{1} \right) + \left(\frac{5}{3} - \frac{\sqrt{10}}{3} \right) \right]$$

$$= 3 \left[\ln \frac{1}{2(\sqrt{10} - 3)} + \frac{5 - \sqrt{10}}{3} \right]$$

$$= 3 \ln \frac{\sqrt{10} + 3}{2} + \frac{5 - \sqrt{10}}{1} = 5.21$$

Exercises 9–3

In Exercises 1 through 16, integrate each of the given functions.

1. $\displaystyle\int \frac{\sqrt{1-x^2}}{x^2}\,dx$

2. $\displaystyle\int \frac{dx}{(x^2+9)^{3/2}}$

3. $\displaystyle\int \frac{dx}{\sqrt{x^2-4}}$

4. $\displaystyle\int \frac{\sqrt{x^2-25}}{x}\,dx$

5. $\displaystyle\int \frac{dx}{x^2\sqrt{x^2+9}}$

6. $\displaystyle\int \frac{dx}{x\sqrt{4-x^2}}$

7. $\displaystyle\int \frac{dx}{(4-x^2)^{3/2}}$

8. $\displaystyle\int \frac{x^3\,dx}{\sqrt{9+x^2}}$

9. $\displaystyle\int_0^{0.5} \frac{x^3\,dx}{\sqrt{1-x^2}}$

10. $\displaystyle\int_4^5 \frac{\sqrt{x^2-16}}{x^2}\,dx$

11. $\displaystyle\int \frac{dx}{\sqrt{x^2+2x+2}}$

12. $\displaystyle\int \frac{dx}{\sqrt{x^2+2x}}$

13. $\displaystyle\int \frac{dx}{x\sqrt{4x^2-9}}$

14. $\displaystyle\int \sqrt{16-x^2}\,dx$

15. $\displaystyle\int \frac{dx}{\sqrt{e^{2x}-1}}$

16. $\displaystyle\int \frac{\sec^2x\,dx}{(4-\tan^2x)^{3/2}}$

In Exercises 17 through 24, solve the given problems by integration.

17. Find the area of a circle of radius 1 by integration.

18. Find the area bounded by $y = \dfrac{1}{x^2\sqrt{x^2-1}}$, $x = \sqrt{2}$, $x = \sqrt{5}$, and $y = 0$. See Fig. 9–7.

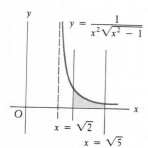

$y = \dfrac{1}{x^2\sqrt{x^2-1}}$

$x = \sqrt{2}$

$x = \sqrt{5}$

Figure 9–7

19. Find the moment of inertia with respect to the y-axis of the first-quadrant area under the circle $x^2 + y^2 = a^2$ in terms of its mass.

20. Find the moment of inertia of a sphere of radius a with respect to its axis in terms of its mass.

21. Find the volume generated by rotating the area bounded by $y = \dfrac{\sqrt{x^2-16}}{x^2}$, $y = 0$, and $x = 5$ about the y-axis.

22. Find the volume generated by rotating the area bounded by $y = x^2\sqrt{x^2-4}$, $y = 0$, and $x = 2.5$ about the y-axis.

23. Find the equation of the curve for which $\dfrac{dy}{dx} = \dfrac{\sqrt{4+x^2}}{x^4}$, if the curve passes through $(2, 1)$

24. Find the length of arc along the curve of $y = \ln x$ from $x = 1$ to $x = 3$.

9–4 Integration by Partial Fractions: Nonrepeated Linear Factors

In algebra we combine fractions into a single fraction by means of addition. However, if we wish to integrate an expression which contains a rational fraction (see Eq. (2–2)), in which both the numerator and the denominator are polynomials, it is often advantageous to reverse the operation of addition and express the rational fraction as the sum of simpler fractions.

Example A

In attempting to integrate $\int \dfrac{7 - x}{x^2 + x - 2}\, dx$ we find that it does not fit any of the standard forms of Chapter 8. However, it is possible to show that

$$\frac{7 - x}{x^2 + x - 2} = \frac{2}{x - 1} - \frac{3}{x + 2}$$

This means that

$$\int \frac{7 - x}{x^2 + x - 2}\, dx = \int \frac{2\, dx}{x - 1} - \int \frac{3\, dx}{x + 2}$$

$$= 2\,\ln|x - 1| - 3\,\ln|x + 2| + C$$

In Example A we see that the integral is readily determined once the rational fraction $(7 - x)/(x^2 + x - 2)$ is replaced by the simpler fractions. In this section and the next we will describe how certain rational fractions can be expressed in terms of simpler fractions and thereby be integrated. *This technique is called the* **method of partial fractions.**

In order to express the rational fraction $f(x)/g(x)$ in terms of simpler partial fractions, *the degree of the numerator f(x) must be less than that of the denominator g(x).* If this is not the case, we divide numerator by denominator until the remainder is of the proper form. Then the denominator $g(x)$ is factored into a product of linear and quadratic factors. The method of determining the partial fractions depends on the factors which are obtained. In advanced algebra the form of the partial fractions is shown (but we shall not show the proof here).

There are four cases for the types of factors of the denominator. They are (1) *nonrepeated linear factors,* (2) *repeated linear factors,* (3) *nonrepeated quadratic factors,* and (4) *repeated quadratic factors.* In this section we consider the case of nonrepeated linear factors, and the other cases are discussed in the next section.

For the case of nonrepeated linear factors we use the fact that *corresponding to each linear factor ax + b, occurring once in the denominator, there will be a partial fraction of the form*

$$\frac{A}{ax + b}$$

where A is a constant to be determined. The following examples illustrate the method.

Example B

Integrate: $\int \dfrac{7 - x}{x^2 + x - 2}\, dx$

This is the same integral as in Example A. Here we will see how the partial fractions are found.

First we note that the degree of the numerator is 1 (the highest power is x) and that of the denominator is 2 (the highest power is x^2). Since the degree of the denominator is higher we may proceed to factoring it. Thus,

$$\frac{7 - x}{x^2 + x - 2} = \frac{7 - x}{(x - 1)(x + 2)}$$

There are two linear factors, $(x - 1)$ and $(x + 2)$, in the denominator and they are different. This means that there are two partial fractions. Therefore we write

$$\frac{7 - x}{(x - 1)(x + 2)} = \frac{A}{x - 1} + \frac{B}{x + 2} \tag{1}$$

The constants A and B are to be determined so that the equation is an identity. To assist in finding the values of A and B we clear Eq. (1) of fractions by multiplying both sides by $(x - 1)(x + 2)$. This gives us

$$7 - x = A(x + 2) + B(x - 1) \tag{2}$$

Equation (2) is also an identity, which means that there are two ways of determining the values of A and B.

 Solution by substitution: Since Eq. (2) is an identity, it is true for any value of x. Thus, in turn we pick $x = -2$ and $x = 1$, for each of these values makes a factor on the right equal to zero, and the values of B and A are easily found.
Therefore,

for $x = -2$: $\quad 7 - (-2) = A(-2 + 2) + B(-2 - 1)$

$$9 = -3B \qquad\qquad\qquad B = -3$$

for $x = 1$: $\quad 7 - 1 = A(1 + 2) + B(1 - 1)$

$$6 = 3A \qquad\qquad\qquad A = 2$$

Solution by equating coefficients: Since Eq. (2) is an identity, another way of finding the constants A and B is to equate coefficients of like powers of x from each side. Thus, writing Eq. (2) as

$$7 - x = (2A - B) + (A + B)x$$

we have

$$2A - B = 7 \qquad \text{(equating constants: } x^0 \text{ terms)}$$
$$A + B = -1 \qquad \text{(equating coefficients of } x\text{)}$$

Solving this system, as shown at the left, we have $A = 2$ and $B = -3$.
Now using the values of A and B we have

$$\frac{7 - x}{(x - 1)(x + 2)} = \frac{2}{x - 1} - \frac{3}{x + 2}$$

Therefore, the integral is found as in Example A.

$$\int \frac{7 - x}{x^2 + x - 2}\, dx = \int \frac{7 - x}{(x - 1)(x + 2)}\, dx = \int \frac{2\, dx}{x - 1} - \int \frac{3\, dx}{x + 2}$$

$$= 2 \ln|x - 1| - 3 \ln|x + 2| + C$$

Using the properties of logarithms, we may write this as

$$\int \frac{7 - x}{x^2 + x - 2}\, dx = \ln\left|\frac{(x - 1)^2}{(x + 2)^3}\right| + C$$

(left margin work)

$2A - B = 7$
$A + B = -1$
$\overline{3A = 6}$
$A = 2$
$2(2) - B = 7$
$-B = 3$
$B = -3$

In Example B we showed two ways of finding the values of A and B. The method of substitution is generally easier to use for linear factors. However, as we will see in the next section, the method of equating coefficients can also be very useful.

Example C

Integrate: $\displaystyle\int \frac{6x^2 - 14x - 11}{(x + 1)(x - 2)(2x + 1)}\, dx$

First we note that the denominator is factored and that it is of degree 3 (when the denominator is mutiplied out, the highest power of x is x^3). Thus we have three nonrepeated linear factors. This means that

$$\frac{6x^2 - 14x - 11}{(x + 1)(x - 2)(2x + 1)} = \frac{A}{x + 1} + \frac{B}{x - 2} + \frac{C}{2x + 1}$$

Multiplying through by $(x + 1)(x - 2)(2x + 1)$, we have

$$6x^2 - 14x - 11$$
$$= A(x - 2)(2x + 1) + B(x + 1)(2x + 1) + C(x + 1)(x - 2)$$

We now substitute the values of 2, $-\frac{1}{2}$, and -1 for x. Again, these are chosen because they make factors of the coefficients of A, B, or C equal to zero, although any values may be chosen. Therefore,

for $x = 2$: $6(4) - 14(2) - 11 = A(0)(5) + B(3)(5) + C(3)(0)$

$$-15 = 15B \qquad\qquad B = -1$$

for $x = -\dfrac{1}{2}$: $6\left(\dfrac{1}{4}\right) - 14\left(-\dfrac{1}{2}\right) - 11 =$

$$A\left(-\frac{5}{2}\right)(0) + B\left(\frac{1}{2}\right)(0) + C\left(\frac{1}{2}\right)\left(-\frac{5}{2}\right)$$

$$-\frac{5}{2} = -\frac{5}{4}C \qquad\qquad C = 2$$

for $x = -1$: $6(1) - 14(-1) - 11 =$

$$A(-3)(-1) + B(0)(-1) + C(0)(-3)$$

$$9 = 3A \qquad\qquad A = 3$$

Therefore,

$$\int \frac{6x^2 - 14x - 11}{(x + 1)(x - 2)(2x + 1)}\, dx = \int \frac{3\, dx}{x + 1} - \int \frac{dx}{x - 2} + \int \frac{2\, dx}{2x + 1}$$

$$= 3\ln|x + 1| - \ln|x - 2| + \ln|2x + 1| + C_1$$

$$= \ln\left|\frac{(2x + 1)(x + 1)^3}{x - 2}\right| + C_1$$

Here we have let the constant of integration be C_1 since we used C as the numerator of the third partial fraction.

Example D

Integrate: $\int \dfrac{2x^4 - x^3 - 9x^2 + x - 12}{x^3 - x^2 - 6x}\, dx$

Since the numerator is of a higher degree than the denominator, we must first divide the numerator by the denominator. This gives

$$\frac{2x^4 - x^3 - 9x^2 + x - 12}{x^3 - x^2 - 6x} = 2x + 1 + \frac{4x^2 + 7x - 12}{x^3 - x^2 - 6x}$$

We must now express this rational fraction in terms of its partial fractions.

$$\frac{4x^2 + 7x - 12}{x^3 - x^2 - 6x} = \frac{4x^2 + 7x - 12}{x(x + 2)(x - 3)} = \frac{A}{x} + \frac{B}{x + 2} + \frac{C}{x - 3}$$

Clearing fractions, we have

$$4x^2 + 7x - 12 = A(x + 2)(x - 3) + Bx(x - 3) + Cx(x + 2)$$

Now using values of x of -2, 3, and 0 for substitution we obtain the values of A, B, and C. Therefore,

for $x = -2$: $4(4) + 7(-2) - 12 = B(-2)(-5)$

$-10 = 10B \qquad\qquad B = -1$

for $x = 3$: $4(9) + 7(3) - 12 = C(3)(5)$

$45 = 15C \qquad\qquad C = 3$

for $x = 0$: $4(0) + 7(0) - 12 = A(2)(-3)$

$-12 = -6A \qquad\qquad A = 2$

Therefore,

$$\int \frac{2x^4 - x^3 - 9x^2 + x - 12}{x^3 - x^2 - 6x}\, dx = \int \left(2x + 1 + \frac{2}{x} - \frac{1}{x + 2} + \frac{3}{x - 3}\right) dx$$

$$= x^2 + x + 2\,\ln|x| - \ln|x + 2|$$
$$+ 3\,\ln|x - 3| + C_1$$

$$= x^2 + x + \ln\left|\frac{x^2(x - 3)^3}{x + 2}\right| + C_1$$

Exercises 9–4

In Exercises 1 through 16, integrate the given functions.

1. $\displaystyle\int \frac{x + 3}{(x + 1)(x + 2)}\, dx$

2. $\displaystyle\int \frac{x + 2}{x(x + 1)}\, dx$

3. $\displaystyle\int \frac{dx}{x^2 - 4}$

4. $\displaystyle\int \frac{x - 9}{2x^2 - 3x + 4}\, dx$

5. $\displaystyle\int \frac{x^2 + 3}{x^2 + 3x}\, dx$

6. $\displaystyle\int \frac{x^3}{x^2 + 3x + 2}\, dx$

7. $\displaystyle\int_0^1 \frac{2x + 4}{3x^2 + 5x + 2}\, dx$

8. $\displaystyle\int_1^3 \frac{x - 1}{4x^2 + x}\, dx$

9. $\displaystyle\int \frac{4x^2 - 10}{x(x + 1)(x - 5)}\, dx$

10. $\displaystyle\int \frac{4x^2 + 21x + 6}{(x + 2)(x - 3)(x + 4)}\, dx$

11. $\displaystyle\int \frac{6x^2 - 2x - 1}{4x^3 - x}\, dx$

12. $\displaystyle\int_2^3 \frac{dx}{x^3 - x}$

13. $\displaystyle\int_1^2 \frac{x^3 + 7x^2 + 9x + 2}{x(x^2 + 3x + 2)}\, dx$

14. $\displaystyle\int \frac{2x^3 + x - 1}{x^3 + x^2 - 4x - 4}\, dx$

15. $\displaystyle\int \frac{dx}{(x^2 - 4)(x^2 - 9)}$

16. $\displaystyle\int \frac{5x^3 - 2x^2 - 15x + 24}{x^4 - 2x^3 - 11x^2 + 12x}\, dx$

In Exercises 17 through 22, solve the given problems by integration.

17. The general expression for the slope of a curve is $(3x + 5)/(x^2 + 5x)$. Find the equation of the curve if it passes through $(1, 0)$.

18. The current i, in amperes, as a function of the time, in seconds, in a certain electric circuit is given by $i = (4t + 3)/(2t^2 + 3t + 1)$. Find the total charge that passes a given point in the circuit during the first second.

19. The force F, in newtons, applied by a stamping machine in making a certain computer part is given by $F = 4x/(x^2 + 3x + 2)$, where x is the distance, in centimeters, through which the force acts. Find the work done by the force from $x = 0$ to $x = 0.500$ cm.

20. Under specified conditions the time t, in minutes, required to form x grams of a substance during a chemical reaction is given by $t = \int dx/[(4 - x)(2 - x)]$. Find the equation relating t and x if $x = 0$ g when $t = 0$ min.

21. Find the first quadrant area bounded by $y = 1/(x^3 + 3x^2 + 2x)$, $x = 1$, and $x = 3$. See Fig. 9–8.

22. Find the volume generated if the area of Exercise 21 is rotated about the y-axis.

$y = \dfrac{1}{x^3 + 3x^2 + 2x}$

$x = 1$ $x = 3$

Figure 9–8

9–5 Integration by Partial Fractions: Other Cases

In the previous section we introduced the method of partial fractions and considered the case of nonrepeated linear factors. In this section we develop the use of partial fractions for the cases of repeated linear factors and nonrepeated quadratic factors. We also briefly discuss the case of repeated quadratic factors.

For the case of repeated linear factors we use the fact that *corresponding to each linear factor ax + b that occurs n times in the denominator there will be n partial fractions*

$$\frac{A_1}{ax + b} + \frac{A_2}{(ax + b)^2} + \cdots + \frac{A_n}{(ax + b)^n}$$

where $A_1, A_2, \ldots, A_n$ *are constants to be determined.*

Example A

Integrate: $\displaystyle\int \frac{dx}{x(x + 3)^2}$

Here we see that the denominator has a factor of x and two factors of $x + 3$. For the factor of x we use a partial fraction as in the previous section, for it is a nonrepeated factor. For the factor $x + 3$ we need two partial fractions, one with a denominator of $x + 3$ and the other with a denominator of $(x + 3)^2$. Thus, we write

$$\frac{1}{x(x + 3)^2} = \frac{A}{x} + \frac{B}{x + 3} + \frac{C}{(x + 3)^2}$$

Multiplying each side by $x(x + 3)^2$, we have

$$1 = A(x + 3)^2 + Bx(x + 3) + Cx \qquad\qquad (1)$$

Using the values of x of -3 and 0, we have

for $x = -3$: $\quad 1 = A(0^2) + B(-3)(0) + (-3)C$

$\qquad\qquad\qquad 1 = -3C \qquad\qquad\qquad\qquad C = -\dfrac{1}{3}$

for $x = 0$: $\quad 1 = A(3^2) + B(0)(3) + C(0)$

$\qquad\qquad\qquad 1 = 9A \qquad\qquad\qquad\qquad A = \dfrac{1}{9}$

Since there are no other values which make a factor in Eq. (1) equal to zero, we must either choose some other value of x or equate coefficients of some power of x in Eq. (1). We will illustrate both methods.

Since Eq. (1) is an indentity, we may choose any value of x. Choosing $x = 1$, we have

$$1 = A(4^2) + B(1)(4) + C(1)$$
$$1 = 16A + 4B + C$$

Using the known values of A and C, we have

$$1 = 16\left(\frac{1}{9}\right) + 4B - \frac{1}{3}$$
$$9 = 16 + 36B - 3, \qquad 36B = -4, \qquad B = -\frac{1}{9}$$

To equate powers of x we first write Eq. (1) as

$$1 = A(x^2 + 6x + 9) + B(x^2 + 3x) + Cx$$

or

$$1 = (A + B)x^2 + (6A + 3B + C)x + 9A$$

Using coefficients of x^2, we have

$$0 = A + B, \qquad B = -A, \qquad B = -\frac{1}{9}$$

(*continued on next page*)

This means that

$$\frac{1}{x(x + 3)^2} = \frac{\frac{1}{9}}{x} + \frac{-\frac{1}{9}}{x + 3} + \frac{-\frac{1}{3}}{(x + 3)^2}$$

or

$$\int \frac{dx}{x(x + 3)^2} = \frac{1}{9} \int \frac{dx}{x} - \frac{1}{9} \int \frac{dx}{x + 3} - \frac{1}{3} \int \frac{dx}{(x + 3)^2}$$

$$= \frac{1}{9} \ln|x| - \frac{1}{9} \ln|x + 3| - \frac{1}{3}\left(\frac{1}{-1}\right)(x + 3)^{-1} + C_1$$

$$= \frac{1}{9} \ln\left|\frac{x}{x + 3}\right| + \frac{1}{3(x + 3)} + C_1$$

Example B   Integrate: $\displaystyle\int \frac{3x^3 + 15x^2 + 21x + 15}{(x - 1)(x + 2)^3} \, dx$

First we set up the partial fractions as

$$\frac{3x^3 + 15x^2 + 21x + 15}{(x - 1)(x + 2)^3} = \frac{A}{x - 1} + \frac{B}{x + 2} + \frac{C}{(x + 2)^2} + \frac{D}{(x + 2)^3}$$

Next we clear fractions.

$$3x^3 + 15x^2 + 21x + 15 = A(x + 2)^3 + B(x - 1)(x + 2)^2$$
$$+ C(x - 1)(x + 2) + D(x - 1) \tag{1}$$

For $x = 1$: $3 + 15 + 21 + 15 = 27A$, $54 = 27A$, $A = 2$

for $x = -2$: $3(-8) + 15(4) + 21(-2) + 15 = -3D$,

$9 = -3D$, $D = -3$

To find B and C we equate coefficients of powers of x. Therefore, we write Eq. (1) as

$$3x^3 + 15x^2 + 21x + 15 = (A + B)x^3 + (6A + 3B + C)x^2$$
$$+ (12A + C + D)x + (8A - 4B - 2C - D)$$

Coefficients of x^3: $3 = A + B$, $3 = 2 + B$, $B = 1$

Coefficients of x^2: $15 = 6A + 3B + C$, $15 = 12 + 3 + C$, $C = 0$

Therefore,

$$\frac{3x^3 + 15x^2 + 21x + 15}{(x - 1)(x + 2)^3} = \frac{2}{x - 1} + \frac{1}{x + 2} + \frac{0}{(x + 2)^2} + \frac{-3}{(x + 2)^3}$$

$$\int \frac{3x^3 + 15x^2 + 21x + 15}{(x - 1)(x + 2)^3} \, dx = 2 \int \frac{dx}{x - 1} + \int \frac{dx}{x + 2} - 3 \int \frac{dx}{(x + 2)^3}$$

$$= 2 \ln|x - 1| + \ln|x + 2| - 3\left(\frac{1}{-2}\right)(x + 2)^{-2} + C_1$$

$$= \ln|(x - 1)^2(x + 2)| + \frac{3}{2(x + 2)^2} + C_1$$

If there is one repeated factor in the denominator and it is the only factor present in the denominator, a substitution is easier and more convenient than partial fractions. This is illustrated in the following example.

Example C

Integrate: $\int \dfrac{x \, dx}{(x-2)^3}$

This could be integrated by first setting up the appropriate partial fractions. However, the solution is more easily found by use of the substitution $u = x - 2$. Using this we have

$$u = x - 2, \qquad x = u + 2, \qquad dx = du$$

$$\int \frac{x \, dx}{(x-2)^3} = \int \frac{(u+2)(du)}{u^3}$$

$$= \int \frac{du}{u^2} + 2 \int \frac{du}{u^3}$$

$$= \int u^{-2} \, du + 2 \int u^{-3} \, du$$

$$= \frac{1}{-u} + \frac{2}{-2} u^{-2} + C$$

$$= -\frac{1}{u} - \frac{1}{u^2} + C$$

$$= -\frac{u+1}{u^2} + C$$

$$= -\frac{x-2+1}{(x-2)^2} + C$$

$$= \frac{1-x}{(x-2)^2} + C$$

For the case of nonrepeated quadratic factors we use the fact that *corresponding to each irreducible (cannot be further factored) quadratic factor* $ax^2 + bx + c$ *that occurs once in the denominator there is a partial fraction of the form*

$$\frac{Ax + B}{ax^2 + bx + c}$$

where A and B are constants to be determined.

Example D

Integrate: $\displaystyle\int \frac{4x + 4}{x^3 + 4x}\, dx$

In setting up the partial fractions, we note that the denominator factors as $x^3 + 4x = x(x^2 + 4)$. Here the factor $x^2 + 4$ cannot be further factored. Thus, we have

$$\frac{4x + 4}{x^3 + 4x} = \frac{4x + 4}{x(x^2 + 4)} = \frac{A}{x} + \frac{Bx + C}{x^2 + 4}$$

Clearing fractions, we have

$$\begin{aligned} 4x + 4 &= A(x^2 + 4) + Bx^2 + Cx \\ &= (A + B)x^2 + Cx + 4A \end{aligned}$$

Equating coefficients of powers of x gives us

for x^2: $0 = A + B$

for x: $4 = C$

for constants: $4 = 4A$,   or $A = 1$

Therefore we easily find that $B = -1$ from the first equation. This means that

$$\frac{4x + 4}{x^3 + 4x} = \frac{1}{x} + \frac{-x + 4}{x^2 + 4}$$

and

$$\int \frac{4x + 4}{x^3 + 4x}\, dx = \int \frac{1}{x}\, dx + \int \frac{-x + 4}{x^2 + 4}\, dx = \int \frac{1}{x}\, dx - \int \frac{x\, dx}{x^2 + 4} + \int \frac{4\, dx}{x^2 + 4}$$

$$= \ln|x| - \frac{1}{2}\ln|x^2 + 4| + 2\,\text{Arctan}\,\frac{x}{2} + C_1$$

Example E

Integrate: $\displaystyle\int \frac{x^3 + 3x^2 + 2x + 4}{x^2(x^2 + 2x + 2)}\, dx$

In the denominator we have a repeated linear factor, x^2, and a quadratic factor. Therefore,

$$\frac{x^3 + 3x^2 + 2x + 4}{x^2(x^2 + 2x + 2)} = \frac{A}{x} + \frac{B}{x^2} + \frac{Cx + D}{x^2 + 2x + 2}$$

$$\begin{aligned} x^3 + 3x^2 + 2x + 4 &= Ax(x^2 + 2x + 2) + B(x^2 + 2x + 2) + Cx^3 + Dx^2 \\ &= (A + C)x^3 + (2A + B + D)x^2 + (2A + 2B)x + 2B \end{aligned}$$

Equating coefficients, we have

for constants: $2B = 4$, $B = 2$

for x: $2A + 2B = 2$, $A + B = 1$, $A = -1$

for x^2: $2A + B + D = 3$, $-2 + 2 + D = 3$, $D = 3$

for x^3: $A + C = 1$, $-1 + C = 1$, $C = 2$

Therefore,

$$\frac{x^3 + 3x^2 + 2x + 4}{x^2(x^2 + 2x + 2)} = -\frac{1}{x} + \frac{2}{x^2} + \frac{2x + 3}{x^2 + 2x + 2}$$

$$\int \frac{x^3 + 3x^2 + 2x + 4}{x^2(x^2 + 2x + 2)}\,dx = -\int \frac{dx}{x} + 2\int \frac{dx}{x^2} + \int \frac{2x + 3}{x^2 + 2x + 2}\,dx$$

$$= -\ln|x| - 2\left(\frac{1}{x}\right) + \int \frac{2x + 2 + 1}{x^2 + 2x + 2}\,dx$$

$$= -\ln|x| - \frac{2}{x} + \int \frac{2x + 2}{x^2 + 2x + 2}\,dx$$

$$+ \int \frac{dx}{(x^2 + 2x + 1) + 1}$$

$$= -\ln|x| - \frac{2}{x} + \ln|x^2 + 2x + 2| + \int \frac{dx}{(x + 1)^2 + 1}$$

$$= -\ln|x| - \frac{2}{x} + \ln|x^2 + 2x + 2|$$

$$+ \operatorname{Arctan}(x + 1) + C_1$$

▶ Note the manner in which the integral for the quadratic denominator was handled. First the numerator, $2x + 3$, was changed to the form $(2x + 2) + 1$ so that we could fit the logarithmic form with the $2x + 2$. Then we completed the square in the denominator of the final integral so that it fitted an inverse tangent form.

Finally, for the case of repeated quadratic factors we use the fact that *corresponding to each irreducible quadratic factor $ax^2 + bx + c$ that occurs n times in the denominator there will be n partial fractions*

$$\frac{A_1 x + B_1}{ax^2 + bx + c} + \frac{A_2 x + B_2}{(ax^2 + bx + c)^2} + \cdots + \frac{A_n x + B_n}{(ax^2 + bx + c)^n}$$

where $A_1, A_2, \ldots, A_n, B_1, B_2, \ldots, B_n$ are constants to be determined. The procedures which lead to the solution are the same as those for the other cases. Exercises 15 and 16 in the following set are solved by using the above types of partial fractions.

Exercises 9–5

In Exercises 1 through 16, integrate each of the given functions.

1. $\int \dfrac{x - 8}{x^3 - 4x^2 + 4x}\,dx$

2. $\int \dfrac{dx}{x^3 - x^2}$

3. $\int \dfrac{2\,dx}{x^2(x^2 - 1)}$

4. $\int_1^3 \dfrac{3x^3 + 8x^2 + 10x + 2}{x(x + 1)^3}\,dx$

5. $\int_1^2 \dfrac{2x\,dx}{(x - 3)^3}$

6. $\int \dfrac{x\,dx}{(x + 2)^4}$

7. $\int \dfrac{x^3 - 2x^2 - 7x + 28}{(x + 1)^2(x - 3)^2}\,dx$ 8. $\int \dfrac{4\,dx}{(x + 1)^2(x - 1)^2}$

9. $\int_0^2 \dfrac{x^2 + x + 5}{(x + 1)(x^2 + 4)}\,dx$ 10. $\int \dfrac{x^2 + x - 1}{(x^2 + 1)(x - 2)}\,dx$

11. $\int \dfrac{5x^2 + 8x + 16}{x^2(x^2 + 4x + 8)}\,dx$ 12. $\int \dfrac{2x^2 + x + 3}{(x^2 + 2)(x - 1)}\,dx$

13. $\int \dfrac{10x^3 + 40x^2 + 22x + 7}{(4x^2 + 1)(x^2 + 6x + 10)}\,dx$ 14. $\int_3^4 \dfrac{5x^3 - 4x}{x^4 - 16}\,dx$

15. $\int \dfrac{2x^3}{(x^2 + 1)^2}\,dx$ 16. $\int \dfrac{-x^3 + x^2 + x + 3}{(x + 1)(x^2 + 1)^2}\,dx$

In Exercises 17 through 24, solve the given problems by integration.

17. Find the area bounded by $y = (x - 3)/(x^3 + x^2)$, $y = 0$, and $x = 1$.

18. Find the first-quadrant area bounded by $y = (3x^2 + 2x + 9)/[(x^2 + 9)(x + 1)]$ and $x = 2$.

19. Find the volume generated by rotating the first-quadrant area bounded by $y = 4/(x^4 + 6x^2 + 5)$ and $x = 2$ about the y-axis.

20. Find the volume generated by rotating the first-quadrant area bounded by $y = x/(x + 3)^2$ and $x = 3$ about the x-axis.

21. Under certain conditions the velocity v, in meters per second, of an object moving along a straight line as a function of time, in seconds, is given by

$$v = \frac{t^2 + 14t + 27}{(2t + 1)(t + 5)^2}$$

Find the distance traveled by the object during the first 2.00 s.

22. By a computer analysis the electric current i, in amperes, in a certain integrated circuit is given by

$$i = \frac{0.001(7t^2 + 16t + 48)}{(t + 4)(t^2 + 16)}$$

where t is the time in seconds. Find the total charge that passes a point in the circuit in the first 0.250 s.

23. Find the x-coordinate of the centroid of the first-quadrant area bounded by $y = 4/(x^3 + x)$, $x = 1$, and $x = 2$. See Fig. 9–9.

24. The slope of a curve is given by $(29x^2 + 36)/(4x^4 + 9x^2)$. Find the equation of the curve if it passes through $(1, 5)$.

Figure 9–9

9–6 Integration by Use of Tables

In this chapter we have introduced certain methods of reducing other integrals to the basic forms. Often this transformation and integration require a number of steps to be performed, and therefore integrals are tabulated for reference. The integrals found in tables have been derived by using the methods introduced thus far, as well as many other methods which can be used. Therefore, an understanding of the basic forms and some of the basic methods is very useful in finding integrals from tables. Such an understanding forms a basis for proper recognition of the forms which are used in the tables, as well as the types of results which may be expected. Therefore, *the use of the tables depends on the proper recognition of the form, and the variables and constants of the integral.* The following examples illustrate the use of the table of integrals found in Table 5 in Appendix D.

Example A

Integrate: $\int \dfrac{x \, dx}{\sqrt{2 + 3x}}$.

We first note that this integral fits the form of Formula 6 of Table 5, with $u = x$, $a = 2$, and $b = 3$. Therefore,

$$\int \frac{x \, dx}{\sqrt{2 + 3x}} = -\frac{2(4 - 3x)\sqrt{2 + 3x}}{27} + C$$

Example B

Integrate: $\int \dfrac{\sqrt{4 - 9x^2}}{x} \, dx$.

This fits the form of Formula 18, with proper identification of constants; $u = 3x$, $du = 3 \, dx$, $a = 2$. Hence,

$$\int \frac{\sqrt{4 - 9x^2}}{x} \, dx = \int \frac{\sqrt{4 - 9x^2}}{3x} \, 3 \, dx$$

$$= \sqrt{4 - 9x^2} - 2 \ln\left(\frac{2 + \sqrt{4 - 9x^2}}{3x}\right) + C$$

Example C

Integrate: $\int \sec^3 2x \, dx$.

This fits the form of Formula 37; $n = 3$, $u = 2x$, $du = 2 \, dx$. And so

$$\int \sec^3 2x \, dx = \frac{1}{2} \int \sec^3 2x (2 \, dx)$$

For reference, Eq. (8–12) is

$\int \sec u \, du =$

$\ln |\sec u + \tan u| + C$

$$= \frac{1}{2} \frac{\sec 2x \tan 2x}{2} + \frac{1}{2} \cdot \frac{1}{2} \int \sec 2x (2 \, dx)$$

[To complete this integral, we must use the basic form of Eq. (8–12).] Thus, we complete it by

$$\int \sec^3 2x \, dx = \frac{\sec 2x \tan 2x}{4} + \frac{1}{4} \ln |\sec 2x + \tan 2x| + C$$

Example D

Evaluate: $\int_1^e x^2 \ln 2x \, dx$.

This integral fits the form of Formula 46 if $u = 2x$. Thus, we have

$$\frac{1}{8} \int_1^e (2x)^2 \ln 2x (2 \, dx) = \frac{1}{8} (2x)^3 \left[\frac{\ln 2x}{3} - \frac{1}{9} \right]_1^e$$

$$= e^3 \left(\frac{\ln 2e}{3} - \frac{1}{9} \right) - \left(\frac{\ln 2}{3} - \frac{1}{9} \right)$$

$$= e^3 \left(\frac{\ln 2}{3} + \frac{\ln e}{3} - \frac{1}{9} \right) - \left(\frac{\ln 2}{3} - \frac{1}{9} \right)$$

$$= \frac{\ln 2}{3} (e^3 - 1) + \frac{1}{9} (2e^3 + 1) = 8.984$$

The proper identification of u and du is the key step in the use of tables. Therefore, for the integrals in the following example the proper u and du, along with the appropriate formula from the table, are identified, but the integrations are not performed.

Example E

(a) $\int \dfrac{dx}{16x^2 - 9}$ $u = 4x$, $du = 4 \, dx$,   Formula 9

(b) $\int x \sqrt{1 - x^4} \, dx$ $u = x^2$, $du = 2x \, dx$,   Formula 15

(c) $\int \dfrac{(4x^6 - 9)^{3/2}}{x} \, dx$ $u = 2x^3$, $du = 6x^2 \, dx$,   Formula 22
(introduce a factor of x^2 into the numerator and the denominator)

(d) $\int x^3 \sin x^2 \, dx$ $u = x^2$, $du = 2x \, dx$,   Formula 47
$[x^3 \, dx = x^2(x \, dx)]$

We summarize briefly here the approach to integrating a function. There are two methods of approaching a problem in order to obtain the exact result for the integral.

1. Write the integral such that it fits an integral form. Either a basic form as developed in this chapter or a form from a table of integrals may be used.
2. Use a technique of transforming the integral such that an integral form may be used. Techniques include those covered in this chapter and others which may be found in other sources.

Also, definite integrals may be approximated by methods such as the trapezoidal rule or Simpson's rule. There are other approximation methods, one of which is developed in Example E of Section 13–2.

Exercises 9–6

In Exercises 1 through 32, integrate each function by use of Table 5 in Appendix D.

1. $\displaystyle\int \frac{3x\,dx}{2 + 5x}$

2. $\displaystyle\int \frac{4x\,dx}{(1 + x)^2}$

3. $\displaystyle\int 5x\sqrt{2 + 3x}\,dx$

4. $\displaystyle\int \frac{dx}{x^2 - 4}$

5. $\displaystyle\int \sqrt{4 - x^2}\,dx$

6. $\displaystyle\int_0^{\pi/3} \sin^3 x\,dx$

7. $\displaystyle\int \sin 2x \sin 3x\,dx$

8. $\displaystyle\int \text{Arcsin } 3x\,dx$

9. $\displaystyle\int \frac{\sqrt{4x^2 - 9}}{x}\,dx$

10. $\displaystyle\int \frac{(9x^2 + 16)^{3/2}}{x}\,dx$

11. $\displaystyle\int \cos^5 4x\,dx$

12. $\displaystyle\int \tan^2 x\,dx$

13. $\displaystyle\int \text{Arctan } x^2(x\,dx)$

14. $\displaystyle\int xe^{4x}\,dx$

15. $\displaystyle\int_1^2 (4 - x^2)^{3/2}\,dx$

16. $\displaystyle\int \frac{dx}{9 - 16x^2}$

17. $\displaystyle\int \frac{dx}{x\sqrt{4x^2 + 1}}$

18. $\displaystyle\int \frac{\sqrt{4 + x^2}}{x}\,dx$

19. $\displaystyle\int \frac{dx}{x\sqrt{1 - 4x^2}}$

20. $\displaystyle\int \frac{dx}{x(1 + 4x)^2}$

21. $\displaystyle\int \sin x \cos 5x\,dx$

22. $\displaystyle\int_0^2 x^2 e^{3x}\,dx$

23. $\displaystyle\int x^5 \cos x^3\,dx$

24. $\displaystyle\int \sin^3 x \cos^2 x\,dx$

25. $\displaystyle\int \frac{x\,dx}{(1 - x^4)^{3/2}}$

26. $\displaystyle\int \frac{dx}{x(1 - 4x)}$

27. $\displaystyle\int_1^3 \frac{\sqrt{3 + 5x^2}\,dx}{x}$

28. $\displaystyle\int_0^1 \frac{\sqrt{9 - 4x^2}}{x}\,dx$

29. $\displaystyle\int x^3 \ln x^2\,dx$

30. $\displaystyle\int \frac{x\,dx}{x^2\sqrt{x^4 - 9}}$

31. $\displaystyle\int \frac{x^2\,dx}{(x^6 - 1)^{3/2}}$

32. $\displaystyle\int x^7\sqrt{x^4 + 4}\,dx$

In Exercises 33 through 40, evaluate the integrals by use of Table 5.

33. Find the length of arc of the curve $y = x^2$ from $x = 0$ to $x = 1$.

34. Find the moment of inertia with respect to its axis of the volume generated by rotating the area bounded by $y = \ln x$, $x = e$, and the x-axis about the y-axis.

35. Find the force on the area bounded by $x = 1/\sqrt{1 + y}$, $y = 0$, $y = 3$, and the y-axis, if the surface of the water is at the upper edge of the area and the units are feet. See Fig. 9–10.

36. Under certain conditions the velocity as a function of time for a particle is given by $v = t/\sqrt{1 + 2t}$. Determine the displacement as a function of time if $s = 6$ when $t = 0$.

37. Find the area bounded by $y = \text{Arctan } 2x$, $x = 2$, and the x-axis.

38. Find the area bounded by $y = (x^2 + 4)^{-3/2}$, $x = 3$, and the axes.

39. Find the volume of the solid generated by rotating the area bounded by $y = e^x \sin x$ and the x-axis between $x = 0$ and $x = \pi$ about the x-axis.

40. The electric current in a certain circuit as a function of time is given by $i = 2/\sqrt{t^2 + 100}$. Find the expression for the charge which passes a given point if $q = 0$ when $t = 0$.

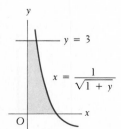

Figure 9–10

9–7 Improper Integrals

In defining the definite integral $\int_a^b f(x)\, dx$ in Chapter 4, we assumed that both limits of integration are finite; we also assumed that the integrand $f(x)$ is continuous for $a \le x \le b$. We now extend the definition of a definite integral to **improper integrals,** *for which one or both limits of integration are infinite, or which have infinite discontinuous integrands.*

First we will discuss the type of improper integral for which one or both of the limits of integration are infinite. To illustrate such an integral, consider the area in the following example.

Example A

Find the area bounded by $y = 1/x^3$, $y = 0$, $x = 1$, and $x = b$, where $b > 1$. See Fig. 9–11.

Figure 9–11

From the figure we see that the area is given by

$$A = \int_1^b \frac{1}{x^3}\, dx = -\frac{1}{2x^2}\Big|_1^b = -\frac{1}{2b^2} + \frac{1}{2}$$
$$= \frac{1}{2}\left(1 - \frac{1}{b^2}\right)$$

If we now let b increase without bound, we see that the area increases, with a limiting value of $\frac{1}{2}$. This we can show as

$$A = \lim_{b \to +\infty} \int_1^b \frac{1}{x^3}\, dx = \lim_{b \to +\infty} \frac{1}{2}\left(1 - \frac{1}{b^2}\right) = \frac{1}{2}$$

Representing this limit by $\int_1^{+\infty} \frac{1}{x^3}\, dx$, we see that we have a definite integral with an infinite limit.

Based on reasoning similar to that of Example A *we have the following definition for improper integrals for which one or both limits of integration are infinite.*

$$\int_a^{+\infty} f(x)\ dx = \lim_{b \to +\infty} \int_a^b f(x)\ dx \qquad (9-2)$$

$$\int_{-\infty}^b f(x)\ dx = \lim_{a \to -\infty} \int_a^b f(x)\ dx \qquad (9-3)$$

$$\int_{-\infty}^{+\infty} f(x)\ dx = \lim_{a \to -\infty} \int_a^0 f(x)\ dx + \lim_{b \to +\infty} \int_0^b f(x)\ dx \qquad (9-4)$$

▶ The improper integrals in Eqs. (9-2), (9-3), and (9-4) have the indicated values only if the limits exist. In this case, *when the limits exist, the improper integrals are said to be* **convergent**. On the other hand, *if the limits do not exist, we say that the improper integrals are* **divergent**. For Eq. (9-2) we assume that $f(x)$ is continuous for $x \geq a$, for Eq. (9-3) we assume that $f(x)$ is continuous for $x \leq b$, and for Eq. (9-4) we assume that $f(x)$ is continuous for all x.

Example B

Evaluate $\int_{-\infty}^0 e^x\ dx$ if it exists.
Using Eq. (9-3), we have

$$\int_{-\infty}^0 e^x\ dx = \lim_{a \to -\infty} \int_a^0 e^x\ dx = \lim_{a \to -\infty} e^x \Big|_a^0$$

$$= \lim_{a \to -\infty} (1 - e^a) = 1$$

Here the integral converges, since the limit exists.

Example C

A rocket is launched vertically with a velocity v, in miles per hour, as a function of time, in hours, of $v = 30,000/(t + 1)$. Show that the rocket will eventually be as far from earth as may be required.
The distance s from the earth is given by

$$s = \int_0^{+\infty} \frac{30,000\ dt}{t + 1} = \lim_{b \to +\infty} \int_0^b \frac{30,000\ dt}{t + 1}$$

$$= 30,000 \lim_{b \to +\infty} \ln|t + 1| = 30,000 \lim_{b \to +\infty} \ln|b + 1| = +\infty$$

Since the integral diverges and the limit is infinite, the distance traveled will be unbounded.

Example D

Evaluate $\displaystyle\int_{-\infty}^{+\infty} \frac{dx}{x^2 + 4x + 5}$ if it exists.

Using Eq. (9–4), we have

$$\int_{-\infty}^{+\infty} \frac{dx}{x^2 + 4x + 5} = \lim_{a \to -\infty} \int_a^0 \frac{dx}{x^2 + 4x + 5} + \lim_{b \to +\infty} \int_0^b \frac{dx}{x^2 + 4x + 5}$$

$$= \lim_{a \to -\infty} \int_a^0 \frac{dx}{(x + 2)^2 + 1} + \lim_{b \to +\infty} \int_0^b \frac{dx}{(x + 2)^2 + 1}$$

$$= \lim_{a \to -\infty} \operatorname{Arctan}(x + 2)\Big|_a^0 + \lim_{b \to +\infty} \operatorname{Arctan}(x + 2)\Big|_0^b$$

$$= \operatorname{Arctan} 2 - \left(-\frac{\pi}{2}\right) + \frac{\pi}{2} - \operatorname{Arctan} 2$$

$$= \pi$$

Here we used the fact that $\operatorname{Arctan}(x + 2) \to -\dfrac{\pi}{2}$ when $x \to -\infty$ and that $\operatorname{Arctan}(x + 2) \to \dfrac{\pi}{2}$ when $x \to +\infty$.

 An improper integral may diverge without becoming infinite. See Exercise 31 at the end of this section. Also, the reason that it is necessary to split the improper integral for which both limits are infinite [Eq. (9–4)] is illustrated in Exercise 32.

We next consider improper integrals which have discontinuous integrands. To illustrate this type of improper integral, consider the area in the following example.

Example E

Find the first-quadrant area bounded by $y = 1/\sqrt{x}$ and $x = 4$. See Fig. 9–12. We set up the integral for the area as

$$A = \int_0^4 \frac{1}{\sqrt{x}}\, dx$$

However, we note that the integrand $1/\sqrt{x}$ is discontinuous at $x = 0$, which is the lower limit. Therefore, we set up the integral for the area as

$$A = \lim_{h \to 0} \int_h^4 \frac{1}{\sqrt{x}}\, dx$$

where $0 < h < 4$. We can now find the limiting value of the area if it exists. Therefore,

$$A = \lim_{h \to 0} \int_h^4 x^{-1/2}\, dx = \lim_{h \to 0} 2\sqrt{x}\,\Big|_h^4 = \lim_{h \to 0} (4 - 2\sqrt{h}) = 4$$

Since the limit exists, the limiting value of the area is 4.

Figure 9–12

(figure: graph showing $y = \dfrac{1}{\sqrt{x}}$ with vertical lines $x = h$ and $x = 4$)

Based on reasoning similar to that of Example E, we have the following definitions for improper integrals which have infinite discontinuous integrands. *If f(x) is continuous for a < x ≤ b (but is discontinuous for x = a),*

$$\int_a^b f(x)\, dx = \lim_{h \to 0} \int_{a+h}^b f(x)\, dx \tag{9-5}$$

If f(x) is continuous for a ≤ x < b (but is discontinuous for x = b),

$$\int_a^b f(x)\, dx = \lim_{h \to 0} \int_a^{b-h} f(x)\, dx \tag{9-6}$$

If f(x) is continuous for a ≤ x ≤ b, except at c where a < c < b,

$$\int_a^b f(x)\, dx = \lim_{h \to 0} \int_a^{c-h} f(x)\, dx + \lim_{h' \to 0} \int_{c+h'}^b f(x)\, dx \tag{9-7}$$

For Eqs. (9–5), (9–6), and (9–7) we assume that $h > 0$ and $h' > 0$. The improper integrals are convergent if the limits exist and divergent if the limits do not exist.

Example F

Evaluate $\int_0^1 \dfrac{dx}{(x-1)^{1/3}}$ if it is convergent.

We see that the integrand is discontinuous at the upper limit. Therefore, we use Eq. (9–6).

$$\int_0^1 \frac{dx}{(x-1)^{1/3}} = \lim_{h \to 0} \int_0^{1-h} \frac{dx}{(x-1)^{1/3}}$$
$$= \lim_{h \to 0} \left[\frac{3}{2}(x-1)^{2/3} \right]_0^{1-h}$$
$$= \lim_{h \to 0} \left[\frac{3}{2}(-h)^{2/3} - \frac{3}{2}(-1)^{2/3} \right]$$
$$= -\frac{3}{2}$$

Here the integral is convergent and has the value $-\frac{3}{2}$.

Example G

Evaluate $\int_0^2 \dfrac{dx}{(x-1)^2}$ if it is convergent.

Here we see that the integrand is discontinuous at $x = 1$, which is between the limits of integration. Therefore, we use Eq. (9–7).

$$\int_0^2 \frac{dx}{(x-1)^2} = \lim_{h\to 0}\int_0^{1-h} \frac{dx}{(x-1)^2} + \lim_{h'\to 0}\int_{1+h'}^2 \frac{dx}{(x-1)^2}$$

$$= \lim_{h\to 0}\left[-\frac{1}{x-1}\right]_0^{1-h} + \lim_{h'\to 0}\left[-\frac{1}{x-1}\right]_{h'+1}^2$$

$$= \lim_{h\to 0}\left(\frac{1}{h}-1\right) + \lim_{h'\to 0}\left(-1+\frac{1}{h'}\right)$$

▶ Neither of these limits exists. The divergence of *either* of these integrals means that the improper integral diverges.

If we were to integrate this improper integral as an ordinary integral, we would obtain

$$\frac{dx}{(x-1)^2} = \left.-\frac{1}{x-1}\right|_0^2 = -1-1 = -2$$

The negative result cannot be valid, because the integrand is never negative. This shows the need to be careful to handle improper integrals correctly.

Exercises 9–7

In Exercises 1 through 24, determine whether the given improper integral is convergent or divergent. If it is convergent, evaluate it.

1. $\int_1^{+\infty} \dfrac{dx}{(x+2)^2}$

2. $\int_2^{+\infty} \dfrac{dx}{3x+4}$

3. $\int_{-\infty}^0 \dfrac{dx}{\sqrt{1-3x}}$

4. $\int_0^{+\infty} \dfrac{dx}{(1+x)^3}$

5. $\int_1^{+\infty} \dfrac{x\,dx}{1+x^2}$

6. $\int_3^{+\infty} \dfrac{x\,dx}{\sqrt{x^2-4}}$

7. $\int_0^{+\infty} xe^{-x^2}\,dx$

8. $\int_{-\infty}^{-1} \dfrac{-x\,dx}{(1+x^2)^2}$

9. $\int_{-\infty}^0 \dfrac{2\,dx}{(1-x)^3}$

10. $\int_0^{+\infty} \dfrac{x\,dx}{\sqrt{x+1}}$

11. $\int_{-\infty}^{+\infty} e^{-x}\,dx$

12. $\int_{-\infty}^{+\infty} \dfrac{dx}{x^2+4}$

13. $\int_0^9 \dfrac{dx}{\sqrt{9-x}}$

14. $\int_0^1 \dfrac{dx}{x}$

15. $\int_{\pi/4}^{\pi/3} \tan 2x\,dx$

16. $\int_0^3 \dfrac{dx}{\sqrt{9-x^2}}$

17. $\int_0^3 \dfrac{dx}{(x-2)^2}$

18. $\int_{-1}^1 \dfrac{dx}{x^{1/3}}$

19. $\int_0^1 \dfrac{2\,dx}{1-x^2}$

20. $\int_2^3 \dfrac{dx}{\sqrt{x-2}}$

21. $\int_1^2 \dfrac{dx}{\sqrt{x^2-1}}$

22. $\int_1^3 \dfrac{3\,dx}{3x-x^2}$

23. $\int_1^3 \dfrac{dx}{(x-2)^{1/3}}$

24. $\int_{-\pi/2}^{\pi/2} \sec x\,dx$

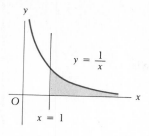

Figure 9–13

In Exercises 25 through 32, solve the given problems by evaluating the appropriate improper integrals (if convergent).

25. (a) Find the area bounded by $y = 1/x$, $y = 0$, and $x = 1$ (to the right of $x = 1$). See Fig. 9–13. (b) Find the volume generated if the area of part (a) is rotated about the x-axis.

26. Compare the area bounded by $y = 1/\sqrt{x}$, $y = 0$, and $x = 4$ (to the right of $x = 4$) with the area of Example E.

27. Find the area bounded by $y = x/\sqrt{x^2 - 1}$, $x = 1$, $x = 2$, and $y = 0$.

28. Find the volume generated by rotating the area bounded by $y = 1/(x - 2)$, $x = 0$, $y = 0$, and $x = 2$ about the x-axis.

29. The gravitational force, in pounds, of the earth on a satellite (the weight of the satellite) is given by $F = 10^{11}/(x + 3960)^2$, where x is the vertical distance, in miles, from the surface of the earth to the satellite. How much work is done in moving the satellite from the earth's surface to an infinite distance from the earth?

30. The electric field intensity E around an infinitely long straight wire carrying an electric current i is given by

$$E = kiam \int_{-\infty}^{+\infty} \frac{dx}{(a^2 + x^2)^{3/2}}$$

where k, i, a, and m are constants. Evaluate the integral.

31. To show that an improper integral may diverge without becoming infinite, evaluate $\int_0^{+\infty} \cos x \, dx$. (Note carefully the values the integral may have.)

32. To show that it is necessary to split the integral $\int_{-\infty}^{+\infty} f(x) \, dx$ into two separate integrals, as in Eq. (9–7), evaluate (a) $\lim_{b \to +\infty} \int_{-b}^{b} x \, dx$ and (b) $\int_{-\infty}^{+\infty} x \, dx$ and compare results.

9–8 Review Exercises for Chapter 9

In Exercises 1 through 24, integrate each of the given functions without the use of Table 5.

1. $\int x \csc^2 2x \, dx$

2. $\int x \operatorname{Arctan} x \, dx$

3. $\int x\sqrt{x - 4} \, dx$

4. $\int \dfrac{2x \, dx}{\sqrt{2x + 5}}$

5. $\int \dfrac{dx}{\sqrt{4x^2 - 9}}$

6. $\int \dfrac{x^2 \, dx}{\sqrt{9 - x^2}}$

7. $\int \dfrac{x + 25}{x^2 - 25} \, dx$

8. $\int \dfrac{2x^2 + 6x + 1}{2x^3 - x^2 - x} \, dx$

9. $\int \dfrac{x^2 - 2x + 3}{(x - 1)^3} \, dx$

10. $\int \dfrac{2x^2 + 3x + 18}{x^3 + 9x} \, dx$

11. $\int_1^{+\infty} \dfrac{dx}{2x + 7}$

12. $\int_{-1}^{1} \dfrac{dx}{\sqrt{x + 1}}$

13. $\int \dfrac{x \, dx}{(x + 3)^{2/3}}$

14. $\int x \tan^2 x \, dx$

15. $\int \dfrac{2 \, dx}{(4x^2 + 1)^{3/2}}$

16. $\int \dfrac{3x^2 - 6x - 2}{3x^3 + x^2} \, dx$

17. $\int \dfrac{x^2 + 3}{x^4 + 3x^2 + 2} \, dx$

18. $\int x^3\sqrt{4 - x^2} \, dx$

19. $\int_0^2 \ln(x + 2)\, dx$ 20. $\int_{-2}^1 \dfrac{x^2\, dx}{\sqrt{2 - x}}$ 21. $\int \dfrac{6(2 - x^2)}{(x^2 - 1)(x^2 - 4)}\, dx$

22. $\int \dfrac{-x^3 + 9x^2 - 24x + 36}{x^2(x - 3)^2}\, dx$ 23. $\int_0^8 \dfrac{x\, dx}{(8 - x)^{1/3}}$

24. $\int_1^{+\infty} \dfrac{dx}{x^2(x^2 + 1)}$

In Exercises 25 through 36, solve the given problems by integration.

25. Find the area bounded by $y = 1/\sqrt{x - 2}$, $x = 2$, $x = 6$, and $y = 0$.

26. Find the area bounded by $y = x/(1 + x)^2$, the x-axis, and the line $x = 4$.

27. Find the area inside the circle $x^2 + y^2 = 25$ and to the right of the line $x = 3$.

28. Find the area bounded by $y = x\sqrt{x + 4}$, $y = 0$, and $x = 5$.

29. Find the volume generated by rotating the area bounded by $y = xe^x$, $y = 0$, and $x = 2$ about the y-axis.

30. Find the volume generated by rotating about the y-axis the area bounded by $y = x + \sqrt{x + 1}$, $x = 3$, and the axes.

31. The force F, in newtons, applied by a stamping machine in punching out a hole 0.500 cm deep is given by $F = 1000x/(2x + 1)^2$, where x is the depth, in centimeters, of the hole at any instant. Find the work done by the machine in punching out the hole.

32. In electrical theory, the integral

$$W = \int_0^\infty Ri^2\, dt$$

occurs. If $i = Ie^{-Rt/L}$, evaluate the integral. Here R, I, and L are constants.

33. Find the centroid of the area bounded by $y = \ln x$, $x = 2$, and the x-axis.

34. The nose cone of a space vehicle is to be covered with a heat shield. The cone is designed such that a cross section x feet from the tip and perpendicular to its axis is a circle of radius $0.125x^2$ feet. Find the surface area of the heat shield if the nose cone is 16.0 ft long. See Fig. 9–14.

35. A window has the shape of a semiellipse, as shown in Fig. 9–15. What is the area of the window?

36. Find the radius of gyration with respect to the y-axis of the first-quadrant area bounded by $y = \sqrt{x + 1}$ and $x = 3$.

$r = 0.125x^2$

16.0 ft

1.25 m

1.12 m

Figure 9–14   Figure 9–15

CHAPTER 10

Introduction to Partial Derivatives and Double Integrals

To this point we have been dealing with functions which have a single independent variable. There are, however, numerous situations in many fields, including science, engineering, and technology, in which functions with more than one independent variable are used. Although some of these applications involve three or more independent variables, many involve only two, and we shall concern ourselves primarily with these.

In this chapter we shall establish the meaning of a function of two variables and then discuss how the graph of this type of function is shown. The remaining sections of the chapter develop some of the basic concepts of the calculus of functions of two variables and show some of the applications in technology.

10–1 Functions of Two Variables

Many familiar formulas express one variable in terms of two or more other variables. The following example illustrates one from geometry.

Example A

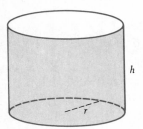

Figure 10–1

The total surface area of a right circular cylinder is a function of the radius and the height of the cylinder. That is, the area will change if either or both of these change. The formula for the total surface area is

$$A = 2\pi r^2 + 2\pi rh$$

We say that A is a function of r and h. See Fig. 10–1.

As we have noted, there are numerous applications of functions of two variables in many fields of study. For example, the voltage in a simple electric circuit depends on the resistance and current in the circuit. The moment of a force depends on the magnitude of the force and the distance of the force from the axis. The pressure of a gas depends on its temperature and volume. The temperature at a point on a flat plate depends on the coordinates of the point.

We define a function of two variables as follows. *If z is uniquely determined for given values of x and y, then z is a function of x and y.* Notation similar to that which is used for one independent variable is employed in this case. This notation is $z = f(x, y)$, where both x and y are independent variables. This is the meaning of a function of two variables.

Similar to the case with one independent variable, $f(a, b)$ *means "the value of the function when x = a and y = b."*

Example B

If $f(x, y) = 3x^2 + 2xy - y^3$, find $f(-1, 2)$. Substituting, we have

$$f(-1, 2) = 3(-1)^2 + 2(-1)(2) - (2)^3$$
$$= 3 - 4 - 8$$
$$= -9$$

Example C

For a certain electric circuit, the current i, in amperes, in terms of the voltage E and resistance R is given by

$$i = \frac{E}{R + 0.25}$$

Find the current for $E = 1.50$ V and $R = 1.20$ Ω, and for $E = 1.60$ V and $R = 1.05$ Ω.

Substituting the first values, we have

$$i = \frac{1.50}{1.20 + 0.25} = 1.03 \text{ A}$$

For the second pair of values we have

$$i = \frac{1.60}{1.05 + 0.25} = 1.23 \text{ A}$$

For this circuit, the current generally changes if either or both of E and R change.

Example D

If $f(x, y) = 2xy^2 - y$, find $f(x, 2x) - f(x, x^2)$.

We note that in each evaluation the x factor remains as x, but that we are to substitute $2x$ for y and subtract the function for which x^2 is substituted for y.

$$\begin{aligned}
f(x, 2x) - f(x, x^2) &= [2x(2x)^2 - (2x)] - [2x(x^2)^2 - x^2] \\
&= [8x^3 - 2x] - [2x^5 - x^2] \\
&= 8x^3 - 2x - 2x^5 + x^2 \\
&= -2x^5 + 8x^3 + x^2 - 2x
\end{aligned}$$

This type of difference of functions is important later in this chapter.

Restricting values of the function to real numbers means that certain restrictions may be placed on the independent variables. *Values of either x or y or both which lead to division by zero or to imaginary values for z are not permissible.* The following example illustrates this point.

Example E

If

$$f(x, y) = \frac{3xy}{(x - y)(x + 3)}$$

all values of x and y are permissible except those for which $x = y$ and $x = -3$. Each of these would indicate division by zero.

If $f(x, y) = \sqrt{4 - x^2 - y^2}$, neither x nor y may be greater than 2 in absolute value, and the sum of their squares may not exceed 4. Otherwise, imaginary values of the function would result.

One of the primary difficulties students have with functions is setting them up from stated conditions. Although many examples of this have been encountered in our previous work, an example of setting up a function of two variables should prove helpful. Although no general rules can be given for this procedure, a careful analysis of the statement should lead to the desired function. The following example illustrates setting up such a function.

Example F

An open rectangular metal box is to be constructed such that it will contain 2 ft³. The cost of sheet metal is c cents per square foot. Express the cost of the sheet metal needed to construct the box as a function of the length and width of the box.

   The cost in question depends on the surface area of the sides of the box. An "open" box is one which has no top. Thus, the surface area of the box is

$$S = lw + 2lh + 2hw$$

$V = 2 \text{ ft}^3$

$h = \dfrac{2}{lw}$

w

l

Figure 10–2

where l is the length, w the width, and h the height of the box (see Fig. 10–2). However, this expression contains three independent variables. Using the condition that the volume of the box is 2 ft³, we have $lwh = 2$. Since we wish to have only l and w, we solve this expression for h, and find that $h = 2/lw$. Substituting this in the expression for the surface area we have

$$S = lw + 2l\left(\frac{2}{lw}\right) + 2\left(\frac{2}{lw}\right)w$$

$$= lw + \frac{4}{w} + \frac{4}{l}$$

Since the cost C is given by $C = cS$, we have the expression for the cost as

$$C = c\left(lw + \frac{4}{w} + \frac{4}{l}\right)$$

Exercises 10–1

r

h

Figure 10–3

In Exercises 1 through 8, determine the indicated functions.

1. Express the volume of a right circular cylinder as a function of the radius and height.
2. Express the volume of a right circular cone as a function of the radius of the base and the height.
3. Express the area of a triangle as a function of the base and the altitude.
4. Express the length of a diagonal of a rectangle as a function of the length and width.
5. A cylindrical can is to be made to contain a volume V. Express the total surface area of the can as a function of V and the radius of the can.
6. A right circular cylinder is to be inscribed in a sphere of radius r. Express the volume of the cylinder as a function of the height h of the cylinder and r. See Fig. 10–3.

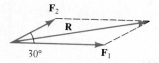

Figure 10-4

7. The angle between two forces F_1 and F_2 is 30°. Express the magnitude of the resultant R in terms of F_1 and F_2. See Fig. 10-4.

8. A computer leasing firm charges a monthly fee F based on the length of time a corporation has used the service plus $100 for every hour the computer is used during the month. Express the total monthly charge T as a function of F and the number of hours h the computer is used.

In Exercises 9 through 24, evaluate the given functions.

9. $f(x, y) = 2x - 6y$, find $f(0, -4)$

10. $F(x, y) = x^2 - 5y + y^2$, find $F(2, -2)$

11. $g(r, s) = r - 2rs - r^2s$, find $g(-2, 1)$

12. $\phi(x, t) = 2x^3 - x^2t - 2t^2$, find $\phi(-3, -1)$

13. $Y(y, t) = \dfrac{2 - 3y}{t - 1} + 2y^2t$, find $Y(y, 2)$

14. $f(r, t) = r^3 - 3r^2t + 3rt^2 - t^3$, find $f(3, t)$

15. $X(x, t) = -6xt + xt^2 - t^3$, find $X(x, -t)$

16. $g(y, z) = 2yz^2 - 6y^2z - y^2z^2$, find $g(y, 2y)$

17. $H(p, q) = p - \dfrac{p - 2q^2 - 5q}{p + q}$, find $H(p, q + k)$

18. $\phi(x, y) = \dfrac{3 - x}{y - x} - xy - xy^2$, find $\phi(x + h, -1)$

19. $f(x, y) = x^2 - 2xy - 4x$, find $f(x + h, y + k) - f(x, y)$

20. $g(y, z) = 4yz - z^3 + 4y$, find $g(y + 1, z + 2) - g(y, z)$

21. $f(x, y) = xy + x^2 - y^2$, find $f(x, x) - f(x, 0)$

22. $f(x, y) = 4x^2 - xy - 2y$, find $f(x, x^2) - f(x, 1)$

23. $g(y, z) = 3y^3 - y^2z + 5z^2$, find $g(3z^2, z) - g(z, z)$

24. $X(x, t) = 2x - \dfrac{t^2 - 2x^2}{x}$, find $X(2t, t) - X(2t^2, t)$

In Exercises 25 through 28, determine which values of x or y, if any, are not permissible.

25. $f(x, y) = \dfrac{\sqrt{y}}{2x}$

26. $f(x, y) = \dfrac{x^2 - 4y^2}{x^2 + 9}$

27. $f(x, y) = \sqrt{x^2 + y^2 - x^2y - y^3}$

28. $f(x, y) = \dfrac{1}{xy - y}$

In Exercises 29 through 38, solve the given problems.

29. The voltage v across a resistor R in an electric circuit is given by $v = iR$, where i is the current. What is the voltage if the current is 3 A and the resistance is 6 Ω?

30. The centripetal acceleration of an object moving in a circular path is $a = v^2/R$, where v is the velocity of the object and R is the radius of the circle. What is the centripetal acceleration of an object moving at 6 ft/s in a circular path whose radius is 4 ft?

31. The pressure of a gas as a function of its volume and temperature is given by $p = nRT/V$, where n and R are constants. If $n = 3$ mol and $R = 8.31$ J/mol · K, find the pressure (in pascals) for $T = 300$ K and $V = 50$ m^3.

32. The power supplied to a resistance R in an electric circuit by a current i is given by $P = i^2R$. Find the power (in watts) supplied by a current of 4 A to a resistance of 2.4 Ω.

33. The reciprocal of the image distance from a lens as a function of the object distance p and the focal length f of the lens is given by

$$\frac{1}{q} = \frac{1}{f} - \frac{1}{p}$$

p = perimeter

Figure 10–5

Find the image distance of an object 20 cm from a lens whose focal length is 5 cm.

34. The pressure exerted by a force on an area is given by $p = F/A$. If a given force is doubled on an area which is $\frac{2}{3}$ a given area, what is the ratio of the initial pressure to the final pressure?

35. A rectangular solar cell panel has a perimeter p and a width w. Express the area A of the panel in terms of p and w and evaluate the area for $p = 250$ cm and $w = 55$ cm. See Fig. 10–5.

36. A gasoline storage tank is in the shape of a right circular cylinder with a hemisphere at each end, as shown in Fig. 10–6. Express the volume V of the tank in terms of r and h and then evaluate the volume for $r = 3.75$ ft and $h = 12.5$ ft.

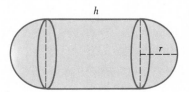

Figure 10–6

37. The crushing load L of a pillar varies as the fourth power of its radius and inversely as the square of its length. Express L as a function of r and l for a pillar 20 ft tall and 1 ft in diameter which is crushed by a load of 20 tons.

38. The resonant frequency of an electric circuit containing an inductance L and capacitance C is inversely proportional to the square root of the product of the inductance and capacitance. If the resonant frequency of a circuit containing a 4-H capacitor and a 64-μF capacitor is 10 Hz, express f as a function of L and C.

10–2 Curves and Surfaces in Three Dimensions

We will now undertake a brief description of the graphical representation of a function of two variables. We shall show first a method of representation in the rectangular coordinate system in two dimensions. The following example illustrates the method.

Example A

To represent $z = 2x^2 + y^2$, we will assume various values of z, and sketch the resulting equation in the xy-plane. For example, if $z = 2$ we have

$$2x^2 + y^2 = 2$$

This we recognize as an ellipse with its major axis along the y-axis and vertices at $(0, \sqrt{2})$ and $(0, -\sqrt{2})$. The ends of the minor axis are $(1, 0)$ and $(-1, 0)$ (see Fig. 10–7a). However, the ellipse $2x^2 + y^2 = 2$ represents the function $z = 2x^2 + y^2$ only if $z = 2$. If $z = 4$, we have $2x^2 + y^2 = 4$, which is another ellipse. In fact, for all positive values of z, an ellipse results. Negative values for z are not possible, since neither x^2 nor y^2 may be negative. Figure 10–7b shows the ellipses obtained by using the indicated values of z.

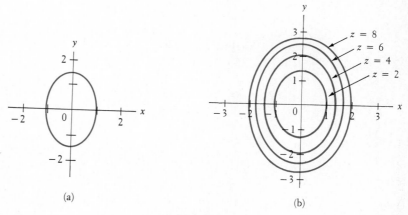

Figure 10–7

The method of representation illustrated in Example A is useful if only a few specific values of z are to be used, or at least if the various curves do not intersect in such a way that they cannot be distinguished. If a general representation of z as a function of x and y is desired, it is necessary to use three coordinate axes, one each for x, y, and z. The most widely applicable system of this kind is to place a third coordinate axis at right angles to each of the x- and y-axes. In this way we employ three dimensions for the representation.

The three mutually perpendicular axes, the x-axis, the y-axis, and the z-axis, are the basis of the **rectangular coordinate system in three dimensions.** Together they form three mutually perpendicular planes in space, the xy-plane, the yz-plane, and the xz-plane. To every point in the space of the coordinate systems is associated the set of numbers (x, y, z). The point at which the axes meet is the origin. The positive directions of the axes are indicated in Fig. 10–8. *That part of space in which all the values are positive is called the first* **octant.** Numbers are not assigned to the other octants.

Figure 10–8

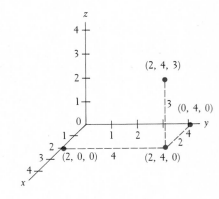

Figure 10–9

Example B

Represent the point (2, 4, 3) in rectangular coordinates.

We first note that a certain distortion is necessary to reasonably represent values of x, since the x-axis "comes out" of the plane of the page. Units $\sqrt{2}/2\ (= 0.7)$ as long as those used on the other axes give a good representation. With this in mind, we draw a line 4 units long from the point (2, 0, 0) on the x-axis in the xy-plane. This locates the point (2, 4, 0). From this point a line 3 units long is drawn vertically upward. This locates the desired point (2, 4, 3). The point may be located by starting from (0, 4, 0), proceeding two units *parallel* to the x-axis to (2, 4, 0), and then proceeding vertically three units to (2, 4, 3). It may also be located by starting from (0, 0, 3) (see Fig. 10–9).

The remainder of this section is devoted to showing certain basic techniques by which three-dimensional figures may be drawn. We start by showing the general equation of a plane.

In Chapter 1 we showed that the graph of the equation $Ax + By + C = 0$ in two dimensions is a straight line. By the following example, we will verify that *the graph of the equation*

$$Ax + By + Cz + D = 0 \qquad\qquad (10\text{–}1)$$

is a **plane** *in three dimensions.*

Example C

Show that the graph of the equation $2x + 3y + z - 6 = 0$ in three dimensions is a plane.

If we let any of the three variables take on a specific value, we obtain a linear equation in the other two variables. For example, the point $(\frac{1}{2}, 1, 2)$ satisfies the equation, and therefore lies on the graph of the equation. For $x = \frac{1}{2}$, we have

$$3y + z - 5 = 0$$

which is the equation of a straight line. This means that all pairs of values of y and z which satisfy this equation, along with $x = \frac{1}{2}$, satisfy the given equation. Thus, for $x = \frac{1}{2}$, the straight line $3y + z - 5 = 0$ lies on the graph of the equation.

For $z = 2$, we have

$$2x + 3y - 4 = 0$$

which is also a straight line. By similar reasoning this line lies on the graph of the equation. Since two lines through a point define a plane, these lines through $(\frac{1}{2}, 1, 2)$ define a plane. This plane is the graph of the equation (see Fig. 10–10).

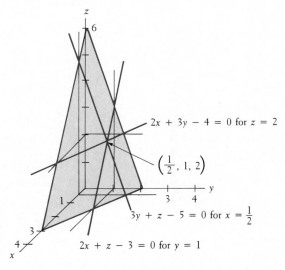

Figure 10–10

To complete the verification, it can be seen that for any point on the graph, there is a straight line parallel to one of the coordinate planes which lies on the graph. Thus, regardless of the point chosen, intersecting straight lines through the point which lies on the graph exist. Thus, the graph is a plane. A similar analysis could be made for any equation of the same form.

Since we know that the graph of an equation of the form of Eq. (10–1) is a plane, its graph can be found by determining its three intercepts, and the plane can then be represented by drawing in the lines between these intercepts. If the plane passes through the origin, by letting two of the variables in turn be zero, two straight lines which define the plane are found.

Example D

Sketch the graph of $3x - y + 2z - 4 = 0$.

The intercepts of the graph of an equation are those points where it crosses the respective axes. Thus, by letting two of the variables at a time equal zero, we obtain the intercepts. For the given equation the intercepts are $(\frac{4}{3}, 0, 0)$, $(0, -4, 0)$ and $(0, 0, 2)$. These points are located (see Fig. 10–11), and lines are drawn between them to represent the plane.

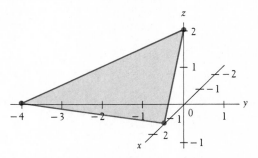

Figure 10–11

In general, the graph of an equation in three variables, which is essentially equivalent to a function with two independent variables, is a **surface** in space. This is seen in the case of the plane, and will be verified for other cases in the examples which follow.

The intersection of two surfaces is a **curve** *in space.* This has been seen in Examples C and D, since the intersection of the given planes and the coordinate planes are lines (which in the general sense are curves). *We define the* **traces** *of a surface to be the curves of intersection of the surface and the coordinate planes.* The traces of a plane are those lines drawn between the intercepts to represent the plane. Many surfaces may be sketched by finding their traces and intercepts.

Example E

Find the intercepts and traces and sketch the graph of the equation $z = 4 - x^2 - y^2$.

The intercepts of the graph of the equation are $(2, 0, 0)$, $(-2, 0, 0)$, $(0, 2, 0)$, $(0, -2, 0)$ and $(0, 0, 4)$.

Since the traces of a surface lie within the coordinate planes, for each trace one of the variables is zero. Thus, by letting each variable in turn be zero, we find the trace of the surface in the plane of the other two variables. Therefore, the traces of this surface are

in the yz-plane: $z = 4 - y^2$ (a parabola)
in the xz-plane: $z = 4 - x^2$ (a parabola)
in the xy-plane: $x^2 + y^2 = 4$ (a circle)

Using the intercepts and sketching the traces (see Fig. 10–12), we obtain the surface of the equation. This figure is called a **circular paraboloid.**

Figure 10-12

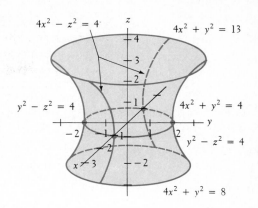

Figure 10-13

There are numerous techniques for analyzing the equation of a surface in order to obtain its graph. Another which we shall discuss here, which is closely associated with a trace, is that of a **section.** *By assuming a specific value of one of the variables, we obtain an equation in two variables, the graph of which lies in a plane parallel to the coordinate plane of the two variables.* The following example illustrates sketching a surface by use of intercepts, traces, and sections.

Example F

Sketch the graph of $4x^2 + y^2 - z^2 = 4$.

The intercepts are $(1, 0, 0)$, $(-1, 0, 0)$, $(0, 2, 0)$, and $(0, -2, 0)$. We note that there are no intercepts on the z-axis, for this would necessitate $z^2 = -4$.

The traces are

$$\text{in the } yz\text{-plane:} \quad y^2 - z^2 = 4 \quad \text{(a hyperbola)}$$
$$\text{in the } xz\text{-plane:} \quad 4x^2 - z^2 = 4 \quad \text{(a hyperbola)}$$
$$\text{in the } xy\text{-plane:} \quad 4x^2 + y^2 = 4 \quad \text{(an ellipse)}$$

The surface is reasonably defined by these curves, but by assuming suitable values of z we may indicate its shape better. For example, if $z = 3$, we have $4x^2 + y^2 = 13$, which is an ellipse. In using it we must remember that it is valid for $z = 3$ and therefore should be drawn 3 units above the xy-plane. If $z = -2$, we have $4x^2 + y^2 = 8$, which is also an ellipse. Thus, we have the following sections:

$$\text{for } z = 3: \quad 4x^2 + y^2 = 13 \quad \text{(an ellipse)}$$
$$\text{for } z = -2: \quad 4x^2 + y^2 = 8 \quad \text{(an ellipse)}$$

Other sections could be found, but these are sufficient to obtain a good sketch of the graph (see Fig. 10-13). The figure is called an **elliptic hyperboloid.**

Now that we have developed the use of the rectangular coordinate system in three dimensions, we can compare the graph of a given function using two dimensions and three dimensions. The following example shows the surface for the function of Example A.

Example G

Sketch the graph of $z = 2x^2 + y^2$.

The only intercept is (0, 0, 0). The traces are

in the yz-plane: $z = y^2$ (a parabola)

in the xz-plane: $z = 2x^2$ (a parabola)

in the xy-plane: the origin

We can see that the trace in the xy-plane is only the point at the origin, since $2x^2 + y^2 = 0$ may be written as $y^2 = -2x^2$, which is true only for $x = 0$ and $y = 0$.

To obtain a better graph we should use at least one positive value for z. As we noted in Example A, negative values of z cannot be used. Since we used $z = 2$, $z = 4$, $z = 6$, and $z = 8$ in Example A, we shall use these values here. Thus,

for $z = 2$: $2x^2 + y^2 = 2$,   for $z = 4$: $2x^2 + y^2 = 4$

for $z = 6$: $2x^2 + y^2 = 6$,   for $z = 8$: $2x^2 + y^2 = 8$

Each of these sections is an ellipse. The surface, called an **elliptic paraboloid,** is shown in Fig. 10–14. We also show Fig. 10–7b again so that the three-dimensional graph can be compared with the two-dimensional graph.

Figure 10–14

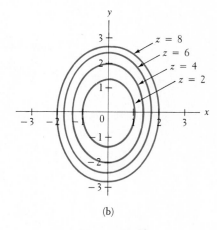

(b)

Figure 10-7b

Example G illustrates how *topographic maps* may be drawn. These maps represent three-dimensional terrain in two dimensions. For example, if Fig. 10–14 represents an excavation in the surface of the earth, then Fig. 10–7b represents the curves of constant elevation, or *contours,* with equally spaced elevations measured from the bottom of the excavation.

An equation with only two variables may represent a surface in space. Since only two variables are included in the equation, the surface is independent of the other variable. Another interpretation is that all sections, for all values of the variable not included, are the same. That is, all sections parallel to the coordinate plane of the included variables are the same as the trace in that plane.

Example H

Sketch the graph of $x + y = 2$ in the rectangular coordinate system in three dimensions and also in two dimensions.

Since z does not appear in the equation, for purposes of graphing in three dimensions we can consider the equation to be $x + y + 0z = 2$. Therefore, for *any value of z* the section is the straight line $x + y = 2$. Thus, the graph is a plane as shown in Fig. 10–15a. The graph as a straight line in two dimensions is shown in Fig. 10–15b.

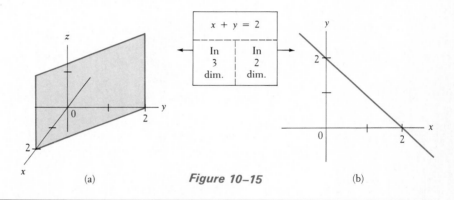

(a)   *Figure 10–15*   (b)

Example I

The graph of $x^2 + 4y^2 = 4$ in three dimensions is a surface whose sections, for all values of z, are given by the ellipse $x^2 + 4y^2 = 4$ (see Fig. 10–16).

The graph of the equation $z = 4 - x^2$ in three dimensions is a surface whose sections, for all values of y, are given by the parabola $z = 4 - x^2$ (see Fig. 10–17).

Figure 10–16

Figure 10–17

Each of the surfaces in Example I is known as a **cylindrical surface.** In general, a cylindrical surface is one which can be generated by a line moving parallel to a fixed line while passing through a plane curve.

It must be realized that most of the figures shown extend beyond the ranges indicated by the traces and sections. However, these traces and sections are convenient for representing and visualizing the surfaces.

A computer can be programmed to draw surfaces by drawing sections which are perpendicular to both the x and y axes or by sections perpendicular to the z-axis. Such computer-drawn surfaces are of great value for all types of surfaces, especially those of a complex nature. Figure 10–18 shows the graph of $z = (x^2 + 3y^2)e^{1 - (x^2 + y^2)}$ drawn by a computer in these two ways.

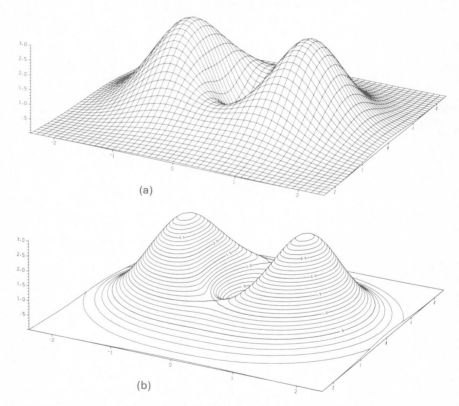

(a)

(b)

Figure 10–18 *(Reprinted by permission from* Calculus *by Marsden and Weinstein, Copyright 1985, Springer-Verlag.)*

Exercises 10–2

In Exercises 1 through 4, use the method of Example A, and draw the graphs of the indicated equations for the given values of z.

1. $z = x^2 + y^2, z = 1, z = 4, z = 9$
2. $z = x - 2y, z = -2, z = 2, z = 5$

See Appendix C for a computer program for graphing a plane.

3. $z = y - x^2$, $z = 0$, $z = 2$, $z = 4$

4. $z = x^2 - 4y^2$, $z = -4$, $z = 1$, $z = 4$

Sketch the graphs of the equations given in Exercises 5 through 24 in the rectangular coordinate system in three dimensions.

5. $x + y + 2z - 4 = 0$ 6. $2x - y - z + 6 = 0$

7. $4x - 2y + z - 8 = 0$ 8. $3x + 3y - 2z - 6 = 0$

9. $z = y - 2x - 2$ 10. $z = x - 4y$

11. $x + 2y = 4$ 12. $2x - 3z = 6$

13. $x^2 + y^2 + z^2 = 4$ 14. $2x^2 + 2y^2 + z^2 = 8$

15. $z = 4 - 4x^2 - y^2$ 16. $z = x^2 + y^2$

17. $z = 2x^2 + y^2 + 2$ 18. $x^2 + y^2 - 4z^2 = 4$

19. $x^2 - y^2 - z^2 = 9$ 20. $z^2 = 9x^2 + 4y^2$

21. $x^2 + y^2 = 16$ 22. $4z = x^2$

23. $y^2 + 9z^2 = 9$ 24. $xy = 2$

In Exercises 25 through 32, sketch the indicated curves and surfaces.

25. Curves which represent a constant temperature are called *isotherms*. The temperature at a point (x, y) of a flat plate is t degrees Celsius, where $t = 4x - y^2$. In two dimensions draw the isotherms for $t = -4$, 0, and 8.

26. At a point (x, y) in the xy-plane the electric potential V is given by $V = y^2 - x^2$, where V is measured in volts. Draw the lines of equal potential for $V = -9$, 0, and 9.

27. Sketch the line in space defined by the intersection of the planes.

$$x + 2y + 3z - 6 = 0 \quad \text{and} \quad 2x + y + z - 4 = 0$$

28. Sketch the curve in space defined by the intersection of the surfaces

$$x^2 + (z - 1)^2 = 1 \quad \text{and} \quad z = 4 - x^2 - y^2$$

29. An electric charge is so distributed that the electric potential at all points on an imaginary surface is the same. Such a surface is called an *equipotential surface*. Sketch the graph of the equipotential surface whose equation is $2x^2 + 2y^2 + 3z^2 = 6$.

30. The pressure, volume, and temperature for a certain gas are related by the equation $p = T/2V$, where p is measured in kilopascals, T in kelvins, and V in cubic meters. Sketch the p-V-T surface, by using the z-axis for p, the x-axis for V, and the y-axis for T. Use units of 100 K for T, 10 kPa for p, and 10 m^3 for V. Sections must be used to sketch this surface, a *thermodynamic surface*, since none of the variables may equal zero.

31. The surface of a small hill can be roughly approximated by the equation $z(2x^2 + y^2 + 100) = 1500$, where the units are meters. Draw the surface of the hill and the contours for $z = 3$ m, $z = 6$ m, $z = 9$ m, $z = 12$ m, and $z = 15$ m.

32. Sketch the graph of $x^2 + y^2 - 2y = 0$ in three dimensions and in two dimensions.

10–3 Partial Derivatives

When we developed the concept of the derivative in Chapter 2, we showed that we were dealing with the rate of change of one variable with respect to another. However, only one independent variable was involved. In order to extend the derivative and its applications to functions of two (or more) variables, we find the derivative of the function with respect to one of the independent variables, while the other is held constant. If $z = f(x, y)$, this is equivalent to saying, for example, that for a specified value of y, we have the derivative of z with respect to x.

Therefore, if $z = f(x, y)$, and **y is kept constant,** *z becomes a function of x alone. The derivative of this function with respect to x is termed the* **partial derivative** *of z with respect to x. Similarly, if $z = f(x, y)$ and* **x is kept constant,** *the derivative of the resulting function with respect to y is the* **partial derivative** *of z with respect to y.*

For the function $z = f(x, y)$, the notations used for the partial derivative of z with respect to x include

$$\frac{\partial z}{\partial x}, \quad \frac{\partial f}{\partial x}, \quad f_x, \quad \frac{\partial}{\partial x} f(x, y), \quad f_x(x, y)$$

Similarly, $\partial z / \partial y$ denotes the partial derivative of z with respect to y. In speaking, this is often shortened to "the partial of z with respect to y."

Example A

If $z = 4x^2 + xy - y^2$, find $\partial z/\partial x$ and $\partial z/\partial y$.

In finding the partial derivative of z with respect to x, we treat y as a constant. Therefore

$$\boxed{\begin{array}{c}\text{Treat as}\\\text{constant}\end{array}}$$
$$\downarrow \quad \downarrow$$
$$z = 4x^2 + xy - y^2$$
$$\frac{\partial z}{\partial x} = 8x + y$$

Similarly, when finding the partial derivative of z with respect to y, we treat x as a constant. Thus

$$\boxed{\begin{array}{c}\text{Treat as}\\\text{constant}\end{array}}$$
$$\downarrow \quad \downarrow$$
$$z = 4x^2 + xy - y^2$$
$$\frac{\partial z}{\partial y} = x - 2y$$

Example B

If $y = e^x \cos t - e^{-x} \sin t$, find $\partial y / \partial x$ and $\partial y / \partial t$.

Treating t as a constant, we find $\partial y / \partial x$.

$$\frac{\partial y}{\partial x} = e^x \cos t + e^{-x} \sin t$$

Now, treating x as a constant, we find $\partial y / \partial t$.

$$\frac{\partial y}{\partial t} = -e^x \sin t - e^{-x} \cos t$$

Example C

If $z = \dfrac{x \ln y}{x^2 + 1}$, find $\partial z / \partial x$ and $\partial z / \partial y$.

$$\frac{\partial z}{\partial x} = \frac{(x^2 + 1)(\ln y) - (x \ln y)(2x)}{(x^2 + 1)^2} = \frac{(1 - x^2) \ln y}{(1 + x^2)^2}$$

$$\frac{\partial z}{\partial y} = \left(\frac{x}{x^2 + 1}\right)\left(\frac{1}{y}\right)$$

$$= \frac{x}{y(x^2 + 1)}$$

We note that in finding $\partial z / \partial x$ it is necessary to use the quotient rule, since x appears in both numerator and denominator. However, when finding $\partial z / \partial y$ the only derivative needed is that of $\ln y$.

Example D

For the function $f(x, y) = x^2 y \sqrt{2 + xy^2}$, find $f_y(2, 1)$.

The notation $f_y(2, 1)$ means the partial derivative of f with respect to y, evaluated for $x = 2$ and $y = 1$. Thus, first finding $f_y(x, y)$, we have

$$f(x, y) = x^2 y (2 + xy^2)^{1/2}$$

$$f_y(x, y) = x^2 y \left(\frac{1}{2}\right)(2 + xy^2)^{-1/2}(2xy) + (2 + xy^2)^{1/2}(x^2)$$

$$= \frac{x^3 y^2}{(2 + xy^2)^{1/2}} + x^2(2 + xy^2)^{1/2}$$

$$= \frac{x^3 y^2 + x^2(2 + xy^2)}{(2 + xy^2)^{1/2}}$$

$$= \frac{2x^2 + 2x^3 y^2}{(2 + xy^2)^{1/2}}$$

$$f_y(2, 1) = \frac{2(4) + 2(8)(1)}{(2 + 2)^{1/2}} = 12$$

To determine the geometric interpretation of a partial derivative, assume that $z = f(x, y)$ is the surface shown in Fig. 10–19. Choosing a point P on the surface, we then draw a plane through P parallel to the xz-plane. On this plane through P, the value of y is constant. The intersection of this plane and the surface is the curve as indicated. *The partial derivative of z with respect to x represents the slope of a line tangent to this curve.* When the values of the coordinates of point P are substituted into the expression, which is the partial derivative, it gives the slope of the tangent line at that point. In the same way, the partial derivative of z with respect to y, evaluated at P, gives the slope of the line tangent to the curve which is found from the intersection of the surface and the plane parallel to the yz-plane through P.

Figure 10–19

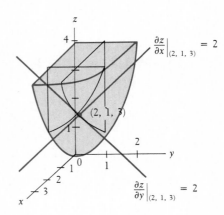

Figure 10–20

Example E

Find the slope of a line tangent to the surface $2z = x^2 + 2y^2$ and parallel to the xz-plane at the point $(2, 1, 3)$. Also find the slope of a line tangent to this surface and parallel to the yz-plane at the same point.

Finding the partial derivative of z with respect to x we have

$$2\frac{\partial z}{\partial x} = 2x \quad \text{or} \quad \frac{\partial z}{\partial x} = x$$

This derivative, evaluated at the point $(2, 1, 3)$, will give us the slope of the line tangent which is also parallel to the xz-plane. Thus, the first required slope is

$$\frac{\partial z}{\partial x}\bigg|_{(2, 1, 3)} = 2$$

The partial derivative of z with respect to y, evaluated at $(2, 1, 3)$ will give us the second required slope. Thus

$$\frac{\partial z}{\partial y} = 2y, \quad \frac{\partial z}{\partial y}\bigg|_{(2, 1, 3)} = 2$$

Therefore, both slopes are 2 (see Fig. 10–20).

The general interpretation of the partial derivative follows that of a derivative of a function with one independent variable. *The partial derivative $f_x(x_0, y_0)$ is the instantaneous rate of change of the function $f(x, y)$ with respect to x, with y held constant at the value of y_0.* This holds regardless of what the variables represent.

Applications of partial derivatives are found in various fields of technology. We present here an application from electricity, and others are found in the exercises.

Example F

A parallel electric circuit in a microwave transmitter has resistances r and R. The current through r can be found from

$$i = \frac{IR}{r + R}$$

where I is the total current for the two branches. Assuming that I is constant at 85.4 mA, find $\partial i / \partial r$ and evaluate it for $R = 0.150 \ \Omega$ and $r = 0.032 \ \Omega$.

Substituting for I and finding the partial derivative, we have the following.

$$i = \frac{85.4R}{r + R} = 85.4R (r + R)^{-1}$$

$$\frac{\partial i}{\partial r} = (-1)(85.4R)(r + R)^{-2}(1)$$

$$= \frac{-85.4R}{(r + R)^2}$$

$$\left. \frac{\partial i}{\partial r} \right|_{\substack{r=0.032 \\ R=0.150}} = \frac{-85.4(0.150)}{(0.032 + 0.150)^2} = -387 \ \text{mA}/\Omega$$

This result tells us that the current is decreasing at the rate of 387 mA per ohm of change in the smaller resistor at the instant when $r = 0.032 \ \Omega$, for a fixed value of $R = 0.150 \ \Omega$.

Since the partial derivatives $\partial f / \partial x$ and $\partial f / \partial y$ are functions of x and y, we can take partial derivatives of each of them. This gives rise to **partial derivatives of higher order,** in a manner similar to the higher derivatives of a function of one independent variable. *The possible* **second-order partial derivatives** *of a function $f(x, y)$ are*

$$\frac{\partial^2 f}{\partial x^2} = \frac{\partial}{\partial x}\left(\frac{\partial f}{\partial x}\right) \qquad \frac{\partial^2 f}{\partial y^2} = \frac{\partial}{\partial y}\left(\frac{\partial f}{\partial y}\right)$$

$$\frac{\partial^2 f}{\partial x \, \partial y} = \frac{\partial}{\partial x}\left(\frac{\partial f}{\partial y}\right) \qquad \frac{\partial^2 f}{\partial y \, \partial x} = \frac{\partial}{\partial y}\left(\frac{\partial f}{\partial x}\right)$$

Example G

Find the second-order partial derivatives of $z = x^3 y^2 - 3xy^3$.
First we find $\partial z / \partial x$ and $\partial z / \partial y$.

$$\frac{\partial z}{\partial x} = 3x^2 y^2 - 3y^3 \qquad \frac{\partial z}{\partial y} = 2x^3 y - 9xy^2$$

Therefore, we have the following second-order partial derivatives.

$$\frac{\partial^2 z}{\partial x^2} = \frac{\partial}{\partial x}\left(\frac{\partial z}{\partial x}\right) = 6xy^2$$

$$\frac{\partial^2 z}{\partial y^2} = \frac{\partial}{\partial y}\left(\frac{\partial z}{\partial y}\right) = 2x^3 - 18xy$$

$$\frac{\partial^2 z}{\partial x\, \partial y} = \frac{\partial}{\partial x}\left(\frac{\partial z}{\partial y}\right) = 6x^2 y - 9y^2$$

$$\frac{\partial^2 z}{\partial y\, \partial x} = \frac{\partial}{\partial y}\left(\frac{\partial z}{\partial x}\right) = 6x^2 y - 9y^2$$

In Example G we note that

$$\frac{\partial^2 z}{\partial x\, \partial y} = \frac{\partial^2 z}{\partial y\, \partial x} \tag{10-2}$$

This is true in general, so long as the function and the partial derivatives are continuous.

Example H

For $f(x, y) = \text{Arctan}\,\dfrac{y}{x^2}$, show that $\dfrac{\partial^2 f}{\partial x\, \partial y} = \dfrac{\partial^2 f}{\partial y\, \partial x}$.

Finding $\partial f / \partial x$ and $\partial f / \partial y$ we have

$$\frac{\partial f}{\partial x} = \frac{1}{1 + \left(\dfrac{y}{x^2}\right)^2}\left(\frac{-2y}{x^3}\right) = \frac{x^4}{x^4 + y^2}\left(-\frac{2y}{x^3}\right) = \frac{-2xy}{x^4 + y^2}$$

$$\frac{\partial f}{\partial y} = \frac{1}{1 + \left(\dfrac{y}{x^2}\right)^2}\left(\frac{1}{x^2}\right) = \frac{x^4}{x^4 + y^2}\left(\frac{1}{x^2}\right) = \frac{x^2}{x^4 + y^2}$$

Now finding $\partial^2 f / \partial x\, \partial y$ and $\partial^2 f / \partial y\, \partial x$ we have

$$\frac{\partial^2 f}{\partial x\, \partial y} = \frac{(x^4 + y^2)(2x) - x^2(4x^3)}{(x^4 + y^2)^2} = \frac{-2x^5 + 2xy^2}{(x^4 + y^2)^2}$$

$$\frac{\partial^2 f}{\partial y\, \partial x} = \frac{(x^4 + y^2)(-2x) - (-2xy)(2y)}{(x^4 + y^2)^2} = \frac{-2x^5 + 2xy^2}{(x^4 + y^2)^2}$$

We see that they are equal.

Exercises 10-3

In Exercises 1 through 24, find the partial derivative of the dependent variable or function with respect to each of the independent variables.

1. $z = 5x + 4x^2y$

2. $z = 3x^2y^3 - 3x + 4y$

3. $z = \dfrac{x^2}{y} - 2xy$

4. $f(x, y) = \dfrac{x + y}{x - y}$

5. $f(x, y) = xe^{2y}$

6. $z = 3y \cos 2x$

7. $f(x, y) = \dfrac{2 + \cos x}{1 - \sec 3y}$

8. $f(x, y) = \dfrac{\text{Arctan } y}{2 + x^2}$

9. $\phi = r\sqrt{1 + 2rs}$

10. $w = uv^2\sqrt{1 - u}$

11. $z = (x^2 + xy^3)^4$

12. $f(x, y) = (2xy - x^2)^5$

13. $z = \sin xy$

14. $y = \text{Arctan}\left(\dfrac{x}{t}\right)$

15. $y = \ln(r^2 + s)$

16. $u = e^{x + 2y}$

17. $f(x, y) = \dfrac{2\sin^3 2x}{1 - 3y}$

18. $f(x, y) = \dfrac{3x + \ln y}{x^2 + y^2}$

19. $z = \dfrac{\text{Arcsin } xy}{3 + x^2}$

20. $z = \dfrac{\sqrt{1 - \tan xy}}{xy + y^2}$

21. $z = \sin x + \cos xy - \cos y$

22. $t = 2re^{rs^2} - \tan(2r + s)$

23. $f(x, y) = e^x\cos xy + e^{-2x}\tan y$

24. $u = \ln\dfrac{y^2}{x - y} + e^{-x}(\sin y - \cos 2y)$

In Exercises 25 through 28, evaluate the indicated partial derivatives at the given points.

25. $z = 3xy - x^2 + y$, $\left.\dfrac{\partial z}{\partial x}\right|_{(1, -2, -9)}$

26. $z = x\sqrt{x^2 - y^2}$, $\left.\dfrac{\partial z}{\partial x}\right|_{(5, 3, 20)}$

27. $z = e^y \ln xy$, $\left.\dfrac{\partial z}{\partial y}\right|_{(e, 1, e)}$

28. $F(r, \phi) = r^2 \sin 4\phi$, $F_\phi\left(2, \dfrac{\pi}{4}\right)$

In Exercises 29 through 32, find all of the second partial derivatives.

29. $z = 2xy^3 - 3x^2y$

30. $z = \dfrac{x}{y} + e^x \sin y$

31. $f(x, y) = y \ln(x + 2y)$

32. $f(x, y) = \dfrac{2 + \cos y}{1 + x^2}$

In Exercises 33 through 48, solve the given problems.

33. Find the slope of a line tangent to the surface $z = 9 - x^2 - y^2$ and parallel to the yz-plane which passes through $(1, 2, 4)$. Repeat the instructions for the line through $(2, 2, 1)$. Draw an appropriate figure.

34. Find $\partial z/\partial y$ for the function $z = 4x^2 - 8$. Explain your result. Draw an appropriate figure.

35. In quality testing, a rectangular sheet of vinyl is stretched. Set up the length of the diagonal d of the sheet as a function of its sides x and y. Find the rate of change of d with respect to x for $x = 6.50$ ft if y remains constant at 4.75 ft.

Figure 10–21

Figure 10–22

36. A metallic machine part contracts while cooling. It is in the shape of a hemisphere attached to a cylinder, as shown in Fig. 10–21. The volume is $V = \frac{1}{3}\pi r^2(2r + 3h)$. Find the rate of change of volume with respect to r when $r = 2.65$ cm and $h = 4.20$ cm.

37. Two resistors R_1 and R_2, placed in parallel, have a combined resistance R_T given by

$$\frac{1}{R_T} = \frac{1}{R_1} + \frac{1}{R_2}$$

Find $\partial R_T/\partial R_1$.

38. For a gas, the isothermal bulk modulus B, a measure of compressibility, is given by

$$B = -V\left(\frac{\partial P}{\partial V}\right)$$

If the pressure, volume, and temperature are related by the van der Waals equation

$$\left(P + \frac{a}{V^2}\right)(V - b) = RT$$

where a, b, and R are constants, find the expression for B.

39. Two masses M and m are attached as shown in Fig. 10–22. If $M > m$, the downward acceleration a of mass M is given by

$$a = \frac{M - m}{M + m}g$$

where g is the acceleration due to gravity. Show that

$$M\frac{\partial a}{\partial M} + m\frac{\partial a}{\partial m} = 0$$

40. If an observer and a source of sound are moving towards or away from each other, the observed frequency of sound is different from that emitted. This is known as the *Doppler effect*. The equation relating the frequency f_o the observer hears and the frequency f_s emitted by the source (a constant) is

$$f_o = f_s\left(\frac{v + v_o}{v - v_s}\right)$$

where v is the velocity of sound in air (a constant), v_o is the velocity of the observer, and v_s is the velocity of the source. Show that

$$f_s\frac{\partial f_o}{\partial v_s} = f_o\frac{\partial f_o}{\partial v_o}$$

41. The *mutual conductance,* measured in reciprocal ohms, of a certain electronic device is defined as $g_m = \partial i_b/\partial e_c$. Under certain circumstances, the current i_b, measured in microamperes, is given by $i_b = 50(e_b + 5e_c)^{1.5}$. Find g_m when $e_b = 200$ V and $e_c = -20$ V.

42. The *amplification factor* of the electronic device of Exercise 41 is defined as $\mu = -\partial e_b/\partial e_c$. For the device of Exercise 41, under the given conditions, find μ.

43. The *coefficient of linear expansion* α of a wire whose length is a function of the tension and temperature is given by

$$\alpha = \frac{1}{L}\left(\frac{\partial L}{\partial T}\right)$$

where L is the length and T is the temperature. If L is a function of T and the tension F given by the relation

$$L = L_0 + k_1 F + k_2 T + k_3 FT^2$$

where k_1, k_2, k_3, and L_0 are constants, find the expression for α.

44. The fundamental frequency of vibration of a string varies directly as the square root of the tension T and inversely as the length L. If a string 60 cm long is under a tension of 65 N and has a fundamental frequency of 30 Hz, find the partial derivative of f with respect to T and evaluate it for the given values.

45. An *isothermal process* is one during which the temperature does not change. If the pressure, volume, and temperature of an ideal gas are related by the equation

$$pV = nRT$$

where n and R are constants, find the expression for $\partial p / \partial V$, which would be the rate of change of pressure with respect to volume for an isothermal process.

46. For the ideal gas equation of Exercise 45, show that

$$\left(\frac{\partial V}{\partial T}\right)\left(\frac{\partial T}{\partial p}\right)\left(\frac{\partial p}{\partial V}\right) = -1$$

47. The temperature u in a metal bar depends on the distance x from one end and the time t. Show that $u(x, t) = 5e^{-t}\sin 4x$ satisfies *the one-dimensional heat conduction equation*

$$\frac{\partial u}{\partial t} = k\frac{\partial^2 u}{\partial x^2}$$

where k is called the *diffusivity*. In this case, $k = \frac{1}{16}$.

48. The displacement y at any point in a taut, flexible string depends on the distance x from one end of the string and the time t. Show that $y(x, t) = 2\sin 2x \cos 4t$ satisfies *the wave equation*

$$\frac{\partial^2 y}{\partial t^2} = a^2\frac{\partial^2 y}{\partial x^2}$$

with $a = 4$.

10–4 Certain Applications of Partial Derivatives

We noted some important areas of application of partial derivatives in the last section. However, just as we found that we can apply the derivative to problems such as related rates, applied maximum and minimum problems, and error calculations by the use of differentials, we can also solve similar problems by the use of partial derivatives. In this section we shall show how to calculate the differential of a function of two variables and apply it to certain applied situations. Then we will discuss applied maximum and minimum problems for functions of two variables.

In Section 4–1 we defined the differential of a function of one variable as

$$dy = f'(x)\,dx$$

In a similar manner, *we define the* **total differential** *of a function* $z = f(x, y)$ *to be*

$$dz = \frac{\partial z}{\partial x}\,dx + \frac{\partial z}{\partial y}\,dy \qquad\qquad (10\text{–}3)$$

It will be noted that this definition is consistent with our original definition of the differential, in that if y is constant, $dz = (\partial z/\partial x)\,dx$. Also, if x is constant, $dz = (\partial z/\partial y)\,dy$. Therefore, each term measures the differential of z with respect to one of the independent variables, assuming the other is constant.

Example A

Find the expression for the total differential of the function $z = x^3y + x^2y^2 - 3xy^3$:

$$\frac{\partial z}{\partial x} = 3x^2y + 2xy^2 - 3y^3$$

$$\frac{\partial z}{\partial y} = x^3 + 2x^2y - 9xy^2$$

Therefore,

$$dz = (3x^2y + 2xy^2 - 3y^3)\,dx + (x^3 + 2x^2y - 9xy^2)\,dy$$

Example B

Find the total differential of $z = e^{-x}\cos 2y - \ln xy$.

First finding $\partial z/\partial x$ and $\partial z/\partial y$, we have

$$\frac{\partial z}{\partial x} = (-e^{-x})\cos 2y - \frac{y}{xy} = -e^{-x}\cos 2y - \frac{1}{x}$$

$$\frac{\partial z}{\partial y} = e^{-x}(-\sin 2y)(2) - \frac{x}{xy} = -2e^{-x}\sin 2y - \frac{1}{y}$$

Therefore,

$$dz = -\left(e^{-x}\cos 2y + \frac{1}{x}\right)dx - \left(2e^{-x}\sin 2y + \frac{1}{y}\right)dy$$

We found that we could apply the differential of a function of one variable to finding changes in the value of the function for small changes in the independent variable. This led to applied problems in error calculations. We may also apply the total differential of a function of two variables in the same manner. The following examples illustrate the method.

Example C

The power delivered to a resistor in an electric circuit is given by $P = i^2R$. Find the approximate change in power if the current changes from 4.0 to 4.1 A, and the resistance from 22.0 to 22.4 Ω.

By finding dP and evaluating the expression for $i = 4.0$, $di = 0.1$, $R = 22.0$, and $dR = 0.4$, we will have the desired result:

$$dP = \frac{\partial P}{\partial i}\, di + \frac{\partial P}{\partial R}\, dR$$

$$= (2iR)\, di + (i^2)\, dR$$

Evaluating, we have

$$dP = [2(4.0)(22.0)](0.1) + (4.0)^2(0.4)$$

$$= 17.6 + 6.4 = 24.0 \text{ W.}$$

Therefore, the approximate change in the power is 24 W.

Example D

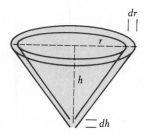

Figure 10-23

A conical funnel, with a very small opening, was to be constructed such that the diameter across the top was 10.0 cm and the depth 6.0 cm. However, in the construction, the diameter was made only 9.8 cm and the depth 5.8 cm. What was the resulting percentage error in the volume of liquid the funnel will hold? See Fig. 10-23.

The volume of a cone as a function of its radius r and height h is

$$V = \frac{1}{3}\pi r^2 h$$

Finding dV, we have

$$dV = \left(\frac{2}{3}\pi rh\right) dr + \left(\frac{1}{3}\pi r^2\right) dh$$

Using the values $r = 5.0$ cm, $dr = -0.1$ cm, $h = 6.0$ cm, and $dh = -0.2$ cm, we have

$$dV = \frac{2}{3}\pi(5.0)(6.0)(-0.1) + \frac{1}{3}\pi(25)(-0.2)$$

$$= -2.0\pi - \frac{1}{3}(5.0\pi) = 11.5 \text{ cm}^3$$

Calculating the intended volume V, we have

$$V = \frac{1}{3}\pi(25)(6) = 50\pi = 157 \text{ cm}^3$$

The percentage error in the volume is

$$100\left(\frac{dV}{V}\right) = 100\left(\frac{11.5}{157}\right) = 7.3\%$$

We will now consider the problem of finding the relative maximum value or relative minimum value of a function of two variables. Considering the geometric interpretation of the relative maximum value of a function of two variables, we observe the surface shown in Fig. 10–24(a). Point M is a relative maximum point, which means that the value of the function at M is greater than at any other point very near M in any direction. It does not mean that the value of the function is greater at M than at any other point *anywhere* on the surface.

At the relative maximum point M, a line tangent to the surface, *in any direction*, must be horizontal. This means that *at a **relative maximum** point*

$$\frac{\partial z}{\partial x} = 0 \quad \text{and} \quad \frac{\partial z}{\partial y} = 0 \qquad \qquad (10\text{–}4)$$

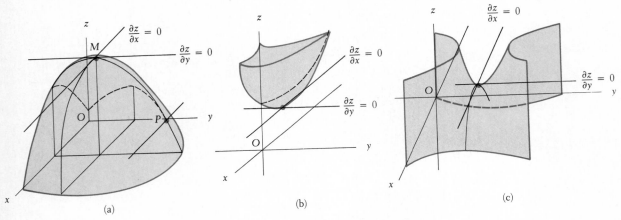

Figure 10–24

From Fig. 10–24(b) we see that Eqs. (10–4) also hold at a relative minimum point. However, it should be noted that it is not sufficient that just one of the partial derivatives in Eqs. (10–4) equal zero to have a relative maximum or relative minimum point. This can be seen by examining point P in Fig. 10–24(a), where $\partial z/\partial x = 0$, but obviously $\partial z/\partial y \neq 0$.

It is possible that Eqs. (10–4) are satisfied but that there is neither a relative maximum nor a relative minimum. See Fig. 10–24(c), where a **saddle point** is illustrated. Therefore, it is clear that Eqs. (10–4) do not provide a complete analysis. Although the proof is beyond the scope of this book, we now state the necessary conditions, which are analogous to the second-derivative test for functions of one independent variable.

If, for a function $f(x, y)$ at a point (a, b),

$$\frac{\partial f}{\partial x} = 0 \quad \text{and} \quad \frac{\partial f}{\partial y} = 0$$

and also

$$D = \frac{\partial^2 f}{\partial x^2}\frac{\partial^2 f}{\partial y^2} - \left(\frac{\partial^2 f}{\partial x\,\partial y}\right)^2$$

then there is a **relative maximum** at (a, b) if $D > 0$ and $\partial^2 f/\partial x^2 < 0$, or a **relative minimum** at (a, b) if $D > 0$ and $\partial^2 f/\partial x^2 > 0$. If $D < 0$, there is neither a maximum nor a minimum, and if $D = 0$, the test fails (no conclusion can be drawn).

Example E

Examine the function $z = x^2 + 2xy + 2y^2 - 4x$ for relative maximum and minimum points.

First, finding the partial derivatives, we have

$$\frac{\partial z}{\partial x} = 2x + 2y - 4$$

and

$$\frac{\partial z}{\partial y} = 2x + 4y$$

Setting these expressions equal to zero, we obtain the equations

$$2x + 2y - 4 = 0$$
$$2x + 4y = 0$$

 Since these equations must hold at the same point, we solve them simultaneously. This gives us $x = 4$ and $y = -2$.

Now finding the second derivatives, we have

$$\frac{\partial^2 z}{\partial x^2} = 2, \qquad \frac{\partial^2 z}{\partial y^2} = 4, \qquad \frac{\partial^2 z}{\partial x\,\partial y} = 2$$

This means that

$$D = 2(4) - 2^2 = 4$$

Since $D > 0$ and $\partial^2 z/\partial x^2 > 0$, there is a relative minimum point at $(4, -2, -8)$.

We shall now show two examples of applied maximum and minimum problems involving functions of two variables.

Example F

A computer analysis shows that the pressure P, in pascals, created by a press at each point of a square plate 8 cm on a side is given by $P = 6xy - x^3 - y^3 + 700$. Find the greatest pressure at any point, if distances are measured from the lower left corner, as shown in Fig. 10–25.

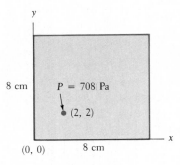

Figure 10–25

Finding the partial derivatives of P, we have

$$\frac{\partial P}{\partial x} = 6y - 3x^2, \qquad \frac{\partial P}{\partial y} = 6x - 3y^2$$

Setting each of these equal to zero, we have

$$6y - 3x^2 = 0, \quad \text{or} \quad 2y - x^2 = 0$$
$$6x - 3y^2 = 0, \quad \text{or} \quad 2x - y^2 = 0$$

Solving the first equation for y and substituting into the second equation gives us

$$y = \frac{1}{2}x^2$$
$$2x - \left(\frac{1}{2}x^2\right)^2 = 0$$
$$8x - x^4 = 0$$
$$x(8 - x^3) = 0$$
$$x = 0, 2$$

For $x = 0$, $y = 0$, and for $x = 2$, $y = 2$. Now finding the second derivatives, we have

$$\frac{\partial^2 P}{\partial x^2} = -6x, \qquad \frac{\partial^2 P}{\partial y^2} = -6y, \qquad \frac{\partial^2 P}{\partial x\, \partial y} = 6$$

At $(0, 0)$, $D = -36$, which means that $(0, 0)$ is neither a relative maximum nor a relative minimum. Actually, $(0, 0)$ is a corner of the plate. For $y = 0$, $\partial P / \partial x = -3x^2$, which means that the pressure is decreasing near $(0, 0)$.
    At $(2, 2)$, $D = (-12)(-12) - 6^2 = 108$ and $\partial^2 P / \partial x^2 = -12$. This means that there is a relative maximum value of $P = 708$ Pa at $(2, 2)$.

Example G

A rectangular box is to be made of material such that the base costs twice as much per square inch as the sides and top. Find the dimensions of the box of minimum cost if it is to hold 324 in.^{3}.

Assuming the length, width, and height of the box to be l, w, and h, respectively, we may write the equations for the volume and cost in terms of these quantities. The equation for the volume is

$$324 = lwh \tag{1}$$

Since the cost per square inch was not actually given, let us assume that the cost for the sides and top is a cents/in.^{2}, which means that of the base is $2a$ cents/in.^{2}. Thus, the expression for the cost is

$$C = 2a(lw) + a(lw + 2wh + 2lh)$$
$$= a(3lw + 2wh + 2lh) \tag{2}$$

Solving Eq. (1) for h and substituting this into Eq. (2), we may express C as a function of l and w:

$$C = a\left(3lw + \frac{648}{l} + \frac{648}{w}\right) \tag{3}$$

Taking the partial derivatives of C with respect to l and w, we may find the values of l and w which give the minimum cost. From these values we then can find h:

$$\frac{\partial C}{\partial w} = a\left(3l - \frac{648}{w^2}\right)$$
$$\frac{\partial C}{\partial l} = a\left(3w - \frac{648}{l^2}\right) \tag{4}$$

Setting each of these expressions equal to zero, we obtain the following equations:

$$l = \frac{216}{w^2}, \qquad w = \frac{216}{l^2}$$

Substituting the expression for l into the solution for w we have

$$w = \frac{216}{\left(\frac{216}{w^2}\right)^2} = \frac{w^4}{216}$$

$$w^3 = 216, \qquad w = 6 \text{ in.}$$

Substituting, we find $l = 6$ in. and $h = 9$ in. Equation (3) shows that the maximum cost is unbounded, because of the l and w in the denominators, which means the solution we obtained must represent a minimum. Also, the second-derivative test will demonstrate that the solution is a minimum.

Exercises 10–4

In Exercises 1 through 12, find the total differential of the given functions.

1. $z = 2x^2 - y^2 + 3x$ 2. $z = xy - 3yx^2 + y^3$ 3. $z = xe^y - y^2$

4. $z = x^3 - y \ln x$ 5. $z = x(y - 2x^2)^5$ 6. $z = y^2\sqrt{1 - 2xy}$

7. $z = \sin xy - y \cos x$ 8. $z = x \tan 2y - x^3$ 9. $z = \dfrac{x - 3y^2}{1 - \sin y}$

10. $z = \dfrac{\sec x^2}{x - y}$ 11. $z = y \operatorname{Arctan} \dfrac{y}{x^2}$ 12. $z = e^{2xy} \ln(x^2 + y^2)$

In Exercises 13 through 20, examine the given functions for relative maximum and minimum points.

13. $z = x^2 + y^2 - 2x - 6y + 10$ 14. $z = x^2 + xy + y^2 + 3x - 3y + 1$

15. $z = x^2 + xy + y^2 - 3y + 2$ 16. $z = x^4 - 4x + y^2$

17. $z = 2y - xy + x^3$ 18. $z = xy + \dfrac{1}{x} + \dfrac{8}{y}$

19. $z = x^2 + 4 - 4x \cos y$ $(0 \le x < 2\pi, 0 \le y < 2\pi)$

20. $z = x - y \ln x$

In Exercises 21 through 34, solve the given problems.

21. The centripetal acceleration of an object moving in a circular path is given by $a = r\omega^2$, where r is the radius of the circle and ω is the angular velocity. Find the approximate change in the centripetal acceleration if the radius changes from 10.0 cm to 10.3 cm and the angular velocity changes from 400 to 406 rad/min.

22. The current i through a resistor R which has a voltage V across it is $i = V/R$. Find the approximate change in current, in amperes, if the voltage changes from 220 V to 225 V and the resistance changes from 20.0 Ω to 21.0 Ω.

23. The index of refraction of a medium is defined as $n = \sin i/\sin r$, where i is the angle of incidence of light on the surface of the medium and r is the angle of refraction. In a particular experiment, n was calculated from values of $i = 45°$ and $r = 30°$. Then an error was discovered, and each angle was 1° too high. What was the approximate percentage error in the calculated value of n?

24. The electric current in a circuit containing a variable resistor R and an inductance of 10 H is a function of R and the time t given by $i = 10(1 - e^{-(Rt/10)})$. Find the approximate percentage change in current as R changes from 2.00 Ω to 2.04 Ω between $t = 10.0$ s and $t = 11.0$ s.

25. Two sides and the included angle of a triangular machine part were measured to be 2.30 cm, 2.10 cm, and 90°, respectively. The length of the third side was calculated from these values. It was then discovered that each of the sides should have been 0.10 cm more. What was the percentage error in the calculation of the third side?

26. If the radius and height of a right circular cylindrical container were measured to be 3.00 ft and 5.00 ft, and then the radius was found to be 3.10 ft and the height 5.20 ft, what was the error in the original value of the volume?

27. Find three positive numbers, whose sum is 120, such that their product is a maximum.

28. Find the point of the plane $2x + 3y - z = 12$ that is closest to the origin. (See Example D of Section 3–7.)

29. The pressure P, in pascals, at each point of a circular plastic sheet of radius 5 cm is given by $P = x^4 + y^2 - 4x + 20$. What is the minimum pressure and where is it located, if distances are measured from the center of the sheet?

30. A flat plate is heated such that the temperature T at any point (x, y) is

$$T = x^2 + 2y^2 - x$$

Find the temperature at the coldest point.

31. An open rectangular box is to be made to contain 500 in.^{3}. What dimensions will require the least material in building the box?

32. A long metal strip 9 in. wide is to be made into a trough by bending up equal edges of length x, making equal angles θ with the bottom. Find the width of the bottom and the angle between the bottom and the edges if the cross-sectional area A of the trough is a maximum. See Fig. 10–26.

9 − 2x

Figure 10–26

33. If x, y, and z are functions of the time t, and we divide each term of Eq. (10–3) by the differential of t, dt, we obtain an equation for the **total derivative** of z with respect to t. Thus, the total derivative is

$$\frac{dz}{dt} = \frac{\partial z}{\partial x}\frac{dx}{dt} + \frac{\partial z}{\partial y}\frac{dy}{dt} \tag{10–5}$$

Using Eq. (10–5), find dz/dt if

$$z = x^2 - xy + y^2, \qquad x = 1 + t^2, \qquad y = 1 - t^2$$

34. If the radius and height of a right circular cylinder are increasing at the rate of 0.05 ft/s and 0.20 ft/s, respectively, use Eq. (10–5) to find the rate at which the volume is increasing when $r = 6.00$ ft and $h = 3.00$ ft.

10–5 Double Integrals

We now turn our attention to integration in the case of a function of two variables. The analysis has similarities to that of partial differentiation, in that an operation is performed while holding one of the independent variables constant.

If $z = f(x, y)$ and we wish to integrate with respect to x and y, we first consider either x or y constant and integrate with respect to the other. After this integral is evaluated, we then integrate with respect to the variable first held constant. We shall now define this type of integral and then give an appropriate geometric interpretation.

*If $z = f(x, y)$, the **double integral** of the function over x and y is defined as*

$$\int_a^b \left[\int_{g(x)}^{G(x)} f(x, y)\, dy \right] dx$$

It will be noted that the limits on the inner integral are functions of x and those on the outer integral are explicit values of x. In performing the integration, x is held constant while the inner integral is found and evaluated. This results in a function of x only. This function is then integrated and evaluated.

It is customary not to include the brackets in stating a double integral. Therefore we write

$$\int_a^b \left[\int_{g(x)}^{G(x)} f(x, y)\, dy \right] dx = \int_a^b \int_{g(x)}^{G(x)} f(x, y)\, dy\, dx \qquad (10\text{--}6)$$

Example A

Evaluate $\int_0^1 \int_{x^2}^x xy\, dy\, dx$.

Integrating the inner integral with y as the variable and x as a constant, we have

Treat as constant

$$\int_{x^2}^x xy\, dy = \left. \left(x\frac{y^2}{2} \right) \right|_{x^2}^x = x\left(\frac{x^2}{2} - \frac{x^4}{2} \right) = \frac{1}{2}(x^3 - x^5)$$

This means

$$\int_0^1 \int_{x^2}^x xy\, dy\, dx = \int_0^1 \frac{1}{2}(x^3 - x^5)\, dx = \left. \frac{1}{2} \left(\frac{x^4}{4} - \frac{x^6}{6} \right) \right|_0^1$$

$$= \frac{1}{2}\left(\frac{1}{4} - \frac{1}{6} \right) - \frac{1}{2}(0) = \frac{1}{24}$$

It is also common to have an integral which is integrated over x first and then over y. The following example illustrates such a case.

Example B

Evaluate $\int_1^4 \int_0^{4y} \sqrt{2x + y}\, dx\, dy$.

 Since the inner differential is dx (the inner limits must then be functions of y, which may be constant), we first integrate with x as the variable and y as a constant.

$$\int_1^4 \int_0^{4y} \sqrt{2x + y}\, dx\, dy = \int_1^4 \left[\frac{1}{2} \int_0^{4y} (2x + y)^{1/2} 2\, dx \right] dy$$

$$= \int_1^4 \left. \left[\frac{1}{3}(2x + y)^{3/2} \right] \right|_0^{4y} dy$$

$$= \frac{1}{3} \int_1^4 \left[(9y)^{3/2} - (y)^{3/2} \right] dy$$

$$= \frac{1}{3} \int_1^4 (27y^{3/2} - y^{3/2})\, dy = \frac{26}{3} \int_1^4 y^{3/2}\, dy$$

$$= \left. \frac{26}{3} \cdot \frac{2}{5} y^{5/2} \right|_1^4 = \frac{52}{15}(32 - 1) = \frac{1612}{15}$$

Example C

Evaluate: $\displaystyle\int_0^{\pi/2} \int_0^{\sin y} e^{2x}\cos y \; dx \; dy$

Here we must first integrate with x as the variable and y as a constant.

$$\int_0^{\pi/2} \int_0^{\sin y} e^{2x}\cos y \; dx \; dy = \int_0^{\pi/2} \left[\frac{1}{2} e^{2x}\cos y \right]_0^{\sin y} dy$$

$$= \frac{1}{2} \int_0^{\pi/2} (e^{2\sin y}\cos y - \cos y) \; dy$$

$$= \frac{1}{2} \left[\frac{1}{2} e^{2\sin y} - \sin y \right]_0^{\pi/2}$$

$$= \frac{1}{2} \left(\frac{1}{2} e^2 - 1 \right) - \frac{1}{2}\left(\frac{1}{2} - 0 \right)$$

$$= \frac{1}{4} e^2 - \frac{1}{2} - \frac{1}{4} = \frac{1}{4}(e^2 - 3) = 1.097$$

To understand the geometric interpretation of a double integral, consider the surface shown in Fig. 10–27(a). Now consider an **element of volume** (as shown), of dimensions dx, dy, and z, which extends from the xy plane to the surface. We now make x a constant and sum (integrate) these elements of volume from the left boundary, $y = g(x)$, to the right boundary, $y = G(x)$. In this way we have the volume of the vertical slice as a function of x, shown as in Fig. 10–27(b). Now, by summing (integrating) the volumes of these slices from $x = a$ ($x = 0$ in the figure) to $x = b$, we have the complete volume, as shown in Fig. 10–27(c).

Therefore, *we may interpret a double integral as the* **volume under a surface,** in the same way as the integral was interpreted as the area of a plane figure. It will be noted that this is not necessarily a volume of revolution, as discussed in Section 5–3. This means we may find the volume of a more general figure than previously possible. The following examples illustrate the use of double integrals to find volumes.

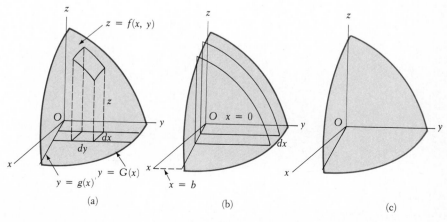

(a)   (b)   (c)

Figure 10–27

Example D

Find the volume in the first octant under the plane

$$x + 2y + 4z - 8 = 0$$

See Fig. 10–28.

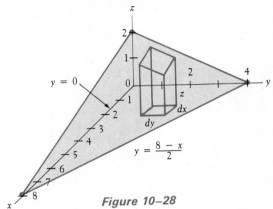

Figure 10–28

This figure is a tetrahedron, for which $V = \frac{1}{3}Bh$. Assuming the base is in the xy-plane, $B = \frac{1}{2}(4)(8) = 16$, and $h = 2$. Therefore, $V = \frac{1}{3}(16)(2) = \frac{32}{3}$ cubic units. We shall use this value to check the one we find by double integration.

To find $z = f(x, y)$, we solve the given equation for z. Thus

$$z = \frac{8 - x - 2y}{4}$$

Next, we must find the limits on y and x. Choosing to integrate over y first, we see that y goes from $y = 0$ to $y = (8 - x)/2$. This last limit is the trace of the surface in the xy-plane. Next we note that x goes from $x = 0$ to $x = 8$. Therefore, we set up and evaluate the integral:

$$\begin{aligned}
V &= \int_0^8 \int_0^{(8-x)/2} \left(\frac{8 - x - 2y}{4}\right) dy\, dx \\
&= \int_0^8 \left[\frac{1}{4}\left(8y - xy - y^2\right)\Big|_0^{(8-x)/2}\right] dx \\
&= \frac{1}{4}\int_0^8 \left[8\left(\frac{8 - x}{2}\right) - x\left(\frac{8 - x}{2}\right) - \left(\frac{8 - x}{2}\right)^2\right] dx \\
&= \frac{1}{4}\int_0^8 \left(32 - 4x - 4x + \frac{x^2}{2} - 16 + 4x - \frac{x^2}{4}\right) dx \\
&= \frac{1}{4}\int_0^8 \left(16 - 4x + \frac{x^2}{4}\right) dx = \frac{1}{4}\left(16x - 2x^2 + \frac{x^3}{12}\right)\Big|_0^8 \\
&= \frac{1}{4}\left(128 - 128 + \frac{512}{12}\right) = \frac{1}{4}\left(\frac{128}{3}\right) = \frac{32}{3} \text{ cubic units}
\end{aligned}$$

We see that the values by the two different methods agree.

Example E

Find the volume above the xy-plane, below the surface $z = xy$, and enclosed by the cylinder $y = x^2$ and the plane $y = x$.

Constructing the figure, shown in Fig. 10–29, we now note that $z = xy$ is the desired function of x and y. Integrating over y first, the limits on y are $y = x^2$ to $y = x$. The corresponding limits on x are $x = 0$ to $x = 1$. Thus, the integral is

$$V = \int_0^1 \int_{x^2}^x xy \; dy \; dx$$

This integral has already been evaluated in Example A of this section, and we can now see the geometric interpretation of that integral. This means that the required volume is $\frac{1}{24}$ cubic unit.

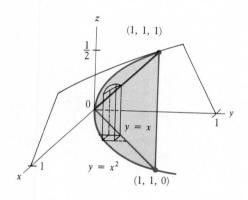

Figure 10–29

Figure 10–30

Example F

Find the volume in the first octant under the surface

$$z = 4 - x^2 - y^2$$

and between the cylinder $x^2 = 3y$ and the plane $y = 1$ (see Fig. 10–30).

Setting up the integration such that we integrate over x first, we have

$$V = \int_0^1 \int_0^{\sqrt{3y}} (4 - x^2 - y^2) \; dx \; dy$$

(*continued on next page*)

Now performing the integration, we have

$$V = \int_0^1 \int_0^{\sqrt{3y}} (4 - x^2 - y^2)\, dx\, dy$$

$$= \int_0^1 \left[4x - \frac{x^3}{3} - y^2 x \right]_0^{\sqrt{3y}} dy$$

$$= \int_0^1 (4\sqrt{3y} - \sqrt{3}y^{3/2} - \sqrt{3}y^{5/2})\, dy$$

$$= \sqrt{3}\left[4\left(\frac{2}{3}\right)y^{3/2} - \frac{2}{5}y^{5/2} - \frac{2}{7}y^{7/2} \right]\Big|_0^1$$

$$= \sqrt{3}\left(\frac{8}{3} - \frac{2}{5} - \frac{2}{7}\right) = \frac{208\sqrt{3}}{105} = 3.431 \text{ cubic units}$$

If we integrate over y first, the integral is

$$V = \int_0^{\sqrt{3}} \int_{x^2/3}^1 (4 - x^2 - y^2)\, dy\, dx$$

Integrating and evaluating this integral, we arrive at the same result,

$$V = \frac{208\sqrt{3}}{105} = 3.431 \text{ cubic units}$$

Exercises 10–5

In Exercises 1 through 16, evaluate the given double integrals.

1. $\int_2^4 \int_0^1 xy^2\, dx\, dy$

2. $\int_0^2 \int_0^1 \frac{y}{(xy+1)^2}\, dx\, dy$

3. $\int_0^1 \int_0^x 2y\, dy\, dx$

4. $\int_0^2 \int_1^x 2x\, dy\, dx$

5. $\int_1^2 \int_0^{y^2} xy^2\, dx\, dy$

6. $\int_0^4 \int_1^{\sqrt{y}} (x - y)\, dx\, dy$

7. $\int_0^1 \int_0^{\sqrt{1-x^2}} y\, dy\, dx$

8. $\int_4^9 \int_0^x \sqrt{x-y}\, dy\, dx$

9. $\int_0^{\pi/6} \int_{\pi/3}^y \sin x\, dx\, dy$

10. $\int_0^{\sqrt{3}} \int_{x^2/3}^1 (4 - x^2 - y^2)\, dy\, dx$

11. $\int_1^e \int_1^y \frac{1}{x}\, dx\, dy$

12. $\int_{-1}^1 \int_1^{e^x} \frac{1}{xy}\, dy\, dx$

13. $\int_1^2 \int_0^x yx^3 e^{xy^2}\, dy\, dx$

14. $\int_0^{\pi/6} \int_0^1 y \sin x\, dy\, dx$

15. $\int_0^{\ln 3} \int_0^x e^{2x+3y}\, dy\, dx$

16. $\int_0^{1/2} \int_y^{y^2} \frac{dx\, dy}{\sqrt{y^2 - x^2}}$

In Exercises 17 through 26, find the indicated volumes by double integration.

17. The first-octant volume under the plane $x + y + z - 4 = 0$.

18. The first-octant volume under the surface $z = y^2$ and bounded by the planes $x = 2$ and $y = 3$.

19. The volume above the xy-plane and under the surface $z = 4 - x^2 - y^2$.

20. The volume above the xy-plane, below the surface $z = x^2 + y^2$, and inside the cylinder $x^2 + y^2 = 4$.

21. The first-octant volume bounded by the xy-plane, the planes $x = y$, $y = 2$, and $z = 2 + x^2 + y^2$.

22. The volume bounded by the planes $x = 0$, $z = 0$, $x + 3y + 2z - 6 = 0$, and $2x = y$.

23. The first-octant volume under the plane $z = x + y$ and inside the cylinder $x^2 + y^2 = 9$.

24. The volume above the xy-plane and bounded by the cylinders $x = y^2$, $y = 8x^2$, and $z = x^2 + 1$. Integrate over y first and then check by integrating over x first.

25. A wedge is to be made in the shape shown in Fig. 10–31 (all vertical cross sections are equal right triangles). By double integration find the volume of the wedge.

26. A circular piece of pipe is cut as shown in Fig. 10–32. Find the volume within the pipe. (*Hint:* Place the z-axis along the axis of the pipe, with the base of the pipe in the xy-plane.)

In Exercises 27 and 28, draw the appropriate figures.

27. Draw an appropriate figure indicating a volume which is found from the integral

$$\int_0^{1/2} \int_{x^2}^1 (4 - x - 2y)\, dy\, dx$$

28. Repeat Exercise 27 for the integral

$$\int_1^2 \int_0^{2-y} \sqrt{1 + x^2 + y^2}\, dx\, dy$$

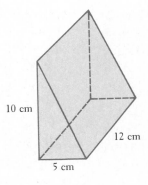

10 cm

12 cm

5 cm

Figure 10–31

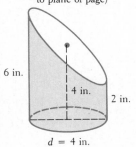

(planes of cuts are perpendicular to plane of page)

6 in.

4 in.

2 in.

$d = 4$ in.

Figure 10–32

10–6 *Centroids and Moments of Inertia by Double Integration*

Among the important applications of double integration are those of finding centroids and moments of inertia of plane areas. In this section we shall show how double integration is applied in these cases.

Considering the area shown in Fig. 10–33, we see that *we may determine the value of this area from the double integral*

$$A = \int_a^b \int_{g(x)}^{G(x)} dy \, dx \tag{10–7}$$

This is the case since, by summing (integrating) the elements of area $dy \, dx$ over y while holding x constant, we find the area of the vertical strip. Then by summing (integrating) from $x = a$ to $x = b$ we find the total area.

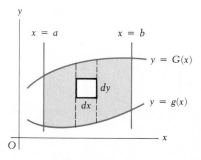

Figure 10–33

In Section 5–4, we showed that the moment of an element of area with respect to the y-axis is $kx \, dA$, where k is the mass per unit area, x is the moment arm, and dA is the element of area. Recalling that the x-coordinate of the centroid is the moment with respect to the y-axis, divided by the area, *we have*

$$\bar{x} = \frac{\int_a^b \int_{g(x)}^{G(x)} x \, dy \, dx}{A} \tag{10–8}$$

as the equation by which we may find the x-coordinate of the centroid by double integration.

Also, *because the moment of the element of area with respect to the x-axis is $ky \, dA$, the y-coordinate of the centroid is found from*

$$\bar{y} = \frac{\int_a^b \int_{g(x)}^{G(x)} y \, dy \, dx}{A} \tag{10–9}$$

For many problems, use of Eqs. (10–8) and (10–9) is more convenient than the equivalent method in Section 5–4. One advantage is that the integration is often simplified. The following example illustrates the use of Eqs. (10–8) and (10–9).

Example A

Find the centroid of the area bounded by the lines $y = 2x$, $y = 3 - x$, and $x = 0$ (see Fig. 10–34).

$$A = \int_0^1 \int_{2x}^{3-x} dy\, dx = \int_0^1 y \Big|_{2x}^{3-x} dx$$

$$= \int_0^1 (3 - 3x)\, dx = 3x - \frac{3x^2}{2}\Big|_0^1 = \frac{3}{2}$$

$$A\bar{x} = \int_0^1 \int_{2x}^{3-x} x\, dy\, dx = \int_0^1 y \Big|_{2x}^{3-x} x\, dx$$

$$= \int_0^1 (3x - 3x^2)\, dx = \frac{3x^2}{2} - x^3 \Big|_0^1 = \frac{1}{2}$$

$$A\bar{y} = \int_0^1 \int_{2x}^{3-x} y\, dy\, dx = \int_0^1 \frac{y^2}{2}\Big|_{2x}^{3-x} dx$$

$$= \frac{1}{2}\int_0^1 (9 - 6x - 3x^2)\, dx = \frac{1}{2}(9x - 3x^2 - x^3)\Big|_0^1 = \frac{5}{2}$$

Therefore, $\bar{x} = \frac{1}{2} \cdot \frac{2}{3} = \frac{1}{3}$ and $\bar{y} = \frac{5}{2} \cdot \frac{2}{3} = \frac{5}{3}$.

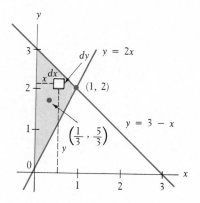

Figure 10–34

In many problems, integrating over x first is definitely preferable. The following example illustrates this type of problem.

Example B

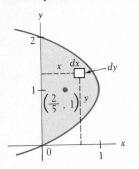

Figure 10–35

Find the centroid of the area bounded by $x = 2y - y^2$ and the y-axis (see Fig. 10–35).

$$A = \int_0^2 \int_0^{2y-y^2} dx\, dy = \int_0^2 (2y - y^2)\, dy = \left[y^2 - \frac{y^3}{3} \right]\Big|_0^2 = \frac{4}{3}$$

$$A\bar{x} = \int_0^2 \int_0^{2y-y^2} x\, dx\, dy = \int_0^2 \frac{x^2}{2}\Big|_0^{2y-y^2}\, dy$$

$$= \frac{1}{2} \int_0^2 (4y^2 - 4y^3 + y^4)\, dy$$

$$= \frac{1}{2}\left(\frac{4}{3}y^3 - y^4 + \frac{y^5}{5} \right)\Big|_0^2 = \frac{8}{15}$$

$$A\bar{y} = \int_0^2 \int_0^{2y-y^2} y\, dx\, dy = \int_0^2 (2y^2 - y^3)\, dy = \left(\frac{2}{3}y^3 - \frac{1}{4}y^4 \right)\Big|_0^2 = \frac{4}{3}$$

Therefore, $\bar{x} = \frac{8}{15} \cdot \frac{3}{4} = \frac{2}{5}$ and $\bar{y} = \frac{4}{3} \cdot \frac{3}{4} = 1$.

From Section 5–5, we recall that the **moment of inertia** of an element of area with respect to the y-axis is $I_y = kx^2\, dA$. Therefore, using double integrals to find the moments of inertia of the area indicated in Fig. 10–36 (which is the same as Fig. 10–33), we have (assuming $k = 1$)

$$I_y = \int_a^b \int_{g(x)}^{G(x)} x^2\, dy\, dx \qquad (10\text{–}10)$$

and

$$I_x = \int_a^b \int_{g(x)}^{G(x)} y^2\, dy\, dx \qquad (10\text{–}11)$$

Also, from Section 5–5, the **radius of gyration** of the area with respect to an axis was defined by $R^2 = I/A$. The following examples illustrate the use of double integrals to find the moment of inertia and radius of gyration of an area.

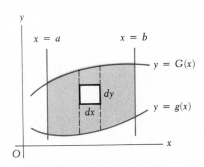

Figure 10–36

Example C

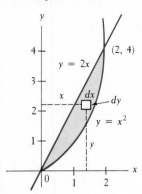

Figure 10–37

Find I_y and I_x for the area bounded by $y = 2x$ and $y = x^2$ (see Fig. (10–37).
Setting up the integrals for integration over y first, we solve for I_y and I_x:

$$I_y = \int_0^2 \int_{x^2}^{2x} x^2 \, dy \, dx = \int_0^2 y \Big|_{x^2}^{2x} x^2 \, dx$$

$$= \int_0^2 (2x^3 - x^4) \, dx = \frac{x^4}{2} - \frac{x^5}{5} \Big|_0^2 = \frac{8}{5}$$

$$I_x = \int_0^2 \int_{x^2}^{2x} y^2 \, dy \, dx = \int_0^2 \frac{y^3}{3} \Big|_{x^2}^{2x} dx$$

$$= \frac{1}{3} \int_0^2 (8x^3 - x^6) \, dx = \frac{1}{3}\left(2x^4 - \frac{x^7}{7}\right)\Big|_0^2 = \frac{32}{7}$$

Example D

Figure 10–38

Find the radius of gyration with respect to the y-axis for the area bounded by $y = x^3$, $x = 2$, and $y = 0$ (see Fig. 10–38).
Since integration is more convenient over y first than over x first, we set up our integrals in this way:

$$A = \int_0^2 \int_0^{x^3} dy \, dx = \int_0^2 y \Big|_0^{x^3} dx = \int_0^2 x^3 \, dx = \frac{x^4}{4} \Big|_0^2 = 4$$

$$I_y = \int_0^2 \int_0^{x^3} x^2 \, dy \, dx = \int_0^2 y \Big|_0^{x^3} x^2 \, dx = \int_0^2 x^5 \, dx = \frac{x^6}{6} \Big|_0^2 = \frac{32}{3}$$

$$R_y^2 = \frac{32}{3} \cdot \frac{1}{4} = \frac{8}{3}$$

$$R_y = \frac{2}{3} \sqrt{6} = 1.63$$

Exercises 10–6

All exercises are to be solved by the use of double integrals.

In Exercises 1 through 4, find the coordinates of the centroid of the area bounded by the given curves.

1. $y = 2 - x$, $y = 0$, $x = 0$ 2. $y = x^2$, $y = 4$, $x = 0$

3. $y = 9 - x^2$, $y = 0$, $x = 0$ 4. $y = 2x$, $x = 1$, $x = 2$

5. Find I_y and I_x for the area of Exercise 1.

6. Find I_y and I_x for the area of Exercise 2.

7. Find R_y for the area of Exercise 3.

Figure 10–39

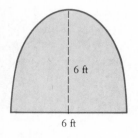

Figure 10–40

8. Find R_x for the area of Exercise 4.

9. Find the coordinates of the centroid and R_y for the area bounded by $y = 1 - x^2$, $x = 1$, and $y = 1$.

10. Find R_y for the area bounded by $y = 3 - x$ and $y = 3 + 2x - x^2$.

11. Find the coordinates of the centroid of a right triangle with legs a and b.

12. Find the coordinates of the centroid of a quarter circle of radius a.

13. Find the moment of inertia of a rectangular area with sides a and b with respect to side a. Express the result in terms of the mass of the area.

14. Find the moment of inertia of an isosceles right triangle with respect to one of the legs a. Express the result in terms of the mass of the area.

15. Find the radius of gyration of a square of side 4 with respect to one of the sides.

16. Find the radius of gyration of a rhombus of side 2, if the angle between two adjacent sides is 45°, with respect to one of the sides.

17. A flat metal plate is in the shape of an isosceles trapezoid, as shown in Fig. 10–39. Where is the centroid of the plate?

18. An aquarium view window has the shape of a parabolic section, as shown in Fig. 10–40. Locate the centroid of the window.

10–7 Review Exercises for Chapter 10

In Exercises 1 through 10, find the partial derivatives of the given functions with respect to each of the independent variables.

1. $z = 5x^3y^2 - 2xy^4$

2. $z = 2x\sqrt{y} - x^2y$

3. $z = \sqrt{x^2 - 3y^2}$

4. $u = \dfrac{r}{(r - 3s)^2}$

5. $z = \dfrac{2x - 3y}{x^2y + 1}$

6. $z = x(y^2 + xy + 2)^4$

7. $u = y \ln \sin(x^2 + 2y)$

8. $q = p \ln(r + 1) - \dfrac{rp}{r + 1}$

9. $z = \text{Arcsin}\sqrt{x + y}$

10. $z = ye^{xy}\sin(2x - y)$

In Exercises 11 and 12, find all of the second partial derivatives of the given functions.

11. $z = 3x^2y - y^3 + 2xy$

12. $z = x\sqrt{2y + 1} + y^2(x - 2)^3$

In Exercises 13 through 20, evaluate each of the given double integrals.

13. $\displaystyle\int_0^2 \int_1^2 (3y + 2xy)\, dx\, dy$

14. $\displaystyle\int_2^7 \int_0^1 x\sqrt{2 + x^2y}\, dx\, dy$

15. $\displaystyle\int_0^3 \int_1^x (x + 2y)\, dy\, dx$

16. $\displaystyle\int_1^2 \int_0^{\pi/4} r\sec^2\theta\, d\theta\, dr$

17. $\displaystyle\int_0^1 \int_0^{2x} x^2 e^{xy}\, dy\, dx$

18. $\displaystyle\int_{\pi/4}^{\pi/2} \int_1^{\sqrt{\cos\theta}} r\sin\theta\, dr\, d\theta$

19. $\displaystyle\int_1^e \int_1^x \dfrac{\ln y}{xy}\, dy\, dx$

20. $\displaystyle\int_1^3 \int_0^x \dfrac{2}{x^2 + y^2}\, dy\, dx$

In Exercises 21 through 24, use the function $z = \sqrt{x^2 + 4y^2}$.

21. Sketch the surface representing the function.

22. Find the equation of a line tangent to the surface of the function at $(2, 1, 2\sqrt{2})$ which is parallel to the yz-plane.

23. Find the approximate change in the function as x changes from 2.00 to 2.04 and y changes from 1.00 to 1.06.

24. Find the expression for the total differential of the function.

In Exercises 25 through 28, use the function $z = e^{x+y}$.

25. Find the equation of the line tangent to the surface of the function at $(1, 1, e^2)$ which is parallel to the xz-plane.

26. Sketch the surface representing the function.

27. Find the volume in the first octant under the surface of the function and inside the planes $x = 1$ and $y = x$.

28. Find the approximate change in the function as x changes from 1.00 to 1.01 and y changes from 1.00 to 1.02.

In Exercises 29 through 48, solve the given problems.

29. In a simple series electric circuit, with two resistors r and R connected across a voltage source E, the voltage v across r is $v = rE/(r + R)$. Assuming E to be constant, find $\partial v/\partial r$ and $\partial v/\partial R$.

30. For a gas, the volume expansivity is defined as

$$\beta = \frac{1}{V}\frac{\partial V}{\partial T}$$

where V is the volume and T the temperature of the gas. If V is a function of T and the pressure p given by

$$V = a + \frac{bT}{p} - \frac{c}{T^2}$$

where a, b and c are constants, find β.

31. In the theory dealing with transistors, the current gain α of a transistor is defined as

$$\alpha = \frac{\partial i_c}{\partial i_e}$$

where i_c is the collector current and i_e is the emitter current. If i_c is a function of i_e and the collector voltage v_c given by

$$i_c = i_e(1 - e^{-2v_c})$$

find α if v_c is 2 V.

32. Young's modulus, which measures the ratio of the stress to strain in a stretched wire, is defined as

$$Y = \frac{L}{A}\frac{\partial F}{\partial L}$$

where L is the length of the wire, A is its cross-sectional area, and F is the tension on the wire. Find Y, in pascals, if

$$F = -\frac{0.01\,T}{L^2}$$

for $L = 1.10$ m, $T = 300$ K, and $A = 10^{-6}$ m^2.

33. The period T of a pendulum as a function of its length l and the acceleration due to gravity g is given by $T = 2\pi \sqrt{l/g}$. Show that

$$\frac{\partial T}{\partial l} = \frac{T}{2l}$$

34. The volume of a right circular cone of radius r and height h is given by $V = \frac{1}{3}\pi r^2 h$. Show that

$$\frac{\partial V}{\partial r} = \frac{2V}{r}$$

35. The image distance q from a lens as a function of the object distance p and the focal length f of the lens is given by

$$q = \frac{pf}{p - f}$$

Find the approximate change in q if f changes from 20 cm to 21 cm and p changes from 100 cm to 105 cm.

36. The impedance Z for an alternating current circuit is given by

$$Z = \sqrt{R^2 + X^2}$$

where R is the resistance and X is the reactance. If R is measured to be 6 Ω and X to be 8 Ω, and each has a maximum possible error of 2%, what is the maximum possible error in the impedance?

37. A flat plate 5 in. square is heated such that the temperature as a function of distances x and y from intersecting edges is

$$T = 2xy - y^2 - 2x^2 + 3y + 2$$

Find the hottest point.

38. A closed rectangular box is made to contain 1000 in.^{3}. What dimensions should it have to make its surface area a minimum?

39. Find the volume in the first octant below the plane $x + y + z - 6 = 0$ and inside the cylinder $y = 4 - x^2$.

40. Find the first-octant volume bounded by $x + z = 4$, $x = 1$, and $y = 1$.

41. Find the first-octant volume bounded by $x^2 + y^2 = 16$ and $x + z = 8$.

42. Find the volume in the first octant below the surface $z = 4 - y^2$ and inside the plane $x + y = 2$.

43. Find, by double integration, the coordinates of the centroid of the area bounded by $y = 2$, $x + y = 4$, $x = 0$, and $y = 0$.

44. Find, by double integration, the coordinates of the centroid of the area bounded by $y = x$ and $y = 4x - x^2$.

45. Find, by double integration, R_y for the area of Exercise 43.

46. Find, by double integration, R_x for the area of Exercise 44.

47. The power delivered to an electric resistor is given by $P = V^2/R$, where V is the voltage across the resistor R. If V is increasing at the rate of 1 V/min and R is increasing at the rate of 2 Ω/min, find the time rate of change of power, in W/min, when $V = 60$ V and $R = 20$ Ω (see Exercise 23 of Section 9–4).

48. The kinetic energy of a moving object is given by $E = \frac{1}{2}mv^2$, where m is the mass of the object and v is its velocity. If m is decreasing at the rate of 0.01 kg/s and v is increasing at the rate of 5 m/s^2, find the time rate of change of E, in joules per second, when $m = 5$ kg and $v = 100$ m/s (see Exercise 33 of Section 9–4).

CHAPTER 11

Polar Coordinates

Thus far we have graphed all plane curves in one coordinate system. This system, the rectangular coordinate system, is probably the most useful and widely applicable system. However, other coordinate systems prove to be better adapted for certain types of curves. One of these coordinate systems, the *polar coordinate system,* is widely used, especially when certain types of applications are involved. We shall discuss this system in this chapter and show some of its applications.

11–1 Polar Coordinates

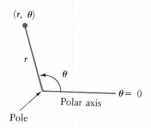

Figure 11–1

Instead of designating a point by its x- and y-coordinates, we can specify its location by its radius vector and the angle the radius vector makes with the x-axis. Thus, the r and θ that are used in the definitions of the trigonometric functions can also be used as the coordinates of points in the plane. The important aspect of choosing coordinates is that, for each set of values, there must be only one point which corresponds to this set. We can see that this condition is satisfied by the use of r and θ as coordinates. *In* **polar coordinates** *the origin is called the* **pole,** *and the half-line for which the angle is zero (equivalent to the positive x-axis) is called the* **polar axis.** The coordinates of a point are designated as (r, θ). We shall use radians when measuring the value of θ. See Fig. 11–1.

When using polar coordinates, we generally label the lines for some of the values of θ; namely, those for $\theta = 0$ (the polar axis), $\theta = \pi/2$ (equivalent to the positive y-axis), $\theta = \pi$ (equivalent to the negative x-axis), $\theta = 3\pi/2$

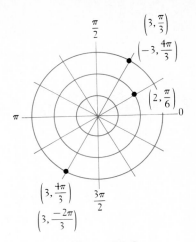

Figure 11–2   *Figure 11–3*

(equivalent to the negative y-axis), and possibly others. In Fig. 11–2 these lines and those for the multiples of $\theta = \pi/6$ are shown. The circles for $r = 1$, $r = 2$, and $r = 3$ are also shown in this figure.

Example A

If $r = 2$ and $\theta = \pi/6$, we have the point as indicated in Fig. 11–3. The coordinates (r, θ) of this point are written as $(2, \pi/6)$ when polar coordinates are used. This point corresponds to $(\sqrt{3}, 1)$ in rectangular coordinates.

One difference between rectangular coordinates and polar coordinates is that, for each point in the plane, there are limitless possibilities for the polar coordinates of that point. For example, the point $(2, \pi/6)$ can also be represented by $(2, 13\pi/6)$ since the angles $\pi/6$ and $13\pi/6$ are coterminal. We also remove one restriction on r that we imposed in the definition of the trigonometric functions. That is, r is allowed to take on positive and negative values. If r is considered

 negative, then the point is found on the opposite side of the pole from that on which it is positive.

Example B

The coordinates $(3, 4\pi/3)$ and $(3, -2\pi/3)$ represent the same point. However, the point $(-3, 4\pi/3)$ is on the opposite side of the pole, three units from the pole. Another possible set of coordinates for $(-3, 4\pi/3)$ is $(3, \pi/3)$ (see Fig. 11–3).

When plotting a point in polar coordinates it is generally easier to *first locate the terminal side of θ and then measure r along this terminal side.* This is illustrated in the following example.

Example C

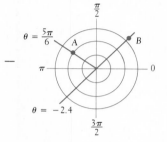

$\theta = \dfrac{5\pi}{6}$

Plot the points $A\left(2, \dfrac{5\pi}{6}\right)$ and $B(-3.2, -2.4)$ in the polar coordinate system.

To locate A we determine the terminal side of $\theta = \dfrac{5\pi}{6}$ and then find $r = 2$. See Fig. 11–4.

To locate B we find the terminal side of $\theta = -2.4$, measuring clockwise from the polar axis (and recalling that $\pi = 3.14 = 180°$). Then we locate $r = -3.2$ on the opposite side of the pole. See Fig. 11–4.

Figure 11–4

Figure 11–5

The relationships between the polar coordinates of a point and the rectangular coordinates of the same point come directly from the definitions of the trigonometric functions. Those most commonly used are (see Fig. 11–5)

$$x = r \cos \theta, \qquad y = r \sin \theta \tag{11–1}$$

and

$$\tan \theta = \frac{y}{x}, \qquad r = \sqrt{x^2 + y^2} \tag{11–2}$$

The following examples show the use of Eqs. (11–1) and (11–2) in changing coordinates in one system to coordinates in the other system. Also, they are used to transform equations from one system to the other.

Example D

Figure 11–6

Using Eqs. (11–1), we can transform the polar coordinates of $(4, \pi/4)$ into the rectangular coordinates $(2\sqrt{2}, 2\sqrt{2})$, since

$$x = 4 \cos \frac{\pi}{4} = 4\left(\frac{\sqrt{2}}{2}\right) = 2\sqrt{2}$$

and

$$y = 4 \sin \frac{\pi}{4} = 4\left(\frac{\sqrt{2}}{2}\right) = 2\sqrt{2}$$

See Fig. 11–6.

Example E

Figure 11–7

Using Eqs. (11–2), we can transform the rectangular coordinates $(3, -5)$ into polar coordinates.

$$\tan \theta = -\frac{5}{3}, \qquad \theta = 5.25 \quad (\text{or } -1.03)$$

$$r = \sqrt{3^2 + (-5)^2} = 5.83$$

We know that θ is a fourth-quadrant angle since x is positive and y is negative. Therefore, the point $(3, -5)$ in rectangular coordinates can be expressed as the point $(5.83, 5.25)$ in polar coordinates (see Fig. 11–7). Other polar coordinates for the point are also possible.

Example F

Figure 11–8

Find the polar equation of the circle $x^2 + y^2 = 2x$.

Here we are given the equation of a circle in rectangular coordinates x and y and are to change it into an equation expressed in polar coordinates r and θ. Since

$$r^2 = x^2 + y^2 \quad \text{and} \quad x = r \cos \theta$$

we have

$$r^2 = 2r \cos \theta$$

or

$$r = 2 \cos \theta$$

as the appropriate polar equation. The circle is shown in Fig. 11–8.

Example G

Figure 11–9

Find the rectangular equation of the *rose* $r = 4 \sin 2\theta$.

Using the relation $2 \sin \theta \cos \theta = \sin 2\theta$, we have $r = 8 \sin \theta \cos \theta$. Then, using Eqs. (11–1) and (11–2), we have

$$\sqrt{x^2 + y^2} = 8\left(\frac{y}{r}\right)\left(\frac{x}{r}\right) = \frac{8xy}{r^2} = \frac{8xy}{x^2 + y^2}$$

Squaring both sides, we obtain

$$x^2 + y^2 = \frac{64x^2y^2}{(x^2 + y^2)^2} \quad \text{or} \quad (x^2 + y^2)^3 = 64x^2y^2$$

From this example we can see that plotting the graph from the rectangular equation would be complicated. However, as we will see in the following section, plotting this graph in polar coordinates is quite simple. The graph is shown in Fig. 11–9.

Exercises 11–1

In Exercises 1 through 12, plot the given polar coordinate points on polar coordinate paper.

1. $\left(3, \dfrac{\pi}{6}\right)$ 2. $(2, \pi)$ 3. $\left(\dfrac{5}{2}, -\dfrac{2\pi}{5}\right)$ 4. $\left(5, -\dfrac{\pi}{3}\right)$

5. $\left(-2, \dfrac{7\pi}{6}\right)$ 6. $\left(-5, \dfrac{\pi}{4}\right)$ 7. $\left(-3, -\dfrac{5\pi}{4}\right)$ 8. $\left(-4, -\dfrac{5\pi}{3}\right)$

9. $\left(0.5, -\dfrac{8\pi}{3}\right)$ 10. $(2.2, -6\pi)$ 11. $(2, 2)$ 12. $(-1, -1)$

In Exercises 13 through 16, find a set of polar coordinates for each of the given points expressed in rectangular coordinates.

13. $(\sqrt{3}, 1)$ 14. $(-1, -1)$ 15. $\left(-\dfrac{\sqrt{3}}{2}, -\dfrac{1}{2}\right)$ 16. $(-5, 4)$

In Exercises 17 through 20, find the rectangular coordinates corresponding to the points for which the polar coordinates are given.

17. $\left(8, \dfrac{4\pi}{3}\right)$ 18. $(-4, -\pi)$ 19. $\left(3, -\dfrac{\pi}{8}\right)$ 20. $(-1, -1)$

In Exercises 21 through 28, find the polar equation of the given rectangular equations.

21. $x = 3$ 22. $y = 2$ 23. $x^2 + y^2 = a^2$

24. $x^2 + y^2 = 4y$ 25. $y^2 = 4x$ 26. $x^2 - y^2 = a^2$

27. $x^2 + 4y^2 = 4$ 28. $y = x^2$

In Exercises 29 through 36, find the rectangular equation of each of the given polar equations.

29. $r = \sin\theta$ 30. $r = 4\cos\theta$ 31. $r\cos\theta = 4$

32. $r\sin\theta = -2$ 33. $r = 2(1 + \cos\theta)$ 34. $r = 1 - \sin\theta$

35. $r^2 = \sin 2\theta$ 36. $r^2 = 16\cos 2\theta$

In Exercises 37 through 40, find the required equations.

37. If we refer back to Eqs. (6–23) and (6–24), we see that the length along the arc of a circle and the area within a circular sector vary with two variables. These variables are those which we refer to as the polar coordinates. Express the arc length s and the area A in terms of rectangular coordinates. (How must we express θ?)

38. Under certain conditions, the x- and y-components of a magnetic field B are given by the equations

$$B_x = \frac{-ky}{x^2 + y^2} \quad \text{and} \quad B_y = \frac{kx}{x^2 + y^2}$$

Write these equations in terms of polar coordinates.

39. Express the equation of the cable (see Exercise 29 of Section 1–6) in polar coordinates.

40. The polar equation of the path of a weather satellite around the earth is

$$r = \frac{4800}{1 + 0.14\cos\theta}$$

where r is measured in miles. Find the rectangular equation of the path of this satellite. The path is an ellipse, with the earth at one of the foci.

11–2 Curves in Polar Coordinates

The basic method for finding a curve in polar coordinates is the same as in rectangular coordinates. We assume values of θ and then find the corresponding values of r. These points are plotted and joined, thus forming the curve which represents the function. However, there are certain basic curves which can be sketched directly from the equation. The following examples illustrate these methods.

Example A

The graph of the polar equation $r = 3$ is a circle of radius 3, with center at the pole. This can be seen to be the case, since $r = 3$ for all possible values of θ. It is not really necessary to find specific points for this circle. See Fig. 11–10.

Example B

The graph of $\theta = \pi/6$ is a straight line through the pole. It represents all points for which $\theta = \pi/6$ for all possible values of r, positive or negative. See Fig. 11–10.

Example C

Plot the graph of $r = 1 + \cos \theta$.
   We find the following table of values of r corresponding to the assumed values of θ.

θ	0	$\frac{\pi}{4}$	$\frac{\pi}{2}$	$\frac{3\pi}{4}$	π	$\frac{5\pi}{4}$	$\frac{3\pi}{2}$	$\frac{7\pi}{4}$	2π
r	2	1.7	1	0.3	0	0.3	1	1.7	2

We now see that the points on the curve start repeating, and it is unnecessary to find additional points. This curve is called a **cardioid** and is shown in Fig. 11–11.

Figure 11–10

Figure 11–11

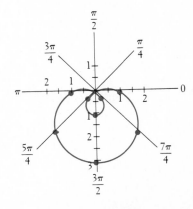

Figure 11–12

Example D

Plot the graph of $r = 1 - 2 \sin \theta$.

θ	0	$\dfrac{\pi}{4}$	$\dfrac{\pi}{2}$	$\dfrac{3\pi}{4}$	π	$\dfrac{5\pi}{4}$	$\dfrac{3\pi}{2}$	$\dfrac{7\pi}{4}$	2π
r	1	-0.4	-1	-0.4	1	2.4	3	2.4	1

 Particular care should be taken in plotting the points for which r is negative. This curve is known as a **limaçon** and is shown in Fig. 11–12.

Example E

Plot the graph of $r = 2 \cos 2\theta$.

θ	0	$\dfrac{\pi}{12}$	$\dfrac{\pi}{6}$	$\dfrac{\pi}{4}$	$\dfrac{\pi}{3}$	$\dfrac{5\pi}{12}$	$\dfrac{\pi}{2}$	$\dfrac{7\pi}{12}$	$\dfrac{2\pi}{3}$	$\dfrac{3\pi}{4}$	$\dfrac{5\pi}{6}$	$\dfrac{11\pi}{12}$	π
r	2	1.7	1	0	-1	-1.7	-2	-1.7	-1	0	1	1.7	2

For the values of θ from π to 2π, the values of r repeat. We have a four-leaf **rose** (Fig. 11–13).

Example F

Plot the graph of $r^2 = 9 \cos 2\theta$.

θ	0	$\dfrac{\pi}{8}$	$\dfrac{\pi}{4}$	$\cdots$	$\dfrac{3\pi}{4}$	$\dfrac{7\pi}{8}$	π
r	± 3	± 2.5	0		0	± 2.5	± 3

There are no values of r corresponding to values of θ in the range $\pi/4 < \theta < 3\pi/4$, since twice these angles are in the second and third quadrants, and the cosine is negative for such angles. The value of r^2 cannot be negative. Also, the values of r repeat for $\theta > \pi$. The figure is called a **lemniscate** (Fig. 11–14).

Figure 11–13

Figure 11–14

See Appendix C for a
computer program for
graphing a curve in polar
coordinates.

Exercises 11–2

In Exercises 1 through 28, plot the given curves in polar coordinates.

1. $r = 4$

2. $r = 2$

3. $\theta = \dfrac{3\pi}{4}$

4. $\theta = \dfrac{5\pi}{3}$

5. $r = 4 \sec \theta$

6. $r = 4 \csc \theta$

7. $r = 2 \sin \theta$

8. $r = 3 \cos \theta$

9. $r = 1 - \cos \theta$ (cardioid)

10. $r = \sin \theta - 1$ (cardioid)

11. $r = 2 - \cos \theta$ (limaçon)

12. $r = 2 + 3 \sin \theta$ (limaçon)

13. $r = 4 \sin 2\theta$ (rose)

14. $r = 2 \sin 3\theta$ (rose)

15. $r^2 = 4 \sin 2\theta$ (lemniscate)

16. $r^2 = 2 \sin \theta$

17. $r = 2^\theta$ (spiral)

18. $r = 10^\theta$ (spiral)

19. $r = -4 \sin 4\theta$ (rose)

20. $r = -\cos 3\theta$ (rose)

21. $r = \dfrac{1}{2 - \cos \theta}$ (ellipse)

22. $r = \dfrac{1}{1 - \cos \theta}$ (parabola)

23. $r = \dfrac{6}{1 - 2 \cos \theta}$ (hyperbola)

24. $r = \dfrac{6}{3 - 2 \sin \theta}$ (ellipse)

25. $r = 4 \cos \tfrac{1}{2}\theta$

26. $r = 2 + \cos 3\theta$

27. $r = 3 - \sin 3\theta$

28. $r = 4 \tan \theta$

In Exercises 29 through 32, sketch the indicated graphs.

29. The charged particle (see Exercise 41 of Section 1–5) is originally traveling along the line $\theta = \pi$ toward the pole. When it reaches the pole it enters the magnetic field. Its path is then described by the polar equation $r = 2a \sin \theta$, where a is the radius of the circle. Sketch the graph of this path.

30. A cam is shaped such that the equation of the edge of the upper "half" is given by $r = 2 + \cos \theta$, and the equation of the edge of the lower "half" by $r = 3/(2 - \cos \theta)$. Plot the curve which represents the shape of the cam.

31. A satellite at a height proper to make one revolution per day around the earth will have for an excellent approximation of the projection on the earth of its path the curve

$$r^2 = R^2 \cos 2\left(\theta + \frac{\pi}{2}\right)$$

where R is the radius of the earth. Sketch the path of the projection.

32. Sketch the graph of the rectangular equation

$$4(x^6 + 3x^4y^2 + 3x^2y^4 + y^6 - x^4 - 2x^2y^2 - y^4) + y^2 = 0$$

[*Hint:* The equation can be written as $4(x^2 + y^2)^3 - 4(x^2 + y^2)^2 + y^2 = 0$. Transform to polar coordinates, and then sketch the curve.]

11–3 Applications of Differentiation and Integration in Polar Coordinates

Polar coordinates are used extensively in applications and theoretical discussions in advanced mechanics, electricity, and certain other fields. In addition, a number of applications of differentiation and integration which we have previously discussed may be solved in polar coordinates. In this section we shall discuss an application of differentiation and one of integration. Certain other applications are in the exercises.

The application of differentiation we shall discuss is that of velocity. In Section 3–3 we were able to determine the resultant velocity of an object if we knew its x- and y-components. Here we shall consider the components of velocity in the direction of increasing r and increasing θ. These components are denoted by v_r and v_θ.

The **radial component** *of velocity v_r and the* **transverse component** *of velocity v_θ are given by*

$$v_r = \frac{dr}{dt} \quad \text{and} \quad v_\theta = r\frac{d\theta}{dt} \tag{11–3}$$

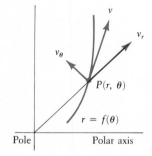

Figure 11–15

See Fig. 11–15. To show that these are proper relations we recall that $x = r\cos\theta$ and $y = r\sin\theta$. Taking derivatives with respect to time, treating both r and θ as functions of time, we have

$$\frac{dx}{dt} = \cos\theta\frac{dr}{dt} - r\sin\theta\frac{d\theta}{dt}$$

$$\frac{dy}{dt} = \sin\theta\frac{dr}{dt} + r\cos\theta\frac{d\theta}{dt}$$

Since $dx/dt = v_x$ and $dy/dt = v_y$, and using Eqs. (11–3), we have

$$v_x = v_r\cos\theta - v_\theta\sin\theta$$
$$v_y = v_r\sin\theta + v_\theta\cos\theta$$

Squaring both sides of each of these equations and adding we have

$$v_x^2 = v_r^2\cos^2\theta - 2v_r v_\theta\cos\theta\sin\theta + v_\theta^2\sin^2\theta$$
$$v_y^2 = v_r^2\sin^2\theta + 2v_r v_\theta\cos\theta\sin\theta + v_\theta^2\cos^2\theta$$

$$v_x^2 + v_y^2 = v_r^2 + v_\theta^2 = v^2 \tag{11–4}$$

 This analysis, along with the fact that dr/dt is the velocity in the r-direction for constant θ, and $r(d\theta/dt)$ is the velocity in the θ-direction for constant r, verifies that these are the correct relations. The following examples illustrate the use of Eqs. (11–3) and (11–4).

Example A

Figure 11–16

Find the radial and transverse components of velocity of an object moving such that $r = t^2 - 2$ and $\theta = \frac{1}{4}t^2$, when $t = 2$. From these values calculate the magnitude of the velocity of v when $t = 2$.

$$v_r = \frac{dr}{dt} = 2t, \qquad v_\theta = r\frac{d\theta}{dt} = (t^2 - 2)\left(\frac{t}{2}\right) = \frac{1}{2}(t^3 - 2t)$$

Evaluating for $t = 2$, we have

$$v_r\big|_{t=2} = 4$$

$$v_\theta\big|_{t=2} = \frac{1}{2}(8 - 4) = 2$$

Thus,

$$v = \sqrt{16 + 4} = \sqrt{20} = 4.47$$

See Fig. 11–16.

Example B

A particle is moving along the curve $r = 3 + 2 \sin \theta$ with a constant angular velocity, $\omega = d\theta/dt = 2$ rad/s (see Fig. 11–17). Find the magnitude of the speed in centimeters per second when $\theta = \pi/3$.

Taking derivatives of $r = 3 + 2 \sin \theta$ with respect to t, we have

$$\frac{dr}{dt} = 2 \cos \theta \frac{d\theta}{dt}$$

Therefore,

$$v_r = \frac{dr}{dt} = 2 \cos \theta \frac{d\theta}{dt}$$

$$v_\theta = r\frac{d\theta}{dt} = (3 + 2 \sin \theta)\frac{d\theta}{dt}$$

Figure 11–17

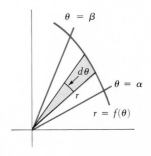

Figure 11–18

Evaluating, we have

$$v_r|_{\theta=\pi/3} = 2\left(\cos\frac{\pi}{3}\right)(2) = 4\left(\frac{1}{2}\right) = 2.00$$

$$v_\theta|_{\theta=\pi/3} = \left[3 + 2\left(\frac{\sqrt{3}}{2}\right)\right](2) = 2(3 + \sqrt{3}) = 9.46$$

$$v = \sqrt{4.00 + 89.5} = \sqrt{93.5} = 9.67 \text{ cm/s}$$

The application of integration in polar coordinates we shall consider is the area bounded by polar curves. We recall Eq. (6–24), which states that the area of a circular sector is $A = \frac{1}{2}r^2\theta$. In Fig. 11–18, consider the element of area with radius r and central angle $d\theta$, as shown. If these elements are summed (integrated) from $\theta = \alpha$ to $\theta = \beta$, the sum will represent the area shown. *Thus, the area is found by*

$$A = \frac{1}{2}\int_\alpha^\beta r^2\,d\theta \tag{11–5}$$

The following examples illustrate the use of Eq. (11–5).

Example C

Find the area enclosed by one loop of the lemniscate $r^2 = \cos 2\theta$.

Employing Eq. (11–5), we substitute $r^2 = \cos 2\theta$. To determine the limits α and β, we note the symmetry of the curve (see Fig. 11–19). By choosing $\alpha = 0$ and β to be the value where r is first zero, we will find the area of the upper half of the loop. Doubling this will give the desired area. Thus, by setting $r^2 = \cos 2\theta = 0$, we find $2\theta = \pi/2$ or $\theta = \pi/4$. This means $\beta = \pi/4$. Therefore, the integral for the area is

$$A = 2\left[\frac{1}{2}\int_0^{\pi/4} \cos 2\theta\,d\theta\right]$$

$$= \frac{1}{2}\int_0^{\pi/4} \cos 2\theta(2\,d\theta) = \frac{1}{2}\sin 2\theta\Big|_0^{\pi/4} = \frac{1}{2}(1 - 0) = \frac{1}{2}$$

Figure 11–19

Figure 11–20

Figure 11–21

Example D

Find the area enclosed by the curve $r = 2 - \cos \theta$.

Using Eq. (11–5), we substitute $r^2 = (2 - \cos \theta)^2 = 4 - 4 \cos \theta + \cos^2 \theta$. To determine the limits α and β, we note the sketch of the curve in Fig. 11–20. Since it is symmetrical to the polar axis, we may let $\alpha = 0$ and $\beta = \pi$ if we also multiply by 2. (Limits of $\alpha = 0$ and $\beta = 2\pi$ would also be proper.) Thus,

$$A = 2\left[\frac{1}{2} \int_0^{\pi} (4 - 4 \cos \theta + \cos^2 \theta) \, d\theta \right]$$

By using Formula 32 of Table 5 and the basic identity $\sin 2\theta = 2 \sin \theta \cos \theta$, we have

$$A = \left[4\theta - 4 \sin \theta + \frac{\theta}{2} + \frac{1}{4} \sin 2\theta \right]_0^{\pi} = \left(4\pi + \frac{\pi}{2} \right) = \frac{9\pi}{2}$$

Example E

Find the area outside the cardioid $r = 1 + \sin \theta$ and inside the circle $r = 3 \sin \theta$.

Drawing the figure, we see that the desired area is the shaded portion of Fig. 11–21. By substituting $r = 3 \sin \theta$ into the equation of the cardioid, we have

$$3 \sin \theta = 1 + \sin \theta$$

$$\sin \theta = \frac{1}{2}$$

$$\theta = \frac{\pi}{6}, \frac{5\pi}{6}$$

This means the points $(3/2, \pi/6)$ and $(3/2, 5\pi/6)$ are common to each curve. Therefore, the area inside the circle between $\theta = \pi/6$ and $\theta = 5\pi/6$ is

$$A_1 = \frac{1}{2} \int_{\pi/6}^{5\pi/6} (3 \sin \theta)^2 \, d\theta$$

The area inside the cardioid between $\theta = \pi/6$ and $\theta = 5\pi/6$ is

$$A_2 = \frac{1}{2} \int_{\pi/6}^{5\pi/6} (1 + \sin \theta)^2 \, d\theta$$

The desired area is the difference of A_1 and A_2. Thus,

$$A = \frac{1}{2} \int_{\pi/6}^{5\pi/6} [(3 \sin \theta)^2 - (1 + \sin \theta)^2] \, d\theta$$

$$= \frac{1}{2} \int_{\pi/6}^{5\pi/6} (8 \sin^2 \theta - 2 \sin \theta - 1) \, d\theta$$

$$= \frac{1}{2} \int_{\pi/6}^{5\pi/6} (4 - 4 \cos 2\theta - 2 \sin \theta - 1) \, d\theta$$

$$= \frac{1}{2} (3\theta - 2 \sin 2\theta + 2 \cos \theta) \Big|_{\pi/6}^{5\pi/6}$$

$$= \frac{1}{2} \left[3 \left(\frac{5\pi}{6} - \frac{\pi}{6} \right) - 2 \left(\sin \frac{5\pi}{3} - \sin \frac{\pi}{3} \right) + 2 \left(\cos \frac{5\pi}{6} - \cos \frac{\pi}{6} \right) \right]$$

$$= \frac{1}{2} \left[2\pi - 2 \left(-\frac{\sqrt{3}}{2} - \frac{\sqrt{3}}{2} \right) + 2 \left(-\frac{\sqrt{3}}{2} - \frac{\sqrt{3}}{2} \right) \right] = \pi$$

Advantage could have been taken of the symmetry by integrating from $\pi/6$ to $\pi/2$ and doubling the result.

Exercises 11–3

In Exercises 1 through 4, r and θ are given as functions of t. Find the magnitudes of the velocities by finding v_r and v_θ for the given values of t.

1. $r = 3t$, $\theta = t^3$, $t = 1$

2. $r = 4t$, $\theta = \sqrt{t + 1}$, $t = 3$

3. $r = e^{0.1t}$, $\theta = \cos 2t$, $t = 2$

4. $r = \sin \dfrac{\pi t}{6}$, $\theta = \dfrac{1}{2t + 1}$, $t = 2$

In Exercises 5 through 12, find the magnitude of the velocity of an object moving along the given curves with a constant angular velocity $\omega = 2$ rad/s for the indicated values of θ with r measured in feet.

5. $r = 2$, $\theta = \pi/3$

6. $r = 4\theta$, $\theta = 2$

7. $r = \sin \theta$, $\theta = \pi/6$

8. $r = \cos 2\theta$, $\theta = \pi/4$

9. $r = 4 - \cos \theta$, $\theta = \pi/2$

10. $r = 5 + 2 \sin 2\theta$, $\theta = \pi/6$

11. $r = \dfrac{1}{2 - \sin \theta}$, $\theta = \pi/6$

12. $r = 1 + \cos^2 \theta$, $\theta = \pi/4$

See Appendix C for a computer program for finding the area bounded by a polar curve.

In Exercises 13 through 28, find the area bounded by the given curves.

13. $\theta = \frac{1}{6}\pi$, $\theta = \frac{2}{3}\pi$, $r = 2$

14. $\theta = 0$, $\theta = \frac{4}{3}\pi$, $r = 3$

15. $\theta = 0$, $\theta = \frac{1}{4}\pi$, $r = \sec \theta$

16. $\theta = \frac{1}{6}\pi$, $\theta = \frac{1}{2}\pi$, $r \sin \theta = 2$

17. $r = 2 \cos \theta$

18. $\theta = 0$, $\theta = \frac{1}{3}\pi$, $r = 4 \sin \theta$

19. $\theta = 0$, $\theta = \frac{1}{2}\pi$, $r = e^{2\theta}$ 20. $r = 1 - \cos\theta$

21. $r = 2\sin 3\theta$ 22. $r = 3\cos 2\theta$

23. $r = 2 + \cos\theta$ 24. $r^2 = 4\sin 2\theta$

25. Inside $r = 1 + \sin\theta$, outside $r = 1$.

26. Inside $r = \sin\theta$, outside $r = 1 + \cos\theta$.

27. Area common to $r = 2\sin\theta$ and $r = 2\cos\theta$.

28. Inside $r = 5\cos\theta$, outside $r = 2 + \cos\theta$.

In Exercises 29 through 36, solve the given problems.

29. A particle moves in a counterclockwise direction on the circle $r = 2\sin\theta$. When $\theta = \pi/3$, $v = 10$ ft/s. Find v_r and v_θ.

30. A particle moves to the left on the parabola $r = 4/(1 + \cos\theta)$ with a constant speed of 4 m/s. Find v_r and v_θ when $\theta = \pi/2$.

31. A cam is in the shape of the curve $r = 6 - 2\cos\theta$. Find the area of one face of the cam in square centimeters.

32. Change $(x^2 + y^2)^2 - (x^2 + y^2) = 2xy$ into polar coordinates and find the area bounded by $\theta = 0$, $\theta = \pi/4$, and the curve.

33. Using $x = r\cos\theta$ and $y = r\sin\theta$, take derivatives of each with respect to θ and treat r as a function of θ. From these results show that

$$\frac{dy}{dx} = \frac{r'\tan\theta + r}{r' - r\tan\theta} \qquad (11\text{–}6)$$

represents the slope of a line tangent to a polar curve. Here, $r' = dr/d\theta$.

34. Use Eq. (11–6) to find the slope of a line tangent to $r = \cos\theta$ at $\theta = \pi/6$.

35. The length of arc along a polar curve is given by

$$s = \int_\alpha^\beta \sqrt{r^2 + (r')^2}\, d\theta \qquad (11\text{–}7)$$

where $r' = dr/d\theta$. Using this equation, find the total perimeter of $r = \cos^2(\theta/2)$.

36. Using Eq. (11–7), find the circumference of the circle $r = \sin\theta + \cos\theta$.

11–4 Review Exercises for Chapter 11

In Exercises 1 through 4, find the polar equation of each of the given rectangular equations.

1. $y = 2x$ 2. $2xy = 1$ 3. $x^2 - y^2 = 16$ 4. $x^2 + y^2 = 7 - 6y$

In Exercises 5 through 8, find the rectangular equation of each of the given polar equations.

5. $r = 2\sin 2\theta$ 6. $r^2 = \sin\theta$

7. $r = \dfrac{4}{2 - \cos \theta}$

8. $r = \dfrac{2}{1 - \sin \theta}$

In Exercises 9 through 20, plot the given curves in polar coordinates.

9. $r = 4(1 + \sin \theta)$

10. $r = 1 + 2 \cos \theta$

11. $r = 4 \cos 3\theta$

12. $r = -3 \sin \theta$

13. $r = \cot \theta$

14. $r = \dfrac{1}{2(\sin \theta - 1)}$

15. $r = 2 \sin\left(\dfrac{\theta}{2}\right)$

16. $r = \theta$

17. $r = 2 + \cos 3\theta$

18. $r = 3 - \cos 2\theta$

19. $r^2 = \theta$

20. $r^2 = 4 \cos \theta$

In Exercises 21 through 28, find the area bounded by the given curves.

21. $r = 4 \cos \theta$, $\theta = 0$, $\theta = \dfrac{\pi}{4}$

22. $r = 2 \csc \theta$, $\theta = \dfrac{\pi}{4}$, $\theta = \dfrac{\pi}{2}$

23. $r = 3 - \sin \theta$

24. $r = \sqrt{\sin \theta}$

25. Inside $r = 6$, to the right of $r \cos \theta = 5$.

26. Smaller loop of $r = 1 - 2 \cos \theta$.

27. Between $r = \theta$ and $\theta = \pi/2$ ($\theta \geq 0$).

28. Between $r = e^{\theta}$, $\theta = 0$, and $\theta = 2\pi$.

In Exercises 29 through 40, solve the given problems.

29. A particle moves such that $r = t^3 - 6t$ and $\theta = t/(t + 1)$, where t is the time. Find the magnitude of the velocity of the particle when $t = 1$.

30. Find the magnitude of the velocity (in meters per second) of an object moving on the curve $r = 2 + 3 \cos 2\theta$ with a constant angular velocity $\omega = 3$ rad/s for $\theta = \pi/12$.

31. Express the equation of the artificial satellite (in Exercise 29 of Section 1–7) in polar coordinates.

32. The x- and y-coordinates of the position of a certain moving object as functions of time are given by $x = 2t$ and $y = \sqrt{8t^2 + 1}$. Find the polar equation of the path of the object.

33. In the theory dealing with an object moving through a liquid, a quantity called the velocity potential is introduced. Find v_r and v_θ if the velocity potential is given by

$$P = \frac{k \cos \theta}{r^2}$$

where k is a constant, and

$$v_r = -\frac{\partial P}{\partial r} \quad \text{and} \quad v_\theta = -\frac{1}{r}\frac{\partial P}{\partial \theta}$$

Express the result in terms of rectangular coordinates.

34. Under certain conditions the electric potential is given by

$$V = \left(\frac{a}{r^2} - br\right) \cos \theta$$

where a and b are constants. The polar components of the electric field are given by

$$E_r = -\frac{\partial V}{\partial r} \quad \text{and} \quad E_\theta = -\frac{1}{r}\frac{\partial V}{\partial \theta}$$

Find the expressions for E_r and E_θ in terms of r and θ.

35. Under a force which varies inversely as the square of the distance from an attracting object (such as the force the sun exerts on the earth), it can be shown that the path that an object follows is described in general by

$$\frac{1}{r} = a + b \cos \theta$$

where a and b are constant for a particular path. By transforming this equation to rectangular coordinates, show that this equation represents one of the conic sections, the particular section depending on the values of a and b. It is through this kind of analysis that we know the paths of the planets and comets are conic sections.

36. The sound produced by a jet engine was measured at a distance of 100 m in all directions. The loudness of the sound (in decibels) was found to be $d = 115 + 10 \cos \theta$, where the $0°$ line for the angle θ is directed in front of the engine. Sketch the graph of d vs. θ in polar coordinates (use d as r).

37. From an observation tower, the terrain which can be effectively observed is within a circle of radius 5 mi. A straight road crosses the observation area, as shown in Fig. 11–22. What is the observation area beyond the road from the tower? The circle can be described by the equation $r = 10 \cos \theta$.

38. The top section of the perimeter of a window is part of a limaçon, and the bottom section is rectangular, as shown in Fig. 11–23. If the limaçon can be described by the equation $r = 4 + \sin \theta$, what is the area of the window?

39. The approximate maximum velocity attained by the artificial satellite of Exercise 40 of Section 11–1 is 5 mi/s. This velocity occurs when the satellite is closest to the earth. Calculate v_r and v_θ for this satellite at this point.

40. The angle ϕ between r and a tangent line to a curve is given by $\tan \phi = r/(dr/d\theta)$. Find the angle between r and a tangent line to the curve $r = \sin(\theta/2)$ for $\theta = \pi/2$.

Figure 11–22

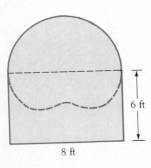

Figure 11–23

CHAPTER 12

Empirical Curve Fitting

A number of methods of graphing functions have been developed in earlier chapters. In this chapter we will show how to start with a set of points and determine a function for which the graph passes through the given points. Our primary aim is to be able to derive an equation which best "fits" given data obtained through experimentation or observation. Such an equation can show a basic relationship between the variables of the experiment and can make it possible to analyze the experiment better.

We will start by introducing a few basic concepts from statistics. Then we will show how to find the equation of the straight line for empirical data (data found through observation and experimentation) which is best described by a linear function. Finally, the method of fitting certain basic nonlinear curves to empirical data will be shown.

12–1 Arithmetic Mean and Standard Deviation

The quality control division of a company which manufactures cables will test these cables to determine the safe load each type of cable can support. In testing the cables and summarizing the results, the basic methods of **statistics** are used. That is, the quality control division (1) collects, (2) analyzes, and (3) interprets data. In this section we will focus on two of the basic statistical measures used in data analysis. These will be used in the following sections when we fit the equations of basic curves to such data.

One way of organizing data in order to develop some understanding of it is to *tabulate the number of occurrences for each particular value within the set. This is called a* **frequency distribution.**

Example A

A test station measured the loudness of the sound of jet aircraft taking off from a certain airport. The decibel readings of the first twenty jets were as follows: 110, 95, 100, 115, 105, 110, 120, 110, 115, 105, 90, 95, 105, 110, 100, 115, 105, 120, 95, 110.

We can see that it is difficult to determine any pattern to the readings. Therefore, we set up the following table to show the frequency distribution.

Decibel reading	90	95	100	105	110	115	120
Frequency	1	3	2	4	5	3	2

We note that the distribution and pattern of readings is clearer in this form.

Just as graphs are useful in representing algebraic functions, so also are they a very convenient method of representing frequency distributions. There are several useful methods of graphing such distributions, among which one of the most important is the **histogram,** which is illustrated in the following example.

Example B

A histogram represents a particular set of data by use of rectangles. The width of the base of each rectangle is the same, and is labeled for one of the readings. The height of the rectangle represents the number of values of a given reading. In Fig. 12–1 a histogram representing the data of the frequency distribution of Example A is shown.

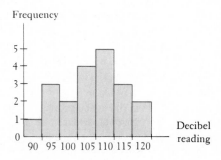

Figure 12–1

Tables of values and graphical representations give a general description of data. However, it is useful to find representative values for the location of the center of a distribution of values and other numbers to give a measure of how the various numbers within the set deviate from this representative central value.

There are several representative central values which may be used. These include *the* **median,** *which is the middle number—that number which has as many values above it as below it in the distribution.* Also, there is *the* **mode,** *which is the value that appears most frequently.* These are important measures, but the one we shall be using for our development in this chapter is the **arithmetic mean.**

The arithmetic mean is calculated by finding the sum of all the values and then dividing this sum by the number of values. (The arithmetic mean, sometimes called simply the *mean,* is the number most people call the "average." However, in statistics the word *average* has the more general meaning of a measure of the center of a distribution.)

Example C

In order to determine the frictional force between two specially designed surfaces, the force to move a block with one surface along an inclined plane with the other surface is measured 10 times. The results, with forces in newtons, are:

2.2, 2.4, 2.1, 2.2, 2.5, 2.2, 2.4, 2.7, 2.1, 2.5

Find the arithmetic mean of these forces.

To find the mean we sum the values of the forces and divide this total by 10. This gives

$$F_{mean} = \frac{2.2 + 2.4 + 2.1 + 2.2 + 2.5 + 2.2 + 2.4 + 2.7 + 2.1 + 2.5}{10}$$

$$= \frac{23.3}{10}$$

$$= 2.3 \text{ N}$$

Thus the mean is 2.3 N.

If we wish to find the arithmetic mean of a large number of values, and if some of them appear more than once, the calculation can be somewhat simplified. By multiplying each number by its frequency, adding these results, and then dividing by the total number of values considered, the mean can be calculated. Letting $\bar{x}$ represent the mean of the values $x_1, x_2, \ldots, x_n$, which occur $f_1, f_2, \ldots, f_n$ times, respectively, we have

$$\bar{x} = \frac{x_1 f_1 + x_2 f_2 + \cdots + x_n f_n}{f_1 + f_2 + \cdots + f_n} \tag{12-1}$$

(This notation can be simplified by letting

$$x_1 f_1 + x_2 f_2 + \cdots + x_n f_n = \sum_{i=1}^{n} x_i f_i$$

and

$$f_1 + f_2 + \cdots + f_n = \sum_{i=1}^{n} f_i$$

This notation is often used to indicate a sum; Σ *is known as the* **summation sign.**)

Example D

Find the arithmetic mean of the measurements of loudness of the sound of jet aircraft of Example A. For reference, we repeat the frequency distribution.

Decibel reading	90	95	100	105	110	115	120
Frequency	1	3	2	4	5	3	2

With d representing the decibel reading, we use Eq. (12–1) to find the mean.

$$\bar{d} = \frac{90(1) + 95(3) + 100(2) + 105(4) + 110(5) + 115(3) + 120(2)}{1 + 3 + 2 + 4 + 5 + 3 + 2}$$

$$= \frac{2130}{20}$$

$$= 107 \text{ dB}$$

The arithmetic mean gives us a measure of the center of a distribution of numbers. However, it does not tell us whether the data are grouped closely together or spread over a large range of values. Therefore, we also need some measure of the deviation, or dispersion, of the values from the median or mean. If the dispersion is small and the numbers are grouped closely together, the measure of central tendency is more reliable and descriptive of the data than the case in which the spread is greater.

There are several measures of dispersion, and we discuss in this section one which is very widely used. It is called the **standard deviation.**

The standard deviation of a set of numbers is given by the equation

$$s = \sqrt{\overline{(x - \bar{x})^2}} \tag{12–2}$$

The definition of s indicates that the following steps are to be taken in computing its value.

1. Find the arithmetic mean $\bar{x}$ of the set of numbers.
2. Subtract the mean from each of the numbers of the set.
3. Square these differences.
4. Find the arithmetic mean of these squares. (Note that a bar over any quantity signifies the mean of the set of values of that quantity.)
5. Find the square root of this last arithmetic mean.

Defined in this way, s must be a positive number, and thus indicates a deviation from the mean, regardless of whether or not individual numbers are greater or less than the mean.

Example E

Find the standard deviation of the numbers 1, 5, 4, 2, 6, 2, 1, 1, 5, 3. The most efficient method of finding s is to make a table in which Steps 1–4 are indicated.

x	$x - \bar{x}$	$(x - \bar{x})^2$
1	-2	4
5	2	4
4	1	1
2	-1	1
6	3	9
2	-1	1
1	-2	4
1	-2	4
5	2	4
3	0	0
$\overline{30}$		$\overline{32}$

$$\bar{x} = \frac{30}{10} = 3$$

$$\overline{(x - \bar{x})^2} = \frac{32}{10} = 3.2$$

$$s = \sqrt{3.2} = 1.8$$

We can see from Example E that a great deal of calculation is required to find the standard deviation of a set of numbers. It is possible to reduce this work somewhat with a different form of the equation for standard deviation. It can be shown that

$$s = \sqrt{\overline{x^2} - \bar{x}^2}$$

(12–3)

Equation (12–3) shows us that the standard deviation s may be found by finding the mean of the squares of the x's, subtracting the square of $\bar{x}$, and then finding the square root. This eliminates the step of subtracting $\bar{x}$ from each x, a step required by the original definition. Obviously, a calculator is very useful in making these calculations.

Example F

By using Eq. (12–3), find s for the numbers in Example A. Again, a table is the most convenient form.

x	x^2
1	1
5	25
4	16
2	4
6	36
2	4
1	1
1	1
5	25
3	9
$\overline{30}$	$\overline{122}$

$$\bar{x} = \frac{30}{10} = 3, \qquad \bar{x}^2 = 9$$

$$\overline{x^2} = \frac{122}{10} = 12.2$$

$$\overline{x^2} - \bar{x}^2 = 12.2 - 9 = 3.2$$

$$s = \sqrt{3.2} = 1.8$$

Example G

In an ammeter, two resistances are connected in parallel. Most of the current passing through the meter goes through the one called the shunt. In order to determine the accuracy of the resistance of shunts being made for ammeters, a manufacturer tested a sample of 100 shunts. The resistance of each, to the nearest hundredth of an ohm, is indicated in the following table. Calculate the standard deviation of the resistances of the shunts.

R (ohms)	f	fR	fR²
0.200	1	0.200	0.0400
0.210	3	0.630	0.1323
0.220	5	1.100	0.2420
0.230	10	2.300	0.5290
0.240	17	4.080	0.9792
0.250	40	10.000	2.5000
0.260	13	3.380	0.8788
0.270	6	1.620	0.4374
0.280	3	0.840	0.2352
0.290	2	0.580	0.1682
	100	24.730	6.1421

$$\overline{R} = \frac{\Sigma fR}{\Sigma f} = \frac{24.73}{100} = 0.2473$$

$$\overline{R}^2 = 0.06115729$$

$$\overline{R^2} = \frac{\Sigma fR^2}{\Sigma f} = \frac{6.142}{100} = 0.06142$$

$$s = \sqrt{\overline{R^2} - \overline{R}^2} = 0.016$$

The arithmetic mean of the resistances is 0.247 Ω, with a standard deviation of 0.016 Ω.

The standard normal distribution curve

Figure 12–2

Many calculators are programmed to give the arithmetic mean and standard deviation when data are entered. Also, these statistical measures are readily programmed to be determined by a computer. Such a program is in Appendix C.

We have seen that the standard deviation is a measure of the dispersion of a set of data. Generally, if we make a great many measurements of a given type, we would expect to find a majority of them near the arithmetic mean. If this is the case, the distribution probably follows, at least to a reasonable extent, the **standard normal distribution curve**. See Fig. 12–2. This curve gives a theoretical distribution about the arithmetic mean, assuming that the number of values in the distribution becomes infinite, and its equation is

$$y = \frac{1}{\sqrt{2\pi}} e^{-x^2/2} \tag{12–4}$$

An important feature of the standard deviation is that *in a normal distribution, about 68% of the values are found within the interval* $\bar{x} - s$ *to* $\bar{x} + s$.

Example H

In Example G, $\bar{R} = 0.247\ \Omega$ and $s = 0.016\ \Omega$. Therefore, $\bar{R} - s = 0.231\ \Omega$ and $\bar{R} + s = 0.263\ \Omega$. From the table in Example G we can see that 70 of the 100 values, or 70%, lie between $0.231\ \Omega$ and $0.263\ \Omega$.

Exercises 12–1

In Exercises 1 through 20, use the following sets of numbers.

A: 3, 6, 4, 2, 5, 4, 7, 6, 3, 4, 6, 4, 5, 7, 3

B: 25, 26, 23, 24, 25, 28, 26, 27, 23, 28, 25

C: 48, 53, 49, 45, 55, 49, 47, 55, 48, 57, 51, 46

D: 105, 108, 103, 108, 106, 104, 109, 104, 110, 108, 108, 104, 113, 106, 107, 106, 107, 109, 105, 111, 109, 108

See Appendix C for a computer program for finding the arithmetic mean and standard deviation.

In Exercises 1 through 4, set up a frequency distribution table, indicating the frequency of each number given in the indicated set.

1. Set A 2. Set B 3. Set C 4. Set D

In Exercises 5 through 8, draw histograms for the data in the given exercise.

5. Exercise 1   6. Exercise 2   7. Exercise 3   8. Exercise 4

In Exercises 9 through 12, determine the arithmetic mean of the numbers of the given set.

9. Set A 10. Set B 11. Set C 12. Set D

In Exercises 13 through 16, use Eq. (12–2) to find the standard deviation s for the numbers of the given set.

13. Set A 14. Set B 15. Set C 16. Set D

In Exercises 17 through 20, use Eq. (12–3) to find the standard deviation s for the numbers of the given set.

17. Set A 18. Set B 19. Set C 20. Set D

In Exercises 21 through 36, find the indicated quantities.

21. A bottling machine is designed to pour 750 mL of soda into each bottle. During the day sample bottles were tested, and the following volumes were recorded (in milliliters.)

    751, 754, 748, 750, 752, 751, 747, 748, 753, 750, 746, 751, 753, 749, 753

    Make a frequency distribution table indicating the frequency of each volume recorded in this data.

22. For the data of Exercise 21, draw a histogram.

23. Find the arithmetic mean of the volumes in Exercise 21.

24. Find the standard deviation of the volumes in Exercise 21.

25. A strobe light is designed to flash every 2.25 s at a certain setting. Sample bulbs were tested with the results in the following table.

Number of bulbs	2	7	18	41	56	32	8	3	3
Time between flashes (seconds)	2.21	2.22	2.23	2.24	2.25	2.26	2.27	2.28	2.29

Draw a histogram for these data.

26. Find the mean of the times in Exercise 25.

27. Find the standard deviation of the times in Exercise 25.

28. One hundred motorists were asked to maintain a miles per gallon record for their cars, all the same model. The records which were reported are shown as follows:

Miles per gallon	19.0–19.4	19.5–19.9	20.0–20.4	20.5–20.9
Number reporting	4	9	14	24

Miles per gallon	21.0–21.4	21.5–21.9	22.0–22.4	22.5–22.9
Number reporting	28	16	4	1

Draw a histogram for these data, using the middle value of each interval to represent the interval.

29. Find the arithmetic mean of the miles per gallon values of Exercise 28, using the middle value of each interval to represent the interval.

30. Find the standard deviation of the miles per gallon values of Exercise 28, using the middle value of each interval to represent the interval.

31. The weekly salaries (in dollars) for the workers in a small factory are as follows:

    250, 350, 275, 225, 175, 300, 200, 350, 275, 400, 300, 225, 250, 300

Find the arithmetic mean of these salaries.

32. Find the standard deviation of the salaries in Exercise 31.

33. In a particular month the electrical usage, rounded to the nearest 100 kWh (kilowatt hours), of 1000 homes in a certain city was summarized as follows:

Number of homes	22	80	106	185	380	122	90	15
Usage	500	600	700	800	900	1000	1100	1200

Find the mean of the electrical usage.

34. Find the standard deviation of the electric usage in Exercise 33.

35. In testing the electric current in a certain microprocessor circuit, the following values (in microamperes) were found on successive trials:

    3.44, 3.48, 3.40, 3.29, 3.46, 3.41, 3.37, 3.45, 3.47,

    3.43, 3.38, 3.50, 3.41, 3.42, 3.44, 3.46, 3.39, 3.47

Find the arithmetic mean of these currents.

36. Find the standard deviation of the currents in Exercise 35.

In Exercises 37 through 44, find the percentage of values in the interval $\bar{x} - s$ to $\bar{x} + s$ for the indicated sets of numbers.

37. Set A   38. Set B   39. Set C   40. Set D

41. The times in Exercise 25.   42. The measurements in Exercise 28.

43. The salaries in Exercise 31.   44. The measurements in Exercise 35.

12–2 Fitting a Straight Line to a Set of Points

In this section we shall show a method of "fitting" a straight line to a given set of points. In the following section fitting suitable nonlinear curves to given sets of points will be discussed. Some of the reasons for doing this are (1) to express a concise relationship between the variables, (2) to use the equation to predict certain fundamental results, (3) to determine the reliability of certain sets of data, and (4) to use the data for testing theoretical concepts.

We shall assume for the examples and exercises of the remainder of this chapter that there is some relationship between the variables. Often when we are analyzing statistics for variables between which we think a relationship might exist, the points are so scattered as to give no reasonable idea regarding the possible functional relationship. We shall assume here that such combinations of variables have been discarded in the analysis.

Example A

Figure 12–3

All the students enrolled in a mathematics course took an entrance test in mathematics. To study the reliability of this test, an instructor tabulated the test scores of 10 students (selected at random), along with their course average, and made a graph of these figures (see table below and Fig. 12–3).

Student	Entrance test score, based on 40	Course average, based on 100
A	29	63
B	33	88
C	22	77
D	17	67
E	26	70
F	37	93
G	30	72
H	32	81
I	23	47
J	30	74

Figure 12–4

We now ask whether or not there is a functional relationship between the test scores and the course grades. Certainly no clear-cut relationship exists, but in general we see that the higher the test score, the higher the course grade. This leads to the possibility that there might be some straight line, from which none of the points would vary too significantly. If such a line could be found, then it could be the basis of predictions as to the possible success a student might have in the course, on the basis of his or her grade on the entrance test. Assuming that such a straight line exists, the problem is to find the equation of this line. Figure 12–4 shows two such possible lines.

There are various methods of determining the line which best fits the given points. We shall employ the one which is most widely used: the **method of least squares.** *The basic principle of this method is that the sum of the squares of the deviation of all points from the best line (in accordance with this method) is the least it can be. By* **deviation** *we mean the difference between the y-value of the line and the y-value for the point (of original data) for a particular value of x.*

Example B

Entrance test

Figure 12–5

In Fig. 12–5 the deviations of some of the points of Example A are shown. The point (67, 17) (student D of Example A) has a deviation of 6 from the indicated line in the figure. Thus we square the value of this deviation to obtain 36. The method of least squares requires that the sum of all such squares be a minimum in order to determine the line which fits best.

Therefore, in applying this method, it is necessary to use the equation of a straight line and the coordinates of the points of the data. The deviations of these are indicated and squared. The problem then arises as to the method of determining the constants m and b in the equation of the line $y = mx + b$ for which the sum of these squares is a minimum. The following example indicates the method employed and forms a basis for determining the equations for finding m and b.

Example C

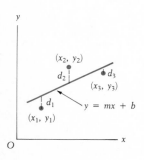

Figure 12–6

Given the points (x_1, y_1), (x_2, y_2), and (x_3, y_3), find m and b in the equation

$$y = mx + b$$

for which the sum of the squares of the deviations is a minimum. See Fig. 12–6.

The deviations are

$$d_1 = y_1 - (mx_1 + b)$$
$$d_2 = y_2 - (mx_2 + b)$$

and

$$d_3 = y_3 - (mx_3 + b)$$

Squaring each of these expressions and finding the sum of the squares, we have

$$S = y_1^2 + y_2^2 + y_3^2 + m^2(x_1^2 + x_2^2 + x_3^2)$$
$$+ 3b^2 - 2m(x_1 y_1 + x_2 y_2 + x_3 y_3)$$
$$- 2b(y_1 + y_2 + y_3) + 2mb(x_1 + x_2 + x_3)$$

We want S to be a minimum with respect to each of m and b. Therefore this is a maximum-minimum problem of the type we studied in Section 10–4. We shall find the partial derivatives of S with respect to m and with respect to b, set each of these equal to zero, and solve the resulting equations simultaneously:

$$\frac{\partial S}{\partial m} = 2m(x_1^2 + x_2^2 + x_3^2) - 2(x_1 y_1 + x_2 y_2 + x_3 y_3)$$
$$+ 2b(x_1 + x_2 + x_3)$$
$$\frac{\partial S}{\partial b} = 6b - 2(y_1 + y_2 + y_3) + 2m(x_1 + x_2 + x_3)$$

Setting the second of these equations equal to zero and solving for b, we have

$$b = \frac{(y_1 + y_2 + y_3) - m(x_1 + x_2 + x_3)}{3}$$

Examination of this equation shows that it may be written as

$$b = \bar{y} - m\bar{x} \qquad (1)$$

where $\bar{y}$ and $\bar{x}$ are the arithmetic means of the coordinates of the given points. Setting $\partial S / \partial m = 0$, we may obtain

$$m(x_1^2 + x_2^2 + x_3^2) = (x_1 y_1 + x_2 y_2 + x_3 y_3) - b(x_1 + x_2 + x_3)$$

Dividing this equation through by 3, we note that it may be written as

$$m\overline{x^2} = \overline{xy} - b\bar{x} \qquad (2)$$

where $\overline{x^2}$ and $\overline{xy}$ are the respective arithmetic means of the coordinates of the given points.

Substituting the expression for b from Eq. (1) in Eq. (2) we have

$$m\overline{x^2} = \overline{xy} - (\bar{y} - m\bar{x})\bar{x}$$

which may be written as

$$m = \frac{\overline{xy} - \bar{x}\,\bar{y}}{\overline{x^2} - \bar{x}^2} \qquad (3)$$

By evaluating the appropriate arithmetic means, m and b may be determined for the least-squares line when actual numerical values are given.

Generalizing on Example C and using Eq. (12-3), the equation of the **least-squares** line

$$\overline{y = mx + b} \qquad (12\text{-}5)$$

can be found by the values

$$\overline{m = \frac{\overline{xy} - \bar{x}\,\bar{y}}{s_x^2}} \qquad (12\text{-}6)$$

and

$$\overline{b = \bar{y} - m\bar{x}} \qquad (12\text{-}7)$$

▶ In Eqs. (12-6) and (12-7), the x's and y's are those of the points in the data and s_x is the standard deviation of the x-values.

Example D

Find the least-squares line of the data of Example A.

Here the y-values will be the entrance-test scores, and the x-values the course averages. The results are best obtained by tabulating the necessary quantities, as we do in the following table.

x	y	xy	x^2
63	29	1,827	3,969
88	33	2,904	7,744
77	22	1,694	5,929
67	17	1,139	4,489
70	26	1,820	4,900
93	37	3,441	8,649
72	30	2,160	5,184
81	32	2,592	6,561
47	23	1,081	2,209
74	30	2,220	5,476
732	279	20,878	55,110

$$\bar{x} = \frac{732}{10} = 73.2, \quad \bar{x}^2 = 5358.24$$

$$\bar{y} = \frac{279}{10} = 27.9, \quad \bar{x}\,\bar{y} = 2042.28$$

$$\overline{xy} = \frac{20,878}{10} = 2087.8$$

$$\overline{x^2} = \frac{55,110}{10} = 5511.0$$

$$s_x^2 = \overline{x^2} - \bar{x}^2 = 152.76$$

$$m = \frac{\overline{xy} - \bar{x}\,\bar{y}}{s_x^2} = \frac{2087.8 - 2042.28}{152.76}$$

$$= 0.2979838$$

$$b = \bar{y} - m\bar{x} = 27.9 - (0.2979838)(73.2) = 6.09$$

$$y = 0.298x + 6.09$$

Thus, the equation of the line is $y = 0.298x + 6.09$. (See Fig. 12–7.)

Figure 12–7

Figure 12–8

We note that when plotting the least-squares line, we know two points on it. These are $(0, b)$ and $(\bar{x}, \bar{y})$. We can see that $(\bar{x}, \bar{y})$ satisfies the equation from the solution for b [Eq. (12–7)].

Example E

In an experiment to determine the relation between the load on a spring and the length of the spring, the following data were found.

Load (pounds)	0.0	1.0	2.0	3.0	4.0	5.0
Length (inches)	10.0	11.2	12.3	13.4	14.6	15.9

Find the least-squares line for these data, which express the length as a function of the load.

x	y	xy	x^2
0.0	10.0	0.0	0.0
1.0	11.2	11.2	1.0
2.0	12.3	24.6	4.0
3.0	13.4	40.2	9.0
4.0	14.6	58.4	16.0
5.0	15.9	79.5	25.0
15.0	77.4	213.9	55.0

$$\bar{x} = \frac{15.0}{6} = 2.50, \quad \bar{x}^2 = 6.25$$

$$\bar{y} = \frac{77.4}{6} = 12.9, \quad \bar{x}\,\bar{y} = 32.25$$

$$\overline{xy} = \frac{213.9}{6} = 35.65$$

$$\overline{x^2} = \frac{55.0}{6} = 9.1666667$$

$$s_x^2 = \overline{x^2} - \bar{x}^2 = 9.1666667 - 6.25 = 2.9166667$$

$$m = \frac{35.65 - 32.25}{2.9166667} = 1.1657143$$

$$b = 12.9 - (1.1657143)(2.50) = 9.985714$$

Therefore, the least-squares line in this case is $y = 1.17x + 9.99$ where y is the length of the spring and x is the load. In Fig. 12–8, the line shown is the least-squares line, and the points are the experimental data points.

Exercises 12–2

See Appendix C for a computer program for finding the least-squares line.

In Exercises 1 through 12, find the equation of the least-squares line for the given data. In each case graph the line and the data points on the same graph.

1. The points in the following table:

x	4	6	8	10	12
y	1	4	5	8	9

2. The points in the following table:

x	1	2	3	4	5	6	7
y	10	17	28	37	49	56	72

3. The points in the following table:

x	20	26	30	38	48	60
y	160	145	135	120	100	90

4. The points in the following table:

x	1	3	6	5	8	10	4	7	3	8
y	15	12	10	8	9	2	11	9	11	7

5. In an electrical experiment, the following data were found for the values of current and voltage for a particular element of the circuit. Find the voltage as a function of the current.

Current (milliamperes)	15.0	10.8	9.30	3.55	4.60
Voltage (volts)	3.00	4.10	5.60	8.00	10.50

6. The velocity of a falling object was found each second by use of an electric device as shown. Find the velocity as a function of time.

Time (seconds)	1.00	2.00	3.00	4.00	5.00	6.00	7.00
Velocity (meters per second)	9.70	19.5	29.5	39.4	49.2	58.9	68.6

7. The altitude *h* of a rocket was measured at several positions at a horizontal distance *x* from the launch site, as shown in the table. Find the least-squares line for *h* as a function of *x*.

x (meters)	0	500	1000	1500	2000	2500
h (meters)	0	1130	2250	3360	4500	5600

8. In testing an air-conditioning system, the temperature *T* in a building was measured during the afternoon hours with the results shown in the table. Find the least-squares line for *T* as a function of the time *t* from noon.

t (hours)	0.0	1.0	2.0	3.0	4.0	5.0
T (degrees Celsius)	20.5	20.6	20.9	21.3	21.7	22.0

9. The pressure *p* was measured along an oil pipeline at different distances from a reference point, with results as shown. Find the least-squares line for *p* as a function of *x*.

x (feet)	0	100	200	300	400
p (lb/in.^{2})	650	630	605	590	570

10. The heat loss *L* per hour through various thicknesses of a particular type of insulation was measured as shown in the table. Find the least-squares line for *L* as a function of *t*.

t (inches)	3.0	4.0	5.0	6.0	7.0
L (Btu)	5900	4800	3900	3100	2450

11. In an experiment on the photoelectric effect, the frequency of the light being used was measured as a function of the stopping potential (the voltage just sufficient to stop the photoelectric current) with the results given below. Find the least-squares line for *V* as a function of *f*. The frequency for *V* = 0 is known as the *threshold frequency*. From the graph, determine the threshold frequency.

f (petahertz)	0.550	0.605	0.660	0.735	0.805	0.880
V (volts)	0.350	0.600	0.850	1.10	1.45	1.80

12. If gas is cooled under conditions of constant volume, it is noted that the pressure falls nearly proportionally as the temperature. If this were to happen until there was no pressure, the theoretical temperature for this case is referred to as *absolute zero*. In an elementary experiment, the following data were found for pressure and temperature for a gas under constant volume.

T (degrees Celsius)	0.0	20	40	60	80	100
P (kilopascals)	133	143	153	162	172	183

Find the least-squares line for *P* as a function of *T*, and, from the graph, determine the value of absolute zero found in this experiment.

The linear coefficient of correlation, a measure of the relatedness of two variables, is defined by $r = m(s_x/s_y)$. Due to its definition, the values of *r* lie in the range $-1 \le r \le 1$. If *r* is near 1, the correlation is considered good. For values of *r* between $-\frac{1}{2}$ and $+\frac{1}{2}$, the correlation is poor. If *r* is near -1, the variables are said to be negatively correlated; that is, one increases while the other decreases.

In Exercises 13 through 16, compute *r* for the given sets of data.

13. Exercise 1   14. Exercise 2   15. Exercise 4   16. Example A

12-3 Fitting Nonlinear Curves to Data

If the experimental points do not appear to be on a straight line, but we recognize them as being approximately on some other type of curve, the method of least squares can be extended to use on these other curves. For example, if the points are apparently on a parabola, we could use the function $y = a + bx^2$. To use the above method, we shall extend the least-squares line to

$$\overline{y = m[f(x)] + b} \qquad (12\text{-}8)$$

▶ Here, $f(x)$ must be calculated first, and then the problem can be treated as a least-squares line to find the values of m and b. Some of the functions $f(x)$ which may be considered for use are x^2, $1/x$, and 10^x.

Example A

Find the least-squares curve $y = mx^2 + b$ for the points in the following table.

y	1	5	12	24	53	76
x	0	1	2	3	4	5

The necessary quantities are tabulated for purposes of calculation.

x	y	$f(x) = x^2$	yx^2	$(x^2)^2$
0	1	0	0	0
1	5	1	5	1
2	12	4	48	16
3	24	9	216	81
4	53	16	848	256
5	76	25	1900	625
	171	55	3017	979

$$\overline{x^2} = \frac{55}{6} = 9.1666667, \qquad \overline{x^2}^2 = 84.027778$$

$$\overline{y} = \frac{171}{6} = 28.5, \qquad \overline{x^2 y} = 261.25$$

$$\overline{yx^2} = \frac{3017}{6} = 502.83333, \qquad \overline{(x^2)^2} = \frac{979}{6} = 163.16667$$

$$s_{(x^2)}^2 = 163.16667 - 84.027778 = 79.138892$$

$$m = \frac{502.83333 - 261.25}{79.138892} = 3.0526499$$

$$b = 28.5 - (3.0526499)(9.1666667)$$
$$= 0.52$$

Therefore, the desired equation is $y = 3.05x^2 + 0.52$. The graph of this equation and the data points are shown in Fig. 12-9.

Figure 12-9

Example B

Figure 12–10

In a physics experiment designed to measure the pressure and volume of a gas at constant temperature, the following data were found. When the points were plotted, they were seen to approximate the hyperbola $y = c/x$. Find the least-squares approximation to this hyperbola $[y = m(1/x) + b]$ (see Fig. 12–10).

P(kPa)	V(cm^3)	$x(= V)$	$y(= P)$	$f(x) = \dfrac{1}{x}$	$y\left(\dfrac{1}{x}\right)$	$\left(\dfrac{1}{x}\right)^2$
120.0	21.0	21.0	120.0	0.0476190	5.7142857	0.0022676
99.2	25.0	25.0	99.2	0.0400000	3.9680000	0.0016000
81.3	31.8	31.8	81.3	0.0314465	2.5566038	0.0009889
60.6	41.1	41.1	60.6	0.0243309	1.4744526	0.0005920
42.7	60.1	60.1	42.7	0.0166389	0.7104825	0.0002769
			403.8	0.1600353	14.4238246	0.0057254

(*Note on calculator use:* The final digits for the values shown may vary, depending on the type of calculator and how values are used. Here all individual values are shown, assuming that the calculator displays eight digits but actually carries more. The value of $1/x$ was found from the value of x, with the eight-digit display indicated. However, the values of $y(1/x)$ and $(1/x)^2$ were found using the extra digits carried by the calculator for $1/x$. The sums were found using the rounded-off values shown.)

$$\overline{\frac{1}{x}} = 0.0320071 \quad \overline{\left(\frac{1}{x}\right)^2} = 0.0010245 \qquad \bar{y} = 80.76 \quad \overline{\frac{1}{x}}\,\bar{y} = 2.5848902$$

$$\overline{y\left(\frac{1}{x}\right)} = 2.8847649 \quad \overline{\left(\frac{1}{x}\right)^2} = 0.0011451 \qquad s^2_{(1/x)} = 0.0011451 - 0.0010245$$
$$= 0.0001206$$

$$m = \frac{2.8847649 - 2.5848902}{0.0001206}$$

$$= 2486.5232$$

$$b = 80.76 - 2486.5232(0.0320071)$$

$$= 1.17$$

Thus the equation of the hyperbola $y = m(1/x) + b$ is

$$y = \frac{2490}{x} + 1.17$$

The graph of this hyperbola and the points representing the data are shown in Fig. 12–11.

$y\ (=P)$

Figure 12–11

Example C

Figure 12–12

It has been found experimentally that the tensile strength of brass (a copper-zinc alloy) increases (within certain limits) with the percentage of zinc. The following table indicates the values which have been found (also see Fig. 12–12).

Tensile strength (10^5 lb/in.^{2})	0.32	0.36	0.40	0.44	0.48
Percentage of zinc	0	5	13	22	34

Fit a curve of the form $y = m(10^x) + b$ to the data. Let x = tensile strength ($\times 10^5$) and y = percentage of zinc.

x	y	$f(x) = 10^x$	$y(10^x)$	$(10^x)^2$
0.32	0	2.0892961	0.000000	4.3651583
0.36	5	2.2908677	11.454338	5.2480746
0.40	13	2.5118864	32.654524	6.3095734
0.44	22	2.7542287	60.593031	7.5857758
0.48	34	3.0199517	102.67836	9.1201084
	74	12.6662306	207.38025	32.6286905

(See the note on calculator use in Example B.)

$$\overline{10^x} = 2.5332461 \qquad \overline{10^x y} = 37.492042$$
$$\overline{10^{x^2}} = 6.4173359 \qquad \overline{y(10^x)} = 41.47605$$
$$\overline{y} = 14.8 \qquad \overline{(10^x)^2} = 6.5257381$$
$$s^2_{(10^x)} = 6.5257381 - 6.4173359 = 0.1084022$$
$$m = \frac{41.47605 - 37.492042}{0.1084022} = 36.752095$$
$$b = 14.8 - (36.752095)(2.5332461) = -78.3$$

The equation of the curve is $y = 36.8(10^x) - 78.3$. It must be remembered that for practical purposes, y must be positive. The graph of the equation is shown in Fig. 12–13, with the solid portion denoting the meaningful part of the curve. The points of the data are also shown.

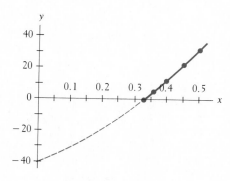

Figure 12–13

Exercises 12–3

In each of the following exercises, find the indicated least-squares curve. Sketch the curve and plot the data points on the same graph.

1. For the points in the following table, find the least-squares curve $y = mx^2 + b$.

x	2	4	6	8	10
y	12	38	72	135	200

2. For the points in the following table, find the least-squares curve $y = m\sqrt{x} + b$.

x	0	4	8	12	16
y	1	9	11	14	15

3. For the points in the following table, find the least-squares curve $y = m(1/x) + b$.

x	1.10	2.45	4.04	5.86	6.90	8.54
y	9.85	4.50	2.90	1.75	1.48	1.30

4. For the points in the following table, find the least-squares curve $y = m(10^x) + b$.

x	0.00	0.200	0.500	0.950	1.325
y	6.00	6.60	8.20	14.0	26.0

5. The following data were found for the distance y that an object rolled down an inclined plane in time t. Determine the least-squares curve $y = mt^2 + b$.

t (seconds)	1.0	2.0	3.0	4.0	5.0
y (centimeters)	6.0	23	55	98	148

6. The increase in length of a certain metallic rod was measured in relation to particular increases in temperature. If y represents the increase in length for the corresponding increase in temperature x, find the least-squares curve $y = m(x^2) + b$ for these data.

x (degrees Celsius)	50.0	100	150	200	250
y (centimeters)		1.00	4.40	9.40	16.4	24.0

7. Use the data of Exercise 6 and determine the least-squares curve $y = m(10^z) + b$ where $z = x/1000$.

8. Measurements were made of the current in an electric circuit as a function of time. The circuit contained a resistance of 5 Ω and an inductance of 10 H. The following data were found.

t (seconds)	0.00	2.00	4.00	6.00	8.00
i (amperes)	0.00	2.52	3.45	3.80	3.92

   Find the least-squares curve $i = m(e^{-0.5t}) + b$ for this data. From the equation, determine the value the current approaches as t approaches infinity.

9. The resonant frequency of an electric circuit containing a 4-μF capacitor was measured as a function of an inductance in the circuit. The following data were found.

L (henrys)	1.0	2.0	4.0	6.0	9.0
f (hertz)	490	360	250	200	170

   Find the least-squares curve $f = m(1/\sqrt{L}) + b$.

10. Measurements of the displacement of a pendulum bob from its equilibrium position as a function of time gave the following results.

Time (seconds)	0.00	0.90	1.60	2.20	3.15
Displacement (centimeters)	0.00	7.80	10.0	8.10	0.00

Find the least-squares curve of the form $y = m(\sin x) + b$, expressing the displacement as y and the time as x.

12–4 Review Exercises for Chapter 12

In Exercises 1 through 4, use the following set of numbers:

2.6, 2.4, 2.9, 2.3, 2.6, 2.6, 2.5, 2.8, 2.7
2.4, 2.6, 2.5, 2.8, 2.4, 2.9

1. Construct a frequency distribution table.

2. Draw a histogram.

3. Find the arithmetic mean of the numbers in the set.

4. Find the standard deviation of the numbers in the set.

In Exercises 5 through 8, use the following set of numbers.

109, 103, 113, 110, 113, 106, 114, 101, 108, 101, 112, 106, 115, 102, 107, 110, 102, 115, 106, 105

5. Construct a frequency distribution table with intervals 101–103, 104–106, and so on.

6. Find the arithmetic mean.

7. Find the standard deviation.

8. Draw a histogram for the table of Exercise 5. Let the middle value of each interval represent the interval.

In Exercises 9 through 12, use the following data:

An important property of oil is its coefficient of viscosity, which gives a measure of how well it flows. In order to determine the viscosity of a certain motor oil, a refinery took samples from 12 different storage tanks and tested them at 50°C. The results (in pascal-seconds) were 0.24, 0.28, 0.29, 0.26, 0.27, 0.26, 0.25, 0.27, 0.28, 0.26, 0.26, 0.25.

9. Construct a frequency distribution table.

10. Make a histogram.

11. Determine the arithmetic mean.

12. Determine the standard deviation.

In Exercises 13 through 15, use the following data:

Two machine parts are considered satisfactorily assembled if their total thickness (to the nearest 0.01 in.) is between or equal to 0.92 and 0.94 in. One hundred sample assemblies are tested, and the thicknesses, to the nearest 0.01 in., are given in the following table:

Total thickness	0.90	0.91	0.92	0.93	0.94	0.95	0.96
Number	3	9	31	38	12	5	2

13. Make a histogram.

14. Determine the arithmetic mean.

15. Determine the standard deviation.

In Exercises 16 through 18, use the following data:

A Geiger counter records the presence of high-energy nuclear particles. Even though no apparent radioactive source is present, a certain number of particles will be recorded. These are primarily cosmic rays, which are caused by very high-energy particles from outer space. In an experiment to measure the amount of cosmic radiation, the number of counts were recorded during 200 5-s intervals. The following table gives the number of counts, and the number of 5-s intervals having this number of counts.

Counts	0	1	2	3	4	5	6	7	8	9	10
Intervals	3	10	25	45	29	39	26	11	7	2	3

16. Make a histogram.

17. Determine the arithmetic mean.

18. Determine the standard deviation.

In Exercises 19 through 26, find the indicated least-squares curves.

19. In a certain experiment, the resistance of a certain resistor was measured as a function of the temperature. The data found were as follows:

T (degrees Celsius)	0.0	20.0	40.0	60.0	80.0	100
R (ohms)	25.0	26.8	28.9	31.2	32.8	34.7

Find the least-squares line for this data, expressing R as a function of T. Sketch the line and data points on the same graph.

20. The solubility of sodium chloride (table salt) (in kilograms per cubic meter of water) as a function of temperature (degrees Celsius) is measured as follows:

Temperature	0.0	20.0	40.0	60.0	80.0
Solubility	357	360	366	373	384

Find the least-squares straight line for these data.

21. An air pollution monitoring station took samples of air each hour during the later morning hours and tested each sample for the number n of parts per million (ppm) of carbon monoxide. The results are shown in the table, where t is the number of hours after 6 A.M. Find the least-squares line for n as a function of t.

t (hours)	0.0	1.0	2.0	3.0	4.0	5.0	6.0
n (ppm)	8.0	8.2	8.8	9.5	9.7	10.0	10.7

22. The *Mach number* of a moving object is the ratio of its speed to the speed of sound (740 mi/h). The following table shows the speed s of a jet aircraft, in terms of Mach numbers, and the time t after it starts to accelerate. Find the least-squares line of s as a function of t.

t (minutes)	0.00	0.60	1.20	1.80	2.40	3.00
s (Mach number)	0.88	0.97	1.03	1.11	1.19	1.25

23. The coefficient of friction of an object sliding down an inclined plane was measured as a function of the angle of incline of the plane. The following results were obtained.

Angle (degrees)	10.0	20.0	30.0	40.0	50.0	60.0	70.0
Coefficient of friction	0.16	0.34	0.55	0.85	1.24	1.82	2.80

Find the least-squares curve of the form $y = m(\tan x) + b$, expressing the coefficient of friction as y and the angle as x.

24. An electric device recorded the total distance (in centimeters) which an object fell at 0.1 s intervals. Following are the data which were found.

Time	0.100	0.200	0.300	0.400	0.500
Distance	4.90	19.5	44.0	78.1	122.0

Find the least-squares curve of the form $y = mx^2 + b$, using y to represent distance and x to represent time.

25. The period of a pendulum as a function of its length was measured, giving the following results.

Length (feet)	1.00	3.00	5.00	7.00	9.00
Period (seconds)	1.10	1.90	2.50	2.90	3.30

Find the least-squares curve of the form $y = m\sqrt{x} + b$, which expresses the period as a function of the length.

26. In an elementary experiment which measured the wavelength of sound as a function of the frequency, the following results were obtained.

Frequency (hertz)	240	320	400	480	560
Wavelength (centimeters)	140	107	81.0	70.0	60.0

Find the least-squares curve of the form $y = m(1/x) + b$ for this data, expressing wavelength as y and frequency as x.

CHAPTER 13

Expansion of Functions in Series

The values of the trigonometric functions can be determined exactly only for a few particular angles. The value of e can only be approximated in decimal form. The question arises as to how these values may be determined, particularly if a specified degree of accuracy is necessary. In this chapter we shall show how a given function may be expressed in terms of a polynomial. Once the polynomial which approximates a given function has been found, we shall be able to evaluate the function to any desired accuracy. A number of applications and other uses of this polynomial form will be shown as well.

In Section 13–6 we shall show how certain types of functions can be represented by a series of sine and cosine terms. Such series are very useful in the study of functions and applications which are periodic in nature.

We begin the chapter by reviewing the meanings and properties of sequences and series.

13–1 Sequences and Series

In algebra certain sequences of numbers are introduced. These include **arithmetic sequences** and **geometric sequences** (also called *progressions*). From algebra we know that in an arithmetic sequence each term after the first is obtained by adding a fixed number to the preceding term. In a geometric sequence each number after the first is obtained by multiplying the preceding term by a fixed number.

Example A

An example of an arithmetic sequence is

2, 5, 8, 11, . . .

where the fixed number is 3.

An example of a geometric sequence is

2, 4, 8, 16, . . .

where the fixed number is 2.

There are many other ways of generating sequences of numbers. The squares of the integers 1, 4, 9, 16, . . . form a sequence. Also, the successive approximations $x_1, x_2, x_3, \ldots$ obtained by using Newton's method in solving a particular equation form a sequence.

In general, *a* **sequence** (*or* **infinite sequence**) *is an infinite succession of numbers. Each of the numbers is a* **term** *of the sequence.* Each term of the sequence is associated with a positive integer, although at times it is convenient to associate the first term with zero (or some specified positive integer). We shall use a_n to designate the term of the sequence corresponding to the integer n.

Example B

Find the first three terms of the sequence for which $a_n = 2n + 1$, $n = 1, 2, 3, \ldots$.

Substituting the values of n, we obtain the values $a_1 = 3, a_2 = 5, a_3 = 7, \ldots$. Therefore, we have the sequence

3, 5, 7, . . .

If we are given $a_n = 2n + 1$ for $n = 0, 1, 2, \ldots$, the sequence is

1, 3, 5, . . .

The indicated sum of the terms of a sequence is called an **infinite series.** Thus, for the sequence

$$a_1, a_2, a_3, \ldots, a_n, \ldots$$

the associated infinite series is

$$a_1 + a_2 + a_3 + \cdots + a_n + \cdots$$

Using the summation sign Σ (see Section 12–1) to indicate the sum we have

$$\sum_{n=1}^{\infty} a_n = a_1 + a_2 + a_3 + \cdots + a_n + \cdots \qquad (13\text{–}1)$$

We must realize that an infinite series, as shown in Eq. (13–1), does not have a sum in the ordinary sense of the word, for it is not possible actually to carry out the addition of infinitely many terms. Therefore, we define the sum for an infinite series in terms of a limit.

For the infinite series of Eq. (13–1) we let S_n represent the sum of the first n terms. Therefore,

$$S_1 = a_1$$
$$S_2 = a_1 + a_2$$
$$S_3 = a_1 + a_2 + a_3$$
$$S_n = a_1 + a_2 + a_3 + \cdots + a_n$$

The numbers $S_1, S_2, S_3, \ldots, S_n, \ldots$ form a sequence. *Each term of this sequence is called a* **partial sum.** *We say that the infinite series, Eq. (13–1), is* **convergent** *and has the sum S given by*

$$S = \lim_{n \to \infty} S_n = \lim_{n \to \infty} \sum_{n=1}^{\infty} a_n \tag{13–2}$$

if this limit exists. If the limit does not exist, the series is **divergent.**

Example C

For the infinite series

$$\sum_{n=0}^{\infty} \frac{1}{5^n} = \frac{1}{5^0} + \frac{1}{5^1} + \frac{1}{5^2} + \cdots + \frac{1}{5^n} + \cdots$$

the first six partial sums are

$$S_0 = 1$$

$$S_1 = 1 + \frac{1}{5} = 1.2$$

$$S_2 = 1 + \frac{1}{5} + \frac{1}{25} = 1.24$$

$$S_3 = 1 + \frac{1}{5} + \frac{1}{25} + \frac{1}{125} = 1.248$$

$$S_4 = 1 + \frac{1}{5} + \frac{1}{25} + \frac{1}{125} + \frac{1}{625} = 1.2496$$

$$S_5 = 1 + \frac{1}{5} + \frac{1}{25} + \frac{1}{125} + \frac{1}{625} + \frac{1}{3125} = 1.24992$$

where the values are conveniently found by use of a calculator. Here it appears that the sequence of partial sums approaches the value 1.25. We therefore conclude that the infinite series converges and that its sum is approximately 1.25.

Example D

The infinite series

$$\sum_{n=1}^{\infty} 5^n = 5 + 5^2 + 5^3 + \cdots + 5^n + \cdots$$

is a divergent series. The first four partial sums are

$$S_1 = 5, \quad S_2 = 30, \quad S_3 = 155, \quad \text{and} \quad S_4 = 780.$$

Obviously they are increasing without bound.

The infinite series

$$\sum_{n=0}^{\infty} (-1)^n = 1 + (-1) + 1 + (-1) + \cdots + (-1)^n + \cdots$$

has as its first five partial sums

$$S_0 = 1, \quad S_1 = 0, \quad S_2 = 1, \quad S_3 = 0, \quad \text{and} \quad S_4 = 1.$$

These values do not approach a limiting value, and therefore the series diverges.

Since convergent series are those which have a value attached to them, they are the ones which are of primary use to us. However, it is not generally easy to determine whether a given series is convergent, and many types of tests have been developed for this purpose.

One important series for which we are able to determine the convergence, and its sum if convergent, is the geometric series. For this series the nth partial sum is

$$S_n = a + ar + ar^2 + \cdots + ar^{n-1} \tag{13–3}$$

where r is the fixed number by which we multiply a given term to get the next term. If we now multiply Eq. (13–3) by r, we obtain

$$rS_n = ar + ar^2 + ar^3 + \cdots + ar^n \tag{13–4}$$

By subtracting rS_n from S_n we have

$$S_n - rS_n = a - ar^n$$

or

$$S_n = \frac{a(1 - r^n)}{1 - r} \quad (r \neq 1) \tag{13–5}$$

If $|r| < 1$, then $\lim_{n \to \infty} r^n = 0$ and

$$S = \lim_{n \to \infty} S_n = \frac{a}{1 - r} \tag{13–6}$$

If $r = 1$, we see that the series is $a + a + a + \cdots + a + \cdots$ and is therefore divergent. If $r = -1$, the series is $a - a + a - a + \cdots$ and is also divergent. If $|r| > 1$, $\lim_{n \to \infty} r^n$ is unbounded. Therefore, *the geometric series is convergent only if $|r| < 1$, and has the value given by Eq. (13–6).*

Example E   Show that the infinite series

$$\sum_{n=0}^{\infty} \frac{1}{5^n} = \frac{1}{5^0} + \frac{1}{5^1} + \frac{1}{5^2} + \cdots + \frac{1}{5^n} + \cdots$$

is convergent and find its sum. This is the same series as in Example C.

We see that this is a geometric series with $r = \frac{1}{5}$. Since $|r| < 1$, the series is convergent. Now using Eq. (13–6), we find its sum to be

$$S = \frac{1}{1 - \frac{1}{5}} = \frac{1}{\frac{4}{5}} = \frac{5}{4} = 1.25$$

We see that this agrees with the conclusion of Example C.

Example F   Find the rational number which has as its repeating decimal form $0.121212 \ldots$.

The decimal form can be considered to be

$$0.12 + 0.0012 + 0.000012 + \cdots$$

which means that we have a geometric series with $a = 0.12$ and $r = 0.01$. Thus,

$$S = \frac{0.12}{1 - 0.01} = \frac{0.12}{0.99} = \frac{4}{33}$$

Thus, $0.121212 \ldots$ and $\frac{4}{33}$ represent the same number.

Exercises 13–1

In Exercises 1 through 4, give the first four terms of the sequences for which a_n is given.

1. $a_n = n^2, \quad n = 1, 2, 3, \ldots$ 2. $a_n = \frac{2}{3^n}, \quad n = 1, 2, 3, \ldots$

3. $a_n = \frac{1}{n + 2}, \quad n = 0, 1, 2, \ldots$ 4. $a_n = \frac{n^2 + 1}{2n + 1}, \quad n = 0, 1, 2, \ldots$

In Exercises 5 through 8, give (a) the first four terms of the sequence for which a_n is given and (b) the first four terms of the infinite series associated with the sequence.

5. $a_n = \left(-\frac{2}{5}\right)^n, \quad n = 1, 2, 3, \ldots$ 6. $a_n = \frac{1}{n} + \frac{1}{n + 1}, \quad n = 1, 2, 3, \ldots$

7. $a_n = 1 + (-1)^n, \quad n = 0, 1, 2, \ldots$

8. $a_n = \frac{1}{n(n + 1)}, \quad n = 2, 3, 4, \ldots$

In Exercises 9 through 12, find the nth term of the given infinite series for $n = 1, 2, 3, \ldots$.

9. $\frac{1}{2} + \frac{1}{3} + \frac{1}{4} + \frac{1}{5} + \cdots$ 10. $\frac{1}{2} + \frac{1}{4} + \frac{1}{8} + \frac{1}{16} + \cdots$

11. $\frac{1}{2 \cdot 3} + \frac{1}{3 \cdot 4} + \frac{1}{4 \cdot 5} + \frac{1}{5 \cdot 6} + \cdots$ 12. $-\frac{2}{3} + \frac{4}{9} - \frac{8}{27} + \frac{16}{81} - \cdots$

See Appendix C for a computer program for finding the first n partial sums.

In Exercises 13 through 20, find the first five partial sums of the given series and determine whether the series appears to be convergent or divergent. If it is convergent, find its approximate sum.

13. $1 + \dfrac{1}{8} + \dfrac{1}{27} + \dfrac{1}{64} + \dfrac{1}{125} + \cdots$

14. $1 + 2 + 5 + 10 + 17 + \cdots$

15. $1 + \dfrac{1}{2} + \dfrac{2}{3} + \dfrac{3}{4} + \dfrac{4}{5} + \cdots$

16. $\dfrac{1}{3} - \dfrac{1}{9} + \dfrac{1}{27} - \dfrac{1}{81} + \dfrac{1}{243} - \cdots$

17. $\displaystyle\sum_{n=0}^{\infty} (-n)$

18. $\displaystyle\sum_{n=1}^{\infty} \dfrac{2}{n(n + 1)}$

19. $\displaystyle\sum_{n=1}^{\infty} \dfrac{2n + 1}{n^2(n + 1)^2}$

20. $\displaystyle\sum_{n=1}^{\infty} \dfrac{n}{2n + 1}$

In Exercises 21 through 28, test each of the given geometric series for convergence or divergence. Find the sum of each convergent series.

21. $1 + 2 + 4 + \cdots + 2^n + \cdots$

22. $1 + \dfrac{1}{2} + \dfrac{1}{4} + \cdots + \dfrac{1}{2^n} + \cdots$

23. $1 - \dfrac{1}{3} + \dfrac{1}{9} - \cdots + \left(-\dfrac{1}{3}\right)^n + \cdots$

24. $1 - \dfrac{3}{2} + \dfrac{9}{4} - \cdots + \left(-\dfrac{3}{2}\right)^n + \cdots$

25. $10 + 9 + 8.1 + 7.29 + 6.561 + \cdots$

26. $4 + 1 + \dfrac{1}{4} + \dfrac{1}{16} + \dfrac{1}{64} + \cdots$

27. $512 - 64 + 8 - 1 + \dfrac{1}{8} - \cdots$

28. $16 + 12 + 9 + \dfrac{27}{4} + \dfrac{81}{16} + \cdots$

In Exercises 29 through 32, find the rational number represented by the given repeating decimal.

29. $0.2222\ldots$

30. $0.070707\ldots$

31. $0.181818\ldots$

32. $0.792792792\ldots$

In Exercises 33 through 38, solve the given problems by using the appropriate infinite series.

33. A ball is dropped from a height of 8.00 ft, and on each rebound it rises $\frac{1}{2}$ of the height it last fell. What is the total distance it will travel in coming to rest?

34. After the power is turned off, an object on a nearly frictionless surface slows down such that it travels 99.9% as far during each second as during the previous second. If it travels 100 cm during the first second after the power is turned off, how far does it travel while stopping?

35. The successive distances traveled by a pendulum bob are 90 cm, 60 cm, 40 cm, $\ldots$. Find the total distance the bob travels before it comes to rest.

36. A square has sides of 20 cm each. Another square is inscribed in the original square by joining the midpoints of the sides. Assuming that such inscribed squares can be formed endlessly, what is the sum of the areas of all of the squares? See Fig. 13–1.

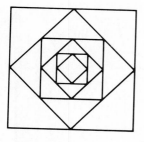

Figure 13–1

37. Use geometric series to show that $\displaystyle\sum_{n=0}^{\infty} x^n = \dfrac{1}{1 - x}$ for $|x| < 1$.

38. Use geometric series to show that $\displaystyle\sum_{n=0}^{\infty} (-1)^n x^n = \dfrac{1}{1 + x}$ for $|x| < 1$.

13–2 Maclaurin Series

In this section we shall develop a very important basic polynomial series form of a function. Before developing the method using calculus, we shall review how this can be done algebraically for some functions.

Example A

$$2 - x \overline{\smash{\big)}\ 2}$$

with work:
$$1 + \frac{x}{2}$$
$$\underline{2 - x}$$
$$x$$
$$\underline{x - \frac{x^2}{2}}$$
$$\frac{x^2}{2}$$

By using long division (as started at left), we have

$$\frac{2}{2 - x} = 1 + \frac{1}{2}x + \frac{1}{4}x^2 + \cdots + \left(\frac{1}{2}x\right)^{n-1} + \cdots \tag{1}$$

where n is the number of the term of the expression on the right. Since x represents a number, the right-hand side of Eq. (1) becomes a geometric series.

From Eq. (13–6) we know that the sum of a geometric series with first term a and common ratio r is

$$S = \frac{a}{1 - r}$$

when $|r| < 1$ and the series converges.

If $x = 1$, the right-hand side of Eq. (1) is

$$1 + \frac{1}{2} + \frac{1}{4} + \cdots + \left(\frac{1}{2}\right)^{n-1} + \cdots$$

For this series, $r = \frac{1}{2}$ and $a = 1$, which means that the series converges and $S = 2$. If $x = 3$, the right-hand side of Eq. (1) is

$$1 + \frac{3}{2} + \frac{9}{4} + \cdots + \left(\frac{3}{2}\right)^{n-1} + \cdots$$

which diverges since $r > 1$. Referring to the left side of Eq. (1), we see that it also equals 2 when $x = 1$. Thus we see that the two sides agree for $x = 1$, but that the series diverges for $x = 3$. In fact, as long as $|x| < 2$, the series will converge to the value of the function on the left. From this we can conclude that the series on the right side properly represents the function on the left, as long as $|x| < 2$.

From Example A, we see that an algebraic function may be properly represented by a function of the form

$$f(x) = a_0 + a_1 x + a_2 x^2 + \cdots + a_n x^n + \cdots \tag{13-7}$$

Equation (13–7) is known as a **power-series expansion** *of the function $f(x)$.* The problem now arises as to whether or not functions in general may be represented in this form. If such a representation were possible, it would provide a means of evaluating the transcendental functions for the purpose of making tables of values. Also, since a power-series expansion is in the form of a polynomial, it makes algebraic operations much simpler due to the properties of polynomials. A further study of calculus shows many other uses of power series.

In Example A we saw that the function could be represented by a power series as long as $|x| < 2$. That is, if we substitute any value of x in this interval into the series and also into the function, the series will converge to the value of the function. *This interval of values for which the series converges is called the* **interval of convergence.**

Example B

In Example A the interval of convergence for the series

$$1 + \frac{1}{2}x + \frac{1}{4}x^2 + \cdots + \left(\frac{1}{2}x\right)^{n-1} + \cdots$$

is $|x| < 2$. We saw that the series converges for $x = 1$, with $S = 2$, and that the value of the function is 2 for $x = 1$. This verifies that $x = 1$ is in the interval of convergence.

Also, we saw that the series diverges for $x = 3$, which verifies that $x = 3$ is not in the interval of convergence.

At this point we shall assume that unless otherwise noted the functions with which we shall be dealing may be properly represented by a power-series expansion (it takes more advanced methods to prove that this is generally possible), for appropriate intervals of convergence. We shall find that the methods of calculus are very useful in developing the method of general representation. Thus, writing a general power series, along with the first few derivatives, we have

$$f(x) = a_0 + a_1x + a_2x^2 + a_3x^3 + a_4x^4 + a_5x^5 + \cdots + a_nx^n + \cdots$$
$$f'(x) = a_1 + 2a_2x + 3a_3x^2 + 4a_4x^3 + 5a_5x^4 + \cdots + na_nx^{n-1} + \cdots$$
$$f''(x) = 2a_2 + 2\cdot3a_3x + 3\cdot4a_4x^2 + 4\cdot5a_5x^3 + \cdots$$
$$+ (n-1)na_nx^{n-2} + \cdots$$
$$f'''(x) = 2\cdot3a_3 + 2\cdot3\cdot4a_4x + 3\cdot4\cdot5a_5x^2 + \cdots$$
$$+ (n-2)(n-1)na_nx^{n-3} + \cdots$$
$$f^{iv}(x) = 2\cdot3\cdot4a_4 + 2\cdot3\cdot4\cdot5a_5x + \cdots$$
$$+ (n-3)(n-2)(n-1)na_nx^{n-4} + \cdots$$

Regardless of the values of the constants a_n for any power series, if $x = 0$, the left and right sides must be equal, and all the terms on the right are zero except the first. Thus, setting $x = 0$ in each of the above equations, we have

$$f(0) = a_0 \qquad f'(0) = a_1$$
$$f''(0) = 2a_2 \qquad f'''(0) = 2\cdot3a_3$$
$$f^{iv}(0) = 2\cdot3\cdot4a_4$$

Solving each of these for the constants a_n, we have

$$a_0 = f(0) \qquad a_1 = f'(0) \qquad a_2 = \frac{f''(0)}{2!} \qquad a_3 = \frac{f'''(0)}{3!} \qquad a_4 = \frac{f^{\text{iv}}(0)}{4!}$$

Substituting these into the expression for $f(x)$, we have

$$f(x) = f(0) + f'(0)x + \frac{f''(0)x^2}{2!} + \frac{f'''(0)x^3}{3!} + \cdots + \frac{f^{(n)}(0)x^n}{n!} + \cdots \qquad (13\text{--}8)$$

Equation $(13\text{--}8)$ *is known as the* **Maclaurin series expansion** *of a function.* For a function to be represented by a Maclaurin expansion, the function and all its derivatives must exist at $x = 0$.

In developing the Maclaurin series, we used the **factorial notation,** $n!$. By definition,

$$n! = n(n - 1)(n - 2)(\cdots)(1) \qquad (13\text{--}9)$$

For example, $4! = 4 \cdot 3 \cdot 2 \cdot 1 = 24$. As we can see, this notation is convenient for use in power-series expansions. It is also used in other fields of mathematics.

As we mentioned earlier, one of the uses we will make of series expansions is that of determining the values of functions for particular values of x. If x is sufficiently small, successive terms become smaller and smaller and the series will converge rapidly. This is considered in the sections which follow.

Example C

Find the first four terms of the Maclaurin series expansion of $f(x) = \dfrac{2}{2 - x}$.
This is written as

$$f(x) = \frac{2}{2 - x} \qquad f(0) = 1$$

$$f'(x) = \frac{2}{(2 - x)^2} \qquad f'(0) = \frac{1}{2}$$

$$f''(x) = \frac{4}{(2 - x)^3} \qquad f''(0) = \frac{1}{2}$$

$$f'''(x) = \frac{12}{(2 - x)^4} \qquad f'''(0) = \frac{3}{4}$$

$$f(x) = 1 + \frac{1}{2}x + \frac{1}{2}\left(\frac{x^2}{2!}\right) + \frac{3}{4}\left(\frac{x^3}{3!}\right) + \cdots$$

or

$$\frac{2}{2 - x} = 1 + \frac{1}{2}x + \frac{1}{4}x^2 + \frac{1}{8}x^3 + \cdots$$

We see that this result agrees with that obtained by direct division.

Example D

Find the first four terms of the Maclaurin series expansion of $f(x) = e^{-x}$.
We write

$$
\begin{aligned}
f(x) &= e^{-x} & f(0) &= 1 \\
f'(x) &= -e^{-x} & f'(0) &= -1 \\
f''(x) &= e^{-x} & f''(0) &= 1 \\
f'''(x) &= -e^{-x} & f'''(0) &= -1
\end{aligned}
$$

$$
f(x) = 1 + (-1)x + 1\left(\frac{x^2}{2!}\right) + (-1)\left(\frac{x^3}{3!}\right) + \cdots
$$

or

$$
e^{-x} = 1 - x + \frac{x^2}{2!} - \frac{x^3}{3!} + \cdots
$$

Example E

Find the first three nonzero terms of the Maclaurin series expansion of $f(x) = \sin 2x$.
We have

$$
\begin{aligned}
f(x) &= \sin 2x & f(0) &= 0 \\
f'(x) &= 2 \cos 2x & f'(0) &= 2 \\
f''(x) &= -4 \sin 2x & f''(0) &= 0 \\
f'''(x) &= -8 \cos 2x & f'''(0) &= -8 \\
f^{iv}(x) &= 16 \sin 2x & f^{iv}(0) &= 0 \\
f^{v}(x) &= 32 \cos 2x & f^{v}(0) &= 32
\end{aligned}
$$

$$
f(x) = 0 + 2x + 0 + (-8)\frac{x^3}{3!} + 0 + 32\frac{x^5}{5!} + \cdots
$$

or

$$
\sin 2x = 2x - \frac{4}{3}x^3 + \frac{4}{15}x^5 - \cdots
$$

This series is called an **alternating series,** *since precisely every other term is negative.*

Example F

Find the first three nonzero terms of the Maclaurin expansion of the function $f(x) = xe^{2x}$.

$$
\begin{aligned}
f(x) &= xe^{2x} & f(0) &= 0 \\
f'(x) &= e^{2x}(1 + 2x) & f'(0) &= 1 \\
f''(x) &= e^{2x}(4 + 4x) & f''(0) &= 4 \\
f'''(x) &= e^{2x}(12 + 8x) & f'''(0) &= 12
\end{aligned}
$$

$$
f(x) = 0 + x + 4\left(\frac{x^2}{2!}\right) + 12\left(\frac{x^3}{3!}\right) + \cdots
$$

or

$$
xe^{2x} = x + 2x^2 + 2x^3 + \cdots
$$

Exercises 13–2

In Exercises 1 through 16, find in each case the first three nonzero terms of the Maclaurin expansion of the given functions.

1. $f(x) = e^x$
         2. $f(x) = \sin x$
         3. $f(x) = \cos x$

4. $f(x) = \ln(1 + x)$
     5. $f(x) = \sqrt{1 + x}$
     6. $f(x) = \dfrac{1}{(1 - x)^{1/3}}$

7. $f(x) = e^{-2x}$
         8. $f(x) = \frac{1}{2}(e^x + e^{-x})$
     9. $f(x) = \cos 4x$

10. $f(x) = e^x \sin x$
     11. $f(x) = \dfrac{1}{1 - x}$
     12. $f(x) = \dfrac{1}{(1 + x)^2}$

13. $\ln(1 - 2x)$
         14. $(1 + x)^{3/2}$

15. $\cos \frac{1}{2}x$
         16. $\ln(1 + 4x)$

In Exercises 17 through 24, find the first two nonzero terms of the Maclaurin expansion of the given functions.

17. $f(x) = \text{Arctan } x$
     18. $f(x) = \cos x^2$
     19. $f(x) = \tan x$

20. $f(x) = \sec x$
         21. $f(x) = \ln \cos x$
     22. $f(x) = xe^{\sin x}$

23. $f(x) = \sin^2 x$
         24. $f(x) = e^{-x^2}$

In Exercises 25 through 28, solve the given problems.

25. Why is it not possible to find a Maclaurin expansion of the following functions?

     (a) $f(x) = \csc x$
         (b) $f(x) = \sqrt{x}$
         (c) $f(x) = \ln x$

26. By finding the Maclaurin expansion, show that

$$(1 + x)^n = 1 + nx + \frac{n(n - 1)}{2!}x^2 + \cdots \qquad (13\text{–}10)$$

This series is the **binomial series,** for which the interval of convergence is $|x| < 1$ for all values of n.

27. If $f(x) = x^3$, show that this function is obtained when a Maclaurin expansion is found.

28. If $f(x) = x^4 + 2x^2$, show that this function is obtained when a Maclaurin expansion is found.

13–3 Certain Operations with Series

The series found in the first four exercises of Section 13–2 are of particular importance. They are used to find values of exponential functions, trigonometric functions, and logarithms. Also they may be used to develop other series. We shall give them here for reference, along with their intervals of convergence:

$$e^x = 1 + x + \frac{x^2}{2!} + \frac{x^3}{3!} + \cdots \quad \text{(all } x) \qquad (13\text{--}11)$$

$$\sin x = x - \frac{x^3}{3!} + \frac{x^5}{5!} - \cdots \quad \text{(all } x) \qquad (13\text{--}12)$$

$$\cos x = 1 - \frac{x^2}{2!} + \frac{x^4}{4!} - \cdots \quad \text{(all } x) \qquad (13\text{--}13)$$

$$\ln(1 + x) = x - \frac{x^2}{2} + \frac{x^3}{3} - \frac{x^4}{4} + \cdots \quad |x| < 1 \qquad (13\text{--}14)$$

In the next section we shall see how we make use of these series in the development of tables. In this section we shall not only see how new series may be developed by using the above basic series, but also we shall see certain other uses of series.

When we discussed functions in Chapter 2, we briefly mentioned functions such as $f(2x)$, $f(x^2)$, and $f(-x)$. *By using functional notation and the preceding series, we can find the series expansions of a great many other series without the use of direct expansion.* This can often save a great deal of time in finding a desired series. The examples given here illustrate the use of series for this purpose.

Example A

Find the Maclaurin expansion of e^{2x}.

From Eq. (13–11), we know the expansion of e^x. Hence,

$$f(x) = 1 + x + \frac{x^2}{2!} + \frac{x^3}{3!} + \cdots$$

Since $e^{2x} = f(2x)$, we have

$$f(2x) = 1 + (2x) + \frac{(2x)^2}{2!} + \frac{(2x)^3}{3!} + \cdots$$

Substitute $2x$ for x in $f(x)$.

or

$$e^{2x} = 1 + 2x + 2x^2 + \frac{4x^3}{3} + \cdots$$

Example B   Find the Maclaurin expansion of $\sin x^2$.

From Eq. (13–12), we know the expansion of $\sin x$. Therefore,

$$f(x) = x - \frac{x^3}{3!} + \frac{x^5}{5!} - \cdots$$

Since $\sin x^2 = f(x^2)$, we have

$$f(x^2) = (x^2) - \frac{(x^2)^3}{3!} + \frac{(x^2)^5}{5!} - \cdots$$

or

$$\sin x^2 = x^2 - \frac{x^6}{3!} + \frac{x^{10}}{5!} - \cdots$$

Direct expansion of this series is quite lengthy.

The basic algebraic operations may be applied to series in the same manner they are applied to polynomials. That is, we may add, subtract, multiply, or divide series in order to obtain other series. The interval of convergence for the resulting series is that which is common to those of the series being used. The multiplication of series is illustrated in the following example.

Example C   Multiply the series for e^x and $\cos x$ in order to obtain the series expansion for $e^x \cos x$.

Using the series expansions for e^x and $\cos x$ as shown in Eqs. (13–11) and (13–13), we have the following indicated multiplication.

$$e^x \cos x = \left(1 + x + \frac{x^2}{2!} + \frac{x^3}{3!} + \frac{x^4}{4!} + \cdots\right)\left(1 - \frac{x^2}{2!} + \frac{x^4}{4!} - \cdots\right)$$

By multiplying the series on the right, we have the following result, considering through the x^4 terms in the product.

$$e^x \cos x = 1 - \frac{x^2}{2} + \frac{x^4}{24} + x - \frac{x^3}{2} + \frac{x^2}{2} - \frac{x^4}{4} + \frac{x^3}{6} + \frac{x^4}{24} + \cdots$$

$$= 1 + x - \frac{1}{3}x^3 - \frac{1}{6}x^4 + \cdots$$

It is also possible to use the operations of differentiation and integration to obtain series expansions, although the proof of this is found in more advanced texts. Consider the following example.

Example D

Show that by differentiating the expansion for $\ln(1 + x)$ term by term, the result is the same as the expansion for $\dfrac{1}{1 + x}$.

The series for $\ln(1 + x)$ is shown in Eq. (13–14) as

$$\ln(1 + x) = x - \frac{x^2}{2} + \frac{x^3}{3} - \frac{x^4}{4} + \cdots$$

Differentiating, we have

$$\frac{1}{1 + x} = 1 - \frac{2x}{2} + \frac{3x^2}{3} - \frac{4x^3}{4} + \cdots = 1 - x + x^2 - x^3 + \cdots$$

Using the binomial expansion for $\dfrac{1}{1 + x} = (1 + x)^{-1}$, we have

$$(1 + x)^{-1} = 1 + (-1)x + \frac{(-1)(-2)}{2!}x^2 + \frac{(-1)(-2)(-3)}{3!}x^3 + \cdots$$

$$= 1 - x + x^2 - x^3 + \cdots$$

We see that the results are the same.

We now demonstrate a very important result in the development of complex numbers in mathematics and in electrical theory. Assuming that the Maclaurin expansions for e^x, $\sin x$, and $\cos x$ are valid for complex numbers (which is shown in advanced calculus), we have ($j = \sqrt{-1}$)

$$e^{j\theta} = 1 + j\theta + \frac{(j\theta)^2}{2!} + \frac{(j\theta)^3}{3!} + \cdots = 1 + j\theta - \frac{\theta^2}{2!} - j\frac{\theta^3}{3!} + \cdots \quad \text{(13–15)}$$

$$j \sin \theta = j\theta - j\frac{\theta^3}{3!} + \cdots \quad \text{(13–16)}$$

$$\cos \theta = 1 - \frac{\theta^2}{2!} + \cdots \quad \text{(13–17)}$$

When we add the terms of Eq. (13–16) to those of Eq. (13–17), the result is the series given in Eq. (13–15). Thus

$$e^{j\theta} = \cos \theta + j \sin \theta \quad \text{(13–18)}$$

The **rectangular form** of a complex number is $a + bj$. The **polar** (or **trigonometric**) **form** is $r(\cos \theta + j \sin \theta)$, and the **exponential form** is $re^{j\theta}$, where $r = \sqrt{a^2 + b^2}$ and $\tan \theta = b/a$. Thus, by these definitions and Eq. (13–18), we have

$$a + bj = r(\cos \theta + j \sin \theta) = re^{j\theta} \quad \text{(13–19)}$$

where

$$r = \sqrt{a^2 + b^2} \qquad \tan \theta = b/a \quad \text{(13–20)}$$

Example E

Express $2e^{4.80j}$ in rectangular form.

Using Eq. (13–19) and the fact that $4.80 = 275°$, we have

$$2e^{4.80j} = 2(\cos 4.80 + j \sin 4.80)$$
$$= 2(\cos 275° + j \sin 275°)$$
$$= 2(0.0872 - j0.9962)$$
$$= 0.1744 - 1.9924j$$

An additional use of power series is now shown. Many integrals which occur in practice cannot be integrated by methods given in the preceding chapters. However, power series can be very useful in giving excellent approximations to some definite integrals.

Example F

Evaluate: $\int_0^1 \dfrac{\sin x}{x}\, dx.$

 We see that this is an improper integral. Therefore,

$$\int_0^1 \frac{\sin x}{x}\, dx = \lim_{h \to 0} \int_h^1 \frac{\sin x}{x}\, dx$$

Now, using the series expansion for $\sin x$, we have

$$\int_0^1 \frac{\sin x}{x}\, dx = \lim_{h \to 0} \int_h^1 \frac{1}{x}\left(x - \frac{x^3}{3!} + \frac{x^5}{5!} - \cdots\right) dx$$

$$= \lim_{h \to 0} \int_h^1 \left(1 - \frac{x^2}{3!} + \frac{x^4}{5!} - \cdots\right) dx$$

$$= \lim_{h \to 0} \left(x - \frac{x^3}{3(3!)} + \frac{x^5}{5(5!)} - \cdots\right)\Big|_h^1$$

$$= 1 - \frac{1}{3(6)} + \frac{1}{5(120)} - \cdots$$

Combining the values indicated, we have $1 - 0.0555556 + 0.0016667 = 0.9461111$. We can see that each of the terms omitted was very small. The correct answer, to 4 decimal places, is 0.9461, which is the value found by using only three terms of the expansion for $\sin x$.

Example G

Evaluate: $\int_0^{0.1} e^{-x^2}\,dx.$

We write

$$e^{-x^2} = 1 + (-x^2) + \frac{(-x^2)^2}{2!} + \cdots$$

Thus,

$$\int_0^{0.1} e^{-x^2}\,dx = \int_0^{0.1}\left(1 - x^2 + \frac{x^4}{2} - \cdots\right) dx$$

$$= \left(x - \frac{x^3}{3} + \frac{x^5}{10} - \cdots\right)\Bigg|_0^{0.1}$$

$$= 0.1 - \frac{0.001}{3} + \frac{0.00001}{10} = 0.0996677$$

This answer is correct to the indicated accuracy.

The question of accuracy now arises. The integrals just evaluated indicate that the more terms used, the greater the accuracy of the result. As an indication of the accuracy involved, Fig. 13–2 shows the graphs of $y = \sin x$ and the graphs of

$$y = x, \qquad y = x - \frac{x^3}{3!}, \quad \text{and} \quad y = x - \frac{x^3}{3!} + \frac{x^5}{5!}$$

which are the first three approximations. We can see that each term added gives a better fit to the curve of $\sin x$. Also, a graphical representation of the meaning of series expansions is indicated.

We have just shown that the more terms included, the more accurate the result. For small values of x, a Maclaurin series can give good accuracy with a very few terms. When this happens, we see that the series *converges* rapidly, as we mentioned earlier. For this reason a Maclaurin series is of particular use if small values of x are being considered. If larger values of x are being considered, usually a function is expanded in a Taylor series (see Section 13–5). Of course, when we omit the later terms in any series, a certain error is inherent in the calculation.

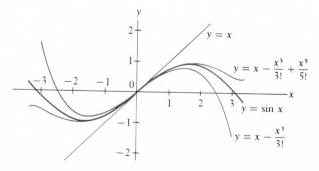

Figure 13–2

Exercises 13–3

In Exercises 1 through 8, find in each case the first four nonzero terms of the Maclaurin expansions of the given functions by using Eqs. (13–11) to (13–14).

1. $f(x) = e^{3x}$ 2. $f(x) = e^{-2x}$ 3. $f(x) = \sin \frac{1}{2}x$
4. $f(x) = \sin x^4$ 5. $f(x) = \cos 4x$ 6. $f(x) = \cos 2x^3$
7. $f(x) = \ln(1 + x^2)$ 8. $f(x) = \ln(1 - x)$

In Exercises 9 through 12, evaluate the given integrals by use of three terms of the appropriate series.

9. $\int_0^1 \sin x^2 \, dx$ 10. $\int_0^{0.1} \frac{e^x - 1}{x} \, dx$
11. $\int_0^{0.2} \cos \sqrt{x} \, dx$ 12. $\int_0^1 \frac{\cos x - 1}{x} \, dx$

In Exercises 13 through 16, change the given complex numbers to the form indicated.

13. $3e^{0.5j}$ (rectangular) 14. $2(\cos 230° + j \sin 230°)$ (exponential)
15. $6 + j$ (polar) 16. $-3 + 2j$ (exponential)

In Exercises 17 through 24, determine the indicated series by the given operation.

17. Find the first four nonzero terms of the expansion of the function $f(x) = \frac{1}{2}(e^x + e^{-x})$ by adding the terms of the appropriate series. The result is the series for cosh x. (See Exercise 38 of Section 7–1.)

18. Find the first four nonzero terms of the expansion of the function $f(x) = \frac{1}{2}(e^x - e^{-x})$ by subtracting the terms of the appropriate series. The result is the series for sinh x. (See Exercise 38 of Section 7–1.)

19. Find the first three terms of the expansion for $e^x \sin x$ by multiplying the proper expansions together, term by term.

20. Find the first three nonzero terms of the expansion for $f(x) = \tan x$, by dividing the series for sin x by that for cos x.

21. Show that by differentiating term by term the expansion for sin x, the result is the expansion for cos x.

22. Show that by differentiating term by term the expansion for e^x, the result is also the expansion for e^x.

23. Show that by integrating term by term the expansion for cos x, the result is the expansion for sin x.

24. Show that by integrating term by term the expansion for $-1/(1 - x)$ (see Exercise 11 of Section 13–2), the result is the expansion for $\ln(1 - x)$.

In Exercises 25 through 32, solve the given problems.

25. Evaluate $\int_0^1 e^x \, dx$ directly and compare the result obtained by using four terms of the series for e^x and then integrating.

26. Evaluate $\lim_{x \to 0} \frac{\sin x}{x}$ by using the series expansion for sin x. Compare the result with Eq. (6–25).

27. Find the approximate value of the area bounded by $y = x^2 e^x$, $x = 0.2$, and the x-axis by use of three terms of the appropriate Maclaurin series.

28. Find the approximate area bounded by $y = e^{-x^2}$, $x = 0$, $x = 1$, and $y = 0$ by use of three terms of the appropriate series.

29. Find the volume generated by rotating the area bounded by $y = \sin x$ from $x = 0$ to $x = \frac{1}{8}\pi$ and the x-axis about the y-axis by use of two terms of the appropriate series.

30. Find the volume generated by rotating the area bounded by $y = e^{-x}$, $y = 0$, $x = 0$, and $x = 0.1$ about the y-axis by use of three terms of the appropriate series.

31. The impedance Z of an alternating current circuit, containing a resistance R, inductance L, and capacitance C, may be defined by the equation $Z = R + j(X_L - X_C)$, where X_L and X_C are known as the inductive reactance and the capacitive reactance, respectively. Show, by use of Eq. (13–19) that the angle between Z and R is given by

$$\tan \theta = \frac{X_L - X_C}{R}$$

32. Find the magnitude of the impedance of a circuit if $R = 8.00 \ \Omega$, $X_L = 7.00 \ \Omega$ and $X_C = 13.0 \ \Omega$. (Finding the magnitude of Z is equivalent to finding r for a complex number).

13–4 Computations by Use of Series Expansions

As mentioned previously, power-series expansions can be used to compute numerical values of transcendental functions. By including a sufficient number of terms, we can calculate the values to any desired degree of accuracy. It is through such calculations that tables of values are made, and values of such numbers as e and π can be found. Also, many of the values determined by a calculator or a computer are found by use of series expansions.

Example A

Calculate the value of $e^{0.1}$.

$$e^x = 1 + x + \frac{x^2}{2!} + \cdots$$

Let $x = 0.1$. Thus,

$$e^{0.1} = 1 + 0.1 + \frac{(0.1)^2}{2} + \cdots$$

$$= 1.105 \qquad \text{(using 3 terms)}$$

Using a calculator, we find that $e^{0.1} = 1.1051709$, which shows that our answer is valid to the accuracy shown.

Example B

Calculate the value of sin 2°.

$$\sin x = x - \frac{x^3}{3!} + \cdots$$

▶ Let $x = \pi/90$ (converting to radians). Hence,

$$\sin 2° = \left(\frac{\pi}{90}\right) - \frac{(\pi/90)^3}{6} + \cdots$$

$$= 0.0348995$$

A calculator also gives the value 0.0348995. Here we note that the second term is much smaller than the first. In fact, a good approximation can be found by use of one term. Thus, we see the reason than $\sin \theta = \theta$ (approximately) for small values of θ.

Example C

Calculate the value of cos 0.5429.
     Since the angle is expressed in radians, using Eq. (13–13) we have

$$\cos 0.5429 = 1 - \frac{0.5429^2}{2} + \frac{0.5429^4}{4!} - \cdots$$

$$= 0.8562495$$

A calculator shows that cos 0.5429 = 0.8562141. Since the angle is not small, additional terms are needed to obtain this accuracy. With one more term, the value 0.8562139 is obtained.

Example D

Calculate the value of ln(1.2).

$$\ln(1 + x) = x - \frac{x^2}{2} + \frac{x^3}{3} - \cdots$$

Let $x = 0.2$. Therefore,

$$\ln(1 + 0.2) = 0.2 - \frac{(0.2)^2}{2} + \frac{(0.2)^3}{3} - \cdots = 0.1827$$

To four significant digits, ln(1.2) = 0.1823. One more term is required to obtain this accuracy.

     Series approximations can also be used to determine errors in calculated values due to errors in measured values. We have already discussed this topic as an application of differentials. A series solution of this kind of problem allows as close a calculation of the error as needed. With differentials, only one term can be found. Of course, for many purposes this is sufficient. The following example illustrates the use of series in error calculations.

Example E

The velocity v attained by an object that has fallen h feet is given by $v = 8\sqrt{h}$. Find the approximate error in the velocity of an object which has been found to drop 100 ft, with a possible error of 2 ft.

▶ If we let $v = 8\sqrt{100 + x}$, where x is the error in h, we may express v as a Maclaurin series in x.

$$f(x) = 8(100 + x)^{1/2} \qquad f(0) = 80$$
$$f'(x) = 4(100 + x)^{-1/2} \qquad f'(0) = 0.4$$
$$f''(x) = -2(100 + x)^{-3/2} \qquad f''(0) = -0.002$$

Therefore,

$$v = 8\sqrt{100 + x} = 80 + 0.4x - 0.001x^2 + \cdots$$

Since the actual calculated value of v, for $x = 0$, is 80, the error e in the value of v is

$$e = 0.4x - 0.001x^2 + \cdots$$

Calculating the error for $x = 2$ is

$$e = 0.4(2) - 0.001(4) = 0.800 - 0.004 = 0.796 \text{ ft/s}$$

The value 0.8 is the same as that which would be found by use of differentials, since it is a result of evaluating the first derivative term of the expansion. The additional terms are corrections to this term. The one additional term in this case indicates that the first term is a good approximation to the error. Although this problem could be done numerically, a series solution, once set up, allows us to calculate the error easily for a given value of x.

Example F

Assuming that the earth is a perfect sphere 4000 mi in radius, find how far the end of a tangent line 2 mi long is from the surface.

From Fig. 13-3 we see that

$$x = 4000 \sec \theta - 4000$$

Finding the series for $\sec \theta$, we have

$$f(\theta) = \sec \theta \qquad f(0) = 1$$
$$f'(\theta) = \sec \theta \tan \theta \qquad f'(0) = 0$$
$$f''(\theta) = \sec^3 \theta + \sec \theta \tan^2 \theta \qquad f''(0) = 1$$

Thus, the first two nonzero terms are $\sec \theta = 1 + (\theta^2/2)$. Therefore,

$$x = 4000(\sec \theta - 1) = 4000\left(1 + \frac{\theta^2}{2} - 1\right) = 2000\ \theta^2$$

The first two terms of the expansion for $\tan \theta$ are $\theta + \theta^3/3$, which means that if θ is small, which it is in this case, then $\tan \theta = \theta$ approximately. From Fig. 13-3, $\tan \theta = 2/4000$. Therefore, $\theta = 1/2000$. Substituting, we have

$$x = 2000\left(\frac{1}{2000}\right)^2 = \frac{1}{2000} = 0.0005 \text{ mi}$$

This means that the end of a 2-mi line drawn tangent to the earth would only be about 2.6 ft from the surface!

Figure 13-3

See Appendix C for a
computer program for finding
values of e^x by use of the
series.

Exercises 13–4

In Exercises 1 through 16, calculate the value of each of the given functions. Use the indicated number of terms of the appropriate series. Compare with the value found directly on a calculator.

1. $e^{0.2}$ (3)
2. $e^{-0.5}$ (3)
3. $\sin 0.1$ (2)
4. $\cos 0.05$ (2)
5. e (7)
6. $\sqrt{e}$ (5)
7. $\cos 3°$ (2)
8. $\sin 4°$ (2)
9. $\ln(1.4)$ (4)
10. $\ln(0.95)$ (4)
11. $\sin 0.3625$ (3)
12. $\cos 0.4072$ (3)
13. $\ln 0.9861$ (3)
14. $\ln 1.0534$ (3)
15. $e^{-0.3165}$ (4)
16. $e^{0.2651}$ (4)

In Exercises 17 through 20, calculate the value of each of the given functions. In Exercises 17 and 18, use the expansion for $\sqrt{1+x}$, and in Exercises 19 and 20, use the expansion for $\sqrt[3]{1+x}$. Use three terms of the appropriate series.

17. $\sqrt{1.1076}$
18. $\sqrt{0.7915}$
19. $\sqrt[3]{0.9628}$
20. $\sqrt[3]{1.1392}$

In Exercises 21 through 24, calculate the maximum error of the values indicated. If a series is alternating (every other term is negative) the maximum possible error in the calculated value is the value of the first term omitted.

21. The value found in Exercise 3.
22. The value found in Exercise 2.
23. The value found in Exercise 7.
24. The value found in Exercise 9.

In Exercises 25 through 30, solve the given problems by the use of series expansions.

25. The electric current in a circuit containing a resistance R, an inductance L, and a battery whose voltage is E is given by

$$i = \frac{E}{R}(1 - e^{-Rt/L})$$

where t is the time. Approximate this expression by using the first three terms of the appropriate exponential series. Under what conditions will this approximation be valid?

26. The image distance q from a certain lens as a function of the object distance p is given by $q = 20p/(p - 20)$. Find the first three nonzero terms of the expansion of the right side. From this expression calculate q for $p = 2$ cm and compare with the value found by substituting 2 in the original expression.

27. From what height can a person see a point 10 mi distant along the earth's surface?

28. The efficiency (in percent) of an internal combustion engine in terms of its compression ratio c is given by $E = 100(1 - c^{-0.4})$. Determine the possible approximate error in the efficiency for a compression ratio measured to be 6.00 with a possible error of 0.50. [*Hint:* Set up a series for $(6 + x)^{-0.4}$.]

29. We can evaluate π by use of the fact that $\frac{1}{4}\pi = \text{Arctan } \frac{1}{2} + \text{Arctan } \frac{1}{3}$ (see Exercise 42 of Section 6–5), along with the use of the series expansion for Arctan x. The first three terms are $\text{Arctan } x = x - \frac{1}{3}x^3 + \frac{1}{5}x^5$. Using these terms, expand Arctan $\frac{1}{2}$ and Arctan $\frac{1}{3}$, and thereby approximate the value of π.

30. Use the fact that $\frac{1}{4}\pi = \text{Arctan } \frac{1}{7} + 2 \text{ Arctan } \frac{1}{3}$ to approximate the value of π. (See Exercise 29.)

13–5 Taylor's Series

If values of certain functions are determined for values of x which are not close to zero, it is usually necessary to use many terms of a Maclaurin expansion to obtain accurate values. Therefore, another type of series, called a **Taylor's series,** can be used for this purpose. Also, functions for which a Maclaurin series may not be found may have a Taylor's series. Actually, a Taylor's series is a more general expansion than a Maclaurin expansion.

The basic assumption in formulating a Taylor's expansion is that a function may be expanded in a polynomial of the form

$$f(x) = c_0 + c_1(x - a) + c_2(x - a)^2 + \cdots \tag{13–21}$$

By following precisely the same line of reasoning that we did in deriving the general Maclaurin expansion, we may find the constants c_0, c_1, c_2, and so forth. That is, derivatives of Eq. (13–21) are taken, and the function and its derivatives are evaluated for $x = a$. This leads to

$$f(x) = f(a) + f'(a)(x - a) + \frac{f''(a)(x - a)^2}{2!} + \cdots \tag{13–22}$$

This series converges rapidly for values of x which are close to a, and this is illustrated in Examples C and D. The following examples illustrate the derivation of Taylor's series.

Example A

Expand $f(x) = e^x$ in a Taylor's series with $a = 1$.

$$f(x) = e^x \qquad f(1) = e \qquad f''(x) = e^x \qquad f''(1) = e$$
$$f'(x) = e^x \qquad f'(1) = e \qquad f'''(x) = e^x \qquad f'''(1) = e$$

$$\vdots$$

$$f(x) = e + e(x - 1) + e\frac{(x - 1)^2}{2!} + e\frac{(x - 1)^3}{3!} + \cdots$$

$$e^x = e\left[1 + (x - 1) + \frac{(x - 1)^2}{2} + \frac{(x - 1)^3}{6} + \cdots\right]$$

This series can be used in evaluating e^x for values of x near 1.

Example B

Expand $f(x) = \sqrt{x}$ in powers of $(x - 4)$.

Another way of stating this is to find the Taylor's series for $f(x) = \sqrt{x}$, with $a = 4$. Thus,

$$f(x) = x^{1/2} \qquad f(4) = 2$$

$$f'(x) = \frac{1}{2x^{1/2}} \qquad f'(4) = \frac{1}{4}$$

$$f''(x) = -\frac{1}{4x^{3/2}} \qquad f''(4) = -\frac{1}{32}$$

$$f'''(x) = \frac{3}{8x^{5/2}} \qquad f'''(4) = \frac{3}{256}$$

$$\vdots$$

$$f(x) = 2 + \frac{1}{4}(x - 4) - \frac{1}{32}\frac{(x - 4)^2}{2!} + \frac{3}{256}\frac{(x - 4)^3}{3!} - \cdots$$

$$\sqrt{x} = 2 + \frac{(x - 4)}{4} - \frac{(x - 4)^2}{64} + \frac{(x - 4)^3}{512} - \cdots$$

This series would be used to evaluate square roots of numbers near 4.

In the last section we evaluated functions by the use of Maclaurin series. In the following examples we shall use Taylor's series to determine certain values of functions.

Example C

By use of a Taylor series, evaluate $\sqrt{4.5}$.

Using the four terms of the series found in Example B, we have

$$\sqrt{4.5} = 2 + \frac{(4.5 - 4)}{4} - \frac{(4.5 - 4)^2}{64} + \frac{(4.5 - 4)^3}{512}$$

$$= 2 + \frac{(0.5)}{4} - \frac{(0.5)^2}{64} + \frac{(0.5)^3}{512}$$

$$= 2.1213379$$

The value found directly on a calculator is 2.1213203. Therefore, the value found by the series expansion is correct to four decimal places.

In Example C we see that successive terms become small rapidly. If a value of x is chosen such that $x - a$ is larger, the successive terms may not become small rapidly and many terms may be required. Therefore, we should choose a

as conveniently close to the x-values which will be used. Also, we should note that a Maclaurin expansion for $\sqrt{x}$ cannot be used since the derivatives of $\sqrt{x}$ are not defined for $x = 0$.

Example D

Calculate the approximate value of sin 29° by means of the appropriate Taylor expansion.

Since the value of sin 30° is known to be $\frac{1}{2}$, if we let $a = \pi/6$ (remember, we must use values expressed in radians), when we evaluate the expansion for $x = 29°$ (when expressed in radians) the quantity $(x - a)$ is $-\pi/180$ (equivalent to $-1°$). This means that its numerical values are small and become smaller when it is raised to higher powers.

Therefore,

$$f(x) = \sin x \qquad f\left(\frac{\pi}{6}\right) = \frac{1}{2}$$

$$f'(x) = \cos x \qquad f'\left(\frac{\pi}{6}\right) = \frac{\sqrt{3}}{2}$$

$$f''(x) = -\sin x \qquad f''\left(\frac{\pi}{6}\right) = -\frac{1}{2}$$

$$f(x) = \frac{1}{2} + \frac{\sqrt{3}}{2}\left(x - \frac{\pi}{6}\right) - \frac{1}{4}\left(x - \frac{\pi}{6}\right)^2 - \cdots$$

or

$$\sin x = \frac{1}{2} + \frac{\sqrt{3}}{2}\left(x - \frac{\pi}{6}\right) - \frac{1}{4}\left(x - \frac{\pi}{6}\right)^2 - \cdots$$

$$\sin 29° = \sin\left(\frac{\pi}{6} - \frac{\pi}{180}\right)$$

$$= \frac{1}{2} + \frac{\sqrt{3}}{2}\left(\frac{\pi}{6} - \frac{\pi}{180} - \frac{\pi}{6}\right) - \frac{1}{4}\left(\frac{\pi}{6} - \frac{\pi}{180} - \frac{\pi}{6}\right)^2 - \cdots$$

$$= \frac{1}{2} + \frac{\sqrt{3}}{2}\left(-\frac{\pi}{180}\right) - \frac{1}{4}\left(-\frac{\pi}{180}\right)^2 - \cdots$$

$$= 0.4848089$$

The value found directly on a calculator is 0.4848096.

Exercises 13-5

In Exercises 1 through 8, evaluate the given functions by use of the series developed in the examples of this section.

1. $e^{1.2}$ (use $e = 2.7183$)
2. $e^{0.7}$
3. $\sqrt{4.2}$
4. $\sqrt{3.5}$
5. sin 31°
6. sin 28°
7. sin 29.53°
8. $\sqrt{3.8527}$

In Exercises 9 through 16, find the first three nonzero terms of the Taylor expansion for the given function and given value of a.

9. e^{-x} $(a = 2)$ 10. $\cos x$ $(a = \pi/4)$ 11. $\sin x$ $(a = \pi/3)$

12. $\ln x$ $(a = 3)$ 13. $\sqrt[3]{x}$ $(a = 8)$ 14. $\dfrac{1}{x}$ $(a = 2)$

15. $\tan x$ $(a = \pi/4)$ 16. $\ln \sin x$ $(a = \pi/2)$

In Exercises 17 through 24, evaluate the given functions by use of three terms of the appropriate Taylor series.

17. $e^{-2.2}$ (Use $e^{-2} = 0.1353$.)   18. $\ln(3.1)$ (Use $\ln 3 = 1.0986$.)

19. $\sqrt{9.3}$ 20. $\sqrt{17}$ 21. $\sqrt[3]{8.3}$

22. $\tan 46°$ 23. $\sin 61°$ 24. $\cos 42°$

In Exercises 25 through 28, solve the given problems.

25. By completing the steps indicated before Eq. (13–22) in the text, complete the derivation of Eq. (13–22).

26. Calculate $e^{0.9}$ by use of four terms of the Maclaurin expansion for e^x. Also calculate $e^{0.9}$ by using the first three terms of the Taylor expansion in Example A, using $e = 2.7183$. Compare the accuracy of the values obtained with that found directly on a calculator.

27. Calculate $\sin 31°$ by use of three terms of the Maclaurin expansion for $\sin x$. Also calculate $\sin 31°$ by using three terms of the Taylor expansion in Example D. Compare the accuracy of the values obtained with that found directly on a calculator.

28. Calculate $\tan 44°$ by use of two terms of the Maclaurin expansion for $\tan x$ (see Exercise 19 of Section 13–2). Also calculate $\tan 44°$ by using two terms of the Taylor expansion (see Exercise 15). Compare the accuracy of the values obtained with that found directly on a calculator.

13–6 Fourier Series

Many problems which are encountered in the various fields of science and technology involve functions which are periodic. *A periodic function is one for which F(x + P) = F(x), where P is the period.* We noted that the trigonometric functions are periodic when we discussed their graphs in Chapter 6. Illustrations of applied problems which involve periodic functions are alternating-current voltages and mechanical oscillations.

Therefore, in this section we use a series made of terms of sines and cosines. This allows us to represent complicated periodic functions in terms of the simpler sines and cosines. It also provides a good approximation over a greater interval than Maclaurin and Taylor series, which give good approximations with a few terms only near a specific value. Illustrations of applications of this type of series are given in Example C and in the exercises.

We shall assume that a function $f(x)$ may be represented by the series of sines and cosines as indicated:

$$f(x) = a_0 + a_1\cos x + a_2\cos 2x + \cdots + a_n \cos nx + \cdots$$
$$+ b_1\sin x + b_2\sin 2x + \cdots + b_n \sin nx + \cdots \quad \textbf{(13–23)}$$

Since all the sines and cosines indicated in this expansion have a period of 2π (the period of any given term may be less than 2π, but all do repeat every 2π units—for example, $\sin 2x$ has a period of π, but it also repeats every 2π), the series expansion indicated in Eq. (13–23) will also have a period of 2π. *This series is called a* **Fourier series.**

The principal problem to be solved is that of finding the coefficients a_n and b_n. Derivatives proved to be useful in finding the coefficients for a Maclaurin expansion. We use the properties of certain integrals to find the coefficients of a Fourier series. To use these properties, we multiply all terms of Eq. (13–23) by $\cos mx$ and then evaluate from $-\pi$ to π (in this way we take advantage of the period 2π). Thus, we have

$$\int_{-\pi}^{\pi} f(x)\cos mx \, dx = \int_{-\pi}^{\pi} (a_0 + a_1\cos x + a_2\cos 2x + \cdots)(\cos mx) \, dx$$

$$+ \int_{-\pi}^{\pi} (b_1\sin x + b_2\sin 2x + \cdots)(\cos mx) \, dx \quad (13\text{–}24)$$

where m is an integer.

Using the methods of integration of Chapters 8 and 9, we have the following:

$$\int_{-\pi}^{\pi} a_0\cos mx \, dx = \frac{a_0}{m}\sin mx \Big|_{-\pi}^{\pi} = \frac{a_0}{m}(0-0) = 0 \quad (13\text{–}25)$$

$$\int_{-\pi}^{\pi} a_n\cos nx \cos mx \, dx =$$

$$a_n\left(\frac{\sin(n-m)x}{2(n-m)} + \frac{\sin(n+m)x}{2(n+m)}\right)\Big|_{-\pi}^{\pi} = 0 \qquad n \neq m; \quad (13\text{–}26)$$

(since the sine of any multiple of π is zero),

$$\int_{-\pi}^{\pi} a_n\cos nx \cos nx \, dx = \int_{-\pi}^{\pi} a_n\cos^2 nx \, dx$$

$$= \left(\frac{a_n x}{2} + \frac{1}{2n}\sin nx \cos nx\right)\Big|_{-\pi}^{\pi} = \frac{a_n x}{2}\Big|_{-\pi}^{\pi} = \pi a_n \quad (13\text{–}27)$$

as well as

$$\int_{-\pi}^{\pi} b_n\sin nx \cos mx \, dx = b_n\left(-\frac{\cos(n-m)x}{2(n-m)} - \frac{\cos(n+m)x}{2(n+m)}\right)\Big|_{-\pi}^{\pi}$$

$$= b_n\left(-\frac{\cos(n-m)\pi}{2(n-m)} - \frac{\cos(n+m)\pi}{2(n+m)}\right.$$

$$\left. + \frac{\cos(n-m)(-\pi)}{2(n-m)} + \frac{\cos(n+m)(-\pi)}{2(n+m)}\right)$$

$$= 0 \; [\text{since } \cos\theta = \cos(-\theta)], \qquad n \neq m \quad (13\text{–}28)$$

$$\int_{-\pi}^{\pi} b_n\sin nx \cos nx \, dx = \frac{b_n}{2n}\sin^2 nx \Big|_{-\pi}^{\pi} = 0 \quad (13\text{–}29)$$

These integrals are seen to be zero, except for the one specific case of $\int_{-\pi}^{\pi} a_n \cos nx \cos mx \, dx$ when $n = m$, for which the result is indicated in Eq.

(13–27). Using these results in Eq. (13–24), we have

$$\int_{-\pi}^{\pi} f(x) \cos nx \, dx = a_n \int_{-\pi}^{\pi} \cos^2 nx \, dx = \pi a_n$$

or

$$a_n = \frac{1}{\pi} \int_{-\pi}^{\pi} f(x) \cos nx \, dx \qquad \textbf{(13–30)}$$

This equation allows us to find the coefficients a_n, except a_0. We find the term a_0 by direct integration of Eq. (13–23) from $-\pi$ to π. When we perform this integration, all the sine and cosine terms integrate to zero, thereby giving the result

$$\int_{-\pi}^{\pi} f(x) \, dx = \int_{-\pi}^{\pi} a_0 \, dx = a_0 x \Big|_{-\pi}^{\pi} = 2\pi a_0$$

or

$$a_0 = \frac{1}{2\pi} \int_{-\pi}^{\pi} f(x) \, dx \qquad \textbf{(13–31)}$$

By multiplying all terms of Eq. (13–23) by $\sin mx$ and then integrating from $-\pi$ to π, we find the coefficients b_n. We obtain the result

$$b_n = \frac{1}{\pi} \int_{-\pi}^{\pi} f(x) \sin nx \, dx \qquad \textbf{(13–32)}$$

We can restate our equations for the Fourier series of a function $f(x)$:

$$f(x) = a_0 + a_1 \cos x + a_2 \cos 2x + \cdots + a_n \cos nx + \cdots$$
$$+ \, b_1 \sin x + b_2 \sin 2x + \cdots + b_n \sin nx + \cdots \qquad \textbf{(13–23)}$$

---

where the coefficients are found by

$$a_0 = \frac{1}{2\pi} \int_{-\pi}^{\pi} f(x) \, dx \qquad \textbf{(13–31)}$$

$$a_n = \frac{1}{\pi} \int_{-\pi}^{\pi} f(x) \cos nx \, dx \qquad \textbf{(13–30)}$$

and

$$b_n = \frac{1}{\pi} \int_{-\pi}^{\pi} f(x) \sin nx \, dx \qquad \textbf{(13–32)}$$

Example A

Find the Fourier series for the square wave function

$$f(x) = \begin{cases} 0 & \text{for } -\pi \le x < 0 \\ 1 & \text{for } \quad 0 \le x < \pi \end{cases}$$

[Many of the functions which we shall expand in Fourier series are discontinuous (not continuous) like this one. See Section 2–2 for a discussion of continuity.]

► Since $f(x)$ is defined differently for the ranges of x indicated, it requires two integrals for each coefficient:

$$a_0 = \frac{1}{2\pi} \int_{-\pi}^{0} 0 \ dx + \frac{1}{2\pi} \int_{0}^{\pi} (1) \ dx = \frac{1}{2}$$

$$a_n = \frac{1}{\pi} \int_{-\pi}^{0} 0 \ dx + \frac{1}{\pi} \int_{0}^{\pi} (1)\cos nx \ dx = \frac{1}{n\pi} (\sin nx) \Big|_{0}^{\pi} = 0$$

for all values of n, since $\sin n\pi = 0$;

$$b_1 = \frac{1}{\pi} \int_{-\pi}^{0} 0 \ dx + \frac{1}{\pi} \int_{0}^{\pi} \sin x \ dx = -\frac{1}{\pi} (\cos x) \Big|_{0}^{\pi}$$

$$= -\frac{1}{\pi} (-1 - 1) = \frac{2}{\pi}$$

$$b_2 = \frac{1}{\pi} \int_{-\pi}^{0} 0 \ dx + \frac{1}{\pi} \int_{0}^{\pi} \sin 2x \ dx = -\frac{1}{2\pi} (\cos 2x) \Big|_{0}^{\pi}$$

$$= -\frac{1}{2\pi} (1 - 1) = 0$$

$$b_3 = \frac{1}{\pi} \int_{-\pi}^{0} 0 \ dx + \frac{1}{\pi} \int_{0}^{\pi} \sin 3x \ dx = -\frac{1}{3\pi} (\cos 3x) \Big|_{0}^{\pi}$$

$$= -\frac{1}{3\pi} (-1 - 1) = \frac{2}{3\pi}$$

► In general we can see that if n is even, $b_n = 0$, and if n is odd, then $b_n = 2/n\pi$. Thus, we have

$$f(x) = \frac{1}{2} + \frac{2}{\pi} \sin x + \frac{2}{3\pi} \sin 3x + \cdots$$

A graph of the function as defined, and the curve found by using the first three terms of the Fourier series, is shown in Fig. 13–4.

Since functions found by Fourier series have a period of 2π, they can represent functions with this period. If the function $f(x)$ were defined to be periodic with period 2π, with the same definitions as originally indicated, we would graph the function as shown in Fig. 13–5. The Fourier series representation would follow it as in Fig. 13–4. If more terms were used, the fit would be closer.

Figure 13–4

Figure 13–5

Example B

Find the Fourier series for the function

$$f(x) = \begin{cases} 1 & \text{for } -\pi \le x < 0 \\ x & \text{for } 0 \le x < \pi \end{cases}$$

For the periodic function, let $f(x + 2\pi) = f(x)$ for all x.

A graph of three periods of this function is shown in Fig. 13–6. The coefficients are

$$a_0 = \frac{1}{2\pi} \int_{-\pi}^{0} dx + \frac{1}{2\pi} \int_{0}^{\pi} x\, dx = \frac{x}{2\pi}\Big|_{-\pi}^{0} + \frac{x^2}{4\pi}\Big|_{0}^{\pi} = \frac{1}{2} + \frac{\pi}{4} = \frac{2 + \pi}{4}$$

$$a_1 = \frac{1}{\pi} \int_{-\pi}^{0} \cos x\, dx + \frac{1}{\pi} \int_{0}^{\pi} x \cos x\, dx$$

$$= \frac{1}{\pi} \sin x \Big|_{-\pi}^{0} + \frac{1}{\pi} (\cos x + x \sin x)\Big|_{0}^{\pi} = -\frac{2}{\pi}$$

$$a_2 = \frac{1}{\pi} \int_{-\pi}^{0} \cos 2x\, dx + \frac{1}{\pi} \int_{0}^{\pi} x \cos 2x\, dx$$

$$= \frac{1}{2\pi} \sin 2x \Big|_{-\pi}^{0} + \frac{1}{4\pi} (\cos 2x + 2x \sin 2x)\Big|_{0}^{\pi} = 0$$

$$a_3 = \frac{1}{\pi} \int_{-\pi}^{0} \cos 3x\, dx + \frac{1}{\pi} \int_{0}^{\pi} x \cos 3x\, dx$$

$$= \frac{1}{3\pi} \sin 3x \Big|_{-\pi}^{0} + \frac{1}{9\pi} (\cos 3x + 3x \sin 3x)\Big|_{0}^{\pi} = -\frac{2}{9\pi}$$

$$b_1 = \frac{1}{\pi} \int_{-\pi}^{0} \sin x\, dx + \frac{1}{\pi} \int_{0}^{\pi} x \sin x\, dx$$

$$= -\frac{1}{\pi} \cos x \Big|_{-\pi}^{0} + \frac{1}{\pi} (\sin x - x \cos x)\Big|_{0}^{\pi} = \frac{\pi - 2}{\pi}$$

$$b_2 = \frac{1}{\pi} \int_{-\pi}^{0} \sin 2x\, dx + \frac{1}{\pi} \int_{0}^{\pi} x \sin 2x\, dx$$

$$= -\frac{\cos 2x}{2\pi} \Big|_{-\pi}^{0} + \frac{\sin 2x - 2x \cos 2x}{4\pi}\Big|_{0}^{\pi} = -\frac{1}{2}$$

Thus

$$f(x) = \frac{2 + \pi}{4} - \frac{2}{\pi} \cos x - \frac{2}{9\pi} \cos 3x - \cdots$$

$$+ \left(\frac{\pi - 2}{\pi}\right) \sin x - \frac{1}{2} \sin 2x + \cdots$$

Figure 13–6

Example C

Certain electronic devices allow an electric current to pass through in only one direction. The result is that when an alternating current is applied, the current exists for only half the cycle. Figure 13–7 is a representation for such a current as a function of time. A device which causes this effect is called a half-wave rectifier. Derive the Fourier series for the half of the rectified wave for which $f(t) = \sin t$ $(0 \leq t \leq \pi)$, and $f(t) = 0$ for the other half of the cycle.

Figure 13-7

We write

$$a_0 = \frac{1}{2\pi} \int_0^\pi \sin t \, dt = \frac{1}{2\pi} (-\cos t) \Big|_0^\pi = \frac{1}{2\pi} (1 + 1) = \frac{1}{\pi}$$

In the previous example we evaluated each of the coefficients individually. Here we shall show how to set up a general expression for a_n and another for b_n. Once we have determined these, we can substitute values of n in the formula to obtain the individual coefficients:

$$a_n = \frac{1}{\pi} \int_0^\pi \sin t \cos nt \, dt = -\frac{1}{2\pi} \left[\frac{\cos(1-n)t}{1-n} + \frac{\cos(1+n)t}{1+n} \right]_0^\pi$$

$$= -\frac{1}{2\pi} \left[\frac{\cos(1-n)\pi}{1-n} + \frac{\cos(1+n)\pi}{1+n} - \frac{1}{1-n} - \frac{1}{1+n} \right]$$

See Formula 40 of Table 5 in Appendix D. This formula is valid for all values of n except $n = 1$. Now we write

$$a_1 = \frac{1}{\pi} \int_0^\pi \sin t \cos t \, dt = \frac{1}{2\pi} \sin^2 t \Big|_0^\pi = 0$$

$$a_2 = -\frac{1}{2\pi} \left(\frac{-1}{-1} + \frac{-1}{3} - \frac{1}{-1} - \frac{1}{3} \right) = -\frac{2}{3\pi}$$

$$a_3 = -\frac{1}{2\pi} \left(\frac{1}{-2} + \frac{1}{4} - \frac{1}{-2} - \frac{1}{4} \right) = 0$$

$$a_4 = -\frac{1}{2\pi} \left(\frac{-1}{-3} + \frac{-1}{5} - \frac{1}{-3} - \frac{1}{5} \right) = -\frac{2}{15\pi}$$

(*continued on next page*)

$$b_n = \frac{1}{\pi} \int_0^\pi \sin t \sin nt \, dt = \frac{1}{2\pi} \left[\frac{\sin(1-n)t}{1-n} - \frac{\sin(1+n)t}{1+n} \right]_0^\pi$$

$$= \frac{1}{2\pi} \left[\frac{\sin(1-n)\pi}{1-n} - \frac{\sin(1+n)\pi}{1+n} \right]$$

See Formula 39 of Table 5. This formula is valid for all values of n except $n = 1$. Thus,

$$b_1 = \frac{1}{\pi} \int_0^\pi \sin t \sin t \, dt$$

$$= \frac{1}{\pi} \int_0^\pi \sin^2 t \, dt$$

$$= \frac{1}{2\pi} (t - \sin t \cos t) \Big|_0^\pi = \frac{1}{2}$$

We see that $b_n = 0$ if $n > 1$, since each is evaluated in terms of the sine of a multiple of π.

Therefore, the Fourier series for the rectified wave is

$$f(t) = \frac{1}{\pi} + \frac{1}{2} \sin t - \frac{2}{\pi} \left(\frac{1}{3} \cos 2t + \frac{1}{15} \cos 4t + \cdots \right)$$

The standard form of a Fourier series is the one we have considered to this point. In this form the function is defined in the interval form $x = -\pi$ to $x = \pi$. At times it may be preferable to consider the form of the series of a function which is defined over a different interval.

Noting that

$$\sin \frac{n\pi}{L} (x + 2L) = \sin n\left(\frac{\pi x}{L} + 2\pi \right) = \sin \frac{n\pi x}{L}$$

we see that $\sin(n\pi x/L)$ has a period of $2L$. Thus, by using $\sin(n\pi x/L)$ and $\cos(n\pi x/L)$ and the same method of derivation, the following equations are found for the coefficients for the Fourier series for the interval from $x = -L$ to $x = L$.

$$a_0 = \frac{1}{2L} \int_{-L}^{L} f(x) \, dx \tag{13–33}$$

$$a_n = \frac{1}{L} \int_{-L}^{L} f(x) \cos \frac{n\pi x}{L} \, dx \tag{13–34}$$

$$b_n = \frac{1}{L} \int_{-L}^{L} f(x) \sin \frac{n\pi x}{L} \, dx \tag{13–35}$$

Example D

Find the Fourier series for the function

$$f(x) = \begin{cases} 0 & \text{for } -4 \le x < 0 \\ 2 & \text{for } 0 \le x < 4 \end{cases}$$

and for which the period is 8.

Figure 13–8

Here $L = 4$. A graph of the function is shown in Fig. 13–8. The coefficients are

$$a_0 = \frac{1}{2(4)} \int_{-4}^{0} 0 \, dx + \frac{1}{2(4)} \int_{0}^{4} 2 \, dx = \frac{1}{8}(2x)\Big|_{0}^{4} = 1$$

$$a_n = \frac{1}{4} \int_{-4}^{0} 0 \cos\frac{n\pi x}{4} \, dx + \frac{1}{4} \int_{0}^{4} 2 \cos\frac{n\pi x}{4} \, dx$$

$$= \frac{1}{2} \cdot \frac{4}{n\pi} \int_{0}^{4} \cos\frac{n\pi x}{4} \left(\frac{n\pi \, dx}{4}\right) = \frac{2}{n\pi} \sin\frac{n\pi x}{4}\Big|_{0}^{4} = 0 \qquad \text{for all } n$$

$$b_n = \frac{1}{4} \int_{-4}^{0} 0 \sin\frac{n\pi x}{4} \, dx + \frac{1}{4} \int_{0}^{4} 2 \sin\frac{n\pi x}{4} \, dx$$

$$= \frac{1}{2} \cdot \frac{4}{n\pi} \int_{0}^{4} \sin\frac{n\pi x}{4} \left(\frac{n\pi \, dx}{4}\right) = -\frac{2}{n\pi} \cos\frac{n\pi x}{4}\Big|_{0}^{4}$$

$$= -\frac{2}{n\pi}(\cos n\pi - \cos 0) = \frac{2}{n\pi}(1 - \cos n\pi)$$

$$b_1 = \frac{2}{\pi}[1 - (-1)] = \frac{4}{\pi}, \qquad b_2 = \frac{2}{2\pi}(1 - 1) = 0$$

$$b_3 = \frac{2}{3\pi}[1 - (-1)] = \frac{4}{3\pi}, \qquad b_4 = \frac{2}{4\pi}(1 - 1) = 0$$

Therefore,

$$f(x) = 1 + \frac{4}{\pi}\sin\frac{\pi x}{4} + \frac{4}{3\pi}\sin\frac{3\pi x}{4} + \cdots$$

All the types of periodic functions included in this section (as well as many others) may actually be seen on an oscilloscope when the proper signal is sent into it. In this way the oscilloscope may be used to analyze the periodic nature of such phenomena as sound waves and electric currents.

Exercises 13–6

In the following exercises, find a few terms of the Fourier series for the given functions, and sketch at least three periods of the function. In Exercises 1 through 10 and 13 through 14, $f(x + 2\pi) = f(x)$.

1. $f(x) = \begin{cases} 1 & -\pi \le x < 0 \\ 0 & 0 \le x < \pi \end{cases}$

2. $f(x) = \begin{cases} -1 & -\pi \le x < 0 \\ 1 & 0 \le x < \pi \end{cases}$

3. $f(x) = \begin{cases} 1 & -\pi \le x < 0 \\ 2 & 0 \le x < \pi \end{cases}$

4. $f(x) = \begin{cases} 0 & -\pi \le x < 0, \dfrac{\pi}{2} < x < \pi \\ 1 & 0 \le x \le \dfrac{\pi}{2} \end{cases}$

5. $f(x) = \begin{cases} 0 & -\pi \le x < 0 \\ x & 0 \le x < \pi \end{cases}$

6. $f(x) = x, \ -\pi \le x < \pi$

7. $f(x) = \begin{cases} -1 & -\pi \le x < 0 \\ 0 & 0 \le x < \dfrac{\pi}{2} \\ 1 & \dfrac{\pi}{2} \le x < \pi \end{cases}$

8. $f(x) = x^2, \ -\pi \le x < \pi$

9. $f(x) = \begin{cases} -x & -\pi \le x < 0 \\ x & 0 \le x < \pi \end{cases}$

10. $f(x) = \begin{cases} 0 & -\pi \le x < 0 \\ x^2 & 0 \le x < \pi \end{cases}$

11. $f(x) = \begin{cases} 5 & -3 \le x < 0 \\ 0 & 0 \le x < 3, \quad \text{period} = 6 \end{cases}$

12. $f(x) = \begin{cases} -x & -4 \le x < 0 \\ x & 0 \le x < 4, \quad \text{period} = 8 \end{cases}$

13. Find the Fourier expansion of the electronic device known as a *full-wave rectifier*. This is found by using as the function for the current $f(t) = -\sin t$ for $-\pi \le t \le 0$ and $f(t) = \sin t$ for $0 < t \le \pi$. See Fig. 13–9. The portion of the curve to the left of the $f(t)$-axis is dashed because from a physical point of view we can give no significance to this part of the wave, although mathematically we can derive the proper form of the expansion by using it.

Figure 13–9

14. The loudness L, in decibels, of a certain siren as a function of time, in seconds, can be described by the function $L = 0$ for $-\pi \le t < 0$, $L = 120t$ for $0 \le t < \dfrac{\pi}{2}$, and $L = 120(\pi - t)$ for $\dfrac{\pi}{2} \le t < \pi$, with a period of 2π seconds (where only positive values of t have physical significance). Find the Fourier expansion for the loudness of the siren.

15. Each pulse of a pulsating force F is 8 N. The force lasts for 1 s, followed by a 3-s pause. Thus, it can be represented by $F = 0$ for $-2 \leq t < 0$ and $1 \leq t \leq 2$, and $F = 8$ for $0 \leq t < 1$, with a period of 4 s (only positive values of t have physical significance). Find the Fourier expansion for the force.

16. A function for which $f(x) = f(-x)$ is called an **even function,** and a function for which $f(x) = -f(-x)$ is called an **odd function.** From our discussion of symmetry in Chapter 1, we see that an even function is symmetrical to the y-axis, and an odd function is symmetrical to the origin. It can be proven that the Fourier expansion of an even function contains no sine terms, and that of an odd function contains no cosine terms. Show that the function in Exercise 9 is an even function, and the function in Exercise 6 is an odd function. Note and compare the Fourier expansions of these functions.

13–7 Review Exercises for Chapter 13

In Exercises 1 through 8, find the first three nonzero terms of the Maclaurin expansion of the given functions.

1. $f(x) = \dfrac{1}{1 + e^x}$ 2. $f(x) = \frac{1}{2}(e^x - e^{-x})$ 3. $f(x) = \sin 2x^2$

4. $f(x) = \dfrac{1}{(1 - x)^2}$ 5. $f(x) = (x + 1)^{1/3}$ 6. $f(x) = \dfrac{2}{2 - x^2}$

7. $f(x) = \text{Arcsin } x$ 8. $f(x) = \dfrac{1}{1 - \sin x}$

In Exercises 9 through 20, calculate the value of each of the given functions. Use three terms of the appropriate series.

9. $e^{-0.2}$ 10. $\ln(1.10)$ 11. $\sqrt[3]{1.3}$ 12. $\sin 3.5°$

13. $\sqrt{1.07}$ 14. $e^{0.4173}$ 15. $\ln 0.8172$ 16. $\cos 0.1376$

17. $\tan 43.62°$ 18. $\sqrt[4]{260}$ 19. $\sqrt{148}$ 20. $\cos 47°$

In Exercises 21 and 22, evaluate the given integrals by using three terms of the appropriate series.

21. $\displaystyle\int_{0.1}^{0.2} \dfrac{\cos x}{\sqrt{x}}\, dx$ 22. $\displaystyle\int_{0}^{0.1} \sqrt[3]{1 + x^2}\, dx$

In Exercises 23 and 24, find the first three nonzero terms of the Taylor expansion for the given function and the given value of a.

23. $\cos x$ $(a = \pi/3)$ 24. $\ln \cos x$ $(a = \pi/4)$

In Exercises 25 through 28, find a few terms of the Fourier series for the given functions with the given periods. Sketch three periods of the function.

25. $f(x) = \begin{cases} 0 & -\pi \le x < -\dfrac{\pi}{2} \quad \text{and} \quad \dfrac{\pi}{2} < x < \pi, \quad \text{period} = 2\pi \\ 1 & -\dfrac{\pi}{2} \le x \le \dfrac{\pi}{2} \end{cases}$

26. $f(x) = \begin{cases} -x & -\pi \le x < 0, \quad \text{period} = 2\pi \\ 0 & 0 \le x < \pi \end{cases}$

27. $f(x) = x, \ -2 \le x < 2, \quad \text{period} = 4$

28. $f(x) = \begin{cases} -2 & -3 \le x < 0, \quad \text{period} = 6 \\ 2 & 0 \le x < 3 \end{cases}$

In Exercises 29 and 30, determine the indicated series by the given operation.

29. If h is small, show that $\sin(x + h) - \sin(x - h) = 2h \cos x$.

30. Find the first three nonzero terms of the Maclaurin expansion of the function $\sin x + x \cos x$ by differentiating the expansion term by term for $x \sin x$.

In Exercises 31 through 40, solve the given problems.

31. On the same diagram plot the graphs of $y = e^x$, $y = 1$, $y = 1 + x$, and $y = 1 + x + \frac{1}{2}x^2$.

32. Show that the Maclaurin expansion for the polynomial $f(x) = ax^2 + bx + c$ is the polynomial itself.

33. Find the first three terms of the Maclaurin expansion for $\sec x$ by finding the reciprocal of the series for $\cos x$.

34. By simplifying the sum of the squares of the Maclaurin series for $\sin x$ and $\cos x$ through the x^4 terms, verify (to this extent) that $\sin^2 x + \cos^2 x = 1$.

35. From the Maclaurin series for $f(x) = \dfrac{1}{1 - x}$ (see Exercise 11 of Section 13–2), find the series for $\dfrac{1}{1 + x}$.

36. Show that the Maclaurin expansions for $\cos x$ and $\cos(-x)$ are the same.

37. Find the approximate area between the curve $y = \dfrac{x - \sin x}{x^2}$ and the x-axis between $x = 0$ and $x = 0.1$.

38. Find the approximate value of the moment of inertia with respect to its axis of the volume generated by rotating the area bounded by $y = \sin x$, $x = 0.3$, and the x-axis about the y-axis. Use two terms of the appropriate series.

39. A certain electric current is pulsating so that the current as a function of time is given by $f(t) = 0$ if $-\pi \le t < 0$ and $\pi/2 < t < \pi$. If $0 < t < \pi/2$, $f(t) = \sin t$. Find the Fourier expansion for this pulsating current, and plot three periods.

40. By use of series, find the greatest distance that a great circle arc 200 mi long on the earth is from its chord.

CHAPTER 14

First-Order Differential Equations

Various relationships which occur in science and technology give rise to equations involving the rate of change of one variable with respect to another. When this type of relationship exists, we express the rate of change as a derivative. This results in an equation which contains a derivative. Such an equation is called a *differential equation*. In this chapter we shall develop methods of solving some of the basic types of differential equations. We shall see that finding the solutions of differential equations is a powerful method of solving many types of applied problems.

14–1 Solutions of Differential Equations

A **differential equation** *is an equation which contains derivatives or differentials*. In this section it is our purpose to introduce the basic meaning of the solution of a differential equation. In the sections that follow, we consider certain methods of finding such solutions.

The basic types of differential equations we shall consider are those which contain first and second derivatives. *If an equation contains only first derivatives, it is called a* **first-order** *differential equation. If the equation contains second derivatives, and possibly first derivatives, it is called a* **second-order** *differential equation. In general, the* **order** *of a differential equation is that of the highest derivative in the equation. The* **degree** *of a differential equation is the highest power of that derivative.*

Example A

The equation $\dfrac{dy}{dx} + x = y$ is a first-order differential equation. The equations

$$\frac{d^2y}{dx^2} + y = 3x^2$$

and

$$\frac{d^2y}{dx^2} + 2\frac{dy}{dx} = x$$

are both second-order equations. The presence of dy/dx in the second equation does not affect the order.

Example B

The equation

$$\frac{d^2y}{dx^2} + \left(\frac{dy}{dx}\right)^4 - y = 6$$

is a differential equation of the second order and the first degree. That is, the highest derivative which appears is the second, and it is raised to the first power. Since the second derivative appears, the fourth power of the first derivative does not affect the degree.

In the sections of this chapter which follow we will solve differential equations of the first order and first degree. In the next chapter we will consider primarily differential equations of the second order and first degree.

A **solution** *of a differential equation is a relation between the variables which satisfies the differential equation.* That is, when this relation is substituted into the differential equation, an algebraic identity results. *A solution containing a number of independent arbitrary constants equal to the order of the differential equation is called the* **general solution** *of the equation. When specific values are given to at least one of these constants, the solution is called a* **particular solution.**

Example C

Any coefficients which are not specified numerically after like terms have been combined are independent arbitrary constants. The expression $c_1x + c_2 + c_3x$ has only two independent constants, in that the two x terms may be combined with one arbitrary constant; $c_2 + c_4x$ is an equivalent expression, with $c_4 = c_1 + c_3$.

Example D

The equation $y = c_1 e^{-x} + c_2 e^{2x}$ is the general solution of the differential equation

$$\frac{d^2 y}{dx^2} - \frac{dy}{dx} = 2y$$

The order of this differential equation is two, and there are two independent arbitrary constants in the equation representing the solution. The equation $y = 4e^{-x}$ is a particular solution. This particular solution can be derived from the general solution by letting $c_1 = 4$ and $c_2 = 0$. Each of these solutions can be shown to satisfy the differential equation by taking two derivatives and substituting.

To solve a differential equation, we have to find some method of transforming the equation so that the terms may be integrated. This will be the topic of principal concern after this section. The purpose here is to show that a given equation is a solution of a differential equation by taking the required derivatives, and to show that an identity results after substitution.

Example E

Show that $y = c_1 \sin x + c_2 \cos x$ is the general solution of the differential equation $y'' + y = 0$.

The function and its first two derivatives are

$$y = c_1 \sin x + c_2 \cos x, \qquad y' = c_1 \cos x - c_2 \sin x$$

and

$$y'' = -c_1 \sin x - c_2 \cos x$$

Substituting these into the differential equation, we have

$$\underset{\downarrow}{y''} \qquad + \qquad \underset{\downarrow}{y} \qquad = 0$$

$$(-c_1 \sin x - c_2 \cos x) + (c_1 \sin x + c_2 \cos x) = 0 \quad \text{or} \quad 0 = 0$$

We know that this must be the general solution, since there are two independent arbitrary constants, and the order of the differential equation is two.

Example F

Show that $y = cx + x^2$ is a solution of the differential equation $xy' - y = x^2$.

Taking one derivative of the function and substituting into the differential equation, we have

$$y = cx + x^2$$

$$xy' - y = x^2$$

$$y' = c + 2x$$

$$x(c + 2x) - (cx + x^2) = x^2 \quad \text{or} \quad x^2 = x^2$$

Exercises 14–1

In the following exercises, show that each equation is a solution of the indicated differential equation.

1. $\dfrac{dy}{dx} = 1$, $y = x + 3$

2. $\dfrac{dy}{dx} = 2x$, $y = x^2 + 1$

3. $\dfrac{dy}{dx} - y = 1$, $y = e^x - 1$

4. $\dfrac{dy}{dx} - 3 = 2x$, $y = x^2 + 3x$

5. $xy' = 2y$, $y = cx^2$

6. $y' = 2xy^2$, $y = -\dfrac{1}{x^2 + c}$

7. $y' + 2y = 2x$, $y = ce^{-2x} + x - \frac{1}{2}$

8. $y' - 3x^2 = 1$, $y = x^3 + x + c$

9. $\dfrac{d^2y}{dx^2} + 4y = 0$, $y = 3\cos 2x$

10. $y'' + 9y = 4\cos x$, $2y = \cos x$

11. $y'' - 4y' + 4y = e^{2x}$, $y = e^{2x}\left(c_1 + c_2 x + \dfrac{x^2}{2}\right)$

12. $\dfrac{d^3y}{dx^3} = \dfrac{d^2y}{dx^2}$, $y = c_1 + c_2 x + c_3 e^x$

13. $x^2 y' + y^2 = 0$, $xy = cx + cy$

14. $xy' - 3y = x^2$, $y = cx^3 - x^2$

15. $x\dfrac{d^2y}{dx^2} + \dfrac{dy}{dx} = 0$, $y = c_1 \ln x + c_2$

16. $\dfrac{d^3y}{dx^3} + 4\dfrac{d^2y}{dx^2} + 4\dfrac{dy}{dx} = 0$, $y = c_1 + c_2 e^{-2x} + xe^{-2x}$

17. $y' + y = 2\cos x$, $y = \sin x + \cos x - e^{-x}$

18. $(x + y) - xy' = 0$, $y = x \ln x - cx$

19. $y'' + y' = 6\sin 2x$, $y = e^{-x} - \frac{3}{5}\cos 2x - \frac{6}{5}\sin 2x$

20. $xy''' + 2y'' = 0$, $y = c_1 x + c_2 \ln x$

21. $\cos x \dfrac{dy}{dx} + \sin x = 1 - y$, $y = \dfrac{x + c}{\sec x + \tan x}$

22. $2xyy' + x^2 = y^2$, $x^2 + y^2 = cx$

23. $(y')^2 + xy' = y$, $y = cx + c^2$

24. $x^4(y')^2 - xy' = y$, $y = c^2 + \dfrac{c}{x}$

14–2 Separation of Variables

The first type of differential equation we shall solve is one of the first order and first degree. There are many methods for solving such equations, a few of which are presented in this and the following two sections. The first of these is *the method of* **separation of variables.**

By the definitions of order and degree, a differential equation of the first order and first degree contains the first derivative to the first power. That is, it may be written as $dy/dx = f(x, y)$. This type of equation is more commonly expressed in its differential form

$$M(x, y)\, dx + N(x, y)\, dy = 0 \tag{14–1}$$

where $M(x, y)$ and $N(x, y)$ may represent constants, functions of either x or y, or functions of x and y.

To find the solution of an equation of the form (14–1), it is necessary to integrate. However, if $M(x, y)$ is a function of x and y, we cannot integrate this term. It is integrable only if $M(x, y)$ is a function of x only. For the same reason, the second term is integrable only if $N(x, y)$ is a function of y only. If it is possible, by some algebraic means, to rewrite Eq. (14–1) in the form

$$A(x)\ dx\ +\ B(y)\ dy\ =\ 0$$

(14–2)

where $A(x)$ is a function of x alone and $B(y)$ is a function of y alone, then *we may find the solution by integrating each term and adding the constant of integration.* (If division is involved, we must remember that the solution is not valid for values which make the divisor zero.) Many elementary differential equations lend themselves to this type of solution.

Example A

Solve the differential equation $dx - 4xy^3\ dy = 0$.

We can write this equation as

(1) $dx\ +\ (-4xy^3)\ dy\ =\ 0$

which means that $M(x, y) = 1$ and $N(x, y) = -4xy^3$.

We have to remove the x from the coefficient of dy, without introducing any factor of y in the coefficient of dx. This can be done by dividing each term of the equation by x, which results in the equation

$$\frac{dx}{x} - 4y^3\ dy = 0$$

It is now possible to integrate each term separately. Performing this integration, we have

$$\ln|x| - y^4 = c$$

which is the desired solution. The c is the constant of integration, which becomes the arbitrary constant of the solution.

In Example A we showed the integration of dx/x as $\ln|x|$, which follows our discussion in Section 8–2. We know that $\ln|x| = \ln x$ if $x > 0$ and $\ln|x| = \ln(-x)$ if $x < 0$. Since we know the values being considered when we find a particular solution, we generally will not use the absolute value notation when integrating logarithmic forms. We would show the integration of dx/x as $\ln x$, with the understanding that we know that $x > 0$. If we are using negative values of x, we could express it as $\ln(-x)$.

Example B

Solve the differential equation $xy\, dx + (x^2 + 1)\, dy = 0$.

In order to integrate each term, it is necessary to divide each term by $y(x^2 + 1)$. When this is done, we have

$$\frac{x\, dx}{x^2 + 1} + \frac{dy}{y} = 0$$

Integrating, we have

$$\tfrac{1}{2} \ln(x^2 + 1) + \ln y = c$$

which is the desired solution.

It is possible to make use of the properties of logarithms to make the form of this solution much neater. If we write the constant of integration as $\ln c_1$, rather than c, we have $\tfrac{1}{2}\ln(x^2 + 1) + \ln y = \ln c_1$. Multiplying through by 2 and using the property of logarithms given by Eq. (7–9), we have $\ln(x^2 + 1) + \ln y^2 = \ln c_1^2$. Next, using the property of logarithms given by Eq. (7–7), we have $\ln(x^2 + 1)\, y^2 = \ln c_1^2$, which means

$$(x^2 + 1)\, y^2 = c_1^2$$

For reference, Eq. (7–9) is $n \log_b x = \log_b x^n$. Eq. (7–7) is $\log_b x + \log_b y = \log_b xy$.

This form of the solution is much more compact, and would generally be preferred. However, it must be pointed out that any expression representing a constant may be chosen as the constant of integration and leads to a proper result. In checking answers, we must remember that a different choice of constant will lead to a different form of the answer. Thus, two different-appearing answers may both be correct. *It often happens that there is more than one reasonable choice of a constant, and different forms of the answer may be expected.*

Example C

Solve the differential equation

$$\frac{dy}{dx} = \frac{y}{x^2 + 4}$$

The variables are separated by writing the equation in the form

$$\frac{dy}{y} = \frac{dx}{x^2 + 4}$$

Integrating, we have

$$\ln y = \frac{1}{2} \operatorname{Arctan} \frac{x}{2} + \frac{c}{2} \quad \text{or} \quad 2 \ln y = \operatorname{Arctan} \frac{x}{2} + c$$

The choice of $\ln c$ as the constant of integration (on the left) is also reasonable. It would lead to the result $2 \ln cy = \operatorname{Arctan} (x/2)$.

In order to separate the variables of a differential equation with exponential functions, it may be necessary to use the properties of exponents. When trigonometric functions are involved, the basic trigonometric identities may be used.

Example D

Solve the differential equation $2e^{3x}\sin y \, dx + e^x \csc y \, dy = 0$.

In the dx-term we want only a function of x, which means that we must divide by $\sin y$. Also, the dy-term indicates that we must divide by e^x. Thus, dividing by $e^x \sin y$, we have

$$\frac{2e^{3x}\sin y \, dx}{e^x \sin y} + \frac{e^x \csc y \, dy}{e^x \sin y} = 0$$

$$2e^{2x} \, dx + \csc^2 y \, dy = 0$$

Since the variables are now separated, we complete the solution.

$$e^{2x}(2 \, dx) + \csc^2 y \, dy = 0$$

$$e^{2x} - \cot y = c$$

The following examples show how particular solutions of a differential equation are found.

Example E

Solve the differential equation $3x^2y^2 \, dx + y^2 \, dx + dy = 0$, subject to the condition that $x = 2$ when $y = 1$.

Dividing through by y^2 and combining terms, we have

$$(3x^2 + 1) \, dx + \frac{dy}{y^2} = 0$$

Integrating, the result is $x^3 + x - (1/y) = c$. Since a specified set of values is given, we can evaluate the constant of integration. Using these values, we have $8 + 2 - 1 = c$, or $c = 9$. Thus, the solution may be written as

$$x^3 + x - \frac{1}{y} = 9 \quad \text{or} \quad yx^3 + yx = 1 + 9y$$

Example F

If the function in Example B is subject to the condition that $x = 0$ when $y = e$, we have

$$\frac{1}{2}\ln(0 + 1) + \ln e = c, \quad \frac{1}{2}\ln 1 + 1 = c \quad \text{or} \quad c = 1$$

The solution is then

$$\frac{1}{2}\ln(x^2 + 1) + \ln y = 1, \quad \ln(x^2 + 1) + 2 \ln y = 2$$

$$\ln y^2(x^2 + 1) = 2 \quad \text{or} \quad y^2(x^2 + 1) = e^2$$

If we use the solution with c_1, we have $(0 + 1)e^2 = c_1^2$, or $c_1^2 = e^2$. The solution is then $y^2(x^2 + 1) = e^2$. We see, therefore, that the choice of the constant does not affect the final result, and therefore is truly arbitrary.

Exercises 14–2

In Exercises 1 through 24, solve the given differential equations.

1. $2x\,dx + dy = 0$

2. $y^2\,dy + x^3\,dx = 0$

3. $y^2\,dx + dy = 0$

4. $y\,dx + x\,dy = 0$

5. $x^2 + (x^3 + 5)y' = 0$

6. $xyy' + \sqrt{1 + y^2} = 0$

7. $e^{x^2}\,dy = x\sqrt{1 - y}\,dx$

8. $(x^3 + x^2)\,dx + (x + 1)\,y\,dy = 0$

9. $e^{x+y}\,dx + dy = 0$

10. $e^{2x}\,dy + e^x\,dx = 0$

11. $y' - y = 4$

12. $y' + 4x = 3$

13. $x\dfrac{dy}{dx} = y^2 + y^2 \ln x$

14. $(yx^2 + y)\dfrac{dy}{dx} = \text{Arctan } x$

15. $y \tan x\,dx + \cos^2 x\,dy = 0$

16. $\sin x \sec y\,dx = dy$

17. $yx^2\,dx = y\,dx - x^2\,dy$

18. $2y(x^3 + 1)\,dy + 3x^2(y^2 - 1)\,dx = 0$

19. $y\sqrt{1 - x^2}\,dy + 2\,dx = 0$

20. $\sqrt{1 + 4x^2}\,dy = y^3 x\,dx$

21. $2 \ln x\,dx + x\,dy = 0$

22. $e^{\sin x}\,dx + \sec x\,dy = 0$

23. $y^2 e^x + (e^x + 1)\dfrac{dy}{dx} = 0$

24. $y + 1 + \sec x(\sin x + 1)\dfrac{dy}{dx} = 0$

In Exercises 25 through 30, in each case find the particular solution of the given differential equations for the indicated conditions.

25. $\dfrac{dy}{dx} + yx^2 = 0; \quad x = 0$ when $y = 1$

26. $\dfrac{dy}{dx} + 2y = 6; \quad x = 0$ when $y = 1$

27. $(xy^2 + x)\dfrac{dy}{dx} = \ln x; \quad x = 1$ when $y = 0$

28. $y' = \sec y; \quad x = 0$ when $y = 0$

29. $y' = (1 - y) \cos x; \quad x = \pi/6$ when $y = 0$

30. $2y \cos y\,dy - \sin y\,dy = y \sin y\,dx; \quad x = 0$ when $y = \pi/2$

14–3 Integrable Combinations

Many differential equations cannot be solved by the method of separation of variables. Many other methods have been developed for solving such equations. One of these methods is based on the fact that *certain combinations of basic differentials can be integrated together as a unit*. The following differentials suggest some of these combinations which may occur.

$$d(xy) = x \, dy + y \, dx \tag{14-3}$$

$$d(x^2 + y^2) = 2(x \, dx + y \, dy) \tag{14-4}$$

$$d\left(\frac{y}{x}\right) = \frac{x \, dy - y \, dx}{x^2} \tag{14-5}$$

$$d\left(\frac{x}{y}\right) = \frac{y \, dx - x \, dy}{y^2} \tag{14-6}$$

Equation (14–3) suggests that, if the combination of $x \, dy + y \, dx$ occurs in a differential equation, we look for functions of xy as solutions. Equation (14–4) suggests that, if the combination $x \, dx + y \, dy$ occurs, we look for functions of $x^2 + y^2$. Equations (14–5) and (14–6) suggest that, if either of the combination $x \, dy - y \, dx$ or $y \, dx - x \, dy$ occurs, we look for functions of y/x or x/y. The following examples illustrate the use of these combinations.

Example A

Solve the differential equation $x \, dy + y \, dx + xy \, dy = 0$.
    By dividing through by xy, we have

$$\frac{x \, dy + y \, dx}{xy} + dy = 0$$

The left term is the differential of xy divided by xy. This means it integrates to $\ln xy$. Thus, we have

$$\frac{d(xy)}{xy} + dy = 0$$

for which the solution is

$$\ln xy + y = c$$

Example B

Solve the differential equation $y \, dx - x \, dy + x \, dx = 0$.
    The combination of $y \, dx - x \, dy$ suggests that this equation might make use of either Eq. (14–5) or (14–6). This would require dividing through by x^2 or y^2. If we divide by y^2, the last term cannot be integrated, but division by x^2 still allows integration of the last term. Performing this division, we obtain

$$\frac{y \, dx - x \, dy}{x^2} + \frac{dx}{x} = 0$$

This left combination is the negative of Eq. (14–5). Thus we have

$$-d\left(\frac{y}{x}\right) + \frac{dx}{x} = 0$$

for which the solution is

$$-\frac{y}{x} + \ln x = c$$

Example C

Solve the differential equation $(x^2 + y^2 + x)\, dx + y\, dy = 0$.

Regrouping the terms of this equation, we have

$$(x^2 + y^2)\, dx + (x\, dx + y\, dy) = 0$$

By dividing through by $x^2 + y^2$, we have

$$dx + \frac{x\, dx + y\, dy}{x^2 + y^2} = 0$$

The right term now can be put in the form of du/u (with $u = x^2 + y^2$) by multiplying each of the terms of the numerator by 2. The result then becomes

$$x + \frac{1}{2}\ln(x^2 + y^2) = \frac{c}{2}$$

or

$$2x + \ln(x^2 + y^2) = c$$

Example D

Find the particular solution of the differential equation

$$(x^3 + xy^2 + 2y)\, dx + (y^3 + x^2 y + 2x)\, dy = 0$$

which satisfies the condition that $x = 1$ when $y = 0$.

Regrouping the terms of the equation, we have

$$x(x^2 + y^2)\, dx + y(x^2 + y^2)\, dy + 2(y\, dx + x\, dy) = 0$$

Factoring $x^2 + y^2$ from each of the first two terms gives

$$(x^2 + y^2)(x\, dx + y\, dy) + 2(y\, dx + x\, dy) = 0$$

The first term can be integrated by multiplying the $x\, dx + y\, dy$ factor by 2. The result is

$$\frac{1}{2}\cdot\frac{1}{2}(x^2 + y^2)^2 + 2xy + \frac{c}{4} = 0,$$

or

$$(x^2 + y^2)^2 + 8xy + c = 0$$

Using the given condition gives $(1 + 0)^2 + 0 + c = 0$, or $c = -1$. The particular solution is then

$$(x^2 + y^2)^2 + 8xy = 1$$

    The use of these integrable combinations depends on proper recognition of the forms. At times it may take two or three arrangements to arrive at the combination which leads to the result. Of course, many problems cannot be so arranged as to give integrable combinations in all terms.

Exercises 14–3

In Exercises 1 through 16, solve the given differential equations.

1. $x\,dy + y\,dx + x\,dx = 0$

2. $(2y + x)\,dy + y\,dx = 0$

3. $y\,dx - x\,dy + x^3\,dx = 2\,dx$

4. $x\,dy - y\,dx + y^2\,dx = 0$

5. $x^3\,dy + x^2y\,dx + y\,dx - x\,dy = 0$

6. $\sec(xy)\,dx + (x\,dy + y\,dx) = 0$

7. $x^3y^4(x\,dy + y\,dx) = dy$

8. $x\,dy + y\,dx + 4xy^3\,dy = 0$

9. $\sqrt{x^2 + y^2}\,dx - 2y\,dy = 2x\,dx$

10. $x\,dx + (x^2 + y^2 + y)\,dy = 0$

11. $\tan(x^2 + y^2)\,dy + x\,dx + y\,dy = 0$

12. $(x^2 + y^3)^2\,dy + 2x\,dx + 3y^2\,dy = 0$

13. $y\,dy - x\,dx + (y^2 - x^2)\,dx = 0$

14. $e^{x+y}(dx + dy) + 4x\,dx = 0$

15. $10x\,dy + 5y\,dx + 3y\,dy = 0$

16. $x^2\,dy + 3xy\,dx + 2\,dx = 0$

In Exercises 17 through 20, find the particular solutions to the given differential equations which satisfy the given conditions.

17. $2(x\,dy + y\,dx) + 3x^2\,dx = 0$; $x = 1$ when $y = 2$

18. $x\,dx + y\,dy = 2(x^2 + y^2)\,dx$; $x = 1$ when $y = 0$

19. $y\,dx - x\,dy = y^3\,dx + y^2x\,dy$; $x = 2$ when $y = 4$

20. $e^{x/y}(x\,dy - y\,dx) = y^4\,dy$; $x = 0$ when $y = 2$

14–4 The Linear Differential Equation of the First Order

There is one type of differential equation of the first order and first degree for which an integrable combination can always be determined. *It is the* **linear differential equation** *of the first order and is of the form*

$$dy + Py\,dx = Q\,dx \tag{14–7}$$

where P and Q are functions of x only. We shall find that this type of equation occurs widely in applications.

If each side of Eq. (14–7) is multiplied by $e^{\int P\,dx}$ it becomes integrable, since the left side becomes of the form du with $u = ye^{\int P\,dx}$ and the right side is a function of x only. This is shown by finding the differential of $ye^{\int P\,dx}$. Thus,

$$d(ye^{\int P\,dx}) = e^{\int P\,dx}(dy + Py\,dx)$$

(In finding the differential of $\int P\,dx$ we use the fact that, by definition, these are reverse processes. Thus, $d(\int P\,dx) = P\,dx$.) Therefore, if each side is multiplied by $e^{\int P\,dx}$, the left side may be immediately integrated to $ye^{\int P\,dx}$ and the right-side integration may be indicated. The solution becomes

$$ye^{\int P\,dx} = \int Qe^{\int P\,dx}\,dx + c \tag{14–8}$$

Example A

Solve the differential equation $dy + \left(\dfrac{2}{x}\right)y \, dx = 4x \, dx$. Here we note that $P = 2/x$ and $Q = 4x$.

The first expression to find is $e^{\int P \, dx}$. In this case this is $e^{\int (2/x) \, dx} = e^{2 \ln x} = e^{\ln x^2} = x^2$ (see below). This means that the left side integrates to yx^2, while the right side becomes $\int 4x(x^2) \, dx$. Thus,

$$y\underbrace{e^{\int P \, dx}}_{\downarrow \; \downarrow} = \int Q\underbrace{e^{\int P \, dx}}_{\downarrow \; \downarrow} \, dx + c$$

$$y(x^2) = \int (4x)(x^2) \, dx + c$$

Therefore,

$$yx^2 = \int 4x^3 \, dx + c$$

or

$$yx^2 = x^4 + c$$

► In finding solutions to this type of differential equation, we often have need of certain basic properties of logarithms and exponential functions. In finding the factor $e^{\int P \, dx}$, we often obtain an expression of the form $e^{\ln u}$. It is easily shown that this equals u.

Let $y = e^{\ln u}$. Writing this in logarithmic form, we have $\ln y = \ln u$. Thus $y = u$ or $u = e^{\ln u}$.

Example B

Solve the differential equation $x \, dy - 3y \, dx = x^3 \, dx$.

This equation can be put in the form of Eq. (14–7) by dividing through by x. This gives $dy - (3/x)y \, dx = x^2 \, dx$. Here, $P = -3/x$, $Q = x^2$, and the factor $e^{\int P \, dx}$ becomes

$$e^{\int (-3/x) \, dx} = e^{-3 \ln x} = e^{\ln x^{-3}} = x^{-3}$$

Therefore,

$$y\underbrace{e^{\int P \, dx}}_{\downarrow \; \downarrow} = \int Q\underbrace{e^{\int P \, dx}}_{\downarrow \; \downarrow} \, dx + c$$

$$yx^{-3} = \int x^2(x^{-3}) \, dx + c$$

This means that

$$yx^{-3} = \int x^{-1} \, dx + c = \ln x + c$$

or

$$y = x^3(\ln x + c)$$

Example C

Solve the differential equation $dy + y\,dx = x\,dx$. Here, $P = 1$, $Q = x$, and

$$e^{\int P\,dx} = e^{\int (1)\,dx} = e^x$$

Therefore,

$$ye^x = \int xe^x\,dx + c = e^x(x - 1) + c$$

or

$$y = x - 1 + ce^{-x}$$

Example D

Solve the differential equation $x^2\,dy + 2xy\,dx = \sin x\,dx$.

This equation is first written in the form of Eq. (14–7). This gives us

$$dy + \left(\frac{2}{x}\right) y\,dx = \frac{1}{x^2}\sin x\,dx$$

Thus,

$$e^{\int P\,dx} = e^{\int (2/x)\,dx} = x^2$$

The solution to the equation then becomes

$$yx^2 = \int \sin x\,dx + c = -\cos x + c$$

or

$$yx^2 + \cos x = c$$

Example E

Solve the differential equation $\cos x\dfrac{dy}{dx} = 1 - y\sin x$.

Writing this in the form of Eq. (14–7), we have

$$dy + y\tan x\,dx = \sec x\,dx$$

Thus,

$$e^{\int P\,dx} = e^{\int \tan x\,dx} = e^{-\ln \cos x} = \sec x$$

The solution is

$$y\sec x = \int \sec^2 x\,dx = \tan x + c$$

or

$$y = \sin x + c\cos x$$

Example F

Find the particular solution of the differential equation

$$dy = (1 - 2y)x \, dx$$

such that $x = 0$ when $y = 2$.

First writing this equation in the form of Eq. (14–7), we have

$$dy + 2xy \, dx = x \, dx$$

Therefore, in this case

$$e^{\int P \, dx} = e^{\int 2x \, dx} = e^{x^2}$$

This leads to

$$ye^{x^2} = \int xe^{x^2} \, dx$$

Integrating the right side, we have

$$ye^{x^2} = \frac{1}{2}e^{x^2} + c$$

Using the condition that $x = 0$ when $y = 2$ gives

$$(2)(e^0) = \frac{1}{2}(e^0) + c \qquad 2 = \frac{1}{2} + c \qquad c = \frac{3}{2}$$

Therefore, the required solution is

$$ye^{x^2} = \frac{1}{2}e^{x^2} + \frac{3}{2} \qquad y = \frac{1}{2}(1 + 3e^{-x^2})$$

Exercises 14–4

In Exercises 1 through 24, solve the given differential equations.

1. $dy + y \, dx = e^{-x} \, dx$

2. $dy + 3y \, dx = e^{-3x} \, dx$

3. $dy + 2y \, dx = e^{-4x} \, dx$

4. $dy + y \, dx = e^{-x} \cos x \, dx$

5. $\dfrac{dy}{dx} - 2y = 4$

6. $2\dfrac{dy}{dx} = 5 - 6y$

7. $x \, dy - y \, dx = 3x \, dx$

8. $x \, dy + 3y \, dx = dx$

9. $2x \, dy + y \, dx = 8x^3 \, dx$

10. $3x \, dy - y \, dx = 9x \, dx$

11. $dy + y \cot x \, dx = dx$

12. $dy + y \tan x \, dx + \sin x \, dx = 0$

13. $\sin x \dfrac{dy}{dx} = 1 - y \cos x$

14. $\dfrac{dy}{dx} - \dfrac{y}{x} = \ln x$

15. $y' + y = 3$

16. $y' + 2y = \sin x$

17. $\dfrac{dy}{dx} = xe^{4x} + 4y$

18. $y' - 2y = 2e^{2x}$

19. $y' = x^3(1 - 4y)$

20. $y' = x^2y + 3x^2$

21. $x\dfrac{dy}{dx} = y + (x^2 - 1)^2$

22. $2x(dy - dx) + y\,dx = 0$

23. $x\,dy + (1 - 3x)y\,dx = 3x^2e^{3x}\,dx$

24. $(1 + x^2)\,dy + xy\,dx = x\,dx$

In Exercises 25 through 30, find the indicated particular solutions of the given differential equations.

25. $\dfrac{dy}{dx} + 2y = e^{-x}$; $\quad x = 0$ when $y = 1$

26. $\dfrac{dy}{dx} - 4y = 2$; $\quad y = 2$ when $x = 0$

27. $x\,dy - 2y\,dx = x^3\cos x\,dx$; $\quad y = \pi^2$ when $x = \pi/2$

28. $y' - 3y = e^{4x}$; $\quad x = 0$ when $y = 2$

29. $y'\sqrt{x} + \dfrac{1}{2}y = e^{\sqrt{x}}$; $\quad x = 1$ when $y = 3$

30. $(\sin x)y' + y = \tan x$; $\quad x = \pi/4$ when $y = 0$

14–5 Elementary Applications

The differential equations of the first order and first degree which we have discussed thus far have a number of applications in geometry and the various fields of technology. In this section we shall illustrate some of these applications, both through the examples and the exercises.

Example A

The slope of a given curve is given by the expression $6xy$. Find the equation of the curve if it passes through the point $(2, 1)$.

Since the slope is $6xy$, we know that $dy/dx = 6xy$. Writing this differential equation as

$$dy = 6xy\,dx$$

we divide through by y to separate the variables. Integrating, we find the general form of the solution.

$$\frac{dy}{y} = 6x\,dx$$

$$\ln y = 3x^2 + c$$

Using the fact that the curve passes through $(2, 1)$, we evaluate the constant.

$$\ln 1 = 3(2^2) + c \qquad 0 = 12 + c \qquad c = -12$$

Thus, the required particular solution is

$$\ln y = 3x^2 - 12$$

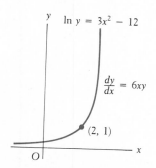

Figure 14–1

See Fig. 14–1.

Example B

A curve which intersects all the members of a family of curves at right angles is called an **orthogonal trajectory** *of the family.* By a family of curves is meant a specified set of curves which satisfy given conditions. Find the equations of the orthogonal trajectories of the parabolas $x^2 = cy$.

Finding the derivative of the given equation, we have $dy/dx = 2x/c$. This equation contains the constant c, which depends on the point (x, y) on the parabola. Eliminating this constant between the equations of the parabola and derivative, we have $dy/dx = 2y/x$. This expression gives us a general expression for the slope of any of the members of the family. For a curve to be perpendicular, its slope must equal the negative reciprocal of this expression, or the slope of the orthogonal trajectories must be given by

$$\left.\frac{dy}{dx}\right|_{OT} = -\frac{x}{2y} \quad \longleftarrow \quad \begin{array}{l}\text{This equation must not}\\ \text{contain the constant } c.\end{array}$$

Solving this differential equation gives the family of orthogonal trajectories:

$$2y\, dy = -x\, dx$$

$$y^2 = -\frac{x^2}{2} + \frac{c'}{2}$$

or

$$2y^2 + x^2 = c'$$

Thus, the orthogonal trajectories are a family of ellipses. Note in Fig. 14–2 that each parabola intersects each ellipse at right angles.

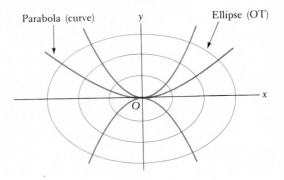

Parabola (curve)   *y*   Ellipse (OT)

Figure 14–2

Example C

Radioactive elements decay at rates proportional to the amount present. A certain isotope of uranium decays so that half of the original amount disappears in 20 days. Determine the equation relating the amount present with the time and the fraction remaining after 50 days.

Let N_0 be the original amount and N the amount present at any time t (in days). Since we know information related to the rate of decay, we express this rate as a derivative. This means that we have the equation

$$\frac{dN}{dt} = kN$$

Separating variables, and integrating, we have

$$\frac{dN}{N} = k\,dt$$

$$\ln N = kt + \ln c$$

Evaluating the constant, we know that $N = N_0$ for $t = 0$. Thus,

$$\ln N_0 = k(0) + \ln c$$

$$c = N_0$$

Therefore, we have the solution

$$\ln N = kt + \ln N_0$$

$$\ln N - \ln N_0 = kt$$

$$\ln\frac{N}{N_0} = kt$$

$$N = N_0 e^{kt}$$

Now using the condition that half this isotope decays in 20 days, we have $N = N_0/2$ when $t = 20$ days. This gives

$$\frac{N_0}{2} = N_0 e^{20k} \quad \text{or} \quad \frac{1}{2} = (e^k)^{20} \quad \text{or} \quad e^k = \left(\frac{1}{2}\right)^{1/20}$$

Therefore, the equation relating N and t is

$$N = N_0\left(\frac{1}{2}\right)^{t/20}$$

In order to determine the fraction remaining after 50 days, we evaluate this equation for $t = 50$. This gives

$$N = N_0\left(\frac{1}{2}\right)^{5/2} = 0.177\,N_0$$

This means that 17.7% remains after 50 days.

Example D

Figure 14–3

The general equation relating the current i, voltage E, inductance L, capacitance C, and resistance R of a simple electric circuit (see Fig. 14–3) is given by

$$L\frac{di}{dt} + Ri + \frac{q}{C} = E$$

where q is the charge on the capacitor. Find the general expression for the current in a circuit containing an inductance, resistance, and voltage source if $i = 0$ when $t = 0$.

The differential equation for this circuit is

$$L\frac{di}{dt} + Ri = E \qquad (14\text{–}9)$$

Using the method of the linear differential equation of the first order, we have the equation

$$di + \frac{R}{L}i\, dt = \frac{E}{L}\, dt$$

The factor $e^{\int P\,dt}$ is $e^{\int (R/L)\,dt} = e^{(R/L)t}$. This gives

$$ie^{(R/L)t} = \frac{E}{L}\int e^{(R/L)t}\, dt = \frac{E}{R}e^{(R/L)t} + c$$

Letting the current be zero for $t = 0$, we have $c = -E/R$. The result is

$$ie^{(R/L)t} = \frac{E}{R}e^{(R/L)t} - \frac{E}{R}$$

or

$$i = \frac{E}{R}(1 - e^{-(R/L)t})$$

(We can see that $i \to E/R$ as $t \to \infty$. In practice the exponential term becomes negligible very quickly.)

Example E

Fifty gallons of brine originally containing 20 lb of salt are in a tank into which 2 gal of water run each minute with the same amount of mixture running out each minute. How much salt remains in the tank after 10 min?

Let $x =$ the number of pounds of salt in the tank at time t. Each gallon of brine contains $x/50$ lb of salt, and in time dt, $2\,dt$ gal of mixture leave the tank with $(x/50)(2\,dt)$ lb of salt. The amount of salt which is leaving may also be written as $-dx$ (the minus sign is included to show that x is decreasing). This means

$$-dx = \frac{2x\,dt}{50}, \quad \text{or} \quad \frac{dx}{x} = -\frac{dt}{25}$$

This leads to $\ln x = -(t/25) + \ln c$. Using the fact that $x = 20$ when $t = 0$, we have $\ln 20 = \ln c$, or $c = 20$. Therefore,

$$x = 20e^{-t/25}$$

is the general expression for the amount of salt in the tank at time t. When $t = 10$ we have

$$x = 20e^{-10/25} = 20e^{-0.4} = 20(0.670) = 13.4$$

There are 13.4 lb of salt in the tank after 10 min.

Example F

An object moving through (or across) a resisting medium often experiences a retarding force approximately proportional to the velocity. Applying Newton's laws of motion to this leads to the equation

$$m\frac{dv}{dt} = F - kv \tag{14–10}$$

where m is the mass of the object, v is the velocity, t is the time, F is the force causing the motion, and k is a constant ($k > 0$). The quantity kv is the retarding force.

We assume these conditions hold for a certain falling object whose mass is 1 slug (this is the unit of mass when force is expressed in pounds) and which experiences a force (its weight) of 32 lb, if the object starts from rest and the retarding force is numerically equal to 0.2 times the velocity (in ft/s).

Substituting in Eq. (14–10), we have

$$\frac{dv}{dt} = 32 - 0.2v$$

Separating variables, we have

$$\frac{dv}{32 - 0.2v} = dt$$

Integrating, we find

$$-5 \ln(32 - 0.2v) = t - 5 \ln c$$

Solving for v, we obtain

$$\ln(32 - 0.2v) = -\frac{t}{5} + \ln c$$

$$32 - 0.2v = ce^{-t/5}$$

$$v = 5(32 - ce^{-t/5})$$

Since the object started from rest, $v = 0$ when $t = 0$. Thus,

$$0 = 5(32 - c), \quad \text{or} \quad c = 32$$

Therefore, the expression giving the velocity as a function of time is

$$v = 160(1 - e^{-t/5})$$

Evaluating v for $t = 5$ s, we have

$$v = 160(1 - e^{-1}) = 160(1 - 0.368) = 101 \text{ ft/s}$$

Exercises 14–5

1. Find the equation of the curve for which the slope is given by $2x/y$ and which passes through $(2, 3)$.

2. Find the equation of the curve for which the slope is given by $-y/(x + y)$ and which passes through $(-1, 3)$.

3. Find the equation of the curve for which the slope is given by $y + x$ and which passes through $(0, 1)$.

4. Find the equation of the curve for which the slope is given by $-2y + e^{-x}$ and which passes through $(0, 2)$.

5. Find the equation of the orthogonal trajectories of the exponential curves $y = ce^x$.

6. Find the equation of the orthogonal trajectories of the cubic curves $y = cx^3$.

7. Find the equation of the orthogonal trajectories of the curves $y = c(\sec x + \tan x)$.

8. Find the equation of the orthogonal trajectories of the family of circles, all with centers at the origin.

9. A certain isotope decays in such a way that half the original amount disappears in 2 h. Find the amount remaining after 3 h, by first finding the relation between the amount present and the time.

10. Radium decays in such a way that half of the original amount disappears in 1800 years. What part of the original amount disappears in 100 years?

11. Carbon-14 has a half-life of about 5600 years. The analysis of the wood in a prehistoric site showed that the concentration of carbon-14 was 60% of that which new wood would have. How old is the wood at the site? (This method, called *carbon dating,* is used to determine the dates of prehistoric events.)

12. After 1 h it is noted that 10% of a certain radioactive material has decayed. Find the half-life of the material.

13. According to Newton's law of cooling, the rate at which a body cools is proportional to the difference in temperature between it and the surrounding medium. How long will it take a cup of hot water, initially at 200°F, to cool to 100°F if the room temperature is 80°F, if it cools to 140°F in 5 min?

14. An object whose temperature is 100°C is placed in a medium whose temperature is 20°C. The temperature of the body falls to 50°C in 10 min. Express the temperature, T, of the body as a function of time. (See Exercise 13.)

15. If interest in a bank account is compounded continuously, the amount grows at a rate which is proportional to the amount which is present. An account earning daily interest very closely approximates this situation. Determine the amount in an account after one year if $1000 is placed in an account which pays 8% interest per year, compounded continuously.

16. If interest is compounded continuously (see Exercise 15), how long will it take a bank account to double in value if the rate of interest is 5% per year?

17. If the current in an RL circuit with a voltage source E is zero when $t = 0$ (see Example D), show that $\lim_{t \to \infty} i = E/R$. See Fig. 14–4.

18. If a circuit contains only an inductance and a resistance, find the expression for the current in terms of the time, if $i = i_0$ when $t = 0$. See Fig. 14–5.

Figure 14–4

Figure 14–5

Figure 14–6

19. For the circuit in Example D, find the current after 1 ms, if $R = 10 \ \Omega$, $L = 0.1$ H, and $E = 100$ V.

20. For the circuit in Exercise 18, what is the value of the current after 1 ms, if $R = 10 \ \Omega$, $L = 0.1$ H, and $i_0 = 2$ A?

21. If a circuit contains only a resistance and a capacitance, find the expression relating the charge on the capacitor in terms of the time, if $i = dq/dt$ and $q = q_0$ when $t = 0$. See Fig. 14–6.

22. If the charge on a capacitor in an RC circuit (see Exercise 21) is 2 mC when $t = 0$, find the charge on the capacitor after 0.01 s if the resistance is 450 Ω and the capacitance is 4 μF.

23. One hundred gallons of brine originally containing 30 lb of salt are in a tank into which 5 gal of water run each minute. The same amount of mixture from the tank leaves each minute. How much salt is in the tank after 20 min?

24. Repeat Exercise 23 with the change that the water entering the tank contains 1 lb of salt per gallon.

25. An object falling under the influence of gravity has a variable acceleration given by $32 - v$, where v represents the velocity. If the object starts from rest, find an expression for the velocity in terms of the time. Also find the limiting value of the velocity. (find $\lim_{t\to\infty} v$).

26. A mass of 2 slugs (weight is 64 lb) is falling under the influence of gravity. Find its velocity after 5 s if it starts from rest and experiences a retarding force numerically equal to 0.1 times the velocity.

27. A boat with a mass of 10 slugs is being towed at 8 mi/h. The tow rope is then cut and a motor which exerts a force of 20 lb on the boat is started. If the water exerts a retarding force which numerically equals twice the velocity (in ft/s), what is the velocity of the boat 3 min later?

28. A parachutist is falling at the rate of 200 ft/s when her parachute opens. If the air resists the fall with a force equal to $0.5v^2$, find the velocity as a function of time. The woman and her equipment have a mass of 5 slugs (weight is 160 lb).

29. A particle moves on the hyperbola $xy = 1$ such that $dx/dt = 2t$. Express x and y in terms of t if $x = 1$, $y = 1$ when $t = 0$.

30. A particle moves on the parabola $y = x^2 + x$, such that $dx/dt = 4t + 1$. Express x and y in terms of t, if both x and y are zero when $t = 0$.

31. The rate of change of air pressure with respect to height is approximately proportional to the pressure. If the pressure is 15 lb/in.^{2} when $h = 0$, and $p = 10$ lb/in.^{2} when $h = 10{,}000$ ft, find the expression relating pressure and height.

32. When a gas undergoes an adiabatic change (no gain or loss of heat), the rate of change of pressure with respect to volume is directly proportional to the pressure and inversely proportional to the volume. Express the pressure in terms of the volume.

33. Assume that the rate of depreciation of an object is proportional to its value at any time t. If a car costs \$8200 new, and its value 3 years later is \$4920, what is its value 11 years after it was purchased?

34. Assume that sugar dissolves at a rate proportional to the undissolved amount. If there are initially 500 g of sugar and 200 g remain after 4 min, how long does it take to dissolve 400 g?

35. Fresh air is being circulated into a room whose volume is 4000 ft^3. Under specified conditions the number of cubic feet x of carbon dioxide present at any time t is found by solving the differential equation $\dfrac{dx}{dt} = 1 - 0.25x$. Find x as a function of t if $x = 12$ ft^3 when $t = 0$ min.

36. The lines of equal potential in a field of force are all at right angles to the lines of force. In an electric field of force caused by charged particles, the lines of force are given by

$$x^2 + y^2 = cx$$

Find the equation of the lines of equal potential. Sketch a few members of the lines of force and those of equal potential.

14–6 Review Exercises for Chapter 14

In Exercises 1 through 20, find the general solution of each given differential equation.

1. $4xy^3\,dx + (x^2 + 1)\,dy = 0$
2. $\dfrac{dy}{dx} = e^{x-y}$
3. $\dfrac{dy}{dx} + 2y = e^{-2x}$
4. $x\,dy + y\,dx = y\,dy$
5. $(x + y)\,dx + (x + y^3)\,dy = 0$
6. $y \ln x\,dx = x\,dy$
7. $x\dfrac{dy}{dx} - 3y = x^2$
8. $dy - 2y\,dx = (x - 2)e^x\,dx$
9. $dy = 2y\,dx + y^2\,dx$
10. $x^2 y\,dy = (1 + x) \csc y\,dx$
11. $y' + 4y = 2$
12. $2xy\,dx = (2y - \ln y)\,dy$
13. $\sin x\dfrac{dy}{dx} + y \cos x + x = 0$
14. $y\,dy = (x^2 + y^2 - x)\,dx$
15. $(x^2 + 2x)\,dy = 2xy\,dx + 5y\,dx$
16. $\dfrac{dy}{dx} = \dfrac{1}{1 + e^x} - y$
17. $x\,dy - y\,dx = x^3 y\,dx$
18. $x\,dy + y\,dx = x^3 y^4\,dy$
19. $y\,dy = 2(x^2 + y^2)^2\,dx - x\,dx$
20. $y' = 2 + xe^{3x}$

In Exercises 21 through 28, find the particular solution of each of the given differential equations.

21. $y' = 2y \cot x$, $x = \pi/2$ when $y = 2$
22. $x\,dy - y\,dx = y^3\,dy$, $x = 1$ when $y = 3$
23. $y' = 4x - 2y$, $x = 0$ when $y = -2$
24. $xy^2\,dx + e^x\,dy = 0$, $x = 0$ when $y = 2$
25. $\dfrac{dy}{dx} = x - xy$, $x = 2$ when $y = 0$

26. $\dfrac{dy}{dx} = 2 - y$, $\quad x = 0$ when $y = 1$

27. $xy' - 2y = x^3 \cos 4x$, $\quad x = \dfrac{\pi}{4}$ when $y = 0$

28. $2xy \, dy + dx = 2x^2 \sin x^2 \, dx$, $\quad x = 1$ when $y = 0$

In Exercises 29 through 42, solve the given problems.

29. The time rate of change of volume of an evaporating substance is proportional to the surface area. Express the radius of a sphere as a function of time. Let $r = r_0$ when $t = 0$. (*Hint:* Express both V and A in terms of the radius r.)

30. An insulated tank is filled with a solution containing radioactive cobalt. Due to the radioactivity, energy is released and the temperature T (in degrees Fahrenheit) of the solution rises with the time t (in hours). The following differential equation expresses the relation between temperature and time for a specific case:

$$56{,}600 = 262(T - 70) + 20{,}200 \frac{dT}{dt}$$

If the initial temperature is 70°F, what is the temperature 24 h later?

31. An object with a mass of 1 kg slides down a long inclined plane. The effective force of gravity is 4 N, and the motion is retarded by a force numerically equal to the velocity. If the object starts from rest, what is its velocity (in meters per second) 4 s later?

32. A 192-lb object falls from rest under the influence of gravity. Find the equation for the velocity at any time t if the air resists the motion with a force numerically equal to twice the velocity.

33. There are initially 100 mg of a certain radioactive material, and after two years there are 95 mg remaining. Find the expression for the mass at any time t.

34. Under proper conditions, the number of bacteria grows at a rate proportional to the number present. In a certain culture there were 10^5 present at a given time, and there were 3.0×10^5 present after 10 h. How many were present after 5 h?

35. Under ideal conditions, the natural law of population change is that it increases at a rate proportional to the population at any time. Under these conditions, project what the population of a city will be in 1990, if its 1970 population was 10,000 and its 1980 population was 12,000.

36. Find the temperature after an hour of an object originally at 100°C, if it cools to 90°C in 5 min in air which is at 20°C. (See Exercise 13 of Section 14–5.)

37. Find the orthogonal trajectories of the family of curves $y = cx^5$.

38. Find the equation of the curves such that their normals at all points are in the same direction as the lines connecting the points and the origin.

39. If a circuit contains a resistance R, capacitance C, and a source of voltage E, express the charge q on the capacitor as a function of time.

40. A 2-H inductor, a 40-Ω resistor, and a 20-V battery are connected in series. Find the current in the circuit as a function of time if the initial current is zero.

$\frac{\pi}{2}$

$r = a \sin \theta$

(r, θ)

π ——————— 0

$\frac{3\pi}{2}$

Figure 14–7

41. The angle ϕ between the radius vector and a curve $r = f(\theta)$ at (r, θ) is given by

$$\frac{dr}{d\theta} = \frac{r}{\tan \phi}$$

Find the polar equation of the orthogonal trajectories of the circles

$$r = a \sin \theta$$

See Fig. 14–7.

42. A differential equation of the form

$$M(x, y)\, dx + N(x, y)\, dy = 0$$

is termed *exact* if

$$\frac{\partial M}{\partial y} = \frac{\partial N}{\partial x}$$

Determine whether or not the equation

$$3x^2 \cos^2 y\, dx - x^3 \sin 2y\, dy = 0$$

is exact.

CHAPTER 15

Second-Order Differential Equations

Until now we have restricted our attention to differential equations of the first order and first degree. Another type of differential equation which is important in technical applications is the linear differential equation of the second order with constant coefficients. In this chapter we will consider this type of differential equation and its applications.

In the first section we shall briefly describe the general higher-order differential equation and the notation we will be using. In the following sections we shall restrict our attention to the second-order linear differential equation with constant coefficients. Finally, in Section 15–5 we shall consider some of its applications, including simple harmonic motion and electric circuits.

15–1 Linear Differential Equations of Higher Order

The general linear differential equation of the nth order is of the form

$$a_0\frac{d^n y}{dx^n} + a_1\frac{d^{n-1}y}{dx^{n-1}} + \cdots + a_{n-1}\frac{dy}{dx} + a_n y = b \tag{15–1}$$

where the a's and b are either functions of x or are constants.

For convenience of notation, the nth derivative with respect to x will be denoted by D^n. The other derivatives are denoted in a similar manner. In this notation, the general linear differential equation becomes

$$a_0 D^n y + a_1 D^{n-1}y + \cdots + a_{n-1}Dy + a_n y = b \tag{15–2}$$

If b = 0 the general linear equation is called **homogeneous,** *and if b ≠ 0 it is called* **nonhomogeneous.** Both these types have important applications.

501

Although the a's may be functions of x, we shall restrict our attention to the case where they are all constants. We shall, however, consider both homogeneous and nonhomogeneous equations. Since second-order equations occur widely in applied problems, the remainder of the chapter will be devoted to this type of equation. The methods developed, however, may be applied to equations of higher order.

15–2 Second-Order Homogeneous Equations with Constant Coefficients

In accordance with the notation and terminology introduced in Section 15–1, *a second-order, linear, homogeneous differential equation with constant coefficients is one of the form*

$$a_0 D^2 y + a_1 Dy + a_2 y = 0 \tag{15–3}$$

where the a's are now constants. The following example will indicate the kind of solution we should expect for this type of equation.

Example A

Solve the differential equation $D^2 y - Dy - 2y = 0$.

First we put this equation in the form $(D^2 - D - 2)y = 0$. This is another way of saying that we are to take the second derivative of y, subtract the first derivative, and finally subtract twice the function. This may now be factored as $(D - 2)(D + 1)y = 0$. (We shall not develop the algebra of the operator D. However, most such algebraic operations can be shown to be valid.) This formula tells us to find the first derivative of the function and add this to the function. Then twice this result is to be subtracted from the derivative of this result. If we let $z = (D + 1)y$, which is valid since $(D + 1)y$ is a function of x, we have $(D - 2)z = 0$. This equation is easily solved by separation of variables. Thus,

$$\frac{dz}{dx} - 2z = 0, \qquad \frac{dz}{z} - 2\,dx = 0, \qquad \ln z - 2x = \ln c_1$$

$$\ln \frac{z}{c_1} = 2x \quad \text{or} \quad z = c_1 e^{2x}$$

Replacing z by $(D + 1)y$, we have

$$(D + 1)y = c_1 e^{2x}$$

This is a linear equation of the first order. Then

$$dy + y\,dx = c_1 e^{2x}\,dx$$

The factor $e^{\int P\,dx}$ is $e^{\int dx} = e^x$. And so

$$ye^x = \int c_1 e^{3x}\,dx = \frac{c_1}{3}e^{3x} + c_2$$

or

$$y = c_1' e^{2x} + c_2 e^{-x}$$

where $c_1' = \frac{1}{3}c_1$. This example indicates that solutions of the form e^{mx} result for this equation.

Basing our reasoning on the result of this example, let us assume that an equation of the form (15–3) has a particular solution e^{mx}. Substituting this into Eq. (15–3) gives

$$a_0 m^2 e^{mx} + a_1 m e^{mx} + a_2 e^{mx} = 0$$

The exponential function e^{mx} is never zero, which means that the polynomial in m is zero, or

$$a_0 m^2 + a_1 m + a_2 = 0 \tag{15–4}$$

Equation (15–4) is called the **auxiliary equation** *of Eq. (15–3). Note that it may be formed directly by inspection from Eq. (15–3).*

There are two roots of the auxiliary Eq. (15–4) and there are two arbitrary constants in the solution of Eq. (15–3). These factors lead us to *the general solution of Eq. (15–3), which is*

$$y = c_1 e^{m_1 x} + c_2 e^{m_2 x} \tag{15–5}$$

where m_1 and m_2 are the solutions of Eq. (15–4). This is seen to be in agreement with the results of Example A.

Example B

Solve the differential equation $D^2 y - 5Dy + 6y = 0$. The auxiliary equation is $m^2 - 5m + 6 = 0$.

Factoring this equation, we have $(m - 3)(m - 2) = 0$, which gives the roots $m_1 = 3$ and $m_2 = 2$. Thus, the solution is

$$y = c_1 e^{3x} + c_2 e^{2x}$$

Obviously it makes no difference which constant is written with each exponential function.

Example C

Solve the differential equation $D^2y - 6Dy = 0$. Here

$$m^2 - 6m = 0$$
$$m(m - 6) = 0$$

or

$$m_1 = 0 \quad \text{and} \quad m_2 = 6$$

The solution is $y = c_1e^{0x} + c_2e^{6x}$. Since $e^{0x} = 1$, we have

$$y = c_1 + c_2e^{6x}$$

Example D

Solve the differential equation $2y'' + 7y' = 4y$.

We first rewrite this equation using the D notation for derivatives. Also, we want to write it in the proper form of a homogeneous equation. This gives us

$$2\,D^2y + 7\,Dy - 4y = 0$$

Now writing the auxiliary equation and solving it, we have

$$2m^2 + 7m - 4 = 0, \qquad (2m - 1)(m + 4) = 0$$

or

$$m_1 = \frac{1}{2} \quad \text{and} \quad m_2 = -4$$

The solution is

$$y = c_1e^{x/2} + c_2e^{-4x}$$

Example E

Solve the differential equation $2\dfrac{d^2y}{dx^2} + \dfrac{dy}{dx} - 7y = 0$.

First writing this equation with the D notation for derivatives, we have

$$2\,D^2y + Dy - 7y = 0$$

This means that the auxiliary equation is

$$2m^2 + m - 7 = 0$$

Since this equation is not factorable, we solve it by use of the quadratic formula. This gives us

$$m = \frac{-1 \pm \sqrt{1 + 56}}{4} = \frac{-1 \pm \sqrt{57}}{4}$$

Therefore, the solution is

$$y = c_1e^{\frac{-1 + \sqrt{57}}{4}x} + c_2e^{\frac{-1 - \sqrt{57}}{4}x}$$

A somewhat better form can be obtained by observing that $e^{-x/4}$ is a factor of each term. Thus,

$$y = e^{-x/4}(c_1e^{\sqrt{57}x/4} + c_2e^{-\sqrt{57}x/4})$$

Example F

Solve the differential equation $D^2y - 2Dy - 15y = 0$, and find the particular solution which satisfies the conditions that $Dy = 2$, $y = -1$ when $x = 0$. (It is necessary to give two conditions since there are two constants to evaluate.)

We have

$$m^2 - 2m - 15 = 0, \qquad (m - 5)(m + 3) = 0$$

or

$$m_1 = 5 \quad \text{and} \quad m_2 = -3$$
$$y = c_1e^{5x} + c_2e^{-3x}$$

This equation is the general solution. In order to evaluate the constants c_1 and c_2, we use the given conditions to find two simultaneous equations in c_1 and c_2. These are then solved to determine the particular solution. Thus,

$$y' = 5c_1e^{5x} - 3c_2e^{-3x}$$

Using the given conditions in the general solution and its derivative, we have

$$c_1 + c_2 = -1, \qquad 5c_1 - 3c_2 = 2$$

The solution to this system of equations is $c_1 = -\frac{1}{8}$ and $c_2 = -\frac{7}{8}$. The particular solution becomes

$$y = -\frac{1}{8}e^{5x} - \frac{7}{8}e^{-3x}$$

or

$$8y + e^{5x} + 7e^{-3x} = 0$$

To solve the auxiliary equation for differential equations of higher order, we use methods developed for higher-degree equations. With these roots, we form the solutions in the same way as we do those for second-order equations.

Exercises 15–1, 15–2

In Exercises 1 through 20, solve the given differential equations.

1. $\dfrac{d^2y}{dx^2} - \dfrac{dy}{dx} - 6y = 0$

2. $\dfrac{d^2y}{dx^2} + \dfrac{dy}{dx} = 0$

3. $\dfrac{d^2y}{dx^2} + 4\dfrac{dy}{dx} + 3y = 0$

4. $\dfrac{d^2y}{dx^2} - 2\dfrac{dy}{dx} - 8y = 0$

5. $D^2y - Dy = 0$

6. $D^2y + 7\,Dy + 6y = 0$

7. $3\,D^2y + 12y = 20\,Dy$

8. $4\,D^2y + 12\,Dy = 7y$

9. $3y'' + 8y' - 3y = 0$

10. $8y'' + 6y' - 9y = 0$

11. $3y'' + 2y' - y = 0$

12. $2y'' - 7y' + 6y = 0$

13. $2\dfrac{d^2y}{dx^2} - 4\dfrac{dy}{dx} + y = 0$

14. $\dfrac{d^2y}{dx^2} + \dfrac{dy}{dx} - 5y = 0$

15. $4D^2y - 3Dy - 2y = 0$

16. $2D^2y - 3Dy - y = 0$

17. $y'' = 3y' + y$

18. $5y'' - y' = 3y$

19. $y'' + y' = 8y$

20. $8y'' = y' + y$

In Exercises 21 through 24, find the particular solutions of the differential equations which satisfy the given conditions.

21. $D^2y - 4Dy - 21y = 0$; $Dy = 0$ and $y = 2$ when $x = 0$

22. $D^2y - 4Dy = 0$; $Dy = 2$ and $y = 4$ when $x = 0$

23. $D^2y - Dy - 12y = 0$; $y = 0$ when $x = 0$ and $y = 1$ when $x = 1$

24. $2D^2y + 5Dy = 0$; $y = 0$ when $x = 0$ and $y = 2$ when $x = 1$

In Exercises 25 through 28, solve the given differential equations. The auxiliary equations will be third degree, and the solutions will have three arbitrary constants.

25. $y''' - 2y'' - 3y' = 0$

26. $2y''' + 3y'' - 2y' = 0$

27. $y''' - 6y'' + 11y' - 6y = 0$

28. $y''' - 2y'' - y' + 2y = 0$

15–3 Auxiliary Equations with Repeated or Complex Roots

The way to solve a second-order homogeneous linear differential equation was shown in the last section. However, we purposely avoided repeated and complex roots of the auxiliary equation. In this section we shall develop the solutions for such equations. The following example indicates the type of solution which results from the case of repeated roots.

Example A

Solve the differential equation $D^2y - 4Dy + 4y = 0$.

Using the method of Example A of the previous section, we have the following steps:

$$(D^2 - 4D + 4)y = 0, \quad (D - 2)(D - 2)y = 0, \quad (D - 2)z = 0$$

where $z = (D - 2)y$. The solution to $(D - 2)z = 0$ is found by separation of variables. And so

$$\frac{dz}{dx} - 2z = 0, \quad \frac{dz}{z} - 2\,dx = 0$$

$$\ln z - 2x = \ln c_1 \quad \text{or} \quad z = c_1e^{2x}$$

Substituting back, we have $(D - 2)y = c_1e^{2x}$, which is a linear equation of the first order. Then

$$dy - 2y\,dx = c_1e^{2x}\,dx, \quad e^{\int -2\,dx} = e^{-2x}$$

This leads to

$$ye^{-2x} = c_1 \int dx = c_1 x + c_2 \quad \text{or} \quad y = c_1 x e^{2x} + c_2 e^{2x}$$

This example indicates the type of solution which results when the auxiliary equation has repeated roots. If the method of the previous section were to be used, the solution of the above example would be $y = c_1 e^{2x} + c_2 e^{2x}$. This would not be the general solution, since both terms are similar, which means that there is only one independent constant. The constants can be combined to give a solution of the form $y = c e^{2x}$, where $c = c_1 + c_2$. This solution would contain only one constant for a second-order equation.

For reference, Eq. (15–3) is $a_0 D^2 y + a_1 Dy + a_2 y = 0$; Eq. (15–4) is $a_0 m^2 + a_1 m + a_2 = 0$.

Based on the above example, *the solution to Eq. (15–3) when the auxiliary Eq. (15–4) has repeated roots is*

$$\underline{y = e^{mx}(c_1 + c_2 x)} \qquad (15–6)$$

where m is the double root. (In Example A, this double root is 2.)

Example B

Solve the differential equation $(D + 2)^2 y = 0$.

The auxiliary equation is $(m + 2)^2 = 0$, for which the solutions are $m = -2, -2$. Since we have repeated roots, the solution of the differential equation is

$$y = e^{-2x}(c_1 + c_2 x)$$

Example C

Solve the differential equation $\dfrac{d^2 y}{dx^2} - 10\dfrac{dy}{dx} + 25y = 0$.

Using the D notation, this equation becomes

$$D^2 y - 10\,Dy + 25y = 0$$

Therefore, the auxiliary equation is

$$m^2 - 10m + 25 = 0$$

Factoring, we have

$$(m - 5)^2 = 0$$

This means the roots are $m = 5, 5$. We therefore have the solution

$$y = e^{5x}(c_1 + c_2 x)$$

When the auxiliary equation has complex roots, it can be solved by the method developed in the preceding section. However, the solution can be put in a special, more useful form. Let us assume that the quadratic formula is used to find the roots of the auxiliary equation, and that these roots are complex of the form $m = \alpha \pm j\beta$. This means that the solution is of the form

$$y = c_1 e^{(\alpha + j\beta)x} + c_2 e^{(\alpha - j\beta)x} = e^{\alpha x}(c_1 e^{j\beta x} + c_2 e^{-j\beta x})$$

Using the exponential form of a complex number, Eq. (13–19), we have

$$y = e^{\alpha x}[c_1 \cos \beta x + jc_1 \sin \beta x + c_2 \cos(-\beta x) + jc_2 \sin(-\beta x)]$$
$$= e^{\alpha x}(c_1 \cos \beta x + c_2 \cos \beta x + jc_1 \sin \beta x - jc_2 \sin \beta x)$$
$$= e^{\alpha x}(c_3 \cos \beta x + c_4 \sin \beta x)$$

where $c_3 = c_1 + c_2$ and $c_4 = jc_1 - jc_2$.

Summarizing these results, we see that *if the auxiliary equation has complex roots of the form* $\alpha \pm j\beta$, *the solution to Eq. (15–3) is*

$$y = e^{\alpha x}(c_1 \sin \beta x + c_2 \cos \beta x) \qquad (15-7)$$

The c_1 and c_2 here are not the same as those above. They are simply the two arbitrary constants of the solution.

Example D

Solve the differential equation $D^2 y - Dy + y = 0$. Thus,

$$m^2 - m + 1 = 0 \quad \text{or} \quad m = \frac{1 \pm j\sqrt{3}}{2}$$

$$\alpha = \frac{1}{2} \quad \text{and} \quad \beta = \frac{\sqrt{3}}{2}$$

We have

$$y = e^{x/2}\left(c_1 \sin \frac{\sqrt{3}}{2}x + c_2 \cos \frac{\sqrt{3}}{2}x\right)$$

Example E

Solve the differential equation $D^2 y + 4y = 0$. In this case we have $m^2 + 4 = 0$, $m_1 = 2j$, and $m_2 = -2j$; $\alpha = 0$ and $\beta = 2$.
The solution is

$$y = e^{0x}(c_1 \sin 2x + c_2 \cos 2x)$$

or

$$y = c_1 \sin 2x + c_2 \cos 2x$$

Example F

Solve the differential equation $y'' - 2y' + 12y = 0$, if $y' = 2$ and $y = 1$ when $x = 0$.

Rewriting this equation as

$$D^2y - 2\,Dy + 12y = 0$$

we have the auxiliary equation

$$m^2 - 2m + 12 = 0$$

Solving for m, we have

$$m = \frac{2 \pm \sqrt{4 - 48}}{2} = \frac{2 \pm 2j\sqrt{11}}{2} = 1 \pm j\sqrt{11}$$

This means that the general solution is

$$y = e^x(c_1\cos\sqrt{11}x + c_2\sin\sqrt{11}x)$$

Using the condition that $y = 1$ when $x = 0$, we have

$$1 = e^0(c_1\cos 0 + c_2\sin 0) \quad \text{or} \quad c_1 = 1$$

Since the other condition involves the value of y', we find the derivative.

$$y' = e^x(c_1\cos\sqrt{11}x + c_2\sin\sqrt{11}x - \sqrt{11}c_1\sin\sqrt{11}x + \sqrt{11}c_2\cos\sqrt{11}x)$$
$$2 = e^0(\cos 0 + c_2\sin 0 - \sqrt{11}\sin 0 + \sqrt{11}c_2\cos 0)$$
$$2 = 1 + \sqrt{11}c_2$$
$$c_2 = \frac{1}{11}\sqrt{11}$$

Therefore, the solution is

$$y = e^x\left(\cos\sqrt{11}x + \frac{1}{11}\sqrt{11}\,\sin\sqrt{11}x\right)$$

Now that we have determined the various types of possible solutions of second-order homogeneous linear differential equations, we can see that it is possible to find the differential equation if the solution is known. Consider the following example.

Example G

If we know that the solution of a differential equation is $y = c_1e^x + c_2e^{2x}$, we know that the auxiliary equation is $(m - 1)(m - 2) = 0$, since it gives solutions of $m_1 = 1$ and $m_2 = 2$. Thus, the simplest form of the differential equation is $(D^2 - 3D + 2)y = 0$.

For the solution $y = c_1e^{2x} + c_2xe^{2x}$, the auxiliary equation would be $(m - 2)^2 = 0$, for repeated roots are indicated by the second term of the solution. The differential equation would be $(D^2 - 4D + 4)y = 0$.

For the solution $y = c_1\sin 2x + c_2\cos 2x$, the auxiliary equation would be $m^2 + 4 = 0$, for imaginary roots are indicated by the terms $\sin 2x$ and $\cos 2x$. The differential equation would be $(D^2 + 4)y = 0$.

Exercises 15–3

In Exercises 1 through 24, solve the given differential equations.

1. $\dfrac{d^2y}{dx^2} - 2\dfrac{dy}{dx} + y = 0$ 2. $\dfrac{d^2y}{dx^2} - 6\dfrac{dy}{dx} + 9y = 0$

3. $D^2y + 12\,Dy + 36y = 0$ 4. $D^2y + 8\,Dy + 16y = 0$

5. $\dfrac{d^2y}{dx^2} + 9y = 0$ 6. $\dfrac{d^2y}{dx^2} + y = 0$

7. $D^2y + Dy + 2y = 0$ 8. $D^2y - 2\,Dy + 4y = 0$

9. $D^2y = 0$ 10. $4\,D^2y - 12\,Dy + 9y = 0$

11. $4\,D^2y + y = 0$ 12. $9\,D^2y + 4y = 0$

13. $16y'' - 24y' + 9y = 0$ 14. $9y'' - 24y' + 16y = 0$

15. $25y'' + 2y = 0$ 16. $y'' - 4y' + 5y = 0$

17. $2\,D^2y + 5y = 4\,Dy$ 18. $D^2y + 4\,Dy + 6y = 0$

19. $25y'' + 16y = 40y'$ 20. $9y'' + 0.6y' + 0.01y = 0$

21. $2\,D^2y - 3\,Dy - y = 0$ 22. $D^2y - 5\,Dy - 4y = 0$

23. $3\,D^2y + 12\,Dy = 2y$ 24. $36\,D^2y = 25y$

In Exercises 25 through 28, find the particular solutions of the given differential equations which satisfy the stated conditions.

25. $y'' + 2y' + 10y = 0$; $y = 0$ when $x = 0$ and $y = e^{-1}$ when $x = \dfrac{\pi}{6}$

26. $D^2y + 16y = 0$; $Dy = 0$ and $y = 2$ when $x = \dfrac{\pi}{2}$

27. $D^2y - 8\,Dy + 16y = 0$; $Dy = 2$ and $y = 4$ when $x = 0$

28. $4y'' + 20y' + 25y = 0$; $y = 0$ when $x = 0$ and $y = e$ when $x = -\tfrac{2}{5}$

In Exercises 29 through 32, find the simplest form of the second-order homogeneous linear differential equation which has the given solution.

29. $y = c_1 e^{3x} + c_2 e^{-3x}$ 30. $y = c_1 e^{3x} + c_2 x e^{3x}$

31. $y = c_1 \cos 3x + c_2 \sin 3x$ 32. $y = c_1 e^{2x} \cos x + c_2 e^{2x} \sin x$

15–4   Solutions to Nonhomogeneous Equations

To solve a nonhomogeneous linear equation of the form

$$a_0\,D^2y + a_1\,Dy + a_2 y = b \qquad\qquad (15\text{–}8)$$

where b is a function of x or is a constant, the solution must be such that if we substitute it into the left side, we obtain the right side. Solutions obtained from the methods of Sections 15–2 and 15–3 will give zero when substituted into the left side, but they do contain the arbitrary constants necessary in the solution. If we could find some particular solution, which when substituted into the left side produced the expression of the right, it could be added to the solution containing the arbitrary constants. Thus, *the solution is of the form*

$$y = y_c + y_p$$

(15–9)

where y_c, called the **complementary solution,** *is obtained by solving the corresponding homogeneous equation and where y_p is the particular solution necessary to produce the expression b of Eq. (15–8).*

Example A

The differential equation $D^2y - Dy - 6y = e^x$ has the solution $y = c_1e^{3x} + c_2e^{-2x} - \frac{1}{6}e^x$, with $y_c = c_1e^{3x} + c_2e^{-2x}$ and $y_p = -\frac{1}{6}e^x$. The complementary solution y_c is obtained by solving the corresponding homogeneous equation $D^2y - Dy - 6y = 0$, and we shall discuss below the method of finding y_p. We can see that y_p alone will satisfy the differential equation, but since it has no arbitrary constants, it cannot be the general solution.

 By inspection of the form of the expression b on the right side of the equation, we can determine the form which the particular solution must have. Since a combination of the derivatives and the function itself must form the function b, **we assume an expression which contains all possible forms of b and its derivatives will form y_p.** The method used to find the exact form of this expression is called the **method of undetermined coefficients.**

Example B

If the function b on the right of Eq. (15–8) is $4x + e^{2x}$, we would assume the particular solution is of the form $y_p = A + Bx + Ce^{2x}$. The Bx-term is included to account for the presence of the $4x$. The A-term is included to account for any derivatives of the Bx-term which may occur. The Ce^{2x} is included to account for the presence of the e^{2x}-term. Since all the derivatives of Ce^{2x} are of this same form, it is not necessary to include any other terms for the e^{2x}.

Example C

If the expression b is of the form $x^2 + e^{-x}$, we would assume y_p to be of the form $y_p = A + Bx + Cx^2 + Ee^{-x}$.

If b is of the form $xe^{-2x} - 5$, we would assume y_p to be of the form $y_p = Ae^{-2x} + Bxe^{-2x} + C$.

If b is of the form $x \sin x$, we would then assume y_p to be of the form $A \sin x + B \cos x + Cx \sin x + Ex \cos x$. All these terms occur in the derivatives of $x \sin x$.

Example D

If the expression for b is of the form $e^x + xe^x$, we would assume y_p to be of the form $y_p = Ae^x + Bxe^x$. These terms occur for xe^x and its derivatives. Since the form of the e^x-term of b is already included in Ae^x, we do not include another e^x-term in y_p.

In the same way, if b is of the form $2x + 4x^2$, the form of y_p is $y_p = A + Bx + Cx^2$. These are the only forms which occur in either $2x$ or $4x^2$ and their derivatives.

Once we have determined the form of the particular solution, we have to find the numerical values of the coefficients $A, B, \ldots$. *This is done by substituting the form into the differential equation, and equating coefficients of like terms.*

Example E

Solve the differential equation $D^2y - Dy - 6y = e^x$.

In this case the solution of the auxiliary equation $m^2 - m - 6 = 0$ gives us the roots $m_1 = 3$ and $m_2 = -2$. Thus

$$y_c = c_1e^{3x} + c_2e^{-2x}$$

The proper form of y_p is $y_p = Ae^x$. Substituting y_p and its derivatives into the differential equation, we have

$$Ae^x - Ae^x - 6Ae^x = e^x$$

To produce equality, the coefficients of e^x must be the same on each side of the equation. Thus,

$$-6A = 1$$

or

$$A = -\frac{1}{6}$$

This gives the complete solution $y = y_c + y_p$, where $y_p = -\frac{1}{6}e^x$. Thus,

$$y = c_1e^{3x} + c_2e^{-2x} - \frac{1}{6}e^x$$

(See Example A)

Example F

Solve the differential equation $D^2y + 4y = x - 4e^{-x}$.

In this case we have $m^2 + 4 = 0$, which gives us $m_1 = 2j$ and $m_2 = -2j$. Therefore,

$$y_c = c_1 \sin 2x + c_2 \cos 2x$$

The proper form of the particular solution is $y_p = A + Bx + Ce^{-x}$. Finding two derivatives and then substituting gives the following result:

$$y_p' = B - Ce^{-x}, \qquad y_p'' = Ce^{-x}$$
$$(Ce^{-x}) + 4(A + Bx + Ce^{-x}) = x - 4e^{-x}$$
$$Ce^{-x} + 4A + 4Bx + 4Ce^{-x} = x - 4e^{-x}$$
$$(4A) + (4Bx) + (5Ce^{-x}) = 0 + (1)x + (-4e^{-x})$$

▶ Equating the coefficients of the constants, x and e^{-x}, gives $4A = 0$, $4B = 1$, and $5C = -4$. Thus, $A = 0$, $B = \frac{1}{4}$, and $C = -\frac{4}{5}$. This means that the particular solution is $y_p = \frac{1}{4}x - \frac{4}{5}e^{-x}$. In turn this tells us that the complete solution is

$$y = c_1 \sin 2x + c_2 \cos 2x + \frac{1}{4}x - \frac{4}{5}e^{-x}$$

Example G

Solve the differential equation $D^2y - 3Dy + 2y = 2\sin x$.

Thus,

$$m^2 - 3m + 2 = 0, \qquad (m - 1)(m - 2) = 0, \qquad m_1 = 1 \quad \text{and} \quad m_2 = 2$$

We write $y_c = c_1 e^x + c_2 e^{2x}$. Then

$$y_p = A \sin x + B \cos x$$
$$Dy_p = A \cos x - B \sin x$$
$$D^2 y_p = -A \sin x - B \cos x$$
$$(-A \sin x - B \cos x) - 3(A \cos x - B \sin x) + 2(A \sin x + B \cos x) = 2 \sin x$$
$$(A + 3B)\sin x + (B - 3A)\cos x = 2 \sin x$$
$$A + 3B = 2, \qquad -3A + B = 0$$

The solution of the system is $A = \frac{1}{5}$ and $B = \frac{3}{5}$. Thus,

$$y_p = \frac{1}{5} \sin x + \frac{3}{5} \cos x$$

Therefore,

$$y = c_1 e^x + c_2 e^{2x} + \frac{1}{5} \sin x + \frac{3}{5} \cos x$$

Exercises 15–4

In Exercises 1 through 12, solve the given differential equations. The form of y_p is indicated.

1. $D^2y - Dy - 2y = 4$ (Let $y_p = A$.)
2. $D^2y - Dy - 6y = 4x$ (Let $y_p = A + Bx$.)
3. $D^2y + y = x^2$ (Let $y_p = A + Bx + Cx^2$.)
4. $D^2y - y = 2 + x^2$ (Let $y_p = A + Bx + Cx^2$.)
5. $D^2y + 4Dy + 3y = 2 + e^x$ (Let $y_p = A + Be^x$.)
6. $D^2y + Dy + y = e^{2x}$ (Let $y_p = Ae^{2x}$.)
7. $y'' - 3y' = 2e^x + xe^x$ (Let $y_p = Ae^x + Bxe^x$.)
8. $y'' + y' - 2y = 8 + 4x + 2xe^{2x}$ (Let $y_p = A + Bx + Ce^{2x} + Exe^{2x}$.)
9. $D^2y - y = \sin x$ (Let $y_p = A \sin x + B \cos x$.)
10. $D^2y + 4y = \sin x + 4$ (Let $y_p = A + B \sin x + C \cos x$.)
11. $\dfrac{d^2y}{dx^2} + 9y = 9x + 5 \cos 2x + 10 \sin 2x$
    (Let $y_p = A + Bx + C \sin 2x + E \cos 2x$.)
12. $\dfrac{d^2y}{dx^2} - 2\dfrac{dy}{dx} + y = 2x + x^2 + \sin 3x$
    (Let $y_p = A + Bx + Cx^2 + E \sin 3x + F \cos 3x$.)

In Exercises 13 through 24, solve the given differential equations.

13. $\dfrac{d^2y}{dx^2} - \dfrac{dy}{dx} - 30y = 10$
14. $2\dfrac{d^2y}{dx^2} + 11\dfrac{dy}{dx} - 6y = 8x$
15. $3\dfrac{d^2y}{dx^2} + 13\dfrac{dy}{dx} - 10y = 14e^{3x}$
16. $\dfrac{d^2y}{dx^2} + 4y = 2 \sin 3x$
17. $D^2y - 4y = \sin x + 2 \cos x$
18. $2D^2y + 5Dy - 3y = e^x + 4e^{2x}$
19. $D^2y + y = 4 + \sin 2x$
20. $D^2y - Dy + y = x + \sin x$
21. $D^2y + 5Dy + 4y = xe^x + 4$
22. $3D^2y + Dy - 2y = 4 + 2x + e^x$
23. $y'' + 6y' + 9y = e^{2x} - e^{-2x}$
24. $y'' + 8y' - y = x^2 + 4e^{-2x}$

In Exercises 25 through 28, find the particular solution of each differential equation for the given conditions.

25. $D^2y - Dy - 6y = 5 - e^x$; $Dy = 4$ and $y = 2$ when $x = 0$.
26. $3y'' - 10y' + 3y = xe^{-2x}$; $Dy = 0$ and $y = -1$ when $x = 0$.
27. $y'' + y = x + \sin 2x$; $Dy = 1$ and $y = 0$ when $x = \pi$.
28. $D^2y - 2Dy + y = xe^{2x} - e^{2x}$; $Dy = 4$ and $y = -2$ when $x = 0$.

15–5 Applications of Second-Order Equations

Linear differential equations of the second order have many important applications. We shall restrict our attention to two of these. In this section, we shall apply second-order differential equations to solving problems in simple harmonic motion and simple electric circuits.

Example A

Simple harmonic motion may be defined as motion in a straight line for which the acceleration is proportional to the displacement and in the opposite direction. Examples of this type of motion are a weight on a spring, a simple pendulum, and an object bobbing in water. If x represents the displacement, d^2x/dt^2 is the acceleration.

Using the definition of simple harmonic motion, we have

$$\frac{d^2x}{dt^2} = -k^2x$$

(We chose k^2 for convenience of notation in the solution.) Writing this equation in the form

$$D^2x + k^2x = 0 \qquad \left(\text{here, } D = \frac{d}{dt}\right)$$

we find that its solution is

$$x = c_1 \sin kt + c_2 \cos kt$$

This solution indicates an oscillating motion, which is known to be the case. If, for example, $k = 4$ and we know that $x = 2$ and $Dx = 0$ (which means the velocity is zero) for $t = 0$, we have

$$Dx = 4c_1 \cos 4t - 4c_2 \sin 4t$$

or (substituting into the expression for x and that for Dx)

$$2 = c_1(0) + c_2(1)$$

and

$$0 = 4c_1(1) - 4c_2(0)$$

which gives $c_1 = 0$ and $c_2 = 2$. Therefore,

$$x = 2 \cos 4t$$

is the equation relating the displacement and time; Dx is the velocity and D^2x is the acceleration.

Example B

In practice an object will in time cease to move due to unavoidable frictional forces. It is found that a "freely" oscillating object has a force retarding it which is approximately proportional to the velocity. The differential equation for this case is $D^2x = -k^2x - b\,Dx$. This results from applying (from physics) ***Newton's second law of motion,*** which states that *the net force acting on an object equals its mass times its acceleration.* The term D^2x represents the acceleration, the term $-k^2x$ is a measure of the restoring force (of the spring, for example), and the term $-b\,Dx$ is the term which represents the retarding (damping) force. This equation can be written as

$$D^2x + b\,Dx + k^2x = 0$$

The auxiliary equation is $m^2 + bm + k^2 = 0$, for which the roots are

$$m = \frac{-b \pm \sqrt{b^2 - 4k^2}}{2}$$

See Appendix C for a computer program for graphical solutions of these cases.

If $k = 3$ and $b = 4$, $m = -2 \pm j\sqrt{5}$, which means the solution is

$$x = e^{-2t}(c_1\sin\sqrt{5}t + c_2\cos\sqrt{5}t) \tag{1}$$

Here, $4k^2 > b^2$, and this case is called **underdamped.**

If $k = 2$ and $b = 5$, $m = -1, -4$, which means the solution is

$$x = c_1e^{-t} + c_2e^{-4t} \tag{2}$$

Here $4k^2 < b^2$, and the case is called **overdamped.** It will be noted that the motion is not oscillatory, since no sine or cosine terms appear.

If $k = 2$ and $b = 4$, $m = -2, -2$, which means the solution is

$$x = e^{-2t}(c_1 + c_2t) \tag{3}$$

Here $4k^2 = b^2$, and the case is called **critically damped.** Again the motion is not oscillatory. See Fig. 15–1, in which Eqs. (1), (2), and (3) are represented in general. Of course, the actual values depend upon c_1 and c_2, which in turn depend upon the conditions imposed on the motion.

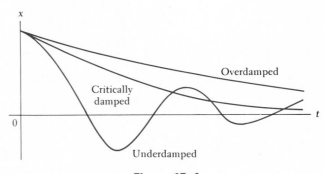

Figure 15–1

Example C

In testing the characteristics of a particular type of spring, it is found that a weight of 16.0 lb stretches the spring 1.60 ft when the weight and spring are placed in a fluid which resists the motion with a force equal to twice the velocity. If the weight is brought to rest and then given a velocity of 12.0 ft/s, find the equation of motion. See Fig. 15–2.

Figure 15–2

In order to find the equation of motion we use Newton's second law of motion (see Example B). The weight (one force) at the end of the spring is offset by the equilibrium position force exerted by the spring, in accordance with Hooke's law (see Section 5–6). Therefore, the net force acting on the weight is the sum of the Hooke's law force due to the displacement from the equilibrium position and the resisting force. Using Newton's second law, we have

mass × acceleration = resisting force + Hooke's law force

$$m\,D^2x = -2.00\,Dx - kx$$

The mass of an object is its weight divided by the acceleration due to gravity. The weight is 16.0 lb and the acceleration due to gravity is 32.0 ft/s^2. Thus, the mass m is

$$m = \frac{16.0\text{ lb}}{32.0\text{ ft/s}^2} = 0.500\text{ slug}$$

where the slug is the unit of mass if the weight is in pounds.

The constant k for the Hooke's law force is found from the fact that the spring stretches 1.60 ft for a force of 16.0 lb. Thus, using Hooke's law,

$$16.0 = k(1.60)$$
$$k = 10.0\text{ lb/ft}$$

This means that the differential equation to be solved is

$$0.500\,D^2x + 2.00\,Dx + 10.0x = 0$$

or

$$1.00\,D^2x + 4.00\,Dx + 20.0x = 0$$

(*continued on next page*)

The auxiliary equation is

$$1.00m^2 + 4.00m + 20.0 = 0$$

for which the solution is

$$m = \frac{-4.00 \pm \sqrt{16.0 - 4(20.0)(1.00)}}{2.00}$$

$$= -2.00 \pm 4.00j$$

Therefore, the general solution is

$$x = e^{-2.00t}(c_1\cos 4.00t + c_2\sin 4.00t)$$

Since the weight started from the equilibrium position with a velocity of 12.0 ft/s, we know that $x = 0$ and $Dx = 12.0$ for $t = 0$. Thus,

$$0 = e^0(c_1 + 0c_2), \quad \text{or} \quad c_1 = 0$$

Thus,

$$x = c_2e^{-2.00t}\sin 4.00t$$

$$Dx = c_2e^{-2.00t}(\cos 4.00t)(4.00) + c_2\sin 4.00t\,(e^{-2.00t})(-2.00)$$

$$12.0 = c_2e^0(1)(4.00) + c_2(0)(e^0)(-2.00)$$

$$c_2 = 3.00$$

This means that the equation of motion is

$$x = 3.00e^{-2.00t}\sin 4.00t$$

This motion is underdamped; the graph is shown in Fig. 15–3.

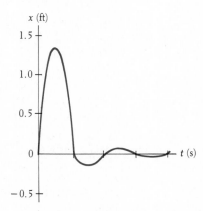

Figure 15–3

In Example C the force was given in pounds and the mass was therefore expressed in slugs. If metric units are used, it is common to give the mass of the object in kilograms, and its weight is therefore in newtons. The weight, which is needed to determine the constant k in the Hooke's law force, is found by multiplying the mass, in kilograms, by the acceleration of gravity, 9.80 m/s^2.

It is possible to have an additional force acting on a weight such as the one in Example C. For example, a vibratory force may be applied to the support of the spring. In such a case, called *forced vibrations,* this additional external force is added to the other net force. This means that the added force $F(t)$ becomes a nonzero function on the right side of the differential equation, and we must then solve a nonhomogeneous equation.

Example D

E or $E_0 \sin \omega t$

Figure 15–4

A basic relation for electric circuits is that the impressed voltage of a circuit equals the sum of the voltages across the components of the circuit. When this principle is applied to a circuit containing a resistance, inductance, capacitance, and a voltage source (see Fig. 15–4), the differential equation is

$$L\frac{d^2q}{dt^2} + R\frac{dq}{dt} + \frac{q}{C} = E \qquad (15\text{–}10)$$

By definition q represents the electric charge, $dq/dt = i$ is the current and d^2q/dt^2 is the time rate of change of current. This equation may be written as $LD^2q + RDq + q/C = E$. The auxiliary equation is $Lm^2 + Rm + 1/C = 0$. The roots are

$$m = \frac{-R \pm \sqrt{R^2 - 4L/C}}{2L} = -\frac{R}{2L} \pm \sqrt{\frac{R^2}{4L^2} - \frac{1}{LC}}$$

If we let $a = R/2L$ and $\omega = \sqrt{1/LC - R^2/4L^2}$, we have (assuming complex roots, which corresponds to realistic values of R, L, and C)

$$q_c = e^{-at}(c_1 \sin \omega t + c_2 \cos \omega t)$$

This indicates an oscillating charge, or an alternating current. However, the exponential term usually is such that the current dies out rapidly unless there is a source of voltage in the circuit.

If there is no source of voltage in the circuit of Example D, we have a homogeneous differential equation to solve. If we have a constant voltage source, the particular solution is of the form $q_p = A$. If there is an alternating voltage source, the particular solution is of the form $q_p = A \sin \omega_1 t + B \cos \omega_1 t$, where ω_1 is the angular velocity of the source. After a very short time, the exponential factor in the complementary solution makes it negligible. For this reason it is referred to as the **transient** term, and *the particular solution is the* **steady-state** *solution.* Therefore, to find the steady-state solution, we need find only the particular solution.

Example E

Find the steady-state solution for the current in a circuit containing the following elements: $C = 400 \ \mu\text{F}$, $L = 1 \ \text{H}$, $R = 10 \ \Omega$, and a voltage source of $500 \sin 100t$.

This means the differential equation to be solved is

$$\frac{d^2q}{dt^2} + 10\frac{dq}{dt} + \frac{10^4}{4}q = 500 \sin 100t$$

Since we wish to find the steady-state solution, we must find q_p, from which we may find i_p by finding a derivative.

$$\frac{d^2q}{dt^2} + 10\frac{dq}{dt} + \frac{10^4}{4}q = 500 \sin 100t$$

$$q_p = A \sin 100t + B \cos 100t$$

$$\frac{dq_p}{dt} = 100A \cos 100t - 100B \sin 100t$$

$$\frac{d^2q_p}{dt^2} = -10^4A \sin 100t - 10^4B \cos 100t$$

Substituting into the differential equation, we have

$$-10^4A \sin 100t - 10^4B \cos 100t + 10^3A \cos 100t - 10^3B \sin 100t$$

$$+ \frac{10^4}{4}A \sin 100t + \frac{10^4}{4}B \cos 100t = 500 \sin 100t$$

This means

$$(-0.75 \times 10^4A - 10^3B)\sin 100t + (-0.75 \times 10^4B + 10^3A)\cos 100t = 500 \sin 100t$$

or

$$-7.5 \times 10^3A - 10^3B = 500, \qquad 10^3A - 7.5 \times 10^3B = 0$$

Solving these equations, we obtain

$$B = -8.73 \times 10^{-3} \quad \text{and} \quad A = -65.5 \times 10^{-3}$$

Therefore,

$$q_p = -65.5 \times 10^{-3}\sin 100t - 8.73 \times 10^{-3}\cos 100t$$

and

$$i_p = \frac{dq_p}{dt} = -6.55 \cos 100t + 0.87 \sin 100t$$

which is the desired solution.

It should be noted that the complementary solutions of the mechanical and electrical cases are of the same identical form. There is also an equivalent mechanical case to that of an impressed sinusoidal voltage source in the electrical case. This arises in the case of forced vibrations, when an external force affecting the vibrations is applied to the system. Thus, we may have transient and steady-state solutions to mechanical and other nonelectrical situations.

Exercises 15–5

1. An object moves with simple harmonic motion according to $D^2x + 100x = 0$; $D = d/dt$. Find the displacement x as a function of the time, subject to the conditions that $x = 4$ and $Dx = 0$ when $t = 0$. Sketch the resulting curve.

2. Solve the problem given in Exercise 1 if the motion is in accordance with the equation $D^2x + 0.2\,Dx + 100x = 0$. Sketch the curve.

3. What must be the value of b so that the motion of an object given by the equation $D^2x + b\,Dx + 100x = 0$ is critically damped?

4. What is the displacement of an object which moves according to the equation $D^2x + 4\,Dx + 6x = 0$ (if $x = 6$ and $Dx = 0$ when $t = 0$) after 2 s?

5. For a certain spring a 4-lb weight stretches it $\frac{1}{8}$ ft. With this weight attached, the spring is pulled 3 in. longer than its equilibrium length and released. Find the equation of the resulting motion, assuming no damping.

6. Find the solution for the spring of Exercise 5 if a damping force numerically equal to the velocity is present.

7. Find the solution for the spring of Exercise 5 if no damping is present, but an external force of $4 \sin 2t$ is acting on the spring.

8. Find the solution for the spring of Exercise 5 if the damping force of Exercise 6 and the impressed force of Exercise 7 are both acting.

9. A mass of 0.820 kg stretches a given spring by 0.250 m. The mass is pulled down 0.150 m below the equilibrium position and released. Find the equation of motion of the mass.

10. Find the equation of motion for the mass of Exercise 9 if a damping force numerically equal to three times the velocity is also present.

11. Find the equation relating the charge and time in an electric circuit with the following elements: $L = 0.2$ H, $R = 8\ \Omega$, $C = 1\ \mu$F, $E = 0$. In this circuit $q = 0$ and $i = 0.5$ A when $t = 0$.

12. For a given electric circuit, $L = 0.5$ H, $R = 0$, $C = 200\ \mu$F, and $E = 0$. Find the equation relating charge and time if $q = 100\ \mu$C and $i = 0$ when $t = 0$.

13. For a given circuit, $L = 0.1$ H, $R = 0$, $C = 100\ \mu$F, and $E = 100$ V. Find the equation relating the charge and time if $q = 0$ and $i = 0$ when $t = 0$.

14. Find the relation between the current and time for the circuit of Exercise 13.

15. For a given circuit, $L = 0.5$ H, $R = 10\ \Omega$, $C = 200\ \mu$F, and $E = 100 \sin 200t$. Find the expression relating charge and time.

16. Find the relation between the current and time for the circuit of Exercise 15.

17. Find the steady-state current for the circuit of Exercise 15.

18. Find the steady-state current for a circuit with $L = 1$ H, $R = 5\ \Omega$, $C = 150\ \mu$F, and $E = 120 \sin 100t$ V.

19. Find the steady-state solution for the charge of the circuit of Exercise 18 if the voltage source is changed to $100 \sin 400t$.

20. Find the steady-state solution for the current in an electric circuit containing the following elements: $C = 20\ \mu$F, $L = 2$ H, $R = 20\ \Omega$, and $E = 200 \sin 10t$.

15-6 Review Exercises for Chapter 15

In Exercises 1 through 20, find the general solution of each of the given differential equations.

1. $2 D^2 y + Dy = 0$

2. $2 D^2 y - 5 Dy + 2y = 0$

3. $y'' + 2y' + y = 0$

4. $y'' + 2y' + 2y = 0$

5. $D^2 y - 2 Dy + 5y = 0$

6. $4 D^2 y - 36 Dy + 81y = 0$

7. $\dfrac{d^2 y}{dx^2} = 0.08y - 0.2\dfrac{dy}{dx}$

8. $\dfrac{d^2 y}{dx^2} = 0.2\dfrac{dy}{dx} - 0.17y$

9. $y'' - y' - 56y = 28$

10. $2\dfrac{d^2 y}{dx^2} + \dfrac{dy}{dx} - 3y = 6$

11. $\dfrac{d^2 y}{dx^2} + 6\dfrac{dy}{dx} + 9y = 3x$

12. $\dfrac{d^2 y}{dx^2} - 25y = 5x$

13. $4y'' + 3y = 3x^2$

14. $y'' - 3y' + 2y = 2x^2 + 1$

15. $9 D^2 y - 18 Dy + 8y = 16 + 4x$

16. $D^2 y + 9y = xe^x$

17. $y'' + y' - y = 2e^x$

18. $y'' + y = 4 \cos 2x$

19. $D^2 y + 9y = \sin x$

20. $y'' + y' = e^x + \cos 2x$

In Exercises 21 through 24, find the indicated particular solutions of the given differential equations.

21. $\dfrac{d^2 y}{dx^2} + \dfrac{dy}{dx} + 4y = 0$, $Dy = \sqrt{15}$, $y = 0$ when $x = 0$

22. $5y'' + 7y' - 6y = 0$, $y' = 10$, $y = 2$ when $x = 0$

23. $(D^2 + 4D + 4)y = 4 \cos x$, $Dy = 1$, $y = 0$ when $x = 0$

24. $y'' - 2y' + y = e^{2x} + x$, $Dy = 0$, $y = 2$ when $x = 0$

In Exercises 25 through 34, solve the given problems.

25. A 64-lb weight stretches a certain spring 2 ft. With this weight suspended on it, the spring is stretched 2 ft beyond the equilibrium position and released. Find the equation of the resulting motion if the medium in which the weight is suspended retards the motion with a force equal to 16 times the velocity. Classify the motion as underdamped, critically damped, or overdamped.

26. Find the solution for the spring of Exercise 25 if there is no force retarding the motion.

27. A mass of 1.25 kg stretches a spring by 0.200 m. From the equilibrium position the mass is given a velocity of 4.50 m/s. Find the equation of motion if there is a resisting force which is numerically equal to four times the velocity.

28. What is the expression for the steady-state displacement x of an object which moves according to the relation $D^2 x - 2 Dx - 8x = 5 \sin 2t$?

29. A 0.5-H inductor, a 6-Ω resistor, and a 20-mF capacitor are connected in series with a generator for which $E = 24 \sin 10t$. Find the charge on the capacitor as a function of time if the initial charge and current are zero.

30. Find the general expression for the charge on the capacitor in a circuit having a resistance of 20 Ω, an inductor of 4 H, and a capacitor of 100 μF if the initial charge on the capacitor is 10 mC and the current is zero when $t = 0$.

31. Find the solution for the current for the circuit of Exercise 30 if a battery of 100 V is placed in the circuit.

32. If a circuit contains an inductor of inductance L, a capacitor of capacitance C, and a sinusoidal source of voltage $E_0 \sin \omega t$, express the charge q on the capacitor as a function of the time. Assume $q = 0$, $i = 0$ when $t = 0$.

33. The approximate differential equation relating the displacement y of a beam at a horizontal distance x from one end is

$$EI \frac{d^2 y}{dx^2} = M$$

where E is the modulus of elasticity, I is the moment of inertia of the cross section of the beam perpendicular to its axis, and M is the bending moment at the cross section. If $M = 2000x - 40x^2$ for a particular beam of length L for which $y = 0$ when $x = 0$ and when $x = L$, express y in terms of x. Consider E and I as constants.

34. When a circular disk of mass m and radius r is suspended by a wire at the center on one of its flat faces and the disk is twisted through an angle θ, torsion in the wire tends to turn the disk back in the opposite direction. The differential equation for this case is

$$\frac{1}{2} mr^2 \frac{d^2 \theta}{dt^2} = -k\theta$$

where k is a constant. Determine the equation of motion if $\theta = \theta_0$ and $d\theta/dt = \omega_0$ when $t = 0$. See Fig. 15–5.

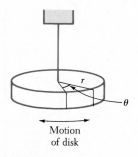

Motion
of disk

Figure 15–5

CHAPTER 16

Other Methods of Solving Differential Equations

In the previous two chapters we presented methods of solving certain types of differential equations. Other types of differential equations, many of which do not have exact solutions, have important applications. Numerous methods and techniques have been developed for solving these equations.

In this chapter we shall develop three of these other methods of solution. The methods developed here are (1) the *increment method,* which is a numerical method of solution, (2) a method of *successive approximations,* and (3) the method of *Laplace transforms,* by which particular solutions can be found by essentially algebraic means.

16-1 Numerical Solutions—the Increment Method

Numerical methods of solving differential equations often require a great deal of time and computation to arrive at the desired solution. For this reason they lend themselves very well to the use of calculators and computers, by which the calculations may be done very rapidly. Also, due to the wide use of computers, numerical methods such as these are important in industry. The **increment method,** which illustrates the basic numerical technique, can be used with nearly any method of computation.

In order to find an approximate solution to a differential equation of the form $dy/dx = f(x, y)$ which passes through a known point (x_1, y_1), we approximate the increment of y, Δy, by the differential of y, dy. From Section 4–1 we recall that this approximation is good if Δx is small. The use of this approximation is illustrated in the following examples.

Example A

Find an approximate solution to the differential equation $dy/dx = x + y$ by the increment method. The curve passes through the point (0, 1).

The differential equation can be written as

$$dy = (x + y)\,\Delta x$$

(since $\Delta x = dx$). We know that $y = 1$ for $x = 0$. Let us choose $\Delta x = 0.1$. With these values we can find a good approximation for dy. Thus,

$$dy = (0 + 1)(0.1) = 0.1$$

This tells us that the curve passes (or nearly passes) through the point (0.1, 1.1). This is seen by adding 0.1 to the original value of x, 0, and then adding 0.1 to the original value of y, 1.0. Assuming this point is correct we use the values $x = 0.1$, $y = 1.1$, and $\Delta x = 0.1$ to find the next approximate point. This time

$$dy = (0.1 + 1.1)(0.1) = 0.12$$

which means that the curve passes through (0.2, 1.22). Continuing this process we find a set of points which would approximately satisfy the function which is the solution of the differential equation. Tabulating the results obtained in this manner, we have the following table.

x	y	$x + y$	dy	Correct value of y
0.0	1.0000	1.0000	0.1000	1.0000
0.1	1.1000	1.2000	0.1200	1.1103
0.2	1.2200	1.4200	0.1420	1.2428
0.3	1.3620	1.6620	0.1662	1.3997
0.4	1.5282	1.9282	0.1928	1.5836
0.5	1.7210	2.2210	0.2221	1.7974
0.6	1.9431	2.5431	0.2543	2.0442

In this table and the others which follow in this section, the values shown have been rounded off, although more digits were carried in the calculations.

In this case we are able to find correct values since the differential equation can be expressed as $dy/dx - y = x$, of which the solution is $y = 2e^x - x - 1$.

(*continued on next page*)

In order to show that better values are obtained by using smaller values of Δx, the following table is presented in which $\Delta x = 0.05$ and $\Delta x = 0.01$ (every fifth value shown) are also used in solving the differential equation.

x	$y(\Delta x = 0.1)$	$y(\Delta x = 0.05)$	$y(\Delta x = 0.01)$	y(correct)
0.00	1.0000	1.0000	1.0000	1.0000
0.05		1.0500	1.0520	1.0525
0.10	1.1000	1.1050	1.1092	1.1103
0.15		1.1653	1.1719	1.1737
0.20	1.2200	1.2310	1.2404	1.2428
0.25		1.3026	1.3149	1.3181
0.30	1.3620	1.3802	1.3957	1.3997
0.35		1.4642	1.4832	1.4881
0.40	1.5282	1.5549	1.5777	1.5836
0.45		1.6527	1.6796	1.6866
0.50	1.7210	1.7578	1.7893	1.7974
0.55		1.8707	1.9070	1.9165
0.60	1.9431	1.9917	2.0334	2.0442

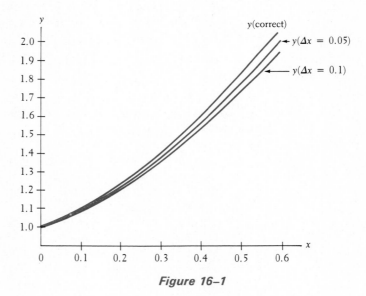

Figure 16–1

See Fig. 16–1 for the graphs using the values of y for $\Delta x = 0.1$ and $\Delta x = 0.05$, as well as the correct values of y.

Example B

By the increment method, solve the differential equation $dy/dx = 2x^2 - y^2$ if the curve passes through the point (1, 2). Use $\Delta x = 0.1$.

By following the method as outlined in Example A, the values in the following table are found.

x	y	$2x^2 - y^2$	dy
1.0	2.0000	-2.0000	-0.2000
1.1	1.8000	-0.8200	-0.0820
1.2	1.7180	-0.0715	-0.0072
1.3	1.7108	0.4530	0.0453
1.4	1.7561	0.8359	0.0836
1.5	1.8397	1.1153	0.1115
1.6	1.9513	1.3125	0.1313

See Fig. 16–2 for the graph of this solution.

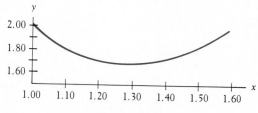

Figure 16–2

Example C

By the increment method, solve the differential equation $dy/dx = \cos x + \sin y$ if the curve passes through the point (0, 0). Use $\Delta x = 0.1$ and construct the graph for $0 \le x \le 0.80$.

By following the method as outlined in Example A, the values in the following table are found.

x	y	$\cos x + \sin y$	dy
0.0	0.0000	1.0000	0.1000
0.1	0.1000	1.0948	0.1095
0.2	0.2095	1.1880	0.1188
0.3	0.3283	1.2778	0.1278
0.4	0.4561	1.3615	0.1361
0.5	0.5922	1.4358	0.1436
0.6	0.7358	1.4965	0.1497
0.7	0.8854	1.5390	0.1539
0.8	1.0393	1.5588	0.1559

Figure 16–3

See Fig. 16–3 for the graph of this solution.

This method does have limitations. The values obtained are not particularly accurate if Δx is large, although reasonable accuracy may be obtained with a calculator, or especially with a computer, and small values of Δx. Also, it is seen from Example A that the values vary more and more from the true values, the farther the values of x are from that of the original given point. However, it is very useful in demonstrating the method of solving a differential equation numerically by successive approximations. Most such methods follow a similar technique.

Exercises 16–1

See Appendix C for a computer program for the increment method.

In Exercises 1 through 8, solve the given differential equations by the increment method for the specified values of x and Δx. Check the results against known values by solving the differential equations exactly. Plot the graphs of the solutions in Exercises 1 through 4.

1. $\dfrac{dy}{dx} = x + 1$, $y = 1$ when $x = 0$, $0 \le x \le 1$, $\Delta x = 0.2$

2. $\dfrac{dy}{dx} = \sqrt{2x + 1}$, $y = 2$ when $x = 0$, $0 \le x \le 1.5$, $\Delta x = 0.3$

3. $\dfrac{dy}{dx} = y(0.4x + 1)$, $y = 2$ when $x = -0.2$, $-0.2 \le x \le 0.5$, $\Delta x = 0.1$

4. $\dfrac{dy}{dx} = y + e^x$, $y = 0$ when $x = 0$, $0 \le x \le 0.5$, $\Delta x = 0.1$

5. Solve the differential equation of Exercise 1 with $\Delta x = 0.1$.

6. Solve the differential equation of Exercise 2 with $\Delta x = 0.1$.

7. Solve the differential equation of Exercise 3 with $\Delta x = 0.05$.

8. Solve the differential equation of Exercise 4 with $\Delta x = 0.05$.

In Exercises 9 through 16, solve the given differential equations by the increment method for the specified values of x and Δx. Plot the graphs of the solutions.

9. $\dfrac{dy}{dx} = xy + 1$, $y = 0$ when $x = 0$, $0 \le x \le 1$, $\Delta x = 0.1$

10. $\dfrac{dy}{dx} = x^2 + y^2$, $y = 1$ when $x = 0$, $0 \le x \le 0.8$, $\Delta x = 0.1$

11. $\dfrac{dy}{dx} = e^{xy}$, $y = 0$ when $x = 0$, $0 \le x \le 1.2$, $\Delta x = 0.2$

12. $\dfrac{dy}{dx} = \sqrt{1 + xy}$, $y = 1$ when $x = 0$, $0 \le x \le 0.5$, $\Delta x = 0.05$

13. $\dfrac{dy}{dx} = \cos(x + y)$, $y = \pi/2$ when $x = 0$, $0 \le x \le 1$, $\Delta x = 0.1$

14. $\dfrac{dy}{dx} = y + \sin x$, $y = 0$ when $x = 0.5$, $0.5 \le x \le 1.5$, $\Delta x = 0.1$

15. The equation of Example A with $\Delta x = 0.02$.

16. The equation of Example B with $\Delta x = 0.02$.

In Exercises 17 through 20, solve the given problems.

17. Evaluate the solution of the differential equation $y' = \ln(xy)$ for $x = 1.7$ if the solution passes through $(1, 1)$. Use $\Delta x = 0.1$.

18. A 3-slug object moves through a resisting medium. A constant force of 10 lb causes the motion, and a retarding force of $2v^{3/2}$ is present, where v is the velocity. The resulting differential equation is

$$3\frac{dv}{dt} = 10 - 2v^{3/2}$$

By the increment method, graph the solution for v as a function of t for $0 \le t \le 2$ s, using $\Delta t = 0.2$ s, if the object starts from rest. The velocity is in feet per second.

19. An electric circuit contains a 1-H inductor, a 2-Ω resistor, and a voltage source of $\sin t$. The resulting differential equation relating current and time is

$$\frac{di}{dt} + 2i = \sin t$$

Find i after 0.5 s by the increment method with $\Delta t = 0.1$ s, if the initial current is zero. Solve the equation exactly and compare the values.

20. An object is being heated such that the rate of change of temperature with respect to time is

$$\frac{dT}{dt} = \sqrt[3]{1 + t^3}$$

Find the temperature T (in degrees Celsius) for $t = 5$ min by using the increment method with $\Delta t = 0.5$ min, if the initial temperature is 0°C.

16–2 A Method of Successive Approximations

The second method of solving a differential equation to be discussed in this chapter is one based on successive approximations. In the previous section we saw how to develop a numerical solution based essentially on successive approximations. In this section we shall demonstrate a method that leads to a function which approximates the solution.

The method generally applies to a differential equation of the form $y' = f(x, y)$. If one point on the solution curve is known, the value of y is substituted into the right side, thereby giving a function of x alone. This equation can then be integrated, yielding y as a function of x, this solution being referred to as the first approximation. By substituting this expression into the right side for y, integrating again, a second approximation may be obtained. This process may be repeated as often as required, the solution for y each time being a better approximation than the last. The following example illustrates the method.

Example A

Obtain the first three approximations to the solution of the differential equation $dy = (x + y)\, dx$, if the solution curve passes through $(0, 1)$.

By substituting $y = 1$ in the right side of the differential equation, we have

$$dy = (x + 1)\, dx$$

Integrating, we obtain the first approximation

$$y_1 = \frac{x^2}{2} + x + 1$$

where the constant of integration has been evaluated as 1 from the fact that the curve passes through $(0, 1)$. Substituting the expression for y_1 in the differential equation, we have

$$dy = \left(x + \frac{x^2}{2} + x + 1\right) dx$$

$$= \left(\frac{x^2}{2} + 2x + 1\right) dx$$

$$y_2 = \frac{x^3}{6} + x^2 + x + 1$$

where the constant of integration again was evaluated to be 1, and y_2 is the second approximation. Substituting the expression for y_2 in the differential equation, we have

$$dy = \left(x + \frac{x^3}{6} + x^2 + x + 1\right) dx$$

$$= \left(\frac{x^3}{6} + x^2 + 2x + 1\right) dx$$

$$y_3 = \frac{x^4}{24} + \frac{x^3}{3} + x^2 + x + 1$$

where, again, the constant of integration is 1 and y_3 is the third approximation.

We note that this differential equation is the same as that of Example A of the previous section. Therefore, the exact solution is

$$y = 2e^x - x - 1$$

Figure 16–4 shows the graph of the exact solution and each of the three approximations. The accuracy of these approximations can be seen on the graph.

Figure 16–4

Example B

Find the first two approximations of the solution of the differential equation $y' = 2 - xy^2$ if the solution curve passes through $(0, 1)$.

Writing the equation in differential form and substituting 1 for y, we have

$$dy = (2 - x)\,dx$$

which when integrated (the constant of integration is 1) gives

$$y_1 = 1 + 2x - \frac{x^2}{2}$$

Substituting this expression for y in the differential equation, we have

$$dy = \left[2 - x\left(1 + 2x - \frac{x^2}{2}\right)^2\right] dx$$

$$= \left(2 - x - 4x^2 - 3x^3 + 2x^4 - \frac{x^5}{4}\right) dx$$

$$y_2 = 1 + 2x - \frac{x^2}{2} - \frac{4x^3}{3} - \frac{3x^4}{4} + \frac{2x^5}{5} - \frac{x^6}{24}$$

where the constant of integration is 1 and y_2 is the desired approximation.

Example C

Find the third approximation to the solution of the differential equation $y' = y + \sin x$, if the solution curve passes through $(0, \frac{1}{2})$. Evaluate the function for $x = 1$.

Following the method as previously illustrated we have the following steps:

$$dy = (\tfrac{1}{2} + \sin x)\, dx$$

$$y_1 = \frac{x}{2} - \cos x + C_1$$

Since the curve passes through $(0, \frac{1}{2})$ we have

$$\frac{1}{2} = 0 - 1 + C_1 \quad \text{or} \quad C_1 = \frac{3}{2}$$

Thus

$$y_1 = \frac{3}{2} + \frac{x}{2} - \cos x$$

$$dy = \left(\frac{3}{2} + \frac{x}{2} - \cos x + \sin x\right) dx$$

$$y_2 = \frac{3x}{2} + \frac{x^2}{4} - \sin x - \cos x + C_2$$

Here $C_2 = \frac{3}{2}$, which means

$$y_2 = \frac{3}{2} + \frac{3x}{2} + \frac{x^2}{4} - \sin x - \cos x$$

$$dy = \left(\frac{3}{2} + \frac{3x}{2} + \frac{x^2}{4} - \cos x\right) dx$$

$$y_3 = \frac{3x}{2} + \frac{3x^2}{4} + \frac{x^3}{12} - \sin x + C_3$$

Here $C_3 = \frac{1}{2}$, which means

$$y_3 = \frac{1}{2} + \frac{3x}{2} + \frac{3x^2}{4} + \frac{x^3}{12} - \sin x$$

Evaluating y_3 for $x = 1$, we have

$$y_3 = \frac{1}{2} + \frac{3}{2} + \frac{3}{4} + \frac{1}{12} - 0.84 = 1.99$$

This function can be integrated exactly by methods developed in Chapter 14. The value found from the exact solution for $x = 1$ is 2.03. Also it should be noted that the constant of integration was not the same for each of the approximations.

The method of this section has the disadvantage that it can be applied only if the resulting differentials can be integrated. This naturally limits the equations which can be solved in this manner. However, for those to which it is applicable, it often provides a quicker and more accurate value for the solution for a specified value of x than the method of the previous section.

Exercises 16–2

In Exercises 1 through 4, find the indicated approximation of the solution of each of the given differential equations for which the solution curve passes through the given point. Solve each exactly and evaluate the solutions for the indicated value of x.

1. $y' = y^2$, y_2, $(0, 1)$, $x = 0.1$

2. $y' = x - y$, y_2, $(0, 1)$, $x = \frac{1}{2}$

3. $y' = 2x(1 + y)$, y_3, $(0, 0)$, $x = 1$

4. $y' = y + e^x$, y_3, $(0, 0)$, $x = 0.2$

In Exercises 5 through 8, find the indicated approximation of the solution of each of the given differential equations for which the solution curve passes through the given point.

5. $y' = x + y^2$, y_3, $(0, 0)$

6. $y' = 1 + xy^2$, y_2, $(0, 0)$

7. $y' = y - \cos 2x$, y_2, $(0, 1)$

8. $y' = x + y \cos x$, y_2, $(0, 1)$

In Exercises 9 through 12, solve the given problems.

9. Substitute the first few terms of the Maclaurin series for e^x in the expression $2e^x - x - 1$ and compare the results with the solution for Example A.

10. Substitute the first few terms for e^{x^2} in the expression for the exact solution to the equation of Exercise 3, and compare the results with the approximate solution.

11. A 1-slug object moves through a resisting medium such that the motion is resisted by a force numerically equal to the cube of the velocity. If a 2-lb force causes the motion, the resulting differential equation is

$$\frac{dv}{dt} = 2 - v^3$$

Find the velocity when $t = 0.3$ s if the object starts from rest. Use the second approximation of the solution.

12. Evaluate i for $t = 0.5$ s for the differential equation in Exercise 19 of Section (16–1) by finding i_2. Compare the results with those found in the indicated exercise.

16–3 Laplace Transforms

The final method of solving differential equations which we shall discuss is by means of **Laplace transforms.** As we shall see, *Laplace transforms provide an algebraic method of obtaining a* **particular** *solution of a differential equation from stated initial conditions.* Since this is often what is desired in practice, Laplace transforms are often preferred for the solution of differential equations in engineering and electronics. In this section we shall discuss the meaning of the transform and its operations. These methods will be applied to solving differential equations in the following section. The treatment in this text is intended only as an introduction to the topic of Laplace transforms.

The Laplace transform of a function $f(t)$ is defined as the function $F(s)$ by the equation

$$F(s) = \int_0^\infty e^{-st}f(t) \, dt \qquad (16\text{–}1)$$

By writing the transform as $F(s)$, we show that the result of integrating and evaluating is a function of s. To denote that we are dealing with "the Laplace transform of the function $f(t)$" the notation $L(f)$ is used. Thus,

$$F(s) = L(f) = \int_0^\infty e^{-st}f(t) \, dt \qquad (16\text{–}2)$$

We shall see that both notations are quite useful.

We note that the integrals in Eqs. (16–1) and (16–2) are improper integrals, which were introduced in Section 9–7. Therefore, it is necessary to evaluate

$$\lim_{h\to\infty} \int_0^h e^{-st}f(t) \, dt$$

in order to find the Laplace transform of the function $f(t)$.

Example A

Find the Laplace transform of the function $f(t) = t,\ t > 0$.

By the definition of the Laplace transform

$$L(f) = L(t) = \int_0^\infty e^{-st}t \, dt$$

This may be integrated by parts or by Formula 44 of Table 5 in Appendix D. Using the formula we have

$$L(t) = \int_0^\infty te^{-st} \, dt = \lim_{h\to\infty}\int_0^h te^{-st} \, dt = \lim_{h\to\infty}\frac{e^{-st}(-st-1)}{s^2}\bigg|_0^h$$

$$= \lim_{h\to\infty}\left[\frac{e^{-sh}(-sh-1)}{s^2}\right] + \frac{1}{s^2}$$

Now, for $s > 0$, as $h \to \infty$, $e^{-sh} \to 0$ and $sh \to \infty$. However, although we cannot prove it here, $e^{-sh} \to 0$ much faster than $sh \to \infty$. We can see that this is reasonable, for $he^{-h} = 4.5 \times 10^{-4}$ for $h = 10$ and $he^{-h} = 3.7 \times 10^{-42}$ for $h = 100$. Thus, the value at the upper limit approaches zero, which means the limit is zero. This means that

$$L(t) = \frac{1}{s^2}$$

As we noted, this transform is defined for $s > 0$.

Example B

Find the Laplace transform of the function $f(t) = \cos at$.
By definition

$$L(f) = L(\cos at) = \int_0^\infty e^{-st}\cos at \, dt$$

Using Formula 50 of Table 5 we have

$$L(\cos at) = \int_0^\infty e^{-st}\cos at \, dt$$

$$= \lim_{h\to\infty} \int_0^h e^{-st}\cos at \, dt$$

$$= \lim_{h\to\infty} \frac{e^{-st}(-s\cos at + a\sin at)}{s^2 + a^2}\bigg|_0^h$$

$$= \lim_{h\to\infty} \frac{e^{-sh}(-s\cos ah + a\sin ah)}{s^2 + a^2} - \left(-\frac{s}{s^2 + a^2}\right)$$

$$= 0 + \frac{s}{s^2 + a^2}$$

$$= \frac{s}{s^2 + a^2} \qquad (s > 0)$$

Therefore, the Laplace transform of the function $\cos at$ is

$$L(\cos at) = \frac{s}{s^2 + a^2}$$

In both examples the resulting transform was an algebraic function of s.

An important property of Laplace transforms is the **linearity property**

$$L[af(t) + bg(t)] = aL(f) + bL(g) \qquad (16\text{–}3)$$

We state this property here since it determines that the transform of a sum of functions is the sum of the transforms. This is of definite importance when dealing with a sum of functions. This property is a direct result of the definition of the Laplace transform.

We now present a short table of Laplace transforms. The transforms given on the following page are sufficient for our remaining work in this chapter. Much more complete tables are available in many standard reference sources.

Table of Laplace Transforms

	$f(t) = L^{-1}(F)$	$L(f) = F(s)$			$f(t) = L^{-1}(F)$	$L(f) = F(s)$
1.	1	$\dfrac{1}{s}$		11.	te^{-at}	$\dfrac{1}{(s+a)^2}$
2.	$\dfrac{t^{n-1}}{(n-1)!}$	$\dfrac{1}{s^n}$ $(n = 1, 2, 3, \ldots)$		12.	$t^{n-1}e^{-at}$	$\dfrac{(n-1)!}{(s+a)^n}$
3.	e^{-at}	$\dfrac{1}{s+a}$		13.	$e^{-at}(1 - at)$	$\dfrac{s}{(s+a)^2}$
4.	$1 - e^{-at}$	$\dfrac{a}{s(s+a)}$		14.	$[(b-a)t + 1]e^{-at}$	$\dfrac{s+b}{(s+a)^2}$
5.	$\cos at$	$\dfrac{s}{s^2+a^2}$		15.	$\sin at - at \cos at$	$\dfrac{2a^3}{(s^2+a^2)^2}$
6.	$\sin at$	$\dfrac{a}{s^2+a^2}$		16.	$t \sin at$	$\dfrac{2as}{(s^2+a^2)^2}$
7.	$1 - \cos at$	$\dfrac{a^2}{s(s^2+a^2)}$		17.	$\sin at + at \cos at$	$\dfrac{2as^2}{(s^2+a^2)^2}$
8.	$at - \sin at$	$\dfrac{a^3}{s^2(s^2+a^2)}$		18.	$t \cos at$	$\dfrac{s^2-a^2}{(s^2+a^2)^2}$
9.	$e^{-at} - e^{-bt}$	$\dfrac{b-a}{(s+a)(s+b)}$		19.	$e^{-at}\sin bt$	$\dfrac{b}{(s+a)^2+b^2}$
10.	$ae^{-at} - be^{-bt}$	$\dfrac{s(a-b)}{(s+a)(s+b)}$		20.	$e^{-at}\cos bt$	$\dfrac{s+a}{(s+a)^2+b^2}$

Another Laplace transform important to the solution of a differential equation is the transform of the derivative of a function. Let us first find the Laplace transform of the first derivative of a function.

By definition

$$L(f') = \int_0^\infty e^{-st}f'(t)\, dt$$

To integrate by parts, let $u = e^{-st}$ and $dv = f'(t)\, dt$, so $du = -se^{-st}\, dt$ and $v = f(t)$ [the integral of the derivative of a function is the function]. Therefore,

$$L(f') = e^{-st}f(t)\Big|_0^\infty + s \int_0^\infty e^{-st}f(t)\, dt$$
$$= 0 - f(0) + sL(f)$$

It is noted that the integral in the second term on the right is the Laplace transform of $f(t)$ by definition. Therefore, *the Laplace transform of the first derivative of a function is*

$$L(f') = sL(f) - f(0) \tag{16-4}$$

Applying the same analysis, we may find *the Laplace transform of the second derivative of a function. It is*

$$L(f'') = s^2 L(f) - sf(0) - f'(0) \tag{16-5}$$

Here it is necessary to integrate by parts twice to derive the result. The transforms of higher derivatives are found in a similar manner.

Equations (16–4) and (16–5) allow us to express the transform of each derivative in terms of s and the transform of the function itself.

Example C

If $f(0) = 0$ and $f'(0) = 1$, express the transform of $f''(t) - 2f'(t)$ in terms of s and the transform of $f(t)$.

By the linearity property, Eq. (16–3), we have

$$L[f''(t) - 2f'(t)] = L(f'') - 2L(f')$$
$$= [s^2 L(f) - s(0) - 1] - 2[sL(f) - 0]$$
$$= (s^2 - 2s)L(f) - 1$$

If the Laplace transform of a function is known, it is then possible to find the function by finding the **inverse transform**

$$L^{-1}(F) = f(t) \tag{16-6}$$

where L^{-1} denotes the inverse transform.

Example D

If $F(s) = \dfrac{s}{s^2 + a^2}$, from Transform 5 of the table we see that

$$L^{-1}(F) = L^{-1}\left(\frac{s}{s^2 + a^2}\right) = \cos at$$

or

$$f(t) = \cos at$$

Example E

If $(s^2 - 2s)L(f) - 1 = 0$, then

$$L(f) = \frac{1}{s^2 - 2s} \quad \text{or} \quad F(s) = \frac{1}{s(s - 2)}$$

Thus,

$$L^{-1}(F) = f(t) = -\frac{1}{2}(1 - e^{2t})$$

from Transform 4 of the table.

Example F

If $F(s) = \dfrac{s + 5}{s^2 + 6s + 10}$, then

$$L^{-1}(F) = L^{-1}\left[\frac{s + 5}{s^2 + 6s + 10}\right]$$

It appears that this function does not fit any of the forms given. However,

$$s^2 + 6s + 10 = (s^2 + 6s + 9) + 1 = (s + 3)^2 + 1$$

By writing $F(s)$ as

$$F(s) = \frac{(s + 3) + 2}{(s + 3)^2 + 1} = \frac{s + 3}{(s + 3)^2 + 1} + \frac{2}{(s + 3)^2 + 1}$$

we can find the inverse of each term. Therefore,

$$L^{-1}(F) = e^{-3t}\cos t + 2e^{-3t}\sin t$$

or

$$f(t) = e^{-3t}(\cos t + 2 \sin t)$$

Partial fractions, which were introduced in Sections 9–4 and 9–5, can also be used to write certain transforms in forms which fit those in the table. Finding the inverse transform by use of partial fractions is illustrated in the following example.

Example G

Find $L^{-1}\left[\dfrac{2s^2 + 1}{s^3 + 2s^2 + s}\right]$

Expressing the transform as

$$\frac{2s^2 + 1}{s^3 + 2s^2 + s} = \frac{2s^2 + 1}{s(s^2 + 2s + 1)} = \frac{2s^2 + 1}{s(s + 1)^2}$$

We see that this transform does not fit any of the forms in the table. However, we have one nonrepeated linear factor and one repeated linear factor in the denominator. Therefore,

$$\frac{2s^2 + 1}{s(s + 1)^2} = \frac{A}{s} + \frac{B}{s + 1} + \frac{C}{(s + 1)^2}$$

Once we determine the constants A, B, and C, we see that each of these forms appears in the table. Multiplying each side by $s(s + 1)^2$, we have

$$2s^2 + 1 = A(s + 1)^2 + Bs(s + 1) + Cs$$

Using the values $s = 0$ and $s = -1$, we have

for $s = 0$: $1 = A$,   or $A = 1$

for $s = -1$: $3 = -C$,   or $C = -3$

Now equating coefficients of s^2, we have

$$2 = A + B, \quad \text{or} \quad B = 1$$

Therefore,

$$\frac{2s^2 + 1}{s^3 + 2s^2 + s} = \frac{1}{s} + \frac{1}{s + 1} - \frac{3}{(s + 1)^2}$$

which means that

$$L^{-1}\left[\frac{2s^2 + 1}{s^3 + 2s^2 + s}\right] = L^{-1}\left[\frac{1}{s} + \frac{1}{s + 1} - \frac{3}{(s + 1)^2}\right]$$

$$= 1 + e^{-t} - 3te^{-t}$$

Exercises 16–3

In Exercises 1 through 4, verify the indicated transforms given in the table.

1. Transform 1
2. Transform 3
3. Transform 6
4. Transform 11

In Exercises 5 through 12, find the transforms of the given functions by use of the table.

5. $f(t) = e^{3t}$
6. $f(t) = 1 - \cos 2t$
7. $f(t) = t^3 e^{-2t}$
8. $f(t) = 2e^{-3t}\sin 4t$
9. $f(t) = \cos 2t - \sin 2t$
10. $f(t) = 2t \sin 3t + e^{-3t}\cos t$
11. $f(t) = 3 + 2t \cos 3t$
12. $f(t) = t^3 - 3te^{-t}$

In Exercises 13 through 16, express the transforms of the given expressions in terms of s and $L(f)$.

13. $y'' + y', f(0) = 0, f'(0) = 0$
14. $y'' - 3y', f(0) = 2, f'(0) = -1$
15. $2y'' - y' + y, f(0) = 1, f'(0) = 0$
16. $y'' - 3y' + 2y, f(0) = -1, f'(0) = 2$

In Exercises 17 through 24, find the inverse transform of the given functions of s.

17. $F(s) = \dfrac{2}{s^3}$
18. $F(s) = \dfrac{3}{s^2 + 4}$
19. $F(s) = \dfrac{1}{s + 5}$
20. $F(s) = \dfrac{3}{s^4 + 4s^2}$
21. $F(s) = \dfrac{1}{s^3 + 3s^2 + 3s + 1}$
22. $F(s) = \dfrac{s^2 - 1}{s^4 + 2s^2 + 1}$
23. $F(s) = \dfrac{s + 2}{(s^2 + 9)^2}$
24. $F(s) = \dfrac{s + 3}{s^2 + 4s + 13}$

In Exercises 25 through 30, use partial fractions to find the inverse transforms of the given functions of s.

25. $F(s) = \dfrac{2s - 1}{s^3 - s}$

26. $F(s) = \dfrac{6s^2}{(s^2 - 4)(s - 4)}$

27. $F(s) = \dfrac{s^2 - 2s + 3}{(s - 1)^2(s + 1)}$

28. $F(s) = \dfrac{8s^2 - 7s + 1}{s(2s - 1)^2}$

29. $F(s) = \dfrac{3s^2 + 2s + 18}{(s + 3)(s^2 + 4)}$

30. $F(s) = \dfrac{6}{(s^2 + 1)(s^2 + 4)}$

16–4 Solving Differential Equations by Laplace Transforms

We will now show how certain differential equations can be solved by the use of Laplace transforms. *It must be remembered that these solutions are the **particular** solutions of the equations subject to the given conditions.* The necessary operations were developed in the preceding section. The following examples illustrate the method.

Example A

Solve the differential equation $2y' - y = 0$ if $y(0) = 1$. (Note that we are using y to denote the function.)

Taking transforms of each term in the equation, we have

$$L(2y') - L(y) = L(0)$$
$$2L(y') - L(y) = 0$$

$L(0) = 0$ by direct use of the definition of the transform. Now using Eq. (16–4), $L(y') = sL(y) - y(0)$, we have

$$2[sL(y) - 1] - L(y) = 0$$

Solving for $L(y)$ we obtain

$$2sL(y) - L(y) = 2 \qquad \qquad L(y)(2s-1)=2$$

$$L(y) = \frac{2}{2s - 1} = \frac{1}{s - \frac{1}{2}}$$

Finding the inverse transform of $F(s) = 1/(s - \frac{1}{2})$, we have

$$y = e^{t/2}$$

 The reader should check this solution with that obtained by methods developed earlier. Also, it should be noted that the solution was essentially an algebraic one. This points out the power and usefulness of Laplace transforms. We are able to translate a differential equation into an algebraic form, which can in turn be translated into the solution of the differential equation. Thus, we are able to solve a differential equation by use of algebra and specific algebraic forms.

Example B

Solve the differential equation $y'' + 2y' + 2y = 0$, if $y(0) = 0$ and $y'(0) = 1$.

Using the same steps as outlined in Example A, we have:

$$L(y'') + 2L(y') + 2L(y) = 0$$
$$[s^2L(y) - sy(0) - y'(0)] + 2[sL(y) - y(0)] + 2L(y) = 0$$
$$s^2L(y) - 1 + 2sL(y) + 2L(y) = 0$$
$$(s^2 + 2s + 2)L(y) = 1$$

$$L(y) = \frac{1}{s^2 + 2s + 2} = \frac{1}{(s + 1)^2 + 1}$$

$$y = e^{-t}\sin t$$

Example C

For reference, Eq. (16–5) is
$L(f'') =$
    $s^2L(f) - sf(0) - f'(0)$.

Solve the differential equation $y'' + y = \cos t$ if $y(0) = 1$ and $y'(0) = 2$.

Taking transforms of each term in the equation, we have

$$L(y'') + L(y) = L(\cos t)$$

The transform on the left is found by use of Eq. (16–5) and the transform on the right is found by direct use of Transform 5 in the table. Therefore, we have

$$[s^2L(y) - s(1) - 2] + L(y) = \frac{s}{s^2 + 1}$$

Solving for $L(y)$, we obtain

$$(s^2 + 1)L(y) = \frac{s}{s^2 + 1} + s + 2$$

$$L(y) = \frac{s}{(s^2 + 1)^2} + \frac{s}{s^2 + 1} + \frac{2}{s^2 + 1}$$

Finding the inverse transform of each expression completes the solution:

$$y = \frac{t}{2}\sin t + \cos t + 2\sin t$$

Example D

Figure 16–5

An electric circuit contains a 1-H inductor, a 10-Ω resistor, and a battery whose voltage is 6 V. Find the current as a function of the time, if the initial current is zero. See Fig. 16–5.

The differential equation for this circuit is

$$\frac{di}{dt} + 10i = 6$$

Following the procedures outlined in the previous examples, we have:

$$L\left(\frac{di}{dt}\right) + 10L(i) = L(6), \qquad [sL(i) - 0] + 10L(i) = \frac{6}{s}$$

$$L(i) = \frac{6}{s(s + 10)} \qquad i = 0.6(1 - e^{-10t})$$

Example E

A spring is stretched 1 ft by a weight of 16 lb (mass of $\frac{1}{2}$ slug). The medium resists the motion of the object with a force of $4v$, where v is the velocity. The differential equation describing the motion is

$$\frac{1}{2}\frac{d^2y}{dt^2} + 4\frac{dy}{dt} + 16y = 0$$

Find y, the displacement of the object, as a function of the time if $y(0) = 1$ and $dy/dt = 0$ when $t = 0$.

Clearing fractions and denoting the derivatives by y'' and y', the differential equation becomes

$$y'' + 8y' + 32y = 0$$

Following the procedure of solving this type of an equation by use of Laplace transforms, we have the following steps:

$$L(y'') + 8L(y') + 32L(y) = 0$$
$$[s^2L(y) - s(1) - 0] + 8[sL(y) - 1] + 32L(y) = 0$$
$$(s^2 + 8s + 32)L(y) = s + 8$$
$$L(y) = \frac{s + 8}{(s + 4)^2 + 4^2} = \frac{s + 4}{(s + 4)^2 + 4^2} + \frac{4}{(s + 4)^2 + 4^2}$$
$$y = e^{-4t}\cos 4t + e^{-4t}\sin 4t$$
$$= e^{-4t}(\cos 4t + \sin 4t)$$

Exercises 16–4

In the following exercises, solve the given differential equations by use of Laplace transforms, where the function is subject to the given conditions.

1. $y' + y = 0$, $y(0) = 1$
2. $y' - 2y = 0$, $y(0) = 2$
3. $2y' - 3y = 0$, $y(0) = -1$
4. $y' + 2y = 1$, $y(0) = 0$
5. $y' + 3y = e^{-3t}$, $y(0) = 1$
6. $y' + 2y = te^{-2t}$, $y(0) = 0$
7. $y'' + 4y = 0$, $y(0) = 0$, $y'(0) = 1$
8. $y'' - 4y = 0$, $y(0) = 2$, $y'(0) = 0$
9. $y'' + 2y' = 0$, $y(0) = 0$, $y'(0) = 2$
10. $y'' + 2y' + y = 0$, $y(0) = 0$, $y'(0) = -2$
11. $y'' - 4y' + 5y = 0$, $y(0) = 1$, $y'(0) = 2$
12. $4y'' + 4y' + y = 0$, $y(0) = 1$, $y'(0) = 0$
13. $y'' + y = 1$, $y(0) = 1$, $y'(0) = 1$
14. $y'' + 4y = 2t$, $y(0) = 0$, $y'(0) = 0$
15. $y'' + 2y' + y = e^{-t}$, $y(0) = 1$, $y'(0) = 2$
16. $y'' + 4y = \sin 2t$, $y(0) = 0$, $y'(0) = 0$

17. A constant force of 6 lb moves a 2-slug mass through a medium which resists the motion with a force equal to v, where v is the velocity. The differential equation relating the velocity and time is

$$2\frac{dv}{dt} = 6 - v$$

Find v as a function of time if the object starts from rest.

18. A 100-Ω resistor and a 400-μF capacitor make up an electric circuit. If the initial charge on the capacitor is 10 mC, find the charge on the capacitor as a function of time after the switch is closed.

19. A 50-Ω resistor, a 4-μF capacitor, and a 40-V battery are connected in series. Find the charge as a function of the time if the initial charge on the capacitor is zero.

20. A 2-H inductor, an 80-Ω resistor, and an 8-V battery are connected in series. Find the current in the circuit as a function of time if the initial current is zero.

21. A 10-H inductor, a 40-μF capacitor, and a voltage supply whose voltage is given by $100 \sin 50t$ are connected in series in an electric circuit. Find the current as a function of time if the initial charge on the capacitor and the initial current are zero.

22. An object moves with simple harmonic motion according to the differential equation $D^2y + 400y = 0$, where $D = d/dt$. Find the displacement y as a function of the time t, if $y = 10$ and $Dy = 0$ when $t = 0$.

23. The weight on a spring undergoes forced vibrations according to the equation $D^2y + 9y = 18 \sin 3t$. Find its displacement y as a function of the time t, if $y = 0$ and $Dy = 0$ when $t = 0$.

24. A spring is such that it is stretched 6 in. by an 8-lb ($\frac{1}{4}$-slug) weight. The spring is stretched 6 in. below the equilibrium position with the weight on it and is given a velocity of 4 ft/s. If the spring is in a medium which resists the motion with a force equal to $4v$, find the displacement y as a function of time.

16–5 Review Exercises for Chapter 16

In Exercises 1 through 8, solve the given differential equations by the increment method for the specified values of x and Δx and for which the solution curve passes through the given point.

1. $y' = x^2 - y^2$, $0 \le x \le 0.5$, $\Delta x = 0.1$, $(0, 1)$

2. $y' = \dfrac{1}{1 + x + y}$, $0 \le x \le 1$, $\Delta x = 0.2$, $(0, 0)$

3. $y' = \sqrt{x + y}$, $0 \le x \le 2$, $\Delta x = 0.2$, $(0, 1)$

4. $y' = x^2 + \sin y$, $0 \le x \le 1$, $\Delta x = 0.1$, $(0, 1)$

5. $y' = \sin x + \tan y$, $0 \le x \le 0.7$, $\Delta x = 0.1$, $(0, 0.5)$

6. $y' = e^{2x} + \sin xy$, $0 \le x \le 1$, $\Delta x = 0.1$, $(0, 0)$

7. The equation of Exercise 1 with $\Delta x = 0.02$.

8. The equation of Exercise 2 with $\Delta x = 0.05$.

In Exercises 9 through 12, find the indicated approximation of the solution of each of the given differential equations for which the solution curve passes through the given point.

9. $y' = y + x^2$, y_3, $(0, 1)$

10. $y' = \cos x - y$, y_3, $(0, 1)$

11. $y' = 1 + x^2 y$, y_2, $(1, 1)$

12. $y' = y^2 - \sin x$, y_2, $(0, 2)$

In Exercises 13 through 20, solve the given differential equations by use of Laplace transforms, where the function is subject to the given conditions.

13. $4y' - y = 0$, $y(0) = 1$

14. $2y' - y = 4$, $y(0) = 1$

15. $y' - 3y = e^t$, $y(0) = 0$

16. $y' + 2y = e^{-2t}$, $y(0) = 2$

17. $y'' + y = 0$, $y(0) = 0$, $y'(0) = -4$

18. $y'' + 4y' + 5y = 0$, $y(0) = 1$, $y'(0) = 1$

19. $y'' + 9y = 3t$, $y(0) = 0$, $y'(0) = -1$

20. $y'' + 4y' + 4y = 2e^{-2t}$, $y(0) = 0$, $y'(0) = 1$

In Exercises 21 through 36, solve the given problems.

21. Find the second approximation to the solution of the differential equation $y' = 1 + ye^{-x}$ if the solution curve passes through $(0, 1)$. Evaluate y_2 for $x = 0.5$.

22. Repeat the instructions of Exercise 21 for $y' = 2x/(x^2 + e^y)$.

23. Solve $y' = 1 + ye^{-x}$ by the increment method for $0 \le x \le 1$ using $\Delta x = 0.1$ if the solution curve passes through $(0, 1)$. Compare the value for $x = 0.5$ with that found in Exercise 21.

24. Repeat the instructions of Exercise 23 for $y' = 2x/(x^2 + e^y)$ and compare the result with that found in Exercise 22.

25. The differential equation relating current and time for a certain electric circuit is

$$2\frac{di}{dt} + i = 12$$

Solve this equation by use of Laplace transforms, given that the initial current is zero. Evaluate the current for $t = 0.3$ s.

26. A 6-H inductor and a 30-Ω resistor are connected in series. Find the current as a function of time if the initial current is 15 A. Use Laplace transforms.

27. A 0.25-H inductor, a 4-Ω resistor, and a 100-μF capacitor are connected in series. If the initial charge on the capacitor is 400 μC and the initial current is zero, find the charge on the capacitor as a function of time. Use Laplace transforms.

28. An inductor of 0.5 H, a resistor of 6 Ω, and a capacitor of 200 μF are connected in series. If the initial charge on the capacitor is 10 mC and the initial current is zero, find the charge on the capacitor as a function of time after the switch is closed. Use Laplace transforms.

29. Use the increment method to find the current for $t = 0.3$ s for the equation of Exercise 25. Use $\Delta t = 0.05$ s.

30. Find the third approximation for i as a function of t for the equation of Exercise 25. Evaluate i for $t = 0.3$ s and compare this value with those found in Exercises 25 and 29.

31. An 8-lb weight ($\frac{1}{4}$-slug mass) stretches a spring 6 in. An external force of $\cos 8t$ is applied to the spring. Express the displacement y of the object as a function of time if the initial displacement and velocity are zero. Use Laplace transforms.

32. Find the displacement y as a function of the time t for the spring and object of Exercise 31 with two conditions changed: the external force is removed and the initial displacement is 9 in.

33. A spring is stretched 1 m by a mass of 5 kg (assume the weight to be 50 N). Find y, the displacement of the object, as a function of time if $y(0) = 1$ and $dy/dt = 0$ when $t = 0$. Use Laplace transforms.

34. Find the solution for the spring system of Exercise 33 if it is in a medium which resists the motion with a force of $10v$, where v is the velocity.

35. Assuming that only Transforms 1 through 6 in the table of transforms are known, derive the inverse transform 9, given $F(s)$ and using partial fractions.

36. Using the conditions and method of Exercise 35, derive the inverse transform 7.

Appendices

Formulas from Geometry

For the indicated figures, the following symbols are used: A = area, B = area of base, c = circumference, S = lateral area, V = volume.

1. **Triangle.** $A = \frac{1}{2}bh$ (Fig. A–1)
2. **Pythagorean theorem.** $c^2 = a^2 + b^2$ (Fig. A–2)
3. **Parallelogram.** $A = bh$ (Fig. A–3)
4. **Trapezoid.** $A = \frac{1}{2}(a + b)h$ (Fig. A–4)
5. **Circle.** $A = \pi r^2$, $c = 2\pi r$ (Fig. A–5)
6. **Rectangular solid.** $A = 2(lw + lh + wh)$, $V = lwh$ (Fig. A–6)
7. **Cube.** $A = 6e^2$, $V = e^3$ (Fig. A–7)
8. **Any cylinder or prism with parallel bases.** $V = Bh$ (Fig. A–8)
9. **Right circular cylinder.** $S = 2\pi rh$, $V = \pi r^2 h$ (Fig. A–9)
10. **Any cone or pyramid.** $V = \frac{1}{3}Bh$ (Fig. A–10)
11. **Right circular cone.** $S = \pi rs$, $V = \frac{1}{3}\pi r^2 h$ (Fig. A–11)
12. **Sphere.** $A = 4\pi r^2$, $V = \frac{4}{3}\pi r^3$ (Fig. A–12)

Figure A–1

Figure A–2

Figure A–3

Figure A–4

Figure A–5

Figure A–6

Figure A–7

Figure A–8

Figure A–9

Figure A–10

Figure A–11

Figure A–12

Units of Measurement

Table B–1. Quantities and Their Associated Units

Quantity	Quantity symbol	Unit				
		U.S. Customary		Metric (SI)		
		Name	*Symbol*	*Name*	*Symbol*	*In terms of other SI units*
Length	s	foot	ft	**meter**	m	
Mass	m	slug		**kilogram**	kg	
Force	F	pound	lb	newton	N	$m \cdot kg/s^2$
Time	t	second	s	**second**	s	
Area	A		ft^2		m^2	
Volume	V		ft^3		m^3	
Capacity	V	gallon	gal	liter	L	$(1\ L = 1\ dm^3)$
Velocity	v		ft/s		m/s	
Acceleration	a		ft/s^2		m/s^2	
Density	d, ρ		lb/ft^3		kg/m^3	
Pressure	p		lb/ft^2	pascal	Pa	N/m^2
Energy, work	E, W		$ft \cdot lb$	joule	J	$N \cdot m$
Power	P	horsepower	hp	watt	W	J/s
Period	T		s		s	
Frequency	f		1/s	hertz	Hz	1/s
Angle	θ	radian	rad	radian	rad	
Electric current	I, i	ampere	A	**ampere**	A	
Electric charge	q	coulomb	C	coulomb	C	$A \cdot s$
Electric potential	V, E	volt	V	volt	V	$J/A \cdot s$
Capacitance	C	farad	F	farad	F	s/Ω
Inductance	L	henry	H	henry	H	$\Omega \cdot s$
Resistance	R	ohm	Ω	ohm	Ω	V/A
Thermodynamic temperature	T			**kelvin**	K	
Temperature	T	Fahrenheit degree	°F	degree Celsius	°C	$(1°C = 1K)$
Quantity of heat	Q	British thermal unit	Btu	joule	J	
Amount of substance	n			**mole**	mol	
Luminous intensity	I	candlepower	cp	**candela**	cd	

Table B–2. Metric Prefixes

Prefix	Factor	Symbol	Prefix	Factor	Symbol
exa	10^{18}	E	deci	10^{-1}	d
peta	10^{15}	P	centi	10^{-2}	c
tera	10^{12}	T	milli	10^{-3}	m
giga	10^{9}	G	micro	10^{-6}	μ
mega	10^{6}	M	nano	10^{-9}	n
kilo	10^{3}	k	pico	10^{-12}	p
hecto	10^{2}	h	femto	10^{-15}	f
deca	10^{1}	da	atto	10^{-18}	a

Table B–3. Reduction and Conversion Factors

1 in. = 2.54 cm (exact)	1 ft^3 = 28.32 L
1 km = 0.6214 mi	1 L = 1.057 qt
1 lb = 453.6 g	1 Btu = 778.0 ft·lb
1 kg = 2.205 lb	1 hp = 550 ft·lb/s (exact)
1 lb = 4.448 N	1 hp = 746.0 W

Special Notes:

1. The SI base units are shown in boldface type.
2. The units symbols shown above are those which are used in the text. Many of them were adopted with the adoption of the SI system. This means, for example, that we use s rather than sec for seconds, and A rather than amp for amperes. Also, other units such as volt are not spelled out, a common practice in the past. When a given unit is used with both systems, we use the SI symbol for the unit.
3. The liter and degree Celsius are not actually SI units. However, they are recognized for use with the SI system due to their practical importance. Also, the symbol for liter has several variations. Presently L is recognized for use in the United States and Canada, l is recognized by the International Committee of Weights and Measures, and ℓ is also recognized for use in several countries.
4. Other units of time, along with their symbols, which are recognized for use with the SI system and are used in this text are: minute, min; hour, h; day, d.
5. There are many additional specialized units which are used with the SI system. However, most of those which appear in this text are shown in the table. A few of the specialized units are noted when used in the text. One which is frequently used is that for revolution, r.
6. Other common U.S. units which are used in the text are as follows: inch, in.; yard, yd; mile, mi; ounce, oz.; ton; quart, qt; acre.
7. There are a number of units which were used with the metric system prior to the development of the SI system. However, many of these are not to be used with the SI system. Among those which were commonly used are the dyne, erg, and calorie.

APPENDIX C

Selected Computer Programs in BASIC

C–1 Introduction

In this appendix there are 20 selected computer programs related to various types of problems throughout the text. Each is referenced to the appropriate text material by use of the computer symbol shown at the left. They are written in the programming language BASIC, which is chosen because of its simplicity and widespread availability on computer systems.

The programs are intended for those who have had some experience with computer programs or who have resources available to learn the fundamentals of programming in BASIC. However, knowledge of only the fundamentals is necessary to use and learn from these programs.

The intention here is to show students how BASIC can be used for calculations and mathematical concepts in direct and reasonably simple programs. Although problems that have lengthy and involved calculations can be solved using them, these programs are intentionally without undue complexity. However, many programs do solve a certain type of problem, and the user must enter certain values and at times the necessary functions.

Many of the programs can be modified and expanded to provide additional use and information. Students are encouraged to experiment with modifications. This can make the programs more versatile and provide insight into the way BASIC is used to develop mathematical concepts. An introductory paragraph accompanies each program, explaining the purpose and possible modifications.

The number of programs that could be used for the problems in this text is essentially limitless. The programs in this appendix constitute only a sample of the possibilities.

C–2 Notes on BASIC

There are many special statements, such as INPUT and PRINT, which are used in BASIC. They provide the computer with the instructions which it must follow. You should have a knowledge of the fundamental BASIC statements before attempting to use these programs. Among those which are necessary are REM, INPUT, LET, PRINT, GOTO, FOR (with NEXT), IF (with THEN), and END. Also, the system commands of NEW, LOAD, LIST, and RUN are necessary.

As well as knowing the fundamental statements, it is necessary to know the symbols for the arithmetic operations and the various mathematical functions. The symbols $+$, $-$, $*$, and $/$ are used for addition, subtraction, multiplication, and division, respectively. For exponentiation, $**$, $\uparrow$, or $\wedge$ may be used, depending on the particular computer. A list of the fundamental mathematical functions is shown in the following table.

Table E–1

BASIC Expression	Function	Comment
SIN(X)	$\sin x$	x in radians
COS(X)	$\cos x$	x in radians
TAN(X)	$\tan x$	x in radians
ATN(X)	Arctan x	Result in radians
LOG(X)	$\ln x$	Natural logarithm
EXP(X)	e^x	Exponential function
SQR(X)	$\sqrt{x}$	Square root
ABS(X)	$|x|$	Absolute value
SGN(X)		Result: $+1$ if $x > 0$ -1 if $x < 0$ 0 if $x = 0$

It should be noted that the trigonometric functions use angles expressed in radians and not degrees. Also, many computers include only the Arctangent and do not have the Arcsine or Arccosine. If a problem involves the use of functions or values which the computer does not have, it may be necessary to define them in the particular program.

There are many variations of BASIC. Therefore, the programs given here (which were written for use on an Apple IIe® computer) may have to be modified to fit the form of BASIC you may be using. However, there are several that may not need modification for use on another computer.

C–3 The Programs

Following are the BASIC programs. There are 20 different programs from 13 different chapters throughout the text. Each program is identified by the chapter in which it is used. A disk containing these programs for both the Apple II and

IBM-PC (and PC compatibles) is available from the publisher. Also, translations of many of these programs for use on computers other than the Apple IIe are found in the Instructor's Guide.

Chapter 1 Graph of Straight Line

This program graphs the equations of straight lines for different slopes and y-intercepts. The values of the slope of the line, M, and the y-intercept, B, are to be entered when requested. The scales on both the x-axis and the y-axis range from -20 to $+20$.

Try keeping B a constant value (for example: $B = 1$) and changing M. Use both small and large positive values for M, and small and large negative values for M. Note the effect of different slopes on the line.

Then, try keeping M a constant value (for example: $M = 1$), change values for B, and note the effect on the graph.

```
10    REM    GRAPH THE EQUATION
20    REM    OF A STRAIGHT LINE
25    REM
30    REM    Y = MX + B
35    REM
40    REM    M AND B TO BE ENTERED
50    REM
100   PRINT
110   PRINT  "ENTER M AND B"
120   PRINT
130   INPUT  "M = ";M: INPUT "B = "
      ;B
140   PRINT
150   GR
160   COLOR= 8
170   FOR X = 0 TO 39
180   PLOT X,20
190   PLOT 20,X
200   NEXT X
210   FOR X = 0 TO 39
220   LET Y = 20 - M * (X - 20) -
      B
230   IF Y > 39 THEN  GOTO 260
240   IF Y < 0 THEN  GOTO 260
250   PLOT X,Y
260   NEXT X
300   END
```

Chapter 2-A Limit of a Function

This program illustrates the limit of a rational function of the form $N(x)/D(x)$ as x approaches a number A. The program prints out values of x and values of the function of x as x approaches A from both the right and left sides. The numerator and denominator of the function are defined in lines 140 and 150 of the program. The function provided in the program is $(x^2 - 9)/(x + 3)$ and can be changed by entering a different numerator function in line 140 and a different denominator function in line 150. The value for A is to be entered when requested.

For example, to find the limit of $(x^2 - 1)/(x - 1)$ as x approaches 1 change the following lines:

```
140  DEF  FN N(X)  =  X*X  -  1
150  DEF  FN D(X)  =   X  -  1
```

and enter 1 as a value for A.

Try running this program for different rational functions, using a value for A that will make the function undefined.

From the values that appear on the screen it should be possible to estimate the limit of the function at $x = A$.

```
10   REM   LIMIT OF FUNCTION
20   REM
30   REM   LIM    N(X)
40   REM          ----
50   REM   X->A   D(X)
60   REM
70   REM   VALUE OF A
75   REM   IS TO BE ENTERED
80   REM
85   REM   ENTER N(X) AND D(X)
90   REM   IN LINES 140, 150
95   REM
100  PRINT
110  PRINT "ENTER A"
120  PRINT
130  INPUT "A= ";A: PRINT
140  DEF  FN N(X) = X * X - 9
150  DEF  FN D(X) = X + 3
160  PRINT "X","FUNCTION"
170  PRINT
180  FOR I = A - .005 TO A - .000
     9 STEP .001
190  PRINT I, FN N(I) /  FN D(I)
200  NEXT I
210  IF  ABS ( FN D(A)) < .0001 THEN
        GOTO 240
220  PRINT A, FN N(A) /  FN D(A)
230  GOTO 250
240  PRINT A,"UNDEFINED"
250  FOR I = A + .001 TO A + .006
       STEP .001

260  PRINT I, FN N(I) /  FN D(I)
270  NEXT I
280  PRINT
290  PRINT "LIMIT = ";(( FN N(A -
     .001) /  FN D(A - .001)) + (
      FN N(A + .001) /  FN D(A +
     .001))) / 2
300  END
```

Chapter 2–B Value of Derivative of Function

This program estimates the value for the derivative of the function $f(x) = x^3 - x^3$ at $x = A$. The value of A is to be entered when requested. Values will be shown as the derivative is approached from the right side and from the left side. Based on these values an estimated value of the derivative will be printed. Note that the estimated value for the derivative will not be valid if the values printed for the right and left sides do not approach the same number.

The value of the derivative of a different function can be estimated by changing the function in line 100. For example, to find the derivative of $f(x) = x^2 - 4x$, change line 100 to the following:

```
100  DEF  FN F(X) = X*X - 4*X
```

Note: The function entered should not be undefined at or near the value for A entered.

```
10   REM   FIND THE VALUE OF
20   REM   THE DERIVATIVE OF
30   REM   A FUNCTION AT X=A
40   REM
50   REM   ENTER FUNCTION
55   REM   IN LINE 100
60   REM
70   REM   VALUE OF A
75   REM   IS TO BE ENTERED
80   REM
100  DEF  FN F(X) = X ^ 3 - X ^ 2
110  PRINT
120  PRINT "ENTER VALUE OF A"
130  INPUT X0
140  LET Y0 =   FN F(X0)
150  PRINT
160  PRINT "RIGHT HAND","LEFT HAN
     D"
170  LET H1 = 1
180  FOR I = 1 TO 10
190  LET H1 = H1 / 2
200  LET Y1 =   FN F(X0 + H1)
210  LET H2 =   - H1
220  LET Y2 =   FN F(X0 + H2)
230  LET D1 = (Y1 - Y0) / H1
240  LET D2 = (Y2 - Y0) / H2
250  PRINT
260  PRINT D1,D2
270  NEXT I
280  PRINT
290  PRINT "F'(A) = ";(D1 + D2) /
     2
300  END
```

Chapter 3–A Newton's Method

This program uses Newton's method to find a solution to the equation $3x^2 - 5x - 2 = 0$. The initial value required to start the iteration process is entered when requested. To solve other equations by this method, the equation must be in the form of a function equal to zero $[f(x) = 0]$. The function is then entered for function F in line 100, and the derivative is entered for function D in line 110. For example, to solve the equation $x^4 - 5 = 0$, lines 100 and 110 would be

```
100  DEF  FN F(X) = X^4 - 5
110  DEF  FN D(X) = 4*X^3
```

Each initial value will lead to only one solution. If other solutions are desired, it will be necessary to start with different values. Several initial values may have to be tried to obtain all the solutions to a particular equation. In the case of the function provided in the program, enter an initial value less than .83, then an initial value greater than .83. Note: An initial value that makes the function D in line 110 equal to zero will cause an error. It is also possible that some initial values may not result in solutions for certain equations. A solution results only if the successive values for x that are printed get closer to a certain number.

In some cases it may be necessary to change the number of iterations to obtain a better solution. This may be done by changing the number 5 in line 160. For example, to perform 10 iterations, change line 160 to:

```
160  for A = 1 to 100
```

```
10   REM   NEWTON'S METHOD
15   REM
20   REM   SOLVE EQUATION
30   REM   F(X) = 0
35   REM
40   REM   VALUE OF X1
50   REM   TO BE ENTERED
55   REM
60   REM   ENTER FUNCTION
70   REM   IN LINE 100
75   REM
80   REM   ENTER DERIVATIVE
90   REM   IN LINE 110
95   REM
100  DEF  FN F(X) = 3 * X * X - 5
          * X - 2
110  DEF  FN D(X) = 6 * X - 5
120  PRINT "ENTER X1"
130  PRINT
140  INPUT "X1 = ";X1
150  PRINT
160  FOR A = 1 TO 5
170  LET R = X1 - ( FN F(X1) /  FN
          D(X1))
180  PRINT "X";A + 1;" = ";R
190  LET X1 = R
200  NEXT A
300  END
```

Chapter 3–B Addition of Two Vectors

This program finds the sum of two vectors, both of which have their initial points at the origin of a rectangular coordinate system. Enter the magnitude, A, and angle, TA, for the first vector, then the magnitude, B, and the angle, TB, for the second vector when requested. The program will print the magnitude and angle of the sum (or resultant) of the two vectors. All angles are measured from the positive x-axis and are in degrees.

The vectors are added by first finding the sum of the x-components (line 180) and the sum of the y-components (line 190), then finding the magnitude, R, and the angle, TR, of the resultant vector (lines 200 and 210). Note: In the special case where the sum of the x-components is exactly zero, an error will occur.

```
10    REM    ADDITION OF TWO VECTORS
20    REM
30    REM    USE ANGLES IN
40    REM    STANDARD POSITION
50    REM
100   PRINT "ENTER MAGNITUDE AND"
105   PRINT "ANGLE OF VECTOR A"
110   PRINT "AND THEN OF VECTOR B"
120   PRINT
130   INPUT "A = ";A: INPUT "TA =
      ";TA: INPUT "B = ";B: INPUT
      "TB = ";TB
140   PRINT
150   LET PI = 3.141592654
160   LET TA = TA * PI / 180
170   LET TB = TB * PI / 180
180   LET RX = A *  COS (TA) + B *
      COS (TB)
190   LET RY = A *  SIN (TA) + B *
      SIN (TB)
200   LET TR = 180 / PI *  ATN (RY
      / RX)
210   LET R =  SQR (RX * RX + RY *
      RY)
220   PRINT "RESULTANT MAGNITUDE =
      ";R
230   IF RX > 0 AND RY > = 0 THEN
      PRINT "RESULTANT ANGLE = ";
      TR
240   IF RX < 0 THEN  PRINT "RESUL
      TANT ANGLE = ";TR + 180
250   IF RX > 0 AND RY < 0 THEN  PRINT
      "RESULTANT ANGLE = ";TR + 36
      0
300   END
```

Chapter 3–C Maximum Value of a Function

This program finds the maximum value of the function $f(x) = 5 - 2x - 3x^2$ over the interval from $x = A$ to $x = B$, where A is less than B. The values of A and B are to be entered when requested. The maximum value of the function and the x-value at which it occurs are printed by the program. The x-value at which the maximum occurs will be determined to a precision of 0.001 units. To find the maximum value to a different precision, change the value of E in line 125. For example, to find the maximum with an x-value precision of 0.0001, change line 125 to the following:

```
125  LET  E  =  0.0001
```

The maximum values of other functions can be determined by changing the function in line 100. For example, to find the maximum value of the function $f(x) = 4x - x^2$ over an interval, replace line 100 by the following:

```
100  DEF  FN  A(X)  =  4*X  —  X*X
```

The minimum value of a function may be found over an interval by changing line 170 to the following:

```
170  IF  M  <=  Y  THEN  GOTO  200
```

In this case the word *maximum* should be changed to *minimum* in lines 10 and 210. Note: Depending on the interval chosen and the value for E, this program may be slow in giving a result.

```
10   REM   FIND THE MAXIMUM VALUE
20   REM   OF A FUNCTION OVER
30   REM   INTERVAL A < X < B
40   REM
50   REM   ENTER FUNCTION
55   REM   IN LINE 100
60   REM
70   REM   VALUES OF A AND B
80   REM   TO BE ENTERED
90   REM
100  DEF   FN A(X)  = 5 - 2 * X - 3
     * X * X
110  PRINT "ENTER VALUES FOR A AN
     D B"
115  PRINT
120  INPUT "A = ";A: INPUT "B = "
     ;B
125  LET E = .001
130  LET M =   FN A(A)
140  LET N = A
150  FOR X = A TO B STEP E
160  LET Y =   FN A(X)

170  IF M >  = Y THEN   GOTO 200
180  LET M = Y
190  LET N = X
200  NEXT X
205  PRINT
210  PRINT "MAXIMUM VALUE OF F(X)
        IS ";M;" WHEN X = ";N
250  END
```

Chapter 4–A Area Under a Curve

This program estimates the area under the curve $y = x^2 + 1$ by summing the areas of rectangles over the region bounded by the curve, the x-axis, and the lines $x = A$ and $x = B$. The values of A and B, as well as the number of rectangles, N, to be used are to be entered when requested. The program prints out the approximate area of the region as the SUM.

The area under a different function can be found by changing the function in line 100. For example, to find the area under $y = x^3 + 2$ change line 100 to:

```
100  DEF  FN F(X) = X^3 + 2
```

Note: If the function crosses the x-axis between A and B, the value printed for SUM will be the algebraic sum of the areas above and below the x-axis.

Try entering several values for N, increasing N each time. The values for SUM should approach a limit (the exact area) as the number of rectangles gets larger.

```
10   REM   AREA UNDER A CURVE
15   REM   FROM X=A TO X=B
30   REM   BY SUMMING THE AREAS
35   REM   OF N RECTANGLES
40   REM
50   REM   ENTER FUNCTION
55   REM   IN LINE 100
60   REM
70   REM   VALUES OF A, B, N
75   REM   TO BE ENTERED
80   REM
100   DEF  FN F(X) = X * X + 1
110   PRINT "ENTER A, B, N"
115   PRINT
120   INPUT "A = ";A: INPUT "B = "
      ;B: INPUT "N = ";N
125   PRINT
130   LET R = 0
140   LET DX = (B - A) / N
150   FOR I = 0 TO N - 1
160   LET X = A + I * DX
170   LET R = R + ( FN F(X)) * DX
180   NEXT I
190   PRINT "SUM = ";R
200   END
```

Chapter 4–B Trapezoidal Rule

This program estimates the area between the graph of the function $f(x) = x^2 + 1$, the x-axis, and the lines $x = A$ and $x = B$ by use of the trapezoidal rule. The values of A and B and the number of intervals, N, to be used are to be entered when requested.

The area for a different function can be estimated by changing the function in line 100. For example, to estimate the area under the graph of the function $f(x) = x^2 - x + 1$, enter the following:

```
100  DEF  FN F(X) = X*X  -  X + 1
```

Note: If the function crosses the x-axis between $x = A$ and $x = B$, the area obtained will be the algebraic sum of the area above and below the x-axis.

Try entering several values for N, increasing N each time. The approximate values obtained for the areas should approach a limit (the exact area) as N increases.

```
10   REM    TRAPEZOIDAL RULE
20   REM
30   REM    FIND THE AREA UNDER
40   REM    Y = F(X)
50   REM    FROM X = A TO X = B
60   REM
70   REM    ENTER FUNCTION
75   REM    IN LINE 100
80   REM
85   REM    VALUES OF A, B, N
90   REM    TO BE ENTERED
95   REM
100   DEF   FN F(X) = X * X + 1
110   PRINT "ENTER A AND B"
115   PRINT
120   INPUT "A = ";A: INPUT "B = "
      ;B
125   PRINT
130   PRINT "ENTER N"
135   PRINT
140   INPUT "N = ";N
145   PRINT
150   LET D = (B - A) / N
160   LET S =   FN F(A) / 2
170   FOR I = 1 TO N - 1
180   LET V =   FN F(A + I * D)
190   LET S = S + V
200   NEXT I
210   LET S = S +   FN F(B) / 2
220   LET T = S * D
230   PRINT "THE AREA BETWEEN ";A;
      " AND ";B;" IS ";T
250   END
```

Chapter 5 Volume of Rotation

This program calculates the exact volume of the solid formed when the area bounded by the curve $y = Cx^N$, the x-axis, and the lines $x = A$ and $x = B$ is rotated about the x-axis. The values for A, B, C, and N are to be entered when requested.

Try keeping A, B, and C constant, increasing the values for N and note the values of the volumes obtained. (For example, let $A = 0$, $B = 1$, $C = 1$, and enter values 1, 2, 3, etc. for N.)

```
10   REM   VOLUME GENERATED BY
20   REM   ROTATING AREA UNDER
30   REM   Y = C*X^N
40   REM   FROM X = A TO X = B
50   REM   ABOUT THE X-AXIS
55   REM
60   REM   VALUES OF A, B, C,
70   REM   AND N TO BE ENTERED
80   REM
100  DEF   FN I(X) = C ^ 2 * X ^ (
     2 * N + 1) / (2 * N + 1)

110  LET PI = 3.141592654
120  PRINT "ENTER A, B, C, N"
130  PRINT
140  INPUT "A = ";A: INPUT "B = "
     ;B: INPUT "C = ";C: INPUT "N
     = ";N
150  PRINT
160  PRINT "VOLUME = ";PI *  FN I
     (B) - PI *  FN I(A)
200  END
```

Chapter 7 Value of e

This program estimates the value for the number e by calculating a sequence of values for the function $(1 + x)^{1/x}$ as x approaches zero. The program prints out five values for x and for the function of x as x gets smaller. These values are then compared with the computer value for e. Additional terms of the sequence can be printed by changing the number 5 in line 140. For example, to obtain 8 values, change line 140 to:

```
140  FOR  N = 1 TO 8
```

Note: Errors in some of the terms of the sequence can occur if too many values are printed as a result of the calculations involved.

```
10   REM   LIMITING VALUE FOR E
20   REM
100  PRINT "VALUE OF (1 + X)^(1/X
     )"
105  PRINT "AS X->0"
110  PRINT
120  PRINT "X","(1+X)^(1/X)"
125  PRINT
130  LET X = 1
140  FOR N = 1 TO 5
150  PRINT X,(1 + X) ^ (1 / X)
160  LET X = X / 10
170  NEXT N
180  PRINT
190  PRINT "ACTUAL VALUE = "; EXP
     (1)
200  END
```

Chapter 8 Integral Forms

This program compares the indefinite integrals of two special rational functions. The first function is $C/(A + Bx^2)$, and the second function is $Cx/(A + Bx^2)$. The values of A, B, and C are to be entered when requested.

Try entering different values for A, B, and C and note the difference in the two integrals.

```
10   REM    INTEGRAL FORMS
15   REM
20   REM    FIND INDEFINITE
25   REM    INTEGRALS OF
30   REM
40   REM      C*DX            C*X*DX
50   REM    --------- AND  ----------
60   REM    A + B*X^2      A + B*X^2
70   REM
80   REM    ENTER POSITIVE VALUES
85   REM    OF A, B, C
90   REM
100  PRINT
110  PRINT "ENTER A, B, C"
115  PRINT
120  INPUT "A = ";A: INPUT "B = "
     ;B: INPUT "C = ";C
125  PRINT : PRINT
130  PRINT "FIRST INTEGRAL IS"
135  PRINT
140  PRINT C /  SQR (A * B);" ARC
     TAN "; SQR (B / A);"X + C'"
145  PRINT : PRINT : PRINT
150  PRINT "SECOND INTEGRAL IS"
155  PRINT
160  PRINT (.5 * C / B)" LN (";A;
     " + ";B;"X^2) + C'"
200  END
```

Chapter 10 Graph of a Plane

This program sketches the traces of the intersection of any plane with the xy-plane, the yz-plane, and the xz-plane in the first octant. The values of the intercepts of the plane with the three axes are to be entered when requested. The intercept with the x-axis is A, the intercept with the y-axis is B, and the intercept with the z-axis is C. The values entered for A, B, and C should all be positive and less than or equal to 10. Marks appear on each of the axes, 8 units from the origin. The x-axis will appear as perpendicular to the screen, the y-axis as a horizontal line, and the z-axis as a vertical line.

Try different values for A, B, and C to see how different planes appear.

```
10    REM    SKETCH GRAPH OF PLANE
15    REM    FROM INTERCEPTS
20    REM    BY TRACES
25    REM
30    REM    USE POSITIVE
35    REM    INTERCEPTS <=10
40    REM
50    REM    VALUES OF INTERCEPTS
55    REM    A, B, C
60    REM    TO BE ENTERED
70    REM
100   PRINT "AXIS MARKINGS AT 8 UN
      ITS"
105   PRINT
110   HGR
115   HCOLOR= 2
120   HPLOT 100,100 TO 220,100
125   HPLOT 100,0 TO 100,100
130   FOR X = 20 TO 100 STEP .5
135   HPLOT X,200 - X
140   NEXT X
150   HPLOT 98,20 TO 102,20
155   HPLOT 180,98 TO 180,102
160   HPLOT 50,148 TO 54,148
170   PRINT "ENTER INTERCEPTS A,B,
      C"
175   PRINT
180   INPUT "A = ";A: INPUT "B = "
      ;B: INPUT "C = ";C
190   FOR X = 100 TO 10 * (10 + B)
      STEP .1
200   HPLOT X,10 * (10 - C) + (C /
      B) * (X - 100)

210   NEXT X
220   FOR X = 100 - 6 * A TO 10 *
      (10 + B) STEP .5
230   HPLOT X,100 + 6 * A - A / (A
      + 1.7 * B) * (X - 100 + 6 *
      A)
240   NEXT X
250   FOR X = 100 - 6 * A TO 100 STEP
      .1
260   HPLOT X,100 + 6 * A - (A + 1
      .6 * C) / A * (X - 100 + 6 *
      A)
270   NEXT X
300   END
```

Chapter 11-A Graph in Polar Coordinates

This program plots a graph of a function in polar coordinates and shows the graph in relation to the x- and y-axes. The function provided in this program is $r = 8 \sin(3t)$. To plot a different function it will be necessary to change the function in line 100. For example, to graph the function $r = 7 \cos(2t)$ change line 100 to:

```
100  DEF  FN  F(T)  =  7*COS(2*T)
```

Only the points for which R is less than or equal to 8 will appear on the screen.

Try graphing different functions of the form $r = A \sin(Bt)$ or $r = A \cos(Bt)$. Note the different graphs that are obtained when A is held constant and B is assigned values 1, 2, 3, etc. Especially note the difference between values of B that are even numbers and values that are odd numbers. Also, note the difference in the graphs obtained when B is held constant and values of A are changed.

```
10    REM    PLOT CURVE IN
20    REM    POLAR COORDINATES
25    REM
30    REM    ONLY VALUES OF R<=8
40    REM    WILL APPEAR
45    REM
50    REM    ENTER FUNCTION IN
60    REM    LINE 100
70    REM
100   DEF   FN F(T) = 8 *  SIN (3 *
      T)
110   HGR
120   HCOLOR= 2
130   HPLOT 40,80 TO 240,80
140   HPLOT 140,0 TO 140,160
150   LET T = 0
160   FOR N = 0 TO 630
170   LET R =   FN F(T)
180   IF   ABS (R) > 8 THEN   GOTO 2
      30
190   LET X = 10 * R *   COS (T)
200   LET Y = 10 * R *   SIN (T)
210   HPLOT 140 + X,80 - Y
220   LET T = T + .01
230   NEXT N
250   END
```

Chapter 11–B Area Bounded by Polar Curve

This program estimates the area bounded by the graph of $r = 4\cos(3t)$ and the lines given by the angles A and B. The values of A and B are to be expressed in radians and are to be entered when requested. The area is estimated by summing the areas of small sectors each with a central angle of 0.005 radians.

To find the area bounded by a different function it is necessary to change the function in line 100. For example, to find the area bounded by $r = 5\sin(2t)$, change line 100 to:

```
100  DEF  FN F(T) = 5*SIN(2*T)
```

For functions of the type provided in the program, try finding the area of one loop, then the area from 0 to 2π radians and compare.

```
10    REM    AREA BOUNDED BY
20    REM    POLAR CURVE
25    REM
30    REM    ENTER INITIAL ANGLE
35    REM    IN RADIANS
40    REM
50    REM    ENTER FINAL ANGLE
55    REM    IN RADIANS
60    REM
70    REM    ENTER FUNCTION
75    REM    IN LINE 100
80    REM
90    REM
100   DEF   FN F(T) = 4 *  COS (3 *
      T)
110   PRINT "ENTER INITIAL ANGLE"
115   PRINT "AND FINAL ANGLE"
120   PRINT
130   INPUT "A = ";A: INPUT "B = "
      ;B
140   LET S = 0
150   LET T = A
160   FOR N = 200 * A TO 200 * B
170   LET R =   FN F(T)
180   LET S = S + .0025 * R * R
190   LET T = T + .005
200   NEXT N
210   PRINT
220   PRINT "AREA = ";S
300   END
```

Chapter 12–A Arithmetic Mean and Standard Deviation

This program calculates the mean and standard deviation of a set of *N* values. The number of values, *N*, is entered, then when requested to "Enter Numbers" each value is entered one at a time. Any number of values can be entered. The computer will print the arithmetic mean of the set of numbers and the standard deviation.

Try values that have a given range but are grouped close together. Then, try values with the same range and nearly the same mean but which are more widely dispersed. Note the differences in the standard deviation.

```
10   REM   ARITHMETIC MEAN AND
20   REM   STANDARD DEVIATION
30   REM   OF N NUMBERS
35   REM
40   REM   N AND NUMBERS
50   REM   TO BE ENTERED
60   REM
100  PRINT "ENTER N"
110  LET SQ = 0
120  LET SUM = 0
125  PRINT
130  INPUT "N = ";N
135  PRINT
140  PRINT "ENTER ";N;" NUMBERS"
145  PRINT
150  FOR C = 1 TO N
160  INPUT "    ";D
170  LET SUM = SUM + D
180  LET SQ = SQ + D * D
190  NEXT C
200  PRINT
210  LET M = SUM / N
215  PRINT
220  LET X = SQ / N
230  LET S =  SQR (X - M * M)
240  PRINT "ARITHMETIC MEAN = ";M

250  PRINT
260  PRINT "STANDARD DEVIATION =
     ";S
300  END
```

Chapter 12–B Least-Squares Line

This program uses the least squares method to find the equation of the best straight line that will "fit" a set of points. The number of points, N, then, the x- and y-coordinates of each point are to be entered when requested. Any number of points may be used. The slope, M, and the y-intercept, B, are calculated by the program. The equation of the line is printed in the form: $y = Mx + B$.

Try plotting the points on a sheet of graph paper, then plot the line on the same graph to see how well the line "fits" the points. Try leaving out one or more of the points to see how this affects the "fit" of the line.

```
10    REM    FIND EQUATION OF
20    REM    LEAST-SQUARES LINE
30    REM    THRU SEVERAL POINTS
35    REM
40    REM    NUMBER OF POINTS
50    REM    AND COORDINATES
60    REM    TO BE ENTERED
70    REM
100   PRINT "ENTER NUMBER OF POINT
      S"
105   PRINT
110   INPUT "N = ";N
120   LET S1 = 0: LET S2 = 0
125   LET S3 = 0: LET S4 = 0
130   FOR I = 1 TO N
135   PRINT
140   PRINT "ENTER POINT NUMBER";I
150   PRINT "ENTER X-COORDINATE"
155   INPUT X
160   PRINT "ENTER Y-COORDINATE"
165   INPUT Y
170   LET S1 = S1 + X
180   LET S2 = S2 + Y
190   LET S3 = S3 + X * X
200   LET S4 = S4 + X * Y
210   NEXT I
220   LET A1 = S1 / N
230   LET A2 = S2 / N
240   LET A3 = S3 / N
250   LET A4 = S4 / N
260   LET D2 = A3 - A1 * A1
270   LET M = (A4 - A1 * A2) / D2
280   LET B = A2 - M * A1
285   PRINT
290   PRINT "LEAST-SQUARES LINE IS
      "
295   PRINT
300   PRINT "Y = ";M;"X + ";B
350   END
```

Chapter 13–A Partial Sums of Infinite Series

This program finds the first N partial sums of an infinite series. The value of N is to be entered when requested. In this program the infinite series has a general term of $(2n + 1)/[n^2(n + 1)^2]$. This is defined as the function F in line 100. To find the partial sums of a series with a different general term it is necessary to change this function. For example: To find the partial sums of the series with the general term of $(-1)^{n+1}/n$, change line 100 to:

```
100  DEF  FN  F(N)  =  (-1)^(N+1)/N
```

Try increasingly larger values of N. By observing the partial sums you can determine if they approach a particular value. If they do, then the infinite series is said to converge to that value.

```
10   REM   FIND THE FIRST N          125   PRINT
20   REM   PARTIAL SUMS              130   INPUT "N= ";I
30   REM   OF INFINITE SERIES        135   PRINT
35   REM                             140   PRINT. "N     TERM
40   REM   NUMBER OF PARTIAL SUMS               PARTIAL SUM"
45   REM   TO BE ENTERED             145   PRINT
50   REM                             150   FOR N = 1 TO I
60   REM   ENTER GENERAL TERM        160   LET S = S +  FN F(N)
65   REM   IN LINE 100               170   LET T =  FN F(N)
70   REM                             180   PRINT N; TAB( 5)T;: HTAB (25
100  DEF  FN F(N) = (2 * N + 1) /          ): PRINT S
     (N * N * (N + 1) ^ 2)          190   NEXT N
110  LET S = 0                       200   END
115  PRINT
120  PRINT "ENTER N"
```

Chapter 13–B Value of e^x

This program uses the partial sums of an infinite series to estimate the value of e^x. The number of terms to be used, N, and the value of X are to be entered when requested. The series used is $1 + x/1! + x^2/2! + x^3/3! + \ldots$. The program compares the result with the computer's value for e^x.

Try various values of N for a given value of X. Determine how many terms are needed for a precision in the sum of .01, .001, .0001, etc. Then, for a given N, try various X-values to determine if the value of X affects the precision obtained.

```
10   REM   FIND THE VALUE OF E^X     150   LET SUM = 1
20   REM   USING FIRST N TERMS       160   FOR A = 1 TO N
30   REM   OF SERIES                 170   LET B = 1
40   REM                             180   FOR C = 1 TO A
50   REM   N AND X                   190   LET B = B * C
60   REM   TO BE ENTERED             200   NEXT C
70   REM                             210   LET SUM = SUM + (X ^ A) / B
100  PRINT "ENTER N AND X"           220   PRINT A,SUM - (X ^ A) / B
105  PRINT                           230   NEXT A
110  INPUT "N = ";N                  235   PRINT
120  INPUT "X = ";X                  240   PRINT "ACTUAL VALUE = "; EXP
130  PRINT                                 (X)
140  PRINT "N","SUM OF N TERMS"      250   END
145  PRINT
```

Chapter 15 Graph of Solution to a Differential Equation

This program gives the graph of the solution to a differential equation of the form: $D^2x + B\,Dx + 4x = 0$ where the initial conditions are $x = 10$ and $Dx = 0$ when $t = 0$. The graph of the solution is given for a value of B that is to be entered when requested. The value of B should be greater than or equal to 0 and less than or equal to 10. The y-axis has a minimum value of -10 and a maximum value of $+10$.

Try various values of B and note the different solution graphs obtained. In particular, note the difference between values of B such that $B = 0$, $B < 4$, $B = 4$, and $B > 4$. Also, note what happens as B becomes larger. (The value B is sometimes referred to as the damping coefficient.)

```
10   REM   GRAPH SOLUTIONS OF
20   REM   DIFFERENTIAL EQUATION
25   REM
30   REM   (D^2)X + BDX + 4X = 0
35   REM
40   REM   FOR INITIAL CONDITIONS
45   REM   X=10 AND DX=0 FOR T=0
50   REM   AND VARIOUS VALUES OF B
60   REM
70   REM   ENTER B SUCH THAT
75   REM   0 <= B <= 10
80   REM
100   PRINT
110   HGR
120   HCOLOR= 2
130   PRINT "ENTER B"
135   INPUT "B = ";B
140   HPLOT 0,80 TO 240,80
150   HPLOT 0,0 TO 0,160
160   IF B > 4 THEN  GOTO 220
170   IF B = 4 THEN  GOTO 280
175   FOR T = 0 TO 240 STEP .5
180   LET X = 80 *  EXP ( - B * T /
      250) *  COS ( SQR (16 - B *
      B) * T / 20)
190   HPLOT T,80 - X
200   NEXT T
210   GOTO 400
220   FOR T = 0 TO 240 STEP .5
230   LET R =  SQR (B * B - 16)
240   LET X = 4 * ((B + R) / R) *
       EXP (( - B + R) / 200 * T) +
      (8 - 4 * (B + R) / R) *  EXP
      (( - B - R) / 200 * T)
250   HPLOT T,80 - 10 * X
260   NEXT T
270   GOTO 400
280   FOR T = 0 TO 240 STEP .5
290   LET X = (80 + 1.6 * T) *  EXP
      ( - .02 * T)
300   HPLOT T,80 - X
310   NEXT T
400   END
```

Chapter 16 Increment Method

This program illustrates a numerical method, called the *increment method,* for estimating the solution to a differential equation of the form: $dy/dx = f(x, y)$. The initial values of x and y, the x-coordinate of the final point, XF, and the increment size, DX, are to be entered when requested. The equation provided in the program is $dy/dx = x + y$. The numerical solution to a different equation can be estimated by changing the function in line 100. For example, to find the numerical solution to the differential equation $dy/dx = x^2 + y^2$ enter the following for line 100:

```
100 DEF FN F(X) = X^2 + Y^2
```

(For this function, use initial conditions of $x = 0$ and $y = 1$. Notice what happens when XF is chosen close to 2.)

A table is printed by the program showing the coordinates of the calculated points, the values of $f(x, y)$ at that point, and values of the increment, dy.

Try keeping the initial values and final value constant and make the increment size smaller. The resulting points should give a better estimate of the solution.

```
10   REM   INCREMENT METHOD
15   REM
20   REM   NUMERICAL SOLUTION
25   REM   OF DY/DX = F(X,Y)
30   REM
35   REM   ENTER FUNCTION
40   REM   IN LINE 100
45   REM
50   REM   COORDINATES OF
55   REM   INITIAL POINT,
60   REM   X-COORDINATE OF
65   REM   FINAL POINT,
70   REM   SIZE OF INCREMENT
75   REM   TO BE ENTERED
80   REM
100  DEF  FN F(X) = 2 * X * X - Y
        * Y
110  PRINT
120  PRINT "ENTER COORDINATES"
125  PRINT "OF INITAL POINT"
130  PRINT
140  INPUT "X = ";X: INPUT "Y = "
        ;Y
145  PRINT

150  PRINT "ENTER X-COORDINATE"
155  PRINT "OF FINAL POINT"
160  PRINT
170  INPUT "XF = ";XF
175  PRINT
180  PRINT "ENTER INCREMENT SIZE"
185  PRINT
190  INPUT "DX = ";DX
195  PRINT
200  PRINT  TAB( 1)"X"; TAB( 6)"Y
        "; TAB( 18)"F(X,Y)"; TAB( 30
        )"DY"
205  PRINT
210  FOR N = X TO XF + .0001 STEP
        DX
220  LET DY = ( FN F(X)) * DX
230  PRINT  TAB( 1)X; TAB( 6)Y; TAB(
        18) FN F(X); TAB( 30)DY
240  LET X = X + DX
250  LET Y = Y + DY
260  NEXT N
300  END
```

APPENDIX D

Tables

Table 1. Powers and Roots

No.	Sq.	Sq. Root	Cube	Cube Root	No.	Sq.	Sq. Root	Cube	Cube Root
1	1	1.000	1	1.000	51	2 601	7.141	132 651	3.708
2	4	1.414	8	1.260	52	2 704	7.211	140 608	3.733
3	9	1.732	27	1.442	53	2 809	7.280	148 877	3.756
4	16	2.000	64	1.587	54	2 916	7.348	157 464	3.780
5	25	2.236	125	1.710	55	3 025	7.416	166 375	3.803
6	36	2.449	216	1.817	56	3 136	7.483	175 616	3.826
7	49	2.646	343	1.913	57	3 249	7.550	185 193	3.849
8	64	2.828	512	2.000	58	3 364	7.616	195 112	3.871
9	81	3.000	729	2.080	59	3 481	7.681	205 379	3.893
10	100	3.162	1 000	2.154	60	3 600	7.746	216 000	3.915
11	121	3.317	1 331	2.224	61	3 721	7.810	226 981	3.936
12	144	3.464	1 728	2.289	62	3 844	7.874	238 328	3.958
13	169	3.606	2 197	2.351	63	3 969	7.937	250 047	3.979
14	196	3.742	2 744	2.410	64	4 096	8.000	262 144	4.000
15	225	3.873	3 375	2.466	65	4 225	8.062	274 625	4.021
16	256	4.000	4 096	2.520	66	4 356	8.124	287 496	4.041
17	289	4.123	4 913	2.571	67	4 489	8.185	300 763	4.062
18	324	4.243	5 832	2.621	68	4 624	8.246	314 432	4.082
19	361	4.359	6 859	2.668	69	4 761	8.307	328 509	4.102
20	400	4.472	8 000	2.714	70	4 900	8.367	343 000	4.121
21	441	4.583	9 261	2.759	71	5 041	8.426	357 911	4.141
22	484	4.690	10 648	2.802	72	5 184	8.485	373 248	4.160
23	529	4.796	12 167	2.844	73	5 329	8.544	389 017	4.179
24	576	4.899	13 824	2.884	74	5 476	8.602	405 224	4.198
25	625	5.000	15 625	2.924	75	5 625	8.660	421 875	4.217
26	676	5.099	17 576	2.962	76	5 776	8.718	438 976	4.236
27	729	5.196	19 683	3.000	77	5 929	8.775	456 533	4.254
28	784	5.292	21 952	3.037	78	6 084	8.832	474 552	4.273
29	841	5.385	24 389	3.072	79	6 241	8.888	493 039	4.291
30	900	5.477	27 000	3.107	80	6 400	8.944	512 000	4.309
31	961	5.568	29 791	3.141	81	6 561	9.000	531 441	4.327
32	1 024	5.657	32 768	3.175	82	6 724	9.055	551 368	4.344
33	1 089	5.745	35 937	3.208	83	6 889	9.110	571 787	4.362
34	1 156	5.831	39 304	3.240	84	7 056	9.165	592 704	4.380
35	1 225	5.916	42 875	3.271	85	7 225	9.220	614 125	4.397
36	1 296	6.000	46 656	3.302	86	7 396	9.274	636 056	4.414
37	1 369	6.083	50 653	3.332	87	7 569	9.327	658 503	4.431
38	1 444	6.164	54 872	3.362	88	7 744	9.381	681 472	4.448
39	1 521	6.245	59 319	3.391	89	7 921	9.434	704 969	4.465
40	1 600	6.325	64 000	3.420	90	8 100	9.487	729 000	4.481
41	1 681	6.403	68 921	3.448	91	8 281	9.539	753 571	4.498
42	1 764	6.481	74 088	3.476	92	8 464	9.592	778 688	4.514
43	1 849	6.557	79 507	3.503	93	8 649	9.644	804 357	4.531
44	1 936	6.633	85 184	3.530	94	8 836	9.695	830 584	4.547
45	2 025	6.708	91 125	3.557	95	9 025	9.747	857 375	4.563
46	2 116	6.782	97 336	3.583	96	9 216	9.798	884 736	4.579
47	2 209	6.856	103 823	3.609	97	9 409	9.849	912 673	4.595
48	2 304	6.928	110 592	3.634	98	9 604	9.899	941 192	4.610
49	2 401	7.000	117 649	3.659	99	9 801	9.950	970 299	4.626
50	2 500	7.071	125 000	3.684	100	10 000	10.000	1 000 000	4.642

Table 2. Common Logarithms

N	0	1	2	3	4	5	6	7	8	9
10	0000	0043	0086	0128	0170	0212	0253	0294	0334	0374
11	0414	0453	0492	0531	0569	0607	0645	0682	0719	0755
12	0792	0828	0864	0899	0934	0969	1004	1038	1072	1106
13	1139	1173	1206	1239	1271	1303	1335	1367	1399	1430
14	1461	1492	1523	1553	1584	1614	1644	1673	1703	1732
15	1761	1790	1818	1847	1875	1903	1931	1959	1987	2014
16	2041	2068	2095	2122	2148	2175	2201	2227	2253	2279
17	2304	2330	2355	2380	2405	2430	2455	2480	2504	2529
18	2553	2577	2601	2625	2648	2672	2695	2718	2742	2765
19	2788	2810	2833	2856	2878	2900	2923	2945	2967	2989
20	3010	3032	3054	3075	3096	3118	3139	3160	3181	3201
21	3222	3243	3263	3284	3304	3324	3345	3365	3385	3404
22	3424	3444	3464	3483	3502	3522	3541	3560	3579	3598
23	3617	3636	3655	3674	3692	3711	3729	3747	3766	3784
24	3802	3820	3838	3856	3874	3892	3909	3927	3945	3962
25	3979	3997	4014	4031	4048	4065	4082	4099	4116	4133
26	4150	4166	4183	4200	4216	4232	4249	4265	4281	4298
27	4314	4330	4346	4362	4378	4393	4409	4425	4440	4456
28	4472	4487	4502	4518	4533	4548	4564	4579	4594	4609
29	4624	4639	4654	4669	4683	4698	4713	4728	4742	4757
30	4771	4786	4800	4814	4829	4843	4857	4871	4886	4900
31	4914	4928	4942	4955	4969	4983	4997	5011	5024	5038
32	5051	5065	5079	5092	5105	5119	5132	5145	5159	5172
33	5185	5198	5211	5224	5237	5250	5263	5276	5289	5302
34	5315	5328	5340	5353	5366	5378	5391	5403	5416	5428
35	5441	5453	5465	5478	5490	5502	5514	5527	5539	5551
36	5563	5575	5587	5599	5611	5623	5635	5647	5658	5670
37	5682	5694	5705	5717	5729	5740	5752	5763	5775	5786
38	5798	5809	5821	5832	5843	5855	5866	5877	5888	5899
39	5911	5922	5933	5944	5955	5966	5977	5988	5999	6010
40	6021	6031	6042	6053	6064	6075	6085	6096	6107	6117
41	6128	6138	6149	6160	6170	6180	6191	6201	6212	6222
42	6232	6243	6253	6263	6274	6284	6294	6304	6314	6325
43	6335	6345	6355	6365	6375	6385	6395	6405	6415	6425
44	6435	6444	6454	6464	6474	6484	6493	6503	6513	6522
45	6532	6542	6551	6561	6571	6580	6590	6599	6609	6618
46	6628	6637	6646	6656	6665	6675	6684	6693	6702	6712
47	6721	6730	6739	6749	6758	6767	6776	6785	6794	6803
48	6812	6821	6830	6839	6848	6857	6866	6875	6884	6893
49	6902	6911	6920	6928	6937	6946	6955	6964	6972	6981
50	6990	6998	7007	7016	7024	7033	7042	7050	7059	7067
51	7076	7084	7093	7101	7110	7118	7126	7135	7143	7152
52	7160	7168	7177	7185	7193	7202	7210	7218	7226	7235
53	7243	7251	7259	7267	7275	7284	7292	7300	7308	7316
54	7324	7332	7340	7348	7356	7364	7372	7380	7388	7396

Table 2. Continued

N	0	1	2	3	4	5	6	7	8	9
55	7404	7412	7419	7427	7435	7443	7451	7459	7466	7474
56	7482	7490	7497	7505	7513	7520	7528	7536	7543	7551
57	7559	7566	7574	7582	7589	7597	7604	7612	7619	7627
58	7634	7642	7649	7657	7664	7672	7679	7686	7694	7701
59	7709	7716	7723	7731	7738	7745	7752	7760	7767	7774
60	7782	7789	7796	7803	7810	7818	7825	7832	7839	7846
61	7853	7860	7868	7875	7882	7889	7896	7903	7910	7917
62	7924	7931	7938	7945	7952	7959	7966	7973	7980	7987
63	7993	8000	8007	8014	8021	8028	8035	8041	8048	8055
64	8062	8069	8075	8082	8089	8096	8102	8109	8116	8122
65	8129	8136	8142	8149	8156	8162	8169	8176	8182	8189
66	8195	8202	8209	8215	8222	8228	8235	8241	8248	8254
67	8261	8267	8274	8280	8287	8293	8299	8306	8312	8319
68	8325	8331	8338	8344	8351	8357	8363	8370	8376	8382
69	8388	8395	8401	8407	8414	8420	8426	8432	8439	8445
70	8451	8457	8463	8470	8476	8482	8488	8494	8500	8506
71	8513	8519	8525	8531	8537	8543	8549	8555	8561	8567
72	8573	8579	8585	8591	8597	8603	8609	8615	8621	8627
73	8633	8639	8645	8651	8657	8663	8669	8675	8681	8686
74	8692	8698	8704	8710	8716	8722	8727	8733	8739	8745
75	8751	8756	8762	8768	8774	8779	8785	8791	8797	8802
76	8808	8814	8820	8825	8831	8837	8842	8848	8854	8859
77	8865	8871	8876	8882	8887	8893	8899	8904	8910	8915
78	8921	8927	8932	8938	8943	8949	8954	8960	8965	8971
79	8976	8982	8987	8993	8998	9004	9009	9015	9020	9025
80	9031	9036	9042	9047	9053	9058	9063	9069	9074	9079
81	9085	9090	9096	9101	9106	9112	9117	9122	9128	9133
82	9138	9143	9149	9154	9159	9165	9170	9175	9180	9186
83	9191	9196	9201	9206	9212	9217	9222	9227	9232	9238
84	9243	9248	9253	9258	9263	9269	9274	9279	9284	9289
85	9294	9299	9304	9309	9315	9320	9325	9330	9335	9340
86	9345	9350	9355	9360	9365	9370	9375	9380	9385	9390
87	9395	9400	9405	9410	9415	9420	9425	9430	9435	9440
88	9445	9450	9455	9460	9465	9469	9474	9479	9484	9489
89	9494	9499	9504	9509	9513	9518	9523	9528	9533	9538
90	9542	9547	9552	9557	9562	9566	9571	9576	9581	9586
91	9590	9595	9600	9605	9609	9614	9619	9624	9628	9633
92	9638	9643	9647	9652	9657	9661	9666	9671	9675	9680
93	9685	9689	9694	9699	9703	9708	9713	9717	9722	9727
94	9731	9736	9741	9745	9750	9754	9759	9763	9768	9773
95	9777	9782	9786	9791	9795	9800	9805	9809	9814	9818
96	9823	9827	9832	9836	9841	9845	9850	9854	9859	9863
97	9868	9872	9877	9881	9886	9890	9894	9899	9903	9908
98	9912	9917	9921	9926	9930	9934	9939	9943	9948	9952
99	9956	9961	9965	9969	9974	9978	9983	9987	9991	9996

Table 3. Trigonometric Functions

Degrees	Radians	Sin	Cos	Tan	Cot		
0	0.0000	0.0000	1.0000	0.0000	—	1.5708	90
1	0.0175	0.0175	0.9998	0.0175	57.29	1.5533	89
2	0.0349	0.0349	0.9994	0.0349	28.64	1.5359	88
3	0.0524	0.0523	0.9986	0.0524	19.08	1.5184	87
4	0.0698	0.0698	0.9976	0.0699	14.30	1.5010	86
5	0.0873	0.0872	0.9962	0.0875	11.43	1.4835	85
6	0.1047	0.1045	0.9945	0.1051	9.514	1.4661	84
7	0.1222	0.1219	0.9925	0.1228	8.144	1.4486	83
8	0.1396	0.1392	0.9903	0.1405	7.115	1.4312	82
9	0.1571	0.1564	0.9877	0.1584	6.314	1.4137	81
10	0.1745	0.1736	0.9848	0.1763	5.671	1.3963	80
11	0.1920	0.1908	0.9816	0.1944	5.145	1.3788	79
12	0.2094	0.2079	0.9781	0.2126	4.705	1.3614	78
13	0.2269	0.2250	0.9744	0.2309	4.331	1.3439	77
14	0.2443	0.2419	0.9703	0.2493	4.011	1.3265	76
15	0.2618	0.2588	0.9659	0.2679	3.732	1.3090	75
16	0.2793	0.2756	0.9613	0.2867	3.487	1.2915	74
17	0.2967	0.2924	0.9563	0.3057	3.271	1.2741	73
18	0.3142	0.3090	0.9511	0.3249	3.078	1.2566	72
19	0.3316	0.3256	0.9455	0.3443	2.904	1.2392	71
20	0.3491	0.3420	0.9397	0.3640	2.747	1.2217	70
21	0.3665	0.3584	0.9336	0.3839	2.605	1.2043	69
22	0.3840	0.3746	0.9272	0.4040	2.475	1.1868	68
23	0.4014	0.3907	0.9205	0.4245	2.356	1.1694	67
24	0.4189	0.4067	0.9135	0.4452	2.246	1.1519	66
25	0.4363	0.4226	0.9063	0.4663	2.145	1.1345	65
26	0.4538	0.4384	0.8988	0.4877	2.050	1.1170	64
27	0.4712	0.4540	0.8910	0.5095	1.963	1.0996	63
28	0.4887	0.4695	0.8829	0.5317	1.881	1.0821	62
29	0.5061	0.4848	0.8746	0.5543	1.804	1.0647	61
30	0.5236	0.5000	0.8660	0.5774	1.732	1.0472	60
31	0.5411	0.5150	0.8572	0.6009	1.664	1.0297	59
32	0.5585	0.5299	0.8480	0.6249	1.600	1.0123	58
33	0.5760	0.5446	0.8387	0.6494	1.540	0.9948	57
34	0.5934	0.5592	0.8290	0.6745	1.483	0.9774	56
35	0.6109	0.5736	0.8192	0.7002	1.428	0.9599	55
36	0.6283	0.5878	0.8090	0.7265	1.376	0.9425	54
37	0.6458	0.6018	0.7986	0.7536	1.327	0.9250	53
38	0.6632	0.6157	0.7880	0.7813	1.280	0.9076	52
39	0.6807	0.6293	0.7771	0.8098	1.235	0.8901	51
40	0.6981	0.6428	0.7660	0.8391	1.192	0.8727	50
41	0.7156	0.6561	0.7547	0.8693	1.150	0.8552	49
42	0.7330	0.6691	0.7431	0.9004	1.111	0.8378	48
43	0.7505	0.6820	0.7314	0.9325	1.072	0.8203	47
44	0.7679	0.6947	0.7193	0.9657	1.036	0.8029	46
45	0.7854	0.7071	0.7071	1.0000	1.000	0.7854	45
		Cos	Sin	Cot	Tan	Radians	Degrees

Table 4. Natural Logarithms and Exponential Functions

x	$\ln x$	e^x	e^{-x}
0.0		1.0000	1.0000
0.1	−2.3026	1.1052	0.9048
0.2	−1.6094	1.2214	0.8187
0.3	−1.2040	1.3499	0.7408
0.4	−0.9163	1.4918	0.6703
0.5	−0.6931	1.6487	0.6065
0.6	−0.5108	1.8221	0.5488
0.7	−0.3567	2.0138	0.4966
0.8	−0.2231	2.2255	0.4493
0.9	−0.1054	2.4596	0.4066
1.0	0.0000	2.7183	0.3679
1.1	0.0953	3.0042	0.3329
1.2	0.1823	3.3201	0.3012
1.3	0.2624	3.6693	0.2725
1.4	0.3365	4.0552	0.2466
1.5	0.4055	4.4817	0.2231
1.6	0.4700	4.9530	0.2019
1.7	0.5306	5.4739	0.1827
1.8	0.5878	6.0496	0.1653
1.9	0.6419	6.6859	0.1496
2.0	0.6931	7.3891	0.1353
2.1	0.7419	8.1662	0.1225
2.2	0.7885	9.0250	0.1108
2.3	0.8329	9.9742	0.1003
2.4	0.8755	11.023	0.0907
2.5	0.9163	12.182	0.0821
2.6	0.9555	13.464	0.0743
2.7	0.9933	14.880	0.0672
2.8	1.0296	16.445	0.0608
2.9	1.0647	18.174	0.0550
3.0	1.0986	20.086	0.0498
3.1	1.1314	22.198	0.0450
3.2	1.1632	24.533	0.0408
3.3	1.1939	27.113	0.0369
3.4	1.2238	29.964	0.0334
3.5	1.2528	33.115	0.0302
3.6	1.2809	36.598	0.0273
3.7	1.3083	40.447	0.0247
3.8	1.3350	44.701	0.0224
3.9	1.3610	49.402	0.0202
4.0	1.3863	54.598	0.0183
4.1	1.4110	60.340	0.0166
4.2	1.4351	66.686	0.0150
4.3	1.4586	73.700	0.0136
4.4	1.4816	81.451	0.0123

x	$\ln x$	e^x	e^{-x}
4.5	1.5041	90.017	0.0111
4.6	1.5261	99.484	0.0101
4.7	1.5476	109.95	0.0091
4.8	1.5686	121.51	0.0082
4.9	1.5892	134.29	0.0074
5.0	1.6094	148.41	0.0067
5.1	1.6292	164.02	0.0061
5.2	1.6487	181.27	0.0055
5.3	1.6677	200.34	0.0050
5.4	1.6864	221.41	0.0045
5.5	1.7047	244.69	0.0041
5.6	1.7228	270.43	0.0037
5.7	1.7405	298.87	0.0033
5.8	1.7579	330.30	0.0030
5.9	1.7750	365.04	0.0027
6.0	1.7918	403.43	0.0025
6.1	1.8083	445.86	
6.2	1.8245	492.75	
6.3	1.8405	544.57	
6.4	1.8563	601.85	
6.5	1.8718	665.14	0.0015
6.6	1.8871	735.10	
6.7	1.9021	812.41	
6.8	1.9169	897.85	
6.9	1.9315	992.27	
7.0	1.9459	1097	0.0009
7.1	1.9601	1212	
7.2	1.9741	1339	
7.3	1.9879	1480	
7.4	2.0015	1636	
7.5	2.0149	1808	0.0006
7.6	2.0281	1998	
7.7	2.0412	2208	
7.8	2.0541	2441	
7.9	2.0669	2697	
8.0	2.0794	2981	0.0003
8.1	2.0919	3294	
8.2	2.1041	3641	
8.3	2.1163	4024	
8.4	2.1282	4447	
8.5	2.1401	4915	0.0002
8.6	2.1518	5432	
8.7	2.1633	6003	
8.8	2.1748	6634	
8.9	2.1861	7332	

x	$\ln x$	e^x	e^{-x}
9.0	2.1972	8,103	0.0001
9.1	2.2083	8,955	
9.2	2.2192	9,897	
9.3	2.2300	10,938	
9.4	2.2407	12,088	
9.5	2.2513	13,360	0.00007
9.6	2.2618	14,765	
9.7	2.2721	16,318	
9.8	2.2824	18,034	
9.9	2.2925	19,930	
10	2.3026	22,026	4.5×10^{-5}
15	2.7081		
20	2.9957	4.9×10^8	2.1×10^{-9}
25	3.2189		
30	3.4012	1.1×10^{13}	9.4×10^{-14}
35	3.5553		
40	3.6889	2.4×10^{17}	4.2×10^{-18}
45	3.8067		
50	3.9120	5.2×10^{21}	1.9×10^{-22}
55	4.0073		
60	4.0943	1.1×10^{26}	8.8×10^{-27}
70	4.2485		
80	4.3820	5.5×10^{34}	1.8×10^{-35}
90	4.4998		
100	4.6052	2.7×10^{43}	3.7×10^{-44}

$\ln 10$	2.3026
$2 \ln 10$	4.6052
$3 \ln 10$	6.9078
$4 \ln 10$	9.2103
$5 \ln 10$	11.5129

Table 5. A Short Table of Integrals

The basic forms of Chapter 8 are not included. The constant of integration is omitted.

Forms containing $a + bu$ and $\sqrt{a + bu}$

1. $\displaystyle\int \frac{u\,du}{a + bu} = \frac{1}{b^2}[(a + bu) - a\ln(a + bu)]$

2. $\displaystyle\int \frac{du}{u(a + bu)} = -\frac{1}{a}\ln\frac{a + bu}{u}$

3. $\displaystyle\int \frac{u\,du}{(a + bu)^2} = \frac{1}{b^2}\!\left(\frac{a}{a + bu} + \ln(a + bu)\right)$

4. $\displaystyle\int \frac{du}{u(a + bu)^2} = \frac{1}{a(a + bu)} - \frac{1}{a^2}\ln\frac{a + bu}{u}$

5. $\displaystyle\int u\sqrt{a + bu}\;du = -\frac{2(2a - 3bu)(a + bu)^{3/2}}{15b^2}$

6. $\displaystyle\int \frac{u\,du}{\sqrt{a + bu}} = -\frac{2(2a - bu)\sqrt{a + bu}}{3b^2}$

7. $\displaystyle\int \frac{du}{u\sqrt{a + bu}} = \frac{1}{\sqrt{a}}\ln\!\left(\frac{\sqrt{a + bu} - \sqrt{a}}{\sqrt{a + bu} + \sqrt{a}}\right),\qquad a > 0$

8. $\displaystyle\int \frac{\sqrt{a + bu}}{u}\,du = 2\sqrt{a + bu} + a\int \frac{du}{u\sqrt{a + bu}}$

Forms containing $\sqrt{u^2 \pm a^2}$ and $\sqrt{a^2 - u^2}$

9. $\displaystyle\int \frac{du}{u^2 - a^2} = \frac{1}{2a}\ln\frac{u - a}{u + a}$

10. $\displaystyle\int \frac{du}{\sqrt{u^2 \pm a^2}} = \ln(u + \sqrt{u^2 \pm a^2})$

11. $\displaystyle\int \frac{du}{u\sqrt{u^2 + a^2}} = -\frac{1}{a}\ln\!\left(\frac{a + \sqrt{u^2 + a^2}}{u}\right)$

12. $\displaystyle\int \frac{du}{u\sqrt{u^2 - a^2}} = \frac{1}{a}\text{Arcsec}\frac{u}{a}$

13. $\displaystyle\int \frac{du}{u\sqrt{a^2 - u^2}} = -\frac{1}{a}\ln\!\left(\frac{a + \sqrt{a^2 - u^2}}{u}\right)$

14. $\displaystyle\int \sqrt{u^2 \pm a^2}\;du = \frac{u}{2}\sqrt{u^2 \pm a^2} \pm \frac{a^2}{2}\ln(u + \sqrt{u^2 \pm a^2})$

15. $\displaystyle\int \sqrt{a^2 - u^2}\;du = \frac{u}{2}\sqrt{a^2 - u^2} + \frac{a^2}{2}\text{Arcsin}\frac{u}{a}$

16. $\displaystyle\int \frac{\sqrt{u^2 + a^2}}{u}\,du = \sqrt{u^2 + a^2} - a\ln\!\left(\frac{a + \sqrt{u^2 + a^2}}{u}\right)$

17. $\displaystyle\int \frac{\sqrt{u^2 - a^2}}{u}\,du = \sqrt{u^2 - a^2} - a\,\text{Arcsec}\frac{u}{a}$

Table 5. Continued

18. $\displaystyle\int \frac{\sqrt{a^2 - u^2}}{u}du = \sqrt{a^2 - u^2} - a \ln\left(\frac{a + \sqrt{a^2 - u^2}}{u}\right)$

19. $\displaystyle\int (u^2 \pm a^2)^{3/2} du = \frac{u}{4}(u^2 \pm a^2)^{3/2} \pm \frac{3a^2 u}{8}\sqrt{u^2 \pm a^2} + \frac{3a^4}{8}\ln(u + \sqrt{u^2 \pm a^2})$

20. $\displaystyle\int (a^2 - u^2)^{3/2} du = \frac{u}{4}(a^2 - u^2)^{3/2} + \frac{3a^2 u}{8}\sqrt{a^2 - u^2} + \frac{3a^4}{8}\text{Arcsin}\frac{u}{a}$

21. $\displaystyle\int \frac{(u^2 + a^2)^{3/2}}{u}du = \frac{1}{3}(u^2 + a^2)^{3/2} + a^2 \sqrt{u^2 + a^2} - a^3\ln\left(\frac{a + \sqrt{u^2 + a^2}}{u}\right)$

22. $\displaystyle\int \frac{(u^2 - a^2)^{3/2}}{u}du = \frac{1}{3}(u^2 - a^2)^{3/2} - a^2 \sqrt{u^2 - a^2} + a^3\text{Arcsec}\frac{u}{a}$

23. $\displaystyle\int \frac{(a^2 - u^2)^{3/2}}{u}du = \frac{1}{3}(a^2 - u^2)^{3/2} - a^2 \sqrt{a^2 - u^2} + a^3\ln\left(\frac{a + \sqrt{a^2 - u^2}}{u}\right)$

24. $\displaystyle\int \frac{du}{(u^2 \pm a^2)^{3/2}} = \pm \frac{u}{a^2 \sqrt{u^2 \pm a^2}}$

25. $\displaystyle\int \frac{du}{(a^2 - u^2)^{3/2}} = \frac{u}{a^2 \sqrt{a^2 - u^2}}$

26. $\displaystyle\int \frac{du}{u(u^2 + a^2)^{3/2}} = \frac{1}{a^2 \sqrt{u^2 + a^2}} - \frac{1}{a^3}\ln\left(\frac{a + \sqrt{u^2 + a^2}}{u}\right)$

27. $\displaystyle\int \frac{du}{u(u^2 - a^2)^{3/2}} = -\frac{1}{a^2 \sqrt{u^2 - a^2}} - \frac{1}{a^3}\text{Arcsec}\frac{u}{a}$

28. $\displaystyle\int \frac{du}{u(a^2 - u^2)^{3/2}} = \frac{1}{a^2 \sqrt{a^2 - u^2}} - \frac{1}{a^3}\ln\left(\frac{a + \sqrt{a^2 - u^2}}{u}\right)$

Trigonometric forms

29. $\displaystyle\int \sin^2 u \; du = \frac{u}{2} - \frac{1}{2}\sin u \cos u$

30. $\displaystyle\int \sin^3 u \; du = -\cos u + \frac{1}{3}\cos^3 u$

31. $\displaystyle\int \sin^n u \; du = -\frac{1}{n}\sin^{n-1} u \cos u + \frac{n-1}{n}\int \sin^{n-2} u \; du$

32. $\displaystyle\int \cos^2 u \; du = \frac{u}{2} + \frac{1}{2}\sin u \cos u$

33. $\displaystyle\int \cos^3 u \; du = \sin u - \frac{1}{3}\sin^3 u$

34. $\displaystyle\int \cos^n u \; du = \frac{1}{n}\cos^{n-1} u \sin u + \frac{n-1}{n}\int \cos^{n-2} u \; du$

35. $\displaystyle\int \tan^n u \; du = \frac{\tan^{n-1} u}{n-1} - \int \tan^{n-2} u \; du$

Table 5. Continued

36. $\displaystyle \int \cot^n u \; du = -\frac{\cot^{n-1} u}{n-1} - \int \cot^{n-2} u \; du$

37. $\displaystyle \int \sec^n u \; du = \frac{\sec^{n-2} u \, \tan u}{n-1} + \frac{n-2}{n-1} \int \sec^{n-2} u \; du$

38. $\displaystyle \int \csc^n u \; du = -\frac{\csc^{n-2} u \, \cot u}{n-1} + \frac{n-2}{n-1} \int \csc^{n-2} u \; du$

39. $\displaystyle \int \sin au \, \sin bu \; du = \frac{\sin(a-b)u}{2(a-b)} - \frac{\sin(a+b)u}{2(a+b)}$

40. $\displaystyle \int \sin au \, \cos bu \; du = -\frac{\cos(a-b)u}{2(a-b)} - \frac{\cos(a+b)u}{2(a+b)}$

41. $\displaystyle \int \cos au \, \cos bu \; du = \frac{\sin(a-b)u}{2(a-b)} + \frac{\sin(a+b)u}{2(a+b)}$

42. $\displaystyle \int \sin^m u \, \cos^n u \; du = \frac{\sin^{m+1} u \, \cos^{n-1} u}{m+n} + \frac{n-1}{m+n} \int \sin^m u \, \cos^{n-2} u \; du$

43. $\displaystyle \int \sin^m u \, \cos^n u \; du = -\frac{\sin^{m-1} u \, \cos^{n+1} u}{m+n} + \frac{m-1}{m+n} \int \sin^{m-2} u \, \cos^n u \; du$

Other forms

44. $\displaystyle \int u e^{au} du = \frac{e^{au}(au-1)}{a^2}$

45. $\displaystyle \int u^2 e^{au} du = \frac{e^{au}}{a^3}(a^2 u^2 - 2au + 2)$

46. $\displaystyle \int u^n \ln u \; du = u^{n+1} \left(\frac{\ln u}{n+1} - \frac{1}{(n+1)^2} \right)$

47. $\displaystyle \int u \sin u \; du = \sin u - u \cos u$

48. $\displaystyle \int u \cos u \; du = \cos u + u \sin u$

49. $\displaystyle \int e^{au} \sin bu \; du = \frac{e^{au}(a \sin bu - b \cos bu)}{a^2 + b^2}$

50. $\displaystyle \int e^{au} \cos bu \; du = \frac{e^{au}(a \cos bu + b \sin bu)}{a^2 + b^2}$

51. $\displaystyle \int \text{Arcsin } u \; du = u \, \text{Arcsin } u + \sqrt{1 - u^2}$

52. $\displaystyle \int \text{Arctan } u \; du = u \, \text{Arctan } u - \frac{1}{2}\ln(1 + u^2)$

Answers to Odd-Numbered Exercises

Exercises 1–1, p. 4

1. (a) Rational, real, (b) Irrational, real, (c) Imaginary, (d) Imaginary **3.** (a) 3, (b) $\frac{7}{2}$, (c) 4
5. (a) < (b) > (c) > **7.** **9.** The points lie on a straight line.

11. On a line parallel to the y-axis, one unit to the right. **13.** 0 **15.** To the right of the y-axis.
17. Above a line two units above x-axis, or below a line two units below the x-axis.
19. In quadrant II above a line parallel to the x-axis and one unit above. **21.** I, III **23.** (5, 4)
25. (a) Positive integer, (b) Negative Integer, (c) Positive rational number (<1) **27.**

Exercises 1–2, p. 9

1. **3.** **5.** **7.** **9.**

11.

13.

15.

17.

19.

21.

23.

25.

27.

29.

31.

33. (a) To y-axis, (b) not to either axis, (c) not to either axis

35. (a) No, (b) yes, $x = 0$, (c) no

39.

41.

43.

45.

47.

49.

Exercises 1–3, p. 16

1. $2\sqrt{29}$ **3.** 3 **5.** $\sqrt{85}$ **7.** 7 **9.** 2.86 **11.** $\frac{5}{2}$ **13.** Undefined **15.** $-\frac{7}{6}$ **17.** 0
19. 0.747 **21.** $\frac{1}{3}\sqrt{3}$ **23.** $-\frac{1}{3}\sqrt{3}$ **25.** 20.0° **27.** 98.5° **29.** Parallel **31.** Perpendicular
33. 8, -2 **35.** -3 **37.** Two sides equal $2\sqrt{10}$ **39.** $m_1 = \frac{1}{4}$, $m_2 = \frac{5}{4}$ **41.** 10 **43.** (1, 5)

Exercises 1–4, p. 21

1. $4x - y + 20 = 0$ **3.** $7x - 6y - 16 = 0$ **5.** $x - y + 2 = 0$ **7.** $y = -3$ **9.** $x = -3$
11. $3x - 2y - 12 = 0$ **13.** $x + 3y + 5 = 0$ **15.** $x + 2y - 4 = 0$ **17.** $4x + 3y + 6 = 0$
19. $2x + 5y + 14 = 0$

21.

23.

25. $y = \frac{3}{2}x - \frac{1}{2}$; $m = \frac{3}{2}$, $b = -\frac{1}{2}$ **27.** $y = \frac{5}{2}x + \frac{5}{2}$; $m = \frac{5}{2}$, $b = \frac{5}{2}$
29. -2 **31.** 1 **33.** $m_1 = m_2 = \frac{3}{2}$ **35.** $m_1 = 2$, $m_2 = -\frac{1}{2}$
37. $x + 2y - 4 = 0$ **39.** $3x + y - 18 = 0$ **41.** $l = 2c + 8$
43. $50x + 60y = 12{,}200$ **45.** $s = 50t + 10$ **47.** $L = \frac{2}{3}F + 15$
49. 3.35 kJ   **51.** $n = \frac{7}{6}t + 10$; at 6:30, $n = 10$; at 8:30, $n = 150$

53.

55.

57.

59.

Exercises 1–5, p. 28

1. $(2, 1)$, $r = 5$ **3.** $(-1, 0)$, $r = 2$ **5.** $x^2 + y^2 = 9$
7. $(x - 2)^2 + (y - 2)^2 = 16$, or $x^2 + y^2 - 4x - 4y - 8 = 0$
9. $(x + 2)^2 + (y - 5)^2 = 5$, or $x^2 + y^2 + 4x - 10y + 24 = 0$
11. $(x + 3)^2 + y^2 = \frac{1}{4}$, or $4x^2 + 4y^2 + 24x + 35 = 0$
13. $(x - 2)^2 + (y - 1)^2 = 8$, or $x^2 + y^2 - 4x - 2y - 3 = 0$
15. $(x + 3)^2 + (y - 5)^2 = 25$, or $x^2 + y^2 + 6x - 10y + 9 = 0$
17. $(x - 2)^2 + (y - 2)^2 = 4$, or $x^2 + y^2 - 4x - 4y + 4 = 0$
19. $(x - 2)^2 + (y - 5)^2 = 25$, or $x^2 + y^2 - 4x - 10y + 4 = 0$; and
$(x + 2)^2 + (y + 5)^2 = 25$, or $x^2 + y^2 + 4x + 10y + 4 = 0$

21. $(0, 3)$
$r = 2$

23. $(-1, 5)$
$r = \frac{9}{2}$

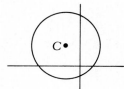

25. $(0, 0)$
$r = 5$

27. $(1, 0)$
$r = 3$

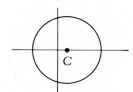

29. $(-4, 5)$
$r = 7$

31. $(1, 2)$
$r = \frac{1}{2}\sqrt{22}$

33. Symmetrical to both axes and origin.   **35.** Symmetrical to y-axis.   **37.** $(7, 0), (-1, 0)$
39. $3x^2 + 3y^2 + 4x + 8y - 20 = 0$, circle
41. $x^2 + y^2 = 2.79$ (Assume center of circle at center of coordinate system.)
43. $x^2 + y^2 = 576, x^2 + y^2 + 36y + 315 = 0$ (in inches)

Exercises 1–6, p. 33
1. $F(1, 0), x = -1$ **3.** $F(-1, 0), x = 1$ **5.** $F(0, 2), y = -2$ **7.** $F(0, -1), y = 1$

9. $F(\frac{1}{2}, 0), x = -\frac{1}{2}$ **11.** $F(0, \frac{1}{4}), y = -\frac{1}{4}$ **13.** $y^2 = 12x$ **15.** $x^2 = 16y$ **17.** $x^2 = 4y$

19. $x^2 = \frac{1}{8}y$

21. $y^2 - 2y - 12x + 37 = 0$ **23.** $x^2 - 2x + 8y - 23 = 0$ **25.** $4x + 3y = 16, y = 4$

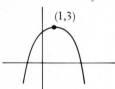

27. 3.68 m   **31.**

29. $x^2 = 100y$

250×10^{-6}

33. $y^2 = 8x$ or $x^2 = 8y$ with vertex midway between island and shore.

Exercises 1–7, p. 39
1. $V(2, 0), V(-2, 0)$
$F(\sqrt{3}, 0), F(-\sqrt{3}, 0)$

3. $V(0, 6), V(0, -6)$
$F(0, \sqrt{11}), F(0, -\sqrt{11})$

5. $V(3, 0), V(-3, 0)$
$F(\sqrt{5}, 0), F(-\sqrt{5}, 0)$

7. $V(0, 7), V(0, -7)$
$F(0, \sqrt{45}), F(0, -\sqrt{45})$

9. $V(0, 4), V(0, -4)$
$F(0, \sqrt{14})$
$F(0, -\sqrt{14})$

11. $V\left(\frac{5}{2}, 0\right), V\left(-\frac{5}{2}, 0\right)$
$F\left(\frac{\sqrt{21}}{2}, 0\right), F\left(\frac{-\sqrt{21}}{2}, 0\right)$

13. $\dfrac{x^2}{225} + \dfrac{y^2}{144} = 1$, or $144x^2 + 225y^2 = 32,400$ **15.** $\dfrac{y^2}{9} + \dfrac{x^2}{5} = 1$, or $9x^2 + 5y^2 = 45$

17. $\dfrac{x^2}{64} + \dfrac{15y^2}{144} = 1$, or $3x^2 + 20y^2 = 192$ **19.** $\dfrac{x^2}{5} + \dfrac{y^2}{20} = 1$, or $4x^2 + y^2 = 20$

21. $16x^2 + 25y^2 - 32x - 50y - 359 = 0$ **23.** $9x^2 + 5y^2 - 18x - 20y - 16 = 0$
25. $2x^2 + 3y^2 - 8x - 4 = 2x^2 + 3(-y)^2 - 8x - 4$ **27.** $9x^2 + 25y^2 = 22,500$
29. $2.70x^2 + 2.76y^2 = 7.45 \times 10^7$ **31.** Major axis $= 8.5$ in., minor axis $= 6.0$ in.
33. $7x^2 + 16y^2 = 112$

Exercises 1–8, p. 45

1. $V(5, 0), V(-5, 0)$
$F(13, 0), F(-13, 0)$

3. $V(0, 3), V(0, -3)$
$F(0, \sqrt{10}), F(0, -\sqrt{10})$

5. $V(\sqrt{2}, 0), V(-\sqrt{2}, 0)$
$F(\sqrt{6}, 0), F(-\sqrt{6}, 0)$

7. $V(0, \sqrt{5}), V(0, -\sqrt{5})$
$F(0, \sqrt{7}), F(0, -\sqrt{7})$

9. $V(0, 2), V(0, -2)$
$F(0, \sqrt{5}), F(0, -\sqrt{5})$

11. $V(2, 0), V(-2, 0)$
$F(\frac{2}{3}\sqrt{13}, 0), F(-\frac{2}{3}\sqrt{13}, 0)$

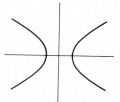

13. $\dfrac{x^2}{9} - \dfrac{y^2}{16} = 1$, or $16x^2 - 9y^2 = 144$ **15.** $\dfrac{y^2}{100} - \dfrac{x^2}{36} = 1$, or $9y^2 - 25x^2 = 900$

17. $\dfrac{x^2}{1} - \dfrac{y^2}{3} = 1$, or $3x^2 - y^2 = 3$ **19.** $\dfrac{x^2}{5} - \dfrac{y^2}{4} = 1$, or $4x^2 - 5y^2 = 20$

21.

23.

25. $9x^2 - 16y^2 - 108x + 64y + 116 = 0$

27. $9x^2 - y^2 - 36x + 27 = 0$

29.

31.

33. dist(rifle to P) $-$ dist(target to P) $=$ constant (related to distance from rifle to target).

Exercises 1–9, p. 50

1. Parabola, $(-1, 2)$ **3.** Hyperbola, $(1, 2)$ **5.** Ellipse, $(-1, 0)$ **7.** Parabola, $(-3, 1)$

9. $(y - 3)^2 = 16(x + 1)$, or $y^2 - 6y - 16x - 7 = 0$

11. $(x + 3)^2 = 4(y - 2)$, or $x^2 + 6x - 4y + 17 = 0$

13. $\dfrac{(x + 2)^2}{25} + \dfrac{(y - 2)^2}{16} = 1$, or $16x^2 + 25y^2 + 64x - 100y - 236 = 0$

15. $\dfrac{(y - 1)^2}{16} + \dfrac{(x + 2)^2}{4} = 1$, or $4x^2 + y^2 + 16x - 2y + 1 = 0$

17. $\dfrac{(y - 2)^2}{1} - \dfrac{(x + 1)^2}{3} = 1$, or $x^2 - 3y^2 + 2x + 12y - 8 = 0$

19. $\dfrac{(x + 1)^2}{9} - \dfrac{(y - 1)^2}{16} = 1$, or $16x^2 - 9y^2 + 32x + 18y - 137 = 0$

21. Parabola $(-1, -1)$ **23.** Ellipse, $(-3, 0)$ **25.** Hyperbola, $(0, 4)$ **27.** Parabola, $(1, 0)$

29. $x^2 - y^2 + 4x - 2y - 22 = 0$ **31.** $y^2 + 4x - 4 = 0$ **33.**

35. $\dfrac{(x - 0.9)^2}{3.7^2} + \dfrac{y^2}{3.6^2} = 1$

Exercises 1–10, p. 55

1. Ellipse   **3.** Hyperbola   **5.** Circle   **7.** Parabola   **9.** Hyperbola   **11.** Circle   **13.** Ellipse

15. Hyperbola   **17.** Ellipse   **19.** Parabola

21. Parabola; $V(-4, 0)$; $F(-4, 2)$ **23.** Hyperbola; $C(+1, -2)$;   **25.** Ellipse; $C(5, 0)$; $V(5, \pm 2\sqrt{2})$

$$V(+1, -2, \pm \sqrt{2})$$

27. Parabola; $V(-\frac{1}{2}, \frac{5}{2})$; $F(\frac{1}{2}, \frac{5}{2})$ **29.** (a) Circle, (b) hyperbola, (c) ellipse

31. The origin   **33.** Hyperbola   **35.** Ellipse

Review Exercises for Chapter 1, p. 56

1. $4x - y - 11 = 0$ **3.** $2x + 3y + 3 = 0$ **5.** $x^2 + y^2 - 2x + 4y - 5 = 0$

7. $y^2 = 12x$ **9.** $9x^2 + 25y^2 = 900$ **11.** $144y^2 - 169x^2 = 24{,}336$

13. $(-3, 0)$, $r = 4$ **15.** $(0, -5)$, $y = 5$ **17.** $V(0, 4)$, $V(0, -4)$
$F(0, \sqrt{15})$, $F(0, -\sqrt{15})$

19. $V(2, 0)$, $V(-2, 0)$
$F(\frac{2}{5}\sqrt{35}, 0)$, $F(-\frac{2}{5}\sqrt{35}, 0)$

21. $V(4, -8)$, $F(4, -7)$

23. $(2, -1)$

25. 4   **27.** 2   **29.** $(1.90, 1.55)$, $(1.90, -1.55)$, $(-1.90, 1.55)$, $(-1.90, -1.55)$

31. $m_1 = -\frac{12}{5}$, $m_2 = \frac{5}{12}$; $d_1^2 = 169$, $d_2^2 = 169$, $d_3^2 = 338$ **33.** $x^2 - 6x - 8y + 1 = 0$ **41.** $x^2 + y^2 = 0.513$

35.   **37.**   **39.**

43. $x^2 = -80y$ **45.**   **47.** $225y = 450x - x^2$

49. (a) 5.8 m, (b) 5.0 m   **51.** $\dfrac{x^2}{1.464 \times 10^6} + \dfrac{y^2}{1.461 \times 10^6} = 1$ **53.**   **55.** 13.5 ft

Exercises 2–1, p. 65

1. $A = s^2$ **3.** $V = \dfrac{8}{3}\pi r^2$ **5.** $d = 55t$ **7.** $p = 100(c - 3)$ **9.** $I = 24t$

11. $n_2 = -\dfrac{2}{3}n_1 + \dfrac{4000}{3}$ **13.** $-18, 70$ **15.** $15, 0$ **17.** $0, 2a^3$ **19.** $-5t + 7, 25t + 42$

21. $\sqrt{x^2 - 2x + 5}$ **23.** $\dfrac{x - 1}{\sqrt{x^2 + 4}}$ **25.**   **27.**

29. $\sqrt[6]{x^6 + 1}$ **31.** $(4x - 5)^{7/2}$ **33.** $\dfrac{3x + 4}{\sqrt{2x + 1}}$ **35.** $\dfrac{1}{(x^2 + 1)^{3/2}}$ **37.**   No

39. Yes, no   **41.** $y \neq 1$ **43.** $y \geq 1$ **45.** (a) -1 (b) $\sqrt{5}$

47.   **49.** 7.8   **51.** $16t^2 + 64t + 64$ ft

53. $\dfrac{(4 + \pi)p^2 - 480p + 14{,}400}{16\pi}$ **55.** $p = 2\left(x + \dfrac{A}{x}\right)$

Exercises 2–2, p. 75

1. Cont. all x **3.** Not cont. $x = -3$, div. by zero   **5.** Cont. $x > 0$ **7.** Cont. all x
9. Not cont. $x = 1$, small change   **11.** Cont. $x < 2$
13. 10.000, 8.500, 7.300, 7.030, 7.003, 4.000, 5.500, 6.700, 6.970, 6.997; $\lim\limits_{x \to 3} f(x) = 7$
15. 1.71, 1.9701, 1.997001, 2.31, 2.0301, 2.003001; $\lim\limits_{x \to 1} f(x) = 2$
17. -0.2484567, -0.2498440, -0.2499850, -0.2515823, -0.2501564, -0.2500155; $\lim\limits_{x \to 2} f(x) = -0.25$
19. 0.4468085, 0.4044266, 0.4004403; $\lim\limits_{x \to \infty} f(x) = 0.4$ **21.** 7   **23.** 1   **25.** 1   **27.** -2 **29.** 2
31. -4 **33.** Does not exist   **35.** 0   **37.** 3   **39.** 0
41. -2.9, -2.99, -2.999, -3.1, -3.01, -3.001; $\lim\limits_{x \to 0} f(x) = -3$
43. 2.1649485, 2.0106032, 2.001006; $\lim\limits_{x \to \infty} f(x) = 2$
45. 1.26×10^{-6}, 1.827×10^{-6}, 1.8837×10^{-6}, 1.8963×10^{-6}, 1.953×10^{-6}, 2.52×10^{-6};
$\lim\limits_{t \to 0.3} B = 1.89 \times 10^{-6}\,T$ **47.** 34.9°C, 0°C

Exercises 2–3, p. 82

1. (slopes) 3.5, 3.9, 3.99,
3.999; $m = 4$

3. (slopes) -3.5, -3.9, -3.99,
-3.999; $m = -4$

5. 4   **7.** -4

9. $m_{\text{tan}} = 2x_1$; 4, -2

11. $m_{\text{tan}} = 2x_1 + 2$; -4, 4

13. $m_{\text{tan}} = 2x_1 + 4$; -2, 8

15. $m_{\tan} = 6 - 2x_1$; 10, 0   **17.** $m_{\tan} = 3x_1^2 - 2$; 1, -2, 1   **19.** $m_{\tan} = 4x_1^3$; 0, 0.5, 4

21. $\dfrac{\Delta y}{\Delta x} = 4.1$, $m_{\tan} = 4$ **23.** $\dfrac{\Delta y}{\Delta x} = -12.61$, $m_{\tan} = -12$

Exercises 2–4, p. 87

1. 3   **3.** -2 **5.** $2x$ **7.** $10x$ **9.** $2x - 7$ **11.** $8 - 4x$ **13.** $3x^2 + 4$ **15.** $-\dfrac{1}{(x + 2)^2}$

17. $1 - \dfrac{1}{x^2}$ **19.** $-\dfrac{4}{x^3}$ **21.** $4x^3 + 3x^2 + 2x + 1$ **23.** $4x^3 + \dfrac{2}{x^2}$ **25.** $6x - 2$; -8

27. $\dfrac{-6}{(x + 3)^2}$; $-\dfrac{1}{6}$ **29.** $\dfrac{1}{2\sqrt{x + 1}}$ **31.** $-\dfrac{3}{2\sqrt{1 - 3x}}$

Exercises 2–5, p. 91

1. 4.00, 4.00, 4.00, 4.00, 4.00; $\lim\limits_{t \to 3} v = 4$ ft/s   **3.** -14, -15, -15.8, -15.98, -15.998; $\lim\limits_{t \to 4} v = -16$ ft/s

5. 4; 4   **7.** $-4t$; -16 **9.** $3 + \dfrac{2}{t^2}$ **11.** $3t^2 - 6$ **13.** $12t - 4$ **15.** -4 **17.** -2

19. $2\pi r$, 6π **21.** $0.285\ \Omega/°C$ **23.** $-83.1\ W/m^2 \cdot h$ **25.** πd^2

27. $-\dfrac{2}{r^3}$, -2×10^6 lx/m, -2×10^3 lx/m

Exercises 2–6, p. 97

1. $5x^4$ **3.** $36x^8$ **5.** $4x^3$ **7.** $2x + 2$ **9.** $15x^2 - 1$ **11.** $8x^7 - 28x^6 - 1$ **13.** $42x^6 - 15x^2$
15. $x^2 + x$ **17.** 16   **19.** 33   **21.** 360   **23.** 14   **25.** $30t^4 - 5$ **27.** $120 - 32t$ **29.** 64
31. 50.6   **33.** 1   **35.** $3\pi r^2$ **37.** 6.0 V/°C   **39.** 300 m/km   **41.** -12 N/cm

Exercises 2–7, p. 102

1. $x^2(3) + (3x + 2)(2x) = 9x^2 + 4x$ **3.** $6x(6x - 5) + (3x^2 - 5x)(6) = 54x^2 - 60x$
5. $(x + 2)(2) + (2x - 5)(1) = 4x - 1$
7. $(x^4 - 3x^2 + 3)(-6x^2) + (1 - 2x^3)(4x^3 - 6x) = -14x^6 + 30x^4 + 4x^3 - 18x^2 - 6x$
9. $(2x - 7)(-2) + (5 - 2x)(2) = -8x + 24$
11. $(x^3 - 1)(4x - 1) + (2x^2 - x - 1)(3x^2) = 10x^4 - 4x^3 - 3x^2 - 4x + 1$

13. $\dfrac{3}{(2x + 3)^2}$ **15.** $\dfrac{-2x}{(x^2 + 1)^2}$ **17.** $\dfrac{6x - 2x^2}{(3 - 2x)^2}$ **19.** $\dfrac{-6x^2 + 6x + 4}{(3x^2 + 2)^2}$ **21.** $\dfrac{-x^2 - 16x - 6}{(x^2 + x + 2)^2}$

23. $\dfrac{-2x^4 + 2x^3 + 5x^2 + 4x}{(x^3 + 2x^2)^2}$ **25.** -107 **27.** 19   **29.** $\dfrac{-12x^3 + 45x^2 - 14x}{(3x - 7)^2}$ **31.** 12   **33.** 1, -1

35. $-\dfrac{1}{16}$ ft/s   **37.** 0.925 V/s   **39.** $\dfrac{E^2(R - r)}{(R + r)^3}$

Exercises 2–8, p. 108

1. $\dfrac{1}{2x^{1/2}}$ **3.** $-\dfrac{2}{x^3}$ **5.** $-\dfrac{1}{x^{4/3}}$ **7.** $\dfrac{3}{2}x^{1/2} + \dfrac{1}{x^2}$ **9.** $10x(x^2 + 1)^4$ **11.** $-192x^2(7 - 4x^3)^7$

13. $\dfrac{2x^2}{(2x^3 - 3)^{2/3}}$ **15.** $\dfrac{8x}{(1 - x^2)^5}$ **17.** $\dfrac{24x^3}{(2x^4 - 5)^{1/4}}$ **19.** $\dfrac{-4x}{(1 - 8x^2)^{3/4}}$ **21.** $\dfrac{3x - 2}{2(x - 1)^{1/2}}$

23. $\dfrac{-16x - 22}{(4x + 3)^{1/2}(8x + 1)^2}$ **25.** $\dfrac{3}{10}$ **27.** $\dfrac{5}{36}$ **29.** $\dfrac{x^3(0) - 1(3x^2)}{x^6} = -3x^{-4}$ **31.** $x = 0$

33. 1   **35.** $\frac{188}{3}\sqrt[3]{60}$ **37.** $\dfrac{-450,000}{V^{5/2}}$, -4.50 kPa/cm^3 **39.** -45.2 W/m$^2 \cdot$h

41. $\dfrac{-kqx}{(x^2 + b^2)^{3/2}}$ **43.** $\dfrac{-3kr}{[r^2 + (l/2)^2]^{5/2}}$

Exercises 2–9, p. 112

1. $-\dfrac{3}{2}$ **3.** $\dfrac{6x + 1}{4}$ **5.** $\dfrac{x}{y}$ **7.** $\dfrac{2x}{5y^4}$ **9.** $\dfrac{2x}{2y + 1}$ **11.** $\dfrac{-3y}{3x + 1}$ **13.** $\dfrac{-2x - y^3}{3xy^2 + 3}$ **15.** $\dfrac{3 - 6xy^2}{6x^2y + 1}$

17. $\dfrac{4(2y - x)^3 - 2x}{8(2y - x)^3 - 1}$ **19.** $\dfrac{-3x(x^2 + 1)^2}{y(y^2 + 1)}$ **21.** 3   **23.** $-\dfrac{108}{157}$ **25.** 1   **27.** $-\dfrac{x}{y}$

Exercises 2–10, p. 116

1. $y' = 3x^2 + 2x$, $y'' = 6x + 2$, $y''' = 6$, $y^{(n)} = 0$ $(n \geq 4)$
3. $f'(x) = 3x^2 - 24x^3$, $f''(x) = 6x - 72x^2$, $f'''(x) = 6 - 144x$, $f^{(4)}(x) = -144$, $f^{(n)}(x) = 0$ $(n \geq 5)$
5. $y' = -8(1 - 2x)^3$, $y'' = 48(1 - 2x)^2$, $y''' = -192(1 - 2x)$, $y^{(4)} = 384$, $y^{(n)} = 0$ $(n \geq 5)$
7. $f'(x) = (8x + 1)(2x + 1)^2$, $f''(x) = 12(2x + 1)(4x + 1)$,
   $f'''(x) = 24(8x + 3)$, $f^{(4)}(x) = 192$, $f^{(n)}(x) = 0$ $(n \geq 5)$ **9.** $84x^5 - 30x^4$

11. $-\dfrac{1}{4x^{3/2}}$ **13.** $-\dfrac{12}{(8x - 3)^{7/4}}$ **15.** $\dfrac{12}{(1 - 2x)^{5/2}}$ **17.** $600(2 - 5x)^2$ **19.** $30(27x^2 - 1)(3x^2 - 1)^3$

21. $\dfrac{4}{(1 - x)^3}$ **23.** $\dfrac{2}{(x + 1)^3}$ **25.** $-\dfrac{9}{y^3}$ **27.** $-\dfrac{6(x^2 - xy + y^2)}{(2y - x)^3}$ **29.** $\dfrac{9}{125}$ **31.** $-\dfrac{13}{384}$

33. -32.2 ft/s^2 **35.** 0.049/m   **37.** $-\dfrac{1.60}{(2t + 1)^{3/2}}$

Review Exercises for Chapter 2, p. 118

1. -4 **3.** $\frac{1}{4}$ **5.** 1   **7.** $\frac{7}{3}$ **9.** $\frac{2}{3}$ **11.** -2 **13.** 5   **15.** $-4x$ **17.** $-\dfrac{4}{x^3}$ **19.** $\dfrac{1}{2\sqrt{x + 5}}$

21. $14x^6 - 6x$ **23.** $\dfrac{2}{x^{1/2}} + \dfrac{1}{x^2}$ **25.** $\dfrac{1}{(1 - x)^2}$ **27.** $-12(2 - 3x)^3$ **29.** $\dfrac{9x}{(5 - 2x^2)^{7/4}}$

31. $\dfrac{-15x^2 + 2x}{(1 - 6x)^{1/2}}$ **33.** $\dfrac{-2x - 3}{2x^2(4x + 3)^{1/2}}$ **35.** $\dfrac{2x - 6(2x - 3y)^2}{1 - 9(2x - 3y)^2}$ **37.** $\dfrac{5}{48}$ **39.** $\dfrac{74}{5}$

41. $36x^2 - \dfrac{2}{x^3}$ **43.** $\dfrac{56}{(1 + 4x)^3}$ **45.** $\frac{1}{2}C_2$, 0   **47.** -31 **49.** 2   **51.** 0.01 m/s   **53.** 0.04 V

55. $-\dfrac{2k}{r^3}$ **57.** $4\pi r - \dfrac{200}{r^2}$ **59.** $\dfrac{a}{\sqrt{v_0^2 + 2as}}$ **61.** $\dfrac{40V_2^{0.4}}{V_1^{1.4}}$ **63.** $4x - x^3$, $4 - 3x^2$

65. At $t = 5$ years, $dV/dt = -\$7500$/year (rate of appreciation—decreasing)
   $d^2V/dt^2 = \$1500$/year^2 (rate at which appreciation changes—increasing)
   (Machinery is depreciating, but depreciation is lessening.)

Exercises 3–1, p. 124

1. $4x - y - 2 = 0$ **3.** $2y + x - 2 = 0$ **5.** $x + 4y - 21 = 0$ **7.** $8x - 4y - 7 = 0$

9. $\sqrt{3}x + 8y - 7 = 0$, $16x - 2\sqrt{3}y - 15\sqrt{3} = 0$ **11.** $2x - 12y + 37 = 0$, $72x + 12y + 37 = 0$
13. $y = 2x - 4$ **15.** $y - 8 = -\frac{1}{24}(x - \frac{3}{2})$, or $2x + 48y - 387 = 0$ **17.** $-\frac{1}{4}$
19. $3x - 5y - 150 = 0$ **21.** $x - \sqrt{3}y + 4 = 0$ **23.** $x + 2y - 3 = 0$, $x = 0$, $x - 2y + 3 = 0$

Exercises 3–2, p. 128

1. 3.4494897   **3.** −0.1804604   **5.** 0.5857864   **7.** 0.3488942   **9.** 2.5615528   **11.** −1.2360680
13. 0.9175433   **15.** 0.6180340   **17.** −1.8557725, 0.6783628, 3.1774097   **19.** 0.0763 m

Exercises 3–3, p. 135

1. 3.16, 341.6°   **3.** 0.479, 211.5°   **5.** $a = 0$ **7.** 0.442, 58.1°   **9.** 301, 86.2°   **11.** 1.106
13. 136 ft/s, 28.1°; 124 ft/s, 345.1°   **15.** 32 ft/s²; 270.0° for both   **17.** 276 m/s, 43.5°; 2090 m/s, 16.7°
19. 21.2 mi/min, 296.6°

Exercises 3–4, p. 139

1. 0.0900 Ω/s   **3.** \$1.22/week   **5.** 0.00401   **7.** 0.375 V/s   **9.** 100 ft²/min   **11.** 10.2 ft/s
13. −3.75 kPa/min   **15.** 2.51 × 10⁶ mm³/s   **17.** 128 km/h   **19.** 12.0 ft/s   **21.** 825 mi/h

Exercises 3–5, p. 146

1. Inc. $x > 1$, dec. $x < 1$ **3.** Inc. $-2 < x < 2$, dec. $x < -2$, $x > 2$ **5.** Min. $(1, -1)$
7. Min. $(-2, -16)$, max. $(2, 16)$ **9.** Conc. up all x **11.** Conc. up $x < 0$, conc. down $x > 0$, infl. $(0, 0)$
13.   **15.**   **17.** Max. $(3, 18)$,   **19.** Max. $(-2, 8)$, min. $(0, 0)$,
                                                                 conc. down all x infl. $(-1, 4)$

21. Max. $(-1, 4)$, min. $(1, -4)$,   **23.** Max. $(1, 1)$, infl. $(0, 0)$,   **25.** Max. $(2000, 1000)$
infl. $(0, 0)$ $(\frac{2}{3}, \frac{16}{27})$

27. Max. $(3, 16)$, min. $(1, 0)$, $(5, 0)$ **29.** $V = 4x^3 - 40x^2 + 96x$ **31.**
max. $(1.57, 67.6)$

Exercises 3–6, p. 150

1. Inc. $x < 0$, dec. $x > 0$
conc. up $x < 0$, $x > 0$
asym. $x = 0$, $y = 0$

3. Dec. $x < -1$, $x > -1$
conc. up $x > -1$, conc. down $x < -1$
int. $(0, 2)$, asym. $x = -1$, $y = 0$

5. Int. $(-\sqrt[3]{2}, 0)$, min. $(1, 3)$,
infl. $(-\sqrt[3]{2}, 0)$, asym. $x = 0$

7. Int. $(1, 0)$, $(-1, 0)$
asym. $x = 0$, $y = x$
conc. up $x < 0$, conc. down $x > 0$

9. Int. $(0, 0)$, max. $(-2, -4)$,
min. $(0, 0)$, asym. $x = -1$

11. Int. $(0, -1)$, max. $(0, -1)$,
asym. $x = 1$, $x = -1$, $y = 0$

13. Asym. $R_T = 5$, inc. $R \geq 0$
Int. $(0, 0)$

15. $A = 2\pi r^2 + \dfrac{40}{r}$, min. $(1.47, 40.8)$
asym. $r = 0$

Exercises 3–7, p. 155

1. 196 ft   **3.** 50 mg   **5.** 0.250 ft/min   **7.** 8, 8   **9.** 5.00×10^5 ft^2 **11.** 17.0 cm

13. $2\sqrt{2}$ by $2\sqrt{2}$ **15.** $5\sqrt{2}$, $5\sqrt{2}$ **17.** 12   **19.** 3 km   **21.** $\dfrac{4}{3}$ in.   **23.** 2 units

25. 0.58 L **27.** width $= \dfrac{16\sqrt{3}}{3}$ in., depth $= \dfrac{16\sqrt{6}}{3}$ in.

Review Exercises for Chapter 3, p. 157

1. $5x - y + 1 = 0$ **3.** $27x - 3y - 26 = 0$ **5.** $x - 2y + 3 = 0$ **7.** 31.0, 359.5°
9. 2.12   **11.** 48.0, 0.3°   **13.** 0.7458983   **15.** 1.9111643
17. Min. $(-2, -16)$
conc. up all x

19. Int. $(0, 0)$, $(\pm 3\sqrt{3}, 0)$
max. $(3, 54)$, min. $(-3, -54)$
infl. $(0, 0)$

21. Min. $(2, -48)$
conc. up $x < 0$, $x > 0$

23. Int. $(0, 0)$, asym. $x = -1$, $y = 1$

25. $2x - y + 1 = 0$
29. 3180 cm/s, 292.2°
33.

27. 0.0 m, 6.527 m
31. -7.44 cm/s
35. 3.27 in./min
37. $r = 1$, $\theta = 2$
39. Int. $(1, 0)$
   max. $(2, \frac{1}{4})$
   infl. $(3, \frac{2}{9})$
   asym. $P = 0$
   $V \geq 1$ (only meaningful values)

41. 13.0 km/h   **43.** 1.94 units

Exercises 4–1, p. 163

1. $(5x^4 + 1)\, dx$ **3.** $\dfrac{-10\, dx}{x^6}$ **5.** $8x(x^2 - 1)^3\, dx$ **7.** $\dfrac{-12x\, dx}{(3x^2 + 1)^2}$ **9.** $x(1 - x)^2(-5x + 2)\, dx$

11. $\dfrac{dx}{(x + 1)^2}$ **13.** 12.28, 12   **15.** 1.71275, 1.675   **17.** -2.4, -2.4730881

19. 0.6257, 0.6264903   **21.** 16 in.²   **23.** 12.2 V   **25.** 2.64 in.³   **27.** $\dfrac{dA}{A} = \dfrac{2\, ds}{s}$

Exercises 4–2, p. 167

1. x^3 **3.** x^6 **5.** $\frac{3}{2}x^4 + x$ **7.** $\frac{2}{3}x^3 - \frac{1}{2}x^2$ **9.** $x^{5/2}$ **11.** $\frac{4}{3}x^{3/2} + 3x$ **13.** $\dfrac{1}{x}$ **15.** $\dfrac{2}{x^3}$

17. $\frac{2}{5}x^5 + x$ **19.** $3x^2 - \dfrac{1}{3x^3}$ **21.** $\frac{1}{3}x^3 + 2x - \dfrac{1}{x}$ **23.** $(2x + 1)^6$ **25.** $(x^2 - 1)^4$

27. $\frac{1}{40}(2x^4 + 1)^5$ **29.** $(6x + 1)^{3/2}$ **31.** $\frac{1}{4}(3x + 1)^{4/3}$

Exercises 4–3, p. 172

1. $x^2 + C$ **3.** $\frac{1}{8}x^8 + C$ **5.** $\frac{2}{5}x^{5/2} + C$ **7.** $-\dfrac{1}{3x^3} + C$ **9.** $\frac{1}{3}x^3 - \frac{1}{6}x^6 + C$

11. $3x^3 + \frac{1}{2}x^2 + 3x + C$ **13.** $-\dfrac{1}{2x^2} + \frac{1}{2}x + C$ **15.** $\frac{2}{7}x^{7/2} - \frac{2}{5}x^{5/2} + C$ **17.** $6x^{1/3} + \frac{1}{9}x + C$

19. $\frac{1}{6}(1 + 2x)^3 + C$ **21.** $\frac{1}{6}(x^2 - 1)^6 + C$ **23.** $\frac{1}{5}(x^4 + 3)^5 + C$ **25.** $\frac{1}{40}(x^5 + 4)^8 + C$
27. $\frac{1}{12}(8x + 1)^{3/2} + C$ **29.** $\frac{1}{6}\sqrt{6x^2 + 1} + C$ **31.** $\sqrt{x^2 - 2x} + C$ **33.** $y = 2x^3 + 2$
35. $y = 5 - \frac{1}{18}(1 - x^3)^6$ **37.** $12y = 83 + (1 - 4x^2)^{3/2}$ **39.** $y = 3x^2 + 2x - 3$

Exercises 4–4, p. 179

1. 9, 12.15   **3.** 1.92, 2.28   **5.** 7.625, 8.208   **7.** 0.464, 0.5995   **9.** 13.5   **11.** $\frac{8}{3}$ **13.** 9
15. 0.8

Exercises 4–5, p. 182

1. 1   **3.** $\frac{254}{7}$ **5.** $6 + 2\sqrt{6} - 2\sqrt{3}$ **7.** $\frac{3}{2}\sqrt[3]{2}$ **9.** $\frac{747}{20}$ **11.** $\frac{33}{20}\sqrt[3]{\frac{11}{5}} - \frac{3}{8}\sqrt[3]{\frac{1}{5}} - \frac{17}{5} = -1.552$
13. $\frac{4}{3}$ **15.** $\frac{81}{4}$ **17.** 2   **19.** $\frac{1}{4}(20.5^{2/3} - 17.5^{2/3}) = 0.1875$ **21.** $\frac{88}{3249} = 0.0271$ **23.** 49
25. $\frac{364}{3}$ **27.** $\frac{110}{3}$

Exercises 4–6, p. 187

1. $\frac{11}{2} = 5.50$, $\frac{16}{3} = 5.33$ **3.** 7.661, $\frac{23}{3} = 7.667$ **5.** 0.2042   **7.** 2.996   **9.** 0.5205   **11.** 21.74
13. 10.58   **15.** 31.70

Exercises 4–7, p. 190

1. (a) 6, (b) 6   **3.** (a) 19.67, (b) 19.67   **5.** 0.2028   **7.** 3.084   **9.** 0.5114   **11.** 10.60

Review Exercises for Chapter 4, p. 191

1. $x^4 - \frac{1}{2}x^2 + C$ **3.** $x^2 - \frac{4}{5}x^{5/2} + C$ **5.** $\frac{20}{3}$ **7.** $\frac{16}{3}$ **9.** $3x - \dfrac{1}{x^2} + C$ **11.** $\sqrt{10} - \sqrt{2} = 1.748$

13. $\dfrac{1}{5(2 - 5x)} + C$ **15.** $-\frac{2}{7}(7 - 2x)^{7/4} + C$ **17.** $\frac{9}{8}(3\sqrt[3]{3} - 1)$ **19.** $-\frac{1}{30}(1 - 2x^3)^5 + C$

21. $-\dfrac{1}{(2x - x^3)} + C$ **23.** $\frac{3350}{3}$ **25.** $\dfrac{-6x\,dx}{(x^2 - 1)^4}$ **27.** $\dfrac{(1 - 4x)\,dx}{(1 - 3x)^{2/3}}$ **29.** 0.061

31. $y = 3x - \frac{1}{3}x^3 + \frac{17}{3}$

33. (a) $x - x^2 + C_1$, (b) $-\frac{1}{4}(1 - 2x)^2 + C_2 = x - x^2 + C_2 - \frac{1}{4}$; $C_1 = C_2 - \frac{1}{4}$ **35.** 22   **37.** 0.842

39. 0.811   **41.** 1.01   **43.** 13.77   **45.** 4.02 mm^3 **47.** $\dfrac{R\,dR}{R^2 + X^2}$

Exercises 5–1, p. 199

1. -96.0 ft/s   **3.** $v = 3t^2 + 5$ **5.** 560 cm   **7.** 17,800 m   **9.** 336 ft   **11.** 13.5 m   **13.** 1.0 mC

15. 5.14 C   **17.** 667 V   **19.** 117 V   **21.** $\theta = 2t^2$ **23.** 66.7 A   **25.** $\dfrac{k}{x_1}$

27. $m = 1002 - 2\sqrt{t + 1}$, 2.51×10^5 min

Exercises 5–2, p. 205

1. 2   **3.** $\frac{8}{3}$ **5.** $\frac{3}{4}$ **7.** $\frac{4}{3}\sqrt{2}$ **9.** $\frac{1}{2}$ **11.** $\frac{1}{6}$ **13.** 12   **15.** $\frac{19}{3}$ **17.** $\frac{15}{4}$ **19.** $\frac{7}{6}$ **21.** $\frac{256}{5}$
23. $\frac{2}{3}$ **25.** $\frac{1}{3}$ **27.** 1   **29.** 18.0 J   **31.** $\frac{128}{5}$ ft   **33.** 4 cm^2 **35.** 0.513 J   **37.** 0.683 m^2

Exercises 5–3, p. 212

1. $\frac{1}{3}\pi$ **3.** $\frac{1}{3}\pi$ **5.** $\frac{348}{5}\pi$ **7.** $\frac{768}{7}\pi$ **9.** 72π **11.** $\frac{16}{3}\pi$ **13.** $\frac{1}{3}\pi$ **15.** $\frac{1}{3}\pi$ **17.** $\frac{2}{5}\pi$
19. $\frac{8}{3}\pi$ **21.** $\frac{16}{15}\pi$ **23.** $\frac{16}{3}\pi$ **25.** $\frac{8}{3}\pi$ **27.** $\frac{1}{3}\pi r^2 h$ **29.** 7.56 mm^3 **31.** 3375π m^3 = 10,600 m^3

Exercises 5–4, p. 219

1. $(\frac{10}{3}, 0)$ **3.** $(\frac{14}{15}, 0)$ **5.** $(-\frac{1}{2}, \frac{1}{2})$ **7.** $(\frac{7}{22}, \frac{5}{22})$ **9.** $(0, \frac{6}{5})$ **11.** $(\frac{4}{3}, \frac{4}{3})$ **13.** $(\frac{3}{5}, \frac{12}{35})$
15. $(0, \frac{5}{6})$ **17.** $(\frac{2}{3}, 0)$ **19.** $(\frac{9}{7}, 0)$ **21.** $(\frac{2}{3}a, \frac{1}{3}b)$ **23.** on the axis, 1.731 in. from larger base

Exercises 5–5, p. 224

1. 128, 4   **3.** 214, 3.27   **5.** $\frac{64}{15}$ **7.** $\frac{2}{3}\sqrt{6}$ **9.** $\frac{1}{6}ma^2$ **11.** $\frac{4}{7}\sqrt{7}$ **13.** $\frac{8}{11}\sqrt{55}$ **15.** $\frac{64}{3}\pi$
17. $\frac{2}{5}\sqrt{10}$ **19.** $\frac{3}{10}mr^2$ **21.** 0.324 g·cm^2

Exercises 5–6, p. 228

1. 37.5 lb-in.   **3.** 90.0 ft-lb   **5.** 2.40 J   **7.** $99k$ J   **9.** 10,000 ft-lb
11. 1800 N·m   **13.** 3.00×10^5 ft·ton   **15.** 8.82×10^4 ft·lb   **17.** 1.26×10^6 N·m = 1.26 MJ

Exercises 5–7, p. 231

1. 2340 lb   **3.** 196 kN   **5.** 6400 lb   **7.** 5320 lb   **9.** 35.9 kN   **11.** 2560 lb   **13.** 11,700 lb

Exercises 5–8, p. 234

1. $\dfrac{38}{3}$ **3.** 21   **5.** $4\pi\sqrt{2}$ **7.** $\dfrac{56\pi}{3}$ **9.** 2.67 A   **11.** 10

Review Exercises for Chapter 5, p. 235

1. 160 ft/s   **3.** 4.22 s   **5.** 9.20 C   **7.** 3640 V   **9.** $y = 20x + \frac{1}{120}x^3$ **11.** $\frac{2}{3}$ **13.** 18
15. $\frac{27}{4}$ **17.** $\frac{48}{5}\pi$ **19.** $\frac{243}{10}\pi$ **21.** $\frac{4}{3}\pi ab^2$ **23.** $(\frac{40}{21}, \frac{10}{3})$ **25.** $(\frac{14}{5}, 0)$ **27.** $\frac{8}{5}$ **29.** $\frac{256}{3}\pi$
31. \$3280   **33.** 107 ft^2 **35.** 9.00 ft·lb   **37.** 8.47×10^6 mi·lb   **39.** 185 kJ   **41.** 3990 lb
43. $\frac{46}{3}$ **45.** 0.286 Ω **47.** 172,000 lb

Exercises 6–1, p. 245

1. 15.2°   **3.** 315°48′   **5.** $\dfrac{\pi}{12}, \dfrac{5\pi}{6}$ **7.** $\dfrac{5\pi}{12}, \dfrac{11\pi}{6}$ **9.** 72°, 270°   **11.** 10°, 315°   **13.** 0.401

15. 43.0°   **17.** -0.8290 **19.** 1.4663   **21.** 0.7071   **23.** 3.372   **25.** -1.732 **27.** -0.1161
29. $\frac{1}{2}\sqrt{2}$ **31.** $\frac{1}{2}$ **33.** 0.3490, 2.7925   **35.** 2.2689, 5.4104

37.

39.

41

43.

45.

47.

49.

Exercises 6–2, p. 253

[*Note:* "Answers" to trigonometric identities are intermediate steps of suggested reductions of the left member.]

1. $\dfrac{\cos x}{\sin x} \cdot \dfrac{1}{\cos x} = \dfrac{1}{\sin x}$ **3.** $\left(\dfrac{\sin y}{\cos y}\right)\left(\dfrac{1}{\sin y}\right) = \dfrac{1}{\cos y}$ **5.** $(1 - \sin^2 x) - \sin^2 x$

7. $\sin x\left(\dfrac{\sin x}{\cos x}\right) + \cos x = \dfrac{\sin^2 x + \cos^2 x}{\cos x} = \dfrac{1}{\cos x}$ **9.** $\tan^2 x + \tan x \cot x = \tan^2 x + 1$

11. $\dfrac{\dfrac{\sec \theta}{1}}{\sec \theta} - \dfrac{\dfrac{\tan \theta}{1}}{\tan \theta} = \sec^2 \theta - \tan^2 \theta$

13. $(\sin x \cos y + \cos x \sin y)(\sin x \cos y - \cos x \sin y) = \sin^2 x \cos^2 y - \cos^2 x \sin^2 y$
$$= \sin^2 x(1 - \sin^2 y) - (1 - \sin^2 x)\sin^2 y$$

15. $\cos x \cos y + \sin x \sin y + \sin x \cos y + \cos x \sin y = \cos x(\cos y + \sin y) + \sin x(\sin y + \cos y)$

17. $2 \sin x + 2 \sin x \cos x = \dfrac{2 \sin x(1 + \cos x)(1 - \cos x)}{1 - \cos x}$

19. $\dfrac{\sin 3x \cos x + \cos 3x \sin x}{\sin x \cos x} = \dfrac{\sin 4x}{\frac{1}{2} \sin 2x} = \dfrac{2 \sin 2x \cos 2x}{\frac{1}{2} \sin 2x}$ **21.** $\dfrac{1 - \cos \alpha}{2\sqrt{\frac{1}{2}(1 - \cos \alpha)}} = \sqrt{\dfrac{1 - \cos \alpha}{2}}$

23. $2\left(\dfrac{1 - \cos x}{2}\right) + \cos x$ **25.** $2A \sin \dfrac{2\pi t}{T} \cos \dfrac{2\pi x}{\lambda}$

27. $\cos \alpha = \dfrac{A}{C}$, $\sin \alpha = \dfrac{B}{C}$:

$$A \sin 2t + B \cos 2t = C\left(\dfrac{A}{C} \sin 2t + \dfrac{B}{C} \cos 2t\right)$$
$$= C(\cos \alpha \sin 2t + \sin \alpha \cos 2t)$$

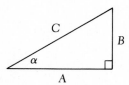

29. 0.262 ft   **31.** 25.7 ft   **33.** 19.6%

Exercises 6–3, p. 258

1. $\cos(x + 2)$ **3.** $4 \cos(2x - 1)$ **5.** $-10 \sin 2x$ **7.** $-6 \sin(3x - 1)$ **9.** $8 \sin 4x \cos 4x = 4 \sin 8x$
11. $-45 \cos^2(5x + 2) \sin(5x + 2)$ **13.** $\sin 3x + 3x \cos 3x$ **15.** $9x^2 \cos 5x - 15x^3 \sin 5x$

17. $2x \cos x^2 \cos 2x - 2 \sin x^2 \sin 2x$ **19.** $-2 \sin 2x \sin x + 4 \cos 2x \cos x$ **21.** $\dfrac{3x \cos 3x - \sin 3x}{x^2}$

23. $-\dfrac{4x^2 \sin x^2 - 2 \cos x^2}{3x^2}$ **25.** $4 \sin 3x(3 \cos 3x \cos 2x - \sin 3x \sin 2x)$

27. $\dfrac{-2 \cos 3x[3 \sin 3x(1 + 2 \sin^2 2x) + 4 \cos 3x \sin 2x \cos 2x]}{(1 + 2 \sin^2 2x)^2}$ **29.** $3 \sin^2 x \cos x + 2 \sin 2x$

31. $1 - 4 \sin^2 4x \cos 4x$ **33.** See table.

35. (a) 0.5403023, value of derivative, (b) 0.5402602, slope of secant line

37. $\dfrac{d \sin x}{dx} = \cos x, \dfrac{d^2 \sin x}{dx^2} = -\sin x, \dfrac{d^3 \sin x}{dx^3} = -\cos x, \dfrac{d^4 \sin x}{dx^4} = \sin x$ **39.** $\sin 2x = 2 \sin x \cos x$

41. $3\sqrt{2}$ **43.** -2.36 **45.** 9.76 ft/s

Exercises 6–4, p. 262

1. $5 \sec^2 5x$ **3.** $2(1 - x) \csc^2(1 - x)^2$ **5.** $6 \sec 2x \tan 2x$ **7.** $\dfrac{3 \csc \sqrt{x} \cot \sqrt{x}}{2\sqrt{x}}$ **9.** $30 \tan 3x \sec^2 3x$

11. $-24 \cot^3 3x \csc^2 3x$ **13.** $2 \tan 4x \sqrt{\sec 4x}$ **15.** $-84 \csc^4 7x \cot 7x$ **17.** $x^2 \sec^2 x + 2x \tan x$

19. $-4 \csc x^2(2x \cos x \cot x^2 + \sin x)$ **21.** $-\dfrac{\csc x(x \cot x + 1)}{x^2}$

23. $\dfrac{-4 \sin 4x - 4 \sin 4x \cot 3x + 3 \cos 4x \csc^2 3x}{(1 + \cot 3x)^2}$ **25.** $\sec^2 x(\tan^2 x - 1)$ **27.** $2 \sec 2x(\sec 2x - \tan 2x)$

29. $24 \tan 3x \sec^2 3x \, dx$ **31.** $4 \sec 4x(\tan^2 4x + \sec^2 4x) \, dx$

33. (a) 3.4255188, value of derivative, (b) 3.4260524, slope of secant line

35. $2 \tan x \sec^2 x = 2 \sec x(\sec x \tan x)$ **37.** -12 **39.** 17.92   **41.** 140 ft/s

Exercises 6–5, p. 267

1. $\dfrac{\pi}{3}$ **3.** 0   **5.** $-\dfrac{\pi}{3}$ **7.** $\dfrac{\pi}{3}$ **9.** $\dfrac{\pi}{6}$ **11.** $-\dfrac{\pi}{4}$ **13.** $\dfrac{\pi}{4}$ **15.** $\frac{1}{2}\sqrt{3}$ **17.** $\frac{1}{2}\sqrt{2}$ **19.** -1

21. -1.3090 **23.** -0.9838 **25.** 1.4413   **27.** 1.4503   **29.** -1.2389

31. -0.2239 **33.** $\dfrac{x}{\sqrt{1 - x^2}}$ **35.** $\dfrac{1}{x}$ **37.** $\dfrac{3x}{\sqrt{9x^2 - 1}}$ **39.** $2x\sqrt{1 - x^2}$

41. $\sin(\text{Arcsin } \frac{3}{5} + \text{Arcsin } \frac{5}{13}) = \frac{3}{5} \cdot \frac{12}{13} + \frac{4}{5} \cdot \frac{5}{13} = \frac{56}{65}$ **43.** $\dfrac{\pi}{6} + \dfrac{\pi}{3} = \dfrac{\pi}{2}$ **45.** $\text{Arcsin}\left(\dfrac{a}{c}\right)$

47. let b = side opp. β; $\tan \beta = \dfrac{b}{1} = b$; $\tan \alpha = a + b = a + \tan \beta$; $\alpha = \text{Arctan}(a + \tan \beta)$

49. Let $\alpha = \text{Arctan} \dfrac{3}{6}$; $\theta + \alpha = \text{Arctan} \dfrac{3}{6 - x}$; $\theta = \text{Arctan} \dfrac{3}{6 - x} - \text{Arctan} \dfrac{1}{2}$

Exercises 6–6, p. 272

1. $\dfrac{2x}{\sqrt{1 - x^4}}$ **3.** $\dfrac{18x^2}{\sqrt{1 - 9x^6}}$ **5.** $-\dfrac{1}{\sqrt{4 - x^2}}$ **7.** $\dfrac{1}{\sqrt{(x - 1)(2 - x)}}$ **9.** $\dfrac{1}{2\sqrt{x}(1 + x)}$ **11.** $-\dfrac{1}{x^2 + 1}$

13. $\dfrac{x}{\sqrt{1 - x^2}} + \text{Arcsin } x$ **15.** $\dfrac{4x}{1 + 4x^2} + 2 \text{ Arctan } 2x$ **17.** $\dfrac{3\sqrt{1 - 4x^2} \text{ Arcsin } 2x - 6x}{\sqrt{1 - 4x^2} \text{ Arcsin}^2 2x}$

19. $\dfrac{2(\text{Arccos } 2x + \text{Arcsin } 2x)}{\sqrt{1 - 4x^2} \text{ Arccos}^2 2x}$ **21.** $\dfrac{-24 \text{ Arccos}^2 4x}{\sqrt{1 - 16x^2}}$ **23.** $\dfrac{8 \text{ Arcsin } 4x}{\sqrt{1 - 16x^2}}$ **25.** $\dfrac{9 \text{ Arctan}^2 x}{1 + x^2}$

27. $\dfrac{-2(2x + 1)^2}{(1 + 4x^2)^2}$ **29.** (a) 1.1547005, value of derivative, (b) 1.1547390, slope of secant line

31. $\dfrac{3 \text{ Arcsin}^2 x \, dx}{\sqrt{1 - x^2}}$ **33.** $\dfrac{-16x}{(1 + 4x^2)^2}$ **35.** Let $y = \text{Arcsec } u$; solve for u; take derivatives; substitute.

37. $-\dfrac{1}{\sqrt{R^2 - R_x^2}} \dfrac{dR_x}{dt}$ **39.** $\dfrac{f}{r\sqrt{fr - f^2}}$

Exercises 6–7, p. 276

1. $d \sin x / dx = \cos x$ and $d \cos x / dx = -\sin x$, and $\sin x = \cos x$ at points of intersection.

3. $\dfrac{1}{x^2 + 1}$ is always positive.   **5.** Dec., $x > 0$, $x < 0$; infl. $(0, 0)$ asym., $x = \pi/2$, $x = -\pi/2$

7. $8\sqrt{2}x + 8y + 4\sqrt{2} - 5\pi\sqrt{2} = 0$

9. 1.9337538   **11.** 10

13. $-\frac{\sqrt{3}}{2}$ ft/s, $-\frac{5}{2}$ ft/s^2

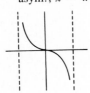

15. 1.05 mm/s   **17.** 2π, 330°   **19.** 4, 225°   **21.** -0.0730 rad/s   **23.** 1.70 unit/s

25. 0.02   **27.** 0.06 cm   **29.** 9.0 in.^{2} **31.** 14.05 ft

Review Exercises for Chapter 6, p. 278

1. $-12 \sin(4x - 1)$ **3.** $-\dfrac{\sec^2\sqrt{3 - x}}{2\sqrt{3 - x}}$ **5.** $\dfrac{9}{9 + x^2}$ **7.** $-6 \csc^2(3x + 2) \cot(3x + 2)$

9. $-24x \cos^3 x^2 \sin x^2$ **11.** -1 **13.** $(-2 \csc 4x)\sqrt{\csc 4x + \cot 4x}$ **15.** $\sec 2x \, (1 + 2x \tan 2x)$

17. $\dfrac{2x(1 + 4x^2)\text{Arctan } 2x - 2x^2}{(1 + 4x^2)(\text{Arctan } 2x)^2}$ **19.** Arccos x **21.** $-\dfrac{2y^2\cos 2x + \sec^2 x}{2y \sin 2x}$ **23.** $\dfrac{2y \sin 2x + \sin 2y}{\cos 2x - 2x \cos 2y}$

25. $2 \sin x \cos x - 2 \cos x \sin x = 0$

27. Infl. $(\frac{1}{2}\pi, \frac{1}{2}\pi)$, $(\frac{3}{2}\pi, \frac{3}{2}\pi)$

29. $7.27x + y - 8.44 = 0$ **31.** 0.5109734   **33.** $-20 \sin 2t$

35. 2π, 330°   **37.** -0.831 units/s   **39.** 0.005934 rad/s

41. 377 ft/s   **43.** 5570 cm^3 **45.** 7.07 in.

Exercises 7–1, p. 286

1. $4 = \log_4 256$ **3.** $16 = 8^{4/3}$ **5.** $\frac{1}{3} = \log_8 2$ **7.** $16 = (0.5)^{-4}$ **9.** $0 = \log_{12} 1$ **11.** $\frac{1}{512} = 8^{-3}$

13. 343   **15.** 0.2   **17.**   **19.**   **21.** $\log_5 3$ **23.** $\log_6 x^{5/2}$

25. $\log_b 4y$ **27.** $\log_2(\frac{3}{5}x^2)$ **29.** 1.099   **31.** 1.921   **33.** $-\dfrac{1}{k} \ln\!\left(\dfrac{p}{p_0}\right) = \ln\!\left(\dfrac{p}{p_0}\right)^{-1/k}$

35. $-\dfrac{L}{R}\ln\left(\dfrac{E - iR}{E}\right)$ **37.** (a) 1.9021, (b) 4.2047, (c) 8.1316, (d) -1.0788; 2.28×10^4

39. $\sinh 0.25 = \frac{1}{2}(e^{0.25} - e^{-0.25}) = 0.2526$; $1.0314^2 - 0.2526^2 = 1.0000$
    $\cosh 0.25 = \frac{1}{2}(e^{0.25} + e^{-0.25}) = 1.0314$

Exercises 7–2, p. 291

1. $\dfrac{2 \log e}{x}$ **3.** $\dfrac{6 \log_5 e}{3x + 1}$ **5.** $\dfrac{-3}{1 - 3x}$ **7.** $\dfrac{4 \sec^2 2x}{\tan 2x} = 4 \sec 2x \csc 2x$ **9.** $\dfrac{1}{2x}$ **11.** $\dfrac{6(x + 1)}{x^2 + 2x}$

13. $1 + \ln x$ **15.** $\dfrac{3(2x + 1)\ln(2x + 1) - 6x}{(2x + 1)[\ln(2x + 1)]^2}$ **17.** $\dfrac{1}{x \ln x}$ **19.** $\dfrac{1}{x^2 + x}$ **21.** $\dfrac{\cos \ln x}{x}$ **23.** $\dfrac{2 \ln 2x}{x}$

25. $\dfrac{x \sec^2 x + \tan x}{x \tan x}$ **27.** $\dfrac{x + 4}{x(x + 2)}$ **29.** 0.5 is value of derivative, 0.4999875 is slope of secant line.

31. See table.   **33.** 2.73   **35.** -1 **37.** $x^x(\ln x + 1)$ **39.** $\dfrac{10 \log e}{I}\dfrac{dI}{dt}$

Exercises 7–3, p. 295

1. $\dfrac{2(3^{2x})}{\log_3 e}$ **3.** $\dfrac{6(4^{6x})}{\log_4 e}$ **5.** $6e^{6x}$ **7.** $\dfrac{e^{\sqrt{x}}}{2\sqrt{x}}$ **9.** $e^x(x + 1)$ **11.** $e^{\sin x}(x \cos x + 1)$ **13.** $\dfrac{3e^{2x}(2x + 1)}{(x + 1)^2}$

15. $e^{-3x}(4 \cos 4x - 3 \sin 4x)$ **17.** $\dfrac{2e^{3x}(12x + 5)}{(4x + 3)^2}$ **19.** $\dfrac{2xe^{x^2}}{e^{x^2} + 4}$ **21.** $16e^{6x}(x \cos x^2 + 3 \sin x^2)$

23. $2(\ln 2x + e^{2x})\left(\dfrac{1}{x} + 2e^{2x}\right)$ **25.** $2(e^{2x} + e^{-2x})$ **27.** $\dfrac{e^{2x}}{x} + 2e^{2x}\ln x$ **29.** $12e^{6x}\cot 2e^{6x}$

31. $\dfrac{4e^{2x}}{\sqrt{1 - e^{4x}}}$ **33.** (a) 2.7182818, value of derivative, (b) 2.7184178, slope of secant line   **35.** -0.724

37. $(-xe^{-x} + e^{-x}) + (xe^{-x}) = e^{-x}$ **39.** $\dfrac{dy}{dx} = e^a$; $y = e^a$ **41.** $10e^{-0.1t}$

43. $\dfrac{d}{dx}\left[\dfrac{1}{2}(e^u - e^{-u})\right] = \dfrac{1}{2}(e^u + e^{-u})\dfrac{du}{dx} = \cosh u \dfrac{du}{dx}$

Exercises 7–4, p. 298

1. Int. (0, 0), max. (0, 0), not defined for $\cos x < 0$, asym. $x = -\frac{1}{2}\pi, \frac{1}{2}\pi, \dots$

3. Int. (0, 0), max. $\left(1, \dfrac{1}{e}\right)$, infl. $\left(2, \dfrac{2}{e^2}\right)$, asym. $y = 0$

5. Int. (0, 0), max. (0, 0), infl. $(-1, -\ln 2)$, $(1, -\ln 2)$

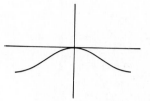

7. Int. (0, 1), max.(0, 1),
infl.($\frac{1}{2}\sqrt{2}, \frac{1}{e}\sqrt{e}$), ($-\frac{1}{2}\sqrt{2}, \frac{1}{e}\sqrt{e}$),
asym. $y = 0$

9. Max.(1, -1), asym. $x = 0$

11. Int.(0, 0), infl.(0, 0),
inc. all x

13. $y = x - 1$ **15.** $2\sqrt{2}x - 2y + 2\sqrt{2} - 3\pi\sqrt{2} = 0$ **17.** 1.3140968   **19.** -0.303 W/day   **21.** 1.21

23. $a = k^2 x$ **25.** $v = -e^{-0.5t}(1.4\cos 6t + 2.3\sin 6t)$, -2.02 **27.** $d\left(\dfrac{T_2}{T_1}\right) = -\dfrac{k}{2}e^{k/\sin(\theta/2)}\csc\dfrac{\theta}{2}\cot\dfrac{\theta}{2}d\theta$

Review Exercises for Chapter 7, p. 299

1. $\dfrac{6x}{x^2 + 1}$ **3.** $2e^{2(x-3)}$ **5.** $-2x\cot x^2$ **7.** $\dfrac{2\cos x \ln(3 + \sin x)}{3 + \sin x}$ **9.** $2(\cos 2x)e^{\sin 2x}$ **11.** $x^2(x + 3)e^x$

13. $\dfrac{2(1 + e^{-x})}{x - e^{-x}}$ **15.** $\dfrac{-\cos x(2e^{3x}\sin x + 3e^{3x}\cos x + 2\sin x)}{(e^{3x} + 1)^2}$ **17.** $x(1 + 2\ln x)$ **19.** $\dfrac{2e^{2x}(x^2 - x + 1)}{(x^2 + 1)^2}$

21. $e^{-x}\sec x(\tan x - 1)$ **23.** $\dfrac{y(1 - 2x\ln y)}{x^2 - y}$ **25.** $1.73x + 3.00y - 0.48 = 0$

27. Int. (0, 0), asym. $x = -1$ **29.** -0.7034674 **31.** $-0.15e^{-0.05t} = 0.05(3 - n)$ **33.** -0.0127 V
Conc. down $x > -1$

35. -0.135 A   **37.** $-\$4.40$ **39.** $\dfrac{ab}{x(b - x)}$ **41.** 0.429

43. -16.0 J/min   **45.** 202 ft/s   **47.** 2.28, 126.9°

Exercises 8–1, p. 304

1. $\frac{1}{5}\sin^5 x + C$ **3.** $-\frac{2}{3}(\cos x)^{3/2} + C$ **5.** $\frac{1}{3}\tan^3 x + C$ **7.** $\frac{1}{8}$ **9.** $\frac{1}{4}(\text{Arcsin } x)^4 + C$

11. $\frac{1}{10}(\text{Arctan } 5x)^2 + C$ **13.** $\frac{1}{3}[\ln(x + 1)]^3 + C$ **15.** 0.179   **17.** $\frac{1}{4}(4 + e^x)^4 + C$

19. $\frac{3}{16}(2e^{2x} - 1)^{4/3} + C$ **21.** $\frac{1}{10}(1 + \sec^2 x)^5 + C$ **23.** $\frac{1}{6}$ **25.** $\frac{1}{2}$ **27.** $y = \frac{1}{3}(\ln x)^3 + 2$ **29.** $\frac{1}{3}mnv^2$

Exercises 8–2, p. 307

1. $\frac{1}{4}\ln|1 + 4x| + C$ **3.** $-\frac{1}{6}\ln|4 - 3x^2| + C$ **5.** $-\frac{1}{2}\ln|\cot 2x| + C$ **7.** $\ln 2 = 0.693$

9. $\ln|1 - e^{-x}| + C$ **11.** $\ln|x + e^x| + C$ **13.** $\frac{1}{4}\ln|1 + 4\sec x| + C$ **15.** $\frac{1}{4}\ln 5 = 0.402$

17. $\ln|\ln x| + C$ **19.** $\ln|2x + \tan x| + C$ **21.** $-\sqrt{1 - 2x} + C$ **23.** $\frac{1}{3}\ln(\frac{5}{4}) = 0.0744$ **25.** 1.10

27. $\pi\ln 2 = 2.18$ **29.** $y = \ln\dfrac{3.5}{3 + \cos x} + 2$ **31.** 0.549 C   **33.** 0.924 ft-lb

Exercises 8–3, p. 311

1. $e^{7x} + C$ **3.** $\frac{1}{2}e^{2x+5} + C$ **5.** $2(e - 1) = 3.44$ **7.** $\frac{1}{3}e^{x^3} + C$ **9.** $2e^{\sqrt{x}} + C$ **11.** $\frac{1}{2}e^{2\sec x} + C$

13. $\dfrac{2e^x - 3}{2e^{2x}} + C$ **15.** $6 - \dfrac{3(e^6 - e^2)}{2} = -588.06$ **17.** $-\dfrac{4}{e^{\sqrt{x}}} + C$ **19.** $e^{\text{Arctan } x} + C$

21. $-\frac{1}{3}e^{\cos 3x} + C$ **23.** 0   **25.** 6.389   **27.** $\pi(e^4 - e) = 163$ **29.** $\frac{1}{8}(e^8 - 1) = 372$

31. $V = \frac{1}{3}(980 - 800e^{-30t})$

Exercises 8–4, p. 315

1. $\frac{1}{2}\sin 2x + C$ **3.** $\frac{1}{3}\tan 3x + C$ **5.** $2\sec\frac{1}{2}x + C$ **7.** $\frac{1}{3}\ln|\sin x^3| + C$ **9.** $\frac{1}{2}\ln|\sec x^2 + \tan x^2| + C$

11. $\cos\left(\frac{1}{x}\right) + C$ **13.** $\frac{1}{2}\sqrt{3}$ **15.** $\frac{1}{5}\sec 5x + C$ **17.** $\frac{1}{2}\ln|\sec 2x + \tan 2x| + C$

19. $\frac{1}{9}\pi + \frac{1}{3}\ln 2 = 0.580$ **21.** $\csc x - \cot x - \ln|\csc x - \cot x| + \ln|\sin x| + C$

23. $\frac{1}{2}(\ln|\sin 2x| + \sin 2x) + C$ **25.** 0.347 **27.** $\pi\sqrt{3} = 5.44$ **29.** $s = \frac{4}{3}\sin 3t$ **31.** 0.741

Exercises 8–5, p. 320

1. $\frac{1}{3}\sin^3 x + C$ **3.** $-\frac{1}{2}\cos 2x + \frac{1}{6}\cos^3 2x + C$ **5.** $\frac{1}{3}\sin^3 x - \frac{1}{5}\sin^5 x + C$ **7.** $\frac{1}{120}(64 - 43\sqrt{2})$

9. $\frac{1}{2}x - \frac{1}{4}\sin 2x + C$ **11.** $\frac{1}{2}x + \frac{1}{12}\sin 6x + C$ **13.** $\frac{1}{2}\tan^2 x + \ln|\cos x| + C$ **15.** $\frac{1}{4}\sec^4 x + C$

17. $\frac{1}{6}\tan^3 2x - \frac{1}{2}\tan 2x + x + C$ **19.** $\frac{1}{15}\sec^5 3x - \frac{1}{9}\sec^3 3x + C$ **21.** $x - \frac{1}{2}\cos 2x + C$

23. $\frac{1}{4}\cot^4 x - \frac{1}{3}\cot^3 x + \frac{1}{2}\cot^2 x - \cot x + C$ **25.** $1 + \frac{1}{2}\ln 2 = 1.347$ **27.** $\frac{1}{5}\tan^5 x + \frac{2}{3}\tan^3 x + \tan x + C$

29. $\frac{1}{2}\pi^2 = 4.935$ **31.** $\frac{4}{3}$ **33.** $\sqrt{2}$

Exercises 8–6, p. 324

1. $\text{Arcsin}\,\frac{1}{2}x + C$ **3.** $\frac{1}{8}\text{Arctan}\frac{1}{8}x + C$ **5.** $\frac{1}{4}\text{Arcsin}\,4x + C$ **7.** $\frac{1}{3}\text{Arctan}\,6 = 0.469$

9. $\frac{2}{5}\sqrt{5}\,\text{Arcsin}\,\frac{1}{2}\sqrt{5}x + C$ **11.** $\frac{4}{9}\ln|9x^2 + 16| + C$ **13.** $\frac{1}{35}\sqrt{35}(\text{Arctan}\frac{2}{7}\sqrt{35} - \text{Arctan}\frac{1}{7}\sqrt{35}) = 0.057$

15. $\text{Arcsin}\,e^x + C$ **17.** $\text{Arctan}(x + 1) + C$ **19.** $4\,\text{Arcsin}\,\frac{1}{2}(x + 2) + C$ **21.** -0.357

23. $2\,\text{Arcsin}(\frac{1}{2}x) + \sqrt{4 - x^2} + C$ **25.** (a) inverse tangent, (b) logarithmic, (c) general power

27. (a) general power, (b) inverse sine, (c) logarithmic   **29.** $\text{Arctan}\,2 = 1.11$ **31.** $\frac{1}{4}\pi\,\text{Arctan}\frac{3}{4} = 0.505$

33. $\text{Arcsin}\,\dfrac{x}{A} = \sqrt{\dfrac{k}{m}}\,t + \text{Arcsin}\dfrac{x_0}{A}$ **35.** 0.22

Review Exercises for Chapter 8, p. 325

1. $-\frac{1}{2}e^{-2x} + C$ **3.** $-\dfrac{1}{\ln 2x} + C$ **5.** $\ln(1 + \sin x) + C$ **7.** $\frac{2}{35}\text{Arctan}\frac{7}{5}x + C$ **9.** 0

11. $\frac{1}{2}\ln 2 = 0.3466$ **13.** $\frac{1}{12}\sec^4 3x + C$ **15.** $-\frac{1}{3}\ln|\cos 3x| + C$ **17.** $\frac{1}{9}\tan^3 3x + \frac{1}{3}\tan 3x + C$

19. $\dfrac{3}{\sqrt{e}} - 2 = -0.1804$ **21.** $\frac{3}{4}\text{Arctan}\dfrac{x^2}{2} + C$ **23.** $\text{Arcsin}\frac{1}{3}e^x + C$ **25.** $\sqrt{e^{2x} + 1} + C$

27. $\frac{1}{6}(3x - \sin 3x \cos 3x) + C$ **29.** $\ln|\tan x - x| + C$ **31.** $\frac{1}{2}\sin e^{2x} + C$ **33.** $\frac{1}{3}$

35. $\frac{1}{2}x^2 - 2x + 3\ln|x + 2| + C$ **37.** $\frac{1}{3}(e^x + 1)^3 + C_1 = \frac{1}{3}e^{3x} + e^{2x} + e^x + C_2; C_2 = C_1 + \frac{1}{3}$

39. $\Delta S = a\ln T + bT + \frac{1}{2}cT^2 + C$ **41.** $v = 320(1 - e^{-t/10})$ **43.** $2(e^3 - 1) = 38.17$

45. $y = \frac{1}{3}\tan^3 x + \tan x$ **47.** 3.47 cm^3 **49.** $\ln 3 = 1.099$

Exercises 9–1, p. 330

1. $\cos x + x\sin x + C$ **3.** $\frac{1}{2}xe^{2x} - \frac{1}{4}e^{2x} + C$ **5.** $x\tan x + \ln|\cos x| + C$

7. $x\,\text{Arctan}\,x - \ln\sqrt{1 + x^2} + C$ **9.** $-2x\sqrt{1 - x} - \frac{4}{3}(1 - x)^{3/2} + C$ **11.** $\frac{1}{2}x^2\ln x - \frac{1}{4}x^2 + C$

13. $\frac{1}{2}x\sin 2x - \frac{1}{4}(2x^2 - 1)\cos 2x + C$ **15.** $\frac{1}{2}(e^{\pi/2} - 1) = 1.91$ **17.** $1 - \dfrac{3}{e^2} = 0.594$

19. $\frac{1}{2}\pi - 1 = 0.571$ **21.** 0.756 **23.** $q = \frac{1}{5}[e^{-2t}(\sin t - 2\cos t) + 2]$

Exercises 9–2, p. 335

1. $\frac{2}{15}(3x - 2)(x + 1)^{3/2} + C$ **3.** $\frac{1}{15}(3x - 1)(2x + 1)^{3/2} + C$ **5.** $\frac{2}{3}(x - 6)(x + 3)^{1/2} + C$

7. $\frac{2}{15}(3x^2 + 8x + 32)(x - 2)^{1/2} + C$ **9.** $\frac{1696}{105}$ **11.** $\frac{3}{28}(4x + 3)(x - 1)^{4/3} + C$

13. $\frac{2}{7}(x - 2)(2x + 3)^{3/4} + C$ **15.** $\frac{3081}{40}$ **17.** $\frac{928}{315} = 2.946$ **19.** $\dfrac{30{,}024\pi}{35} = 2695$

21. $38.5\text{ ft}\cdot\text{lb}$ **23.** $-\dfrac{2}{15}(1 - x)^{3/2}(2 + 3x) + C$

Exercises 9–3, p. 339

1. $-\dfrac{\sqrt{1 - x^2}}{x} - \text{Arcsin } x + C$ **3.** $\ln|x + \sqrt{x^2 - 4}| + C$ **5.** $-\dfrac{\sqrt{x^2 + 9}}{9x} + C$ **7.** $\dfrac{x}{4\sqrt{4 - x^2}} + C$

9. $\dfrac{16 - 9\sqrt{3}}{24} = 0.017$ **11.** $\ln|\sqrt{x^2 + 2x + 2} + x + 1| + C$ **13.** $\frac{1}{3}\text{Arcsec}\frac{2}{3}x + C$

15. $\text{Arcsec } e^x + C$ **17.** π **19.** $\frac{1}{4}ma^2$ **21.** 2.68 **23.** $y = \dfrac{1}{12}\left[12 + 2\sqrt{2} - \dfrac{(4 + x^2)^{3/2}}{x^3}\right]$

Exercises 9–4, p. 343

1. $\ln\left|\dfrac{(x + 1)^2}{x + 2}\right| + C$ **3.** $\dfrac{1}{4}\ln\left|\dfrac{x - 2}{x + 2}\right| + C$ **5.** $x + \ln\left|\dfrac{x}{(x + 3)^4}\right| + C$ **7.** 1.057

9. $\ln\left|\dfrac{x^2(x - 5)^3}{x + 1}\right| + C$ **11.** $\dfrac{1}{4}\ln\left|\dfrac{x^4(2x + 1)^3}{2x - 1}\right| + C$ **13.** $1 + \ln\dfrac{16}{3} = 2.674$

15. $\dfrac{1}{60}\ln\left|\dfrac{(x + 2)^3(x - 3)^2}{(x - 2)^3(x + 3)^2}\right| + C$ **17.** $y = \ln\left|\dfrac{x(x + 5)^2}{36}\right|$ **19.** $0.1633 \text{ N} \cdot \text{cm}$ **21.** $\dfrac{1}{2}\ln\dfrac{5}{4} = 0.1116$

Exercises 9–5, p. 349

1. $\dfrac{3}{x - 2} + 2\ln\left|\dfrac{x - 2}{x}\right| + C$ **3.** $\dfrac{2}{x} + \ln\left|\dfrac{x - 1}{x + 1}\right| + C$ **5.** $-\dfrac{5}{4}$

7. $-\dfrac{2}{x + 1} - \dfrac{1}{x - 3} + \ln|x + 1| + C$ **9.** $\dfrac{1}{8}\pi + \ln 3 = 1.491$ **11.** $-\dfrac{2}{x} + \dfrac{3}{2}\text{Arctan}\dfrac{x + 2}{2} + C$

13. $\dfrac{1}{4}\ln(4x^2 + 1) + \ln|x^2 + 6x + 10| + \text{Arctan}(x + 3) + C$ **15.** $\ln(x^2 + 1) + \dfrac{1}{x^2 + 1} + C$

17. $2 + 4\ln\dfrac{2}{3} = 0.3781$ **19.** $\pi\ln\dfrac{25}{9} = 3.210$ **21.** 0.9190 m **23.** 1.369

Exercises 9–6, p. 353

1. $\frac{2}{25}[2 + 5x - 2\ln|2 + 5x|] + C$ **3.** $-\frac{2}{27}(4 - 9x)(2 + 3x)^{3/2} + C$

5. $\frac{1}{2}x\sqrt{4 - x^2} + 2\text{Arcsin}\frac{1}{2}x + C$ **7.** $\frac{1}{2}\sin x - \frac{1}{10}\sin 5x + C$ **9.** $\sqrt{4x^2 - 9} - 3\text{Arcsec}\left(\dfrac{2x}{3}\right) + C$

11. $\frac{1}{20}\cos^4 4x \sin 4x + \frac{1}{5}\sin 4x - \frac{1}{15}\sin^3 4x + C$ **13.** $\frac{1}{2}x^2\text{Arctan } x^2 - \frac{1}{4}\ln(1 + x^4) + C$

15. $\frac{1}{4}(8\pi - 9\sqrt{3}) = 2.386$ **17.** $-\ln\left(\dfrac{1 + \sqrt{4x^2 + 1}}{2x}\right) + C$ **19.** $-\ln\left(\dfrac{1 + \sqrt{1 - 4x^2}}{2x}\right) + C$

21. $\frac{1}{8}\cos 4x - \frac{1}{12}\cos 6x + C$ **23.** $\frac{1}{3}(\cos x^3 + x^3\sin x^3) + C$ **25.** $\dfrac{x^2}{2\sqrt{1 - x^4}} + C$ **27.** 4.892

29. $\frac{1}{4}x^4(\ln x^2 - \frac{1}{2}) + C$ **31.** $-\dfrac{x^3}{3\sqrt{x^6 - 1}} + C$ **33.** $\frac{1}{4}[2\sqrt{5} + \ln(2 + \sqrt{5})] = 1.479$ **35.** 208 lb

37. $\frac{1}{4}(8 \text{ Arctan } 4 - \ln 17) = 1.943$ **39.** $\frac{1}{8}\pi(e^{2\pi} - 1) = 209.9$

Exercises 9–7, p. 358

1. $\frac{1}{3}$ **3.** Divergent **5.** Divergent **7.** $\frac{1}{2}$ **9.** 1 **11.** Divergent **13.** 6 **15.** Divergent
17. Divergent **19.** Divergent **21.** $\ln(2 + \sqrt{3})$ **23.** 0 **25.** (a) Divergent, (b) π **27.** $\sqrt{3}$
29. $2.53 \times 10^7 \text{ mi} \cdot \text{lb}$ **31.** $\sin x$ takes on all values from -1 to 1 as x increases

Review Exercises for Chapter 9, p. 359

1. $-\frac{1}{2}x\cot 2x + \frac{1}{4}\ln|\sin 2x| + C$ **3.** $\frac{2}{15}(3x + 8)(x - 4)^{3/2} + C$ **5.** $\frac{1}{2}\ln|2x + \sqrt{4x^2 - 9}| + C$

7. $\ln\left|\dfrac{(x - 5)^3}{(x + 5)^2}\right| + C$ **9.** $\ln|x - 1| - \dfrac{1}{(x - 1)^2} + C$ **11.** Divergent

13. $\frac{3}{4}(x + 3)^{1/3}(x - 9) + C$ **15.** $\frac{2x}{\sqrt{4x^2 + 1}} + C$ **17.** $2 \text{ Arctan } x - \frac{1}{2}\sqrt{2} \text{ Arctan } \frac{1}{2}x\sqrt{2} + C$

19. $6 \ln 2 - 2 = 2.159$ **21.** $\ln\left|\frac{(x + 1)(x + 2)}{(x - 1)(x - 2)}\right| + C$ **23.** $\frac{144}{5}$ **25.** 4   **27.** 11.18

29. $4\pi(e^2 - 1) = 80.29$ **31.** $48.3 \text{ N} \cdot \text{cm}$ **33.** $(1.65, 0.244)$ **35.** 1.10 m^2

Exercises 10–1, p. 364

1. $V = \pi r^2 h$ **3.** $A = \frac{1}{2}bh$ **5.** $A = \frac{2V}{r} + 2\pi r^2$ **7.** $R = \sqrt{F_1^2 + F_2^2 + 1.732F_1F_2}$ **9.** 24   **11.** -2

13. $2 - 3y + 4y^2$ **15.** $6xt + xt^2 + t^3$ **17.** $\frac{p^2 + pq + kp - p + 2q^2 + 4kq + 2k^2 + 5q + 5k}{p + q + k}$

19. $2hx - 2kx - 2hy + h^2 - 2hk - 4h$ **21.** 0   **23.** $81z^6 - 9z^5 - 2z^3$ **25.** $x \neq 0, y \geq 0$

27. $y \leq 1$ **29.** 18 V   **31.** 150 Pa   **33.** 6.67 cm   **35.** $A = \frac{pw - 2w^2}{2}$, 3850 cm^2

37. $L = \frac{(1.28 \times 10^5)r^4}{l^2}$

Exercises 10–2, p. 374

1.

3.

5.

7.

9.

11.

13.

15.

17.

19.

21.

23.

25.   **27.**   **29.**

31.

Exercises 10–3, p. 381

1. $\dfrac{\partial z}{\partial x} = 5 + 8xy, \dfrac{\partial z}{\partial y} = 4x^2$ **3.** $\dfrac{\partial z}{\partial x} = \dfrac{2x}{y} - 2y, \dfrac{\partial z}{\partial y} = -\dfrac{x^2}{y^2} - 2x$

5. $\dfrac{\partial f}{\partial x} = e^{2y}, \dfrac{\partial f}{\partial y} = 2xe^{2y}$ **7.** $\dfrac{\partial f}{\partial x} = -\dfrac{\sin x}{1 - \sec 3y}, \dfrac{\partial f}{\partial y} = -\dfrac{3(2 + \cos x)\sec 3y \tan 3y}{(1 - \sec 3y)^2}$

9. $\dfrac{\partial \phi}{\partial r} = \dfrac{1 + 3rs}{\sqrt{1 + 2rs}}, \dfrac{\partial \phi}{\partial s} = \dfrac{r^2}{\sqrt{1 + 2rs}}$ **11.** $\dfrac{\partial z}{\partial x} = 4(2x + y^3)(x^2 + xy^3)^3, \dfrac{\partial z}{\partial y} = 12xy^2(x^2 + xy^3)^3$

13. $\dfrac{\partial z}{\partial x} = y \cos xy, \dfrac{\partial z}{\partial y} = x \cos xy$ **15.** $\dfrac{\partial y}{\partial r} = \dfrac{2r}{r^2 + s}, \dfrac{\partial y}{\partial s} = \dfrac{1}{r^2 + s}$

17. $\dfrac{\partial f}{\partial x} = \dfrac{12 \sin^2 2x \cos 2x}{1 - 3y}, \dfrac{\partial f}{\partial y} = \dfrac{6 \sin^3 2x}{(1 - 3y)^2}$

19. $\dfrac{\partial z}{\partial x} = \dfrac{3y + x^2y - 2x\sqrt{1 - x^2y^2} \; \text{Arcsin } xy}{(3 + x^2)^2\sqrt{1 - x^2y^2}}, \dfrac{\partial z}{\partial y} = \dfrac{x}{(3 + x^2)\sqrt{1 - x^2y^2}}$

21. $\dfrac{\partial z}{\partial x} = \cos x - y \sin xy, \dfrac{\partial z}{\partial y} = -x \sin xy + \sin y$

23. $\dfrac{\partial f}{\partial x} = e^x(\cos xy - y \sin xy) - 2e^{-2x}\tan y, \dfrac{\partial f}{\partial y} = -xe^x \sin xy + e^{-2x} \sec^2 y$ **25.** -8 **27.** $2e$

29. $\dfrac{\partial^2 z}{\partial x^2} = -6y, \dfrac{\partial^2 z}{\partial y^2} = 12xy, \dfrac{\partial^2 z}{\partial x \, \partial y} = 6y^2 - 6x$

31. $\dfrac{\partial^2 f}{\partial x^2} = -\dfrac{y}{(x + 2y)^2}, \dfrac{\partial^2 f}{\partial y^2} = \dfrac{4(x + y)}{(x + 2y)^2}, \dfrac{\partial^2 f}{\partial x \, \partial y} = \dfrac{x}{(x + 2y)^2}$ **33.** $-4, -4$

35. 0.807   **37.** $\left(\dfrac{R_2}{R_1 + R_2}\right)^2$ **39.** $M\left(\dfrac{2mg}{(M + m)^2}\right) + m\left(\dfrac{-2Mg}{(M + m)^2}\right) = 0$ **41.** $3.75 \times 10^{-3} \; 1/\Omega$

43. $\dfrac{k_2 + 2k_3FT}{L_0 + k_1F + k_2T + k_3FT^2}$ **45.** $-\dfrac{nRT}{V^2}$ **47.** $-5e^{-t}\sin 4x = \tfrac{1}{16}(-80e^{-t}\sin 4x)$

Exercises 10–4, p. 390

1. $dz = (4x + 3)\, dx - 2y\, dy$ **3.** $dz = e^y dx + (xe^y - 2y)\, dy$
5. $dz = (y - 22x^2)(y - 2x^2)^4 \, dx + 5x(y - 2x^2)^4 \, dy$
7. $dz = (y \cos xy + y \sin x)\, dx + (x \cos xy - \cos x)\, dy$

9. $dz = \dfrac{dx}{1 - \sin y} + \dfrac{6y \sin y - 6y + x \cos y - 3y^2\cos y}{(1 - \sin y)^2}\, dy$

11. $dz = -\dfrac{2xy^2\, dx}{x^4 + y^2} + \left(\dfrac{yx^2}{x^4 + y^2} + \text{Arctan } \dfrac{y}{x^2}\right) dy$ **13.** Min $(1, 3, 0)$ **15.** Min $(-1, 2, -1)$

17. No max. or min.   **19.** Min. $(2, 0, 0)$ and $(-2, \pi, 0)$ **21.** $9.60 \times 10^4 \text{ rad/min}^2$ **23.** 1.3%
25. 4.1%   **27.** 40, 40, 40   **29.** $P_{min} = 17$ Pa at $(1, 0)$ **31.** 10 in., 10 in., 5 in.   **33.** $6xt - 6yt$

Exercises 10–5, p. 396

1. $\dfrac{28}{3}$ **3.** $\dfrac{1}{3}$ **5.** $\dfrac{127}{14}$ **7.** $\dfrac{1}{3}$ **9.** $\dfrac{\pi - 6}{12}$ **11.** 1   **13.** 495.2   **15.** $\dfrac{74}{5}$ **17.** $\dfrac{32}{3}$ **19.** 8π

21. $\dfrac{28}{3}$ **23.** 18   **25.** 300 cm^3 **27.**

Exercises 10–6, p. 401

1. $\left(\dfrac{2}{3}, \dfrac{2}{3}\right)$ **3.** $\left(\dfrac{9}{8}, \dfrac{18}{5}\right)$ **5.** $\dfrac{4}{3}, \dfrac{4}{3}$ **7.** 1.34   **9.** $\left(\dfrac{3}{4}, \dfrac{7}{10}\right)$, 0.775   **11.** $\left(\dfrac{a}{3}, \dfrac{b}{3}\right)$

13. $\dfrac{mb^2}{3}$ **15.** $\dfrac{4}{3}\sqrt{3}$ **17.** $\left(3, \dfrac{5}{6}\right)$ (origin at lower left corner)

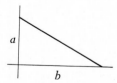

Review Exercises for Chapter 10, p. 402

1. $\dfrac{\partial z}{\partial x} = 15x^2y^2 - 2y^4, \dfrac{\partial z}{\partial y} = 10x^3y - 8xy^3$ **3.** $\dfrac{\partial z}{\partial x} = \dfrac{x}{\sqrt{x^2 - 3y^2}}, \dfrac{\partial z}{\partial y} = \dfrac{-3y}{\sqrt{x^2 - 3y^2}}$

5. $\dfrac{\partial z}{\partial x} = \dfrac{2 - 2x^2y + 6xy^2}{(x^2y + 1)^2}, \dfrac{\partial z}{\partial y} = -\dfrac{3 + 2x^3}{(x^2y + 1)^2}$

7. $\dfrac{\partial u}{\partial x} = 2xy \cot(x^2 + 2y), \dfrac{\partial u}{\partial y} = 2y \cot(x^2 + 2y) + \ln \sin(x^2 + 2y)$ **9.** $\dfrac{\partial z}{\partial x} = \dfrac{\partial z}{\partial y} = \dfrac{1}{2\sqrt{(x + y)(1 - x - y)}}$

11. $\dfrac{\partial^2 z}{\partial x^2} = 6y, \dfrac{\partial^2 z}{\partial y^2} = -6y, \dfrac{\partial^2 z}{\partial x\, \partial y} = 6x + 2$ **13.** 6   **15.** $\dfrac{21}{2}$ **17.** $\dfrac{e^2 - 3}{4} = 1.097$ **19.** $\dfrac{1}{6}$

21.   **23.** 0.113   **25.** $e^2 x - y - e^2 + 1 = 0$ **27.** 1.47

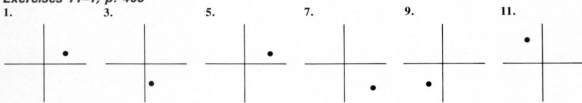

29. $\dfrac{\partial V}{\partial r} = \dfrac{ER}{(r + R)^2}, \dfrac{\partial V}{\partial R} = \dfrac{-rE}{(r + R)^2}$ **31.** 0.982   **33.** $\dfrac{\pi}{2\sqrt{gl}} = \dfrac{2\pi\sqrt{l/g}}{2l}$ **35.** 1.25 cm   **37.** $\left(\dfrac{3}{2}, 3\right)$

39. $\dfrac{292}{15}$ **41.** $32\left(\pi - \dfrac{2}{3}\right) = 79.20$ **43.** $\left(\dfrac{14}{9}, \dfrac{8}{9}\right)$ **45.** 1.82   **47.** -12 W/min

Exercises 11–1, p. 409

1.   **3.**   **5.**   **7.**   **9.**   **11.**

13. $\left(2, \dfrac{\pi}{6}\right)$ **15.** $\left(1, \dfrac{7\pi}{6}\right)$ **17.** $(-4, -4\sqrt{3})$ **19.** $(2.77, -1.15)$ **21.** $r = 3 \sec \theta$ **23.** $r = a$

25. $r = 4 \cot \theta \csc \theta$ **27.** $r^2 = \dfrac{4}{1 + 3 \sin^2\theta}$ **29.** $x^2 + y^2 - y = 0$ **31.** $x = 4$

33. $x^4 + y^4 - 4x^3 + 2x^2 y^2 - 4xy^2 - 4y^2 = 0$ **35.** $(x^2 + y^2)^2 = 2xy$

37. $s = \sqrt{x^2 + y^2} \arctan\left(\dfrac{y}{x}\right), A = \dfrac{1}{2}(x^2 + y^2) \arctan\left(\dfrac{y}{x}\right)$ **39.** $r = 100 \tan \theta \sec \theta$

Exercises 11–2, p. 412

1.   **3.**   **5.**   **7.**   **9.**

11.   **13.**   **15.**   **17.**   **19.**

21.

23.

25.

27.

29.

31.

Exercises 11–3, p. 417

1. 9.49   **3.** 1.85   **5.** 4.0 ft/s   **7.** 2.0 ft/s   **9.** 8.25 ft/s   **11.** 1.54 ft/s   **13.** π **15.** $\frac{1}{2}$
17. π **19.** $\frac{1}{8}(e^{2\pi-1})$ **21.** π **23.** $\frac{9}{2}\pi$ **25.** $\frac{1}{4}(8 + \pi)$ **27.** $\frac{1}{2}(\pi - 2)$ **29.** 5 ft/s, $5\sqrt{3}$ ft/s
31. 119 cm^2 **33.** Use indicated procedure.   **35.** 4

Review Exercises for Chapter 11, p. 418

1. $\theta = $ Arctan $2 = 1.11$ **3.** $r^2\cos 2\theta = 16$ **5.** $(x^2 + y^2)^3 = 16x^2y^2$ **7.** $3x^2 + 4y^2 - 8x - 16 = 0$

9.

11.

13.

15.

17.

19.

21. $\pi + 2$ **23.** $\frac{19}{2}\pi$ **25.** 4.502   **27.** $\frac{1}{48}\pi^3$

29. $\frac{13}{4}$ **31.** $r^2 = \dfrac{7.45 \times 10^7}{2.70 + 0.06 \sin^2\theta}$ **33.** $v_r = \dfrac{2kx}{(x^2 + y^2)^2}, v_\theta = \dfrac{ky}{(x^2 + y^2)^2}$

35. $(a^2 - b^2)x^2 + a^2y^2 + 2bx - 1 = 0$; $a = b$, parabola; $a^2 > b^2$, ellipse; $a^2 < b^2$, hyperbola
37. 9.507 mi^2 **39.** $v_r = 0, v_\theta = 5$ mi/s

Exercises 12–1, p. 427

1.

No.	2	3	4	5	6	7
Freq.	1	3	4	2	3	2

3.

No.	45	46	47	48	49	50	51	52	53	54	55	56	57
Freq.	1	1	1	2	2	0	1	0	1	0	2	0	1

5.   **7.**

9. 4.6   **11.** 50.25   **13.** 1.50   **15.** 3.74   **17.** 1.50   **19.** 3.74

21.

Vol.	746	747	748	749	750	751	752	753	754
No.	1	1	2	1	2	3	1	3	1

23. 750.4 mL   **25.**

27. 0.014 s   **29.** 20.9 mi/gal   **31.** $277   **33.** 862 kW · h

35. 3.43 A   **37.** 60%   **39.** 58%   **41.** 76%   **43.** 64%

Exercises 12–2, p. 433

1. $y = 1.0x - 2.6$ **3.** $y = -1.77x + 191$ **5.** $V = -0.590i + 11.3$

7. $h = 2.24x + 5.2$ **9.** $p = -0.200x + 649$ **11.** $V = 4.32 \times 10^{-15}f - 2.03$
$f_0 = 0.470$ PHz

13. 0.985   **15.** -0.901

Exercises 12–3, p. 438

1. $y = 1.97x^2 + 4.8$ **3.** $y = 10.9/x$ **5.** $y = 5.97t^2 + 0.38$

                    [graph]

7. $y = 35.6(10^t) - 39.9$ **9.** $f = \dfrac{492}{\sqrt{L}} + 4$

[graph]

Review Exercises for Chapter 12, p. 439

1.

No.	2.3	2.4	2.5	2.6	2.7	2.8	2.9
Freq.	1	3	2	4	1	2	2

3. 2.6

5.

Int.	101–103	104–106	107–109	110–112	113–115
Freq.	5	4	3	3	5

7. 4.6

9.

Visc. (Pa · s)	0.24	0.25	0.26	0.27	0.28	0.29
Freq.	1	2	4	2	2	1

11. 0.264 Pa · s

13.

15. 0.012 in.　**17.** 4.2 counts　**19.** $R = 0.0983T + 25.0$

21. $n = 0.45t + 7.9$　**23.** $y = 1.04 \tan x - 0.02$　**25.** $y = 1.10\sqrt{x}$

Exercises 13–1, p. 446

1. 1, 4, 9, 16　**3.** $\frac{1}{2}, \frac{1}{3}, \frac{1}{4}, \frac{1}{5}$　**5.** (a) $-\frac{2}{5}, \frac{4}{25}, -\frac{8}{125}, \frac{16}{625}$　(b) $-\frac{2}{5} + \frac{4}{25} - \frac{8}{125} + \frac{16}{625} - \cdots$

7. (a) 2, 0, 2, 0　(b) $2 + 0 + 2 + 0 + \cdots$　**9.** $a_n = \dfrac{1}{n+1}$　**11.** $a_n = \dfrac{1}{(n+1)(n+2)}$

13. 1, 1.125, 1.1620370, 1.1776620, 1.1856620; convergent; 1.2

15. 1, 1.5, 2.1666667, 2.9166667, 3.7166667; divergent　**17.** 0, −1, −3, −6, −10; divergent

19. 0.75, 0.8888889, 0.9375000, 0.9600000, 0.9722222; convergent; 1　**21.** Divergent

23. Convergent, $S = \frac{3}{4}$　**25.** Convergent, $S = 100$　**27.** Convergent, $S = \frac{4096}{9}$　**29.** $\frac{2}{9}$　**31.** $\frac{2}{11}$

33. 24.0 ft　**35.** 270 cm　**37.** $r = x; S = \dfrac{x^0}{1-x} = \dfrac{1}{1-x}$

Exercises 13–2, p. 452

1. $1 + x + \frac{1}{2}x^2 + \cdots$　**3.** $1 - \frac{1}{2}x^2 + \frac{1}{24}x^4 - \cdots$　**5.** $1 + \frac{1}{2}x - \frac{1}{8}x^2 + \cdots$　**7.** $1 - 2x + 2x^2 - \cdots$

9. $1 - 8x^2 + \frac{32}{3}x^4 - \cdots$　**11.** $1 + x + x^2 + \cdots$　**13.** $-2x - 2x^2 - \frac{8}{3}x^3 - \cdots$

15. $1 - \dfrac{x^2}{8} + \dfrac{x^4}{384} - \cdots$　**17.** $x - \frac{1}{3}x^3 + \cdots$　**19.** $x + \frac{1}{3}x^3 + \cdots$　**21.** $-\frac{1}{2}x^2 - \frac{1}{12}x^4 - \cdots$

23. $x^2 - \frac{1}{3}x^4 + \cdots$　**25.** (a), (c) functions are not defined at $x = 0$; (b) derivatives are not defined at $x = 0$.

27. $f''(0) = 0$ except $f'''(0) = 6$

Exercises 13–3, p. 458

1. $1 + 3x + \frac{9}{2}x^2 + \frac{9}{2}x^3 + \cdots$　**3.** $\dfrac{x}{2} - \dfrac{x^3}{2^3 3!} + \dfrac{x^5}{2^5 5!} - \dfrac{x^7}{2^7 7!} + \cdots$　**5.** $1 - 8x^2 + \frac{32}{3}x^4 - \frac{256}{45}x^6 + \cdots$

7. $x^2 - \frac{1}{2}x^4 + \frac{1}{3}x^6 - \frac{1}{4}x^8 + \cdots$　**9.** 0.3103　**11.** 0.1901

13. $2.63 + 1.44j$　**15.** $6.08(\cos 9.5° + j \sin 9.5°)$　**17.** $1 + \frac{1}{2}x^2 + \frac{1}{24}x^4 + \frac{1}{720}x^6 + \cdots$

19. $x + x^2 + \frac{1}{3}x^3 + \cdots$　**21.** $\dfrac{d}{dx}(x - \frac{1}{6}x^3 + \frac{1}{120}x^5 - \cdots) = 1 - \frac{1}{2}x^2 + \frac{1}{24}x^4 - \cdots$

23. $\int \cos x \, dx = x - \dfrac{x^3}{3!} + \cdots$ **25.** $\int_0^1 e^x \, dx = 1.7182818,\ \int_0^1 (1 + x + \tfrac{1}{2}x^2 + \tfrac{1}{6}x^3)\, dx = 1.7083333$

27. 0.003099   **29.** 0.1249   **31.** Use indicated method.

Exercises 13–4, p. 462
1. 1.22, 1.2214028   **3.** 0.0998333, 0.0998334   **5.** 2.7180556, 2.7182818   **7.** 0.9986292, 0.9986295
9. 0.3349333, 0.3364722   **11.** 0.3546130, 0.3546129   **13.** $-0.0139975, -0.0139975$
15. 0.7283020, 0.7286950   **17.** 1.0523528   **19.** 0.9874462   **21.** 8.3×10^{-8} **23.** 3.1×10^{-7}
25. $i = \dfrac{E}{L}\left(t - \dfrac{Rt^2}{2L}\right)$; small values of t **27.** 66 ft   **29.** 3.146

Exercises 13–5, p. 465
1. 3.32   **3.** 2.049   **5.** 0.5150   **7.** 0.49288   **9.** $e^{-2}\left[1 - (x-2) + \dfrac{(x-2)^2}{2!} - \cdots\right]$
11. $\tfrac{1}{2}\left[\sqrt{3} + (x - \tfrac{1}{3}\pi) - \dfrac{\sqrt{3}}{2!}(x - \tfrac{1}{3}\pi)^2 - \cdots\right]$ **13.** $2 + \tfrac{1}{12}(x-8) - \tfrac{1}{288}(x-8)^2 + \cdots$
15. $1 + 2(x - \tfrac{1}{4}\pi) + 2(x - \tfrac{1}{4}\pi)^2 + \cdots$ **17.** 0.111   **19.** 3.0496   **21.** 2.0247   **23.** 0.87462
25. Use indicated method.   **27.** 0.5150408, 0.5150388, 0.5150381

Exercises 13–6, p. 474
1. $f(x) = \tfrac{1}{2} - \tfrac{2}{\pi}\sin x - \tfrac{2}{3\pi}\sin 3x - \cdots$ **3.** $f(x) = \tfrac{3}{2} + \tfrac{2}{\pi}\sin x + \tfrac{2}{3\pi}\sin 3x + \cdots$

5. $f(x) = \tfrac{\pi}{4} - \tfrac{2}{\pi}(\cos x + \tfrac{1}{9}\cos 3x + \cdots) + (\sin x - \tfrac{1}{2}\sin 2x + \cdots)$

7. $f(x) = -\tfrac{1}{4} - \tfrac{1}{\pi}\cos x + \tfrac{1}{3\pi}\cos 3x - \cdots + \tfrac{3}{\pi}\sin x - \tfrac{1}{\pi}\sin 2x + \tfrac{1}{\pi}\sin 3x - \cdots$

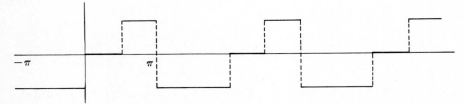

9. $f(x) = \tfrac{\pi}{2} - \tfrac{4}{\pi}\cos x - \tfrac{4}{9\pi}\cos 3x - \cdots$ **11.** $f(x) = \tfrac{5}{2} - \tfrac{10}{\pi}(\sin\tfrac{\pi x}{3} + \tfrac{1}{3}\sin \pi x - \cdots)$

13. $f(t) = \frac{2}{\pi} - \frac{4}{3\pi}\cos 2t - \frac{4}{15\pi}\cos 4t - \cdots$

15. $f(t) = 2 + \frac{8}{\pi}(\cos \frac{\pi}{2}t - \frac{1}{3}\cos \frac{3\pi}{2}t + \cdots + \sin \frac{\pi}{2}t + \sin \pi t + \frac{1}{3}\sin \frac{3\pi}{2}t + \cdots)$

Review Exercises for Chapter 13, p. 475

1. $\frac{1}{2} - \frac{1}{4}x + \frac{1}{48}x^3 - \cdots$ **3.** $2x^2 - \frac{4}{3}x^6 + \frac{4}{15}x^{10} - \cdots$ **5.** $1 + \frac{1}{3}x - \frac{1}{9}x^2 + \cdots$

7. $x + \frac{1}{6}x^3 + \frac{3}{40}x^5 + \cdots$ **9.** 0.82   **11.** 1.09   **13.** 1.0344   **15.** -0.2015 **17.** 0.95299

19. 12.1655   **21.** 0.259   **23.** $\frac{1}{2} - \frac{1}{2}\sqrt{3}(x - \frac{1}{3}\pi) - \frac{1}{4}(x - \frac{1}{3}\pi)^2 + \cdots$

25. $f(x) = \frac{1}{2} + \frac{2}{\pi}(\cos x - \frac{1}{3}\cos 3x + \cdots)$

27. $f(x) = \frac{1}{\pi}(\sin \frac{\pi x}{2} - \frac{1}{2}\sin \pi x + \frac{1}{3}\sin \frac{3\pi x}{2} - \cdots)$

29. $(x + h) - \dfrac{(x + h)^3}{3!} + \cdots - (x - h) + \dfrac{(x - h)^3}{3!} - \cdots = 2h - \dfrac{2hx^2}{2!} + \cdots = 2h\left(1 - \dfrac{x^2}{2!} + \cdots\right)$

31.

$y = 1 + x + \dfrac{x^2}{2}$

$y = 1 + x$

$y = e^x$ $y = 1$

33. $1 + \dfrac{x^2}{2} + \dfrac{5x^4}{24} + \cdots$ **35.** $1 - x + x^2 - \cdots$

37. 0.0008331

39. $f(t) = \frac{1}{2\pi} + \frac{1}{\pi}(\frac{1}{3}\cos t - \frac{1}{3}\cos 2t + \cdots)$
      $+ \frac{1}{4}\sin t + \frac{2}{3\pi}\sin 2t + \cdots$

Exercises 14–1, p. 480

(*Note:* "Answers" in this section are the unsimplified expressions obtained by substituting functions and derivatives.)

1. $1 = 1$ **3.** $e^x - (e^x - 1) = 1$ **5.** $x(2cx) = 2(cx^2)$ **7.** $(-2ce^{-2x} + 1) + 2(ce^{-2x} + x - \frac{1}{2}) = 2x$

9. $(-12 \cos 2x) + 4(3 \cos 2x) = 0$

11. $[e^{2x}(1 + 4c_1 + 4c_2 + 4x + 4c_2x + 2x^2)] - 4[e^{2x}(2c_1 + c_2 + x + 2c_2x + x^2)]$
   $+ 4[e^{2x}(c_1 + c_2x + x^2/2)] = e^{2x}$ **13.** $x^2\left[-\dfrac{c^2}{(x - c)^2}\right] + \left[\dfrac{cx}{(x - c)}\right]^2 = 0$ **15.** $x\left(-\dfrac{c_1}{x^2}\right) + \dfrac{c_1}{x} = 0$

17. $(\cos x - \sin x + e^{-x}) + (\sin x + \cos x - e^{-x}) = 2 \cos x$

19. $(e^{-x} + \frac{12}{5}\cos 2x + \frac{24}{5}\sin 2x) + (-e^{-x} + \frac{6}{5}\sin 2x - \frac{12}{5}\cos 2x) = 6 \sin 2x$

21. $\cos x\left[\dfrac{(\sec x + \tan x) - (x + c)(\sec x \tan x + \sec^2 x)}{(\sec x + \tan x)^2}\right] + \sin x = 1 - \dfrac{x + c}{\sec x + \tan x}$

23. $c^2 + cx = cx + c^2$

Exercises 14–2, p. 484

1. $y = c - x^2$ **3.** $x - \dfrac{1}{y} = c$ **5.** $\ln(x^3 + 5) + 3y = c$ **7.** $4\sqrt{1 - y} = e^{-x^2} + c$

9. $e^x - e^{-y} = c$ **11.** $\ln(y + 4) = x + c$ **13.** $y(1 + \ln x)^2 + cy + 2 = 0$ **15.** $\tan^2 x + 2 \ln y = c$

17. $x^2 + 1 + x \ln y + cx = 0$ **19.** $y^2 + 4 \operatorname{Arcsin} x = c$ **21.** $y = c - (\ln x)^2$

23. $\ln(e^x + 1) - \dfrac{1}{y} = c$ **25.** $3 \ln y + x^3 = 0$ **27.** $\frac{1}{3}y^3 + y = \frac{1}{2}\ln^2 x$

29. $2 \ln(1 - y) = 1 - 2 \sin x$

Exercises 14–3, p. 487

1. $2xy + x^2 = c$ **3.** $x^3 - 2y = cx - 4$ **5.** $x^2 y - y = cx$ **7.** $(xy)^4 = 4 \ln y + c$

9. $2\sqrt{x^2 + y^2} = x + c$ **11.** $y = c - \frac{1}{2}\ln \sin(x^2 + y^2)$ **13.** $\ln(y^2 - x^2) + 2x = c$

15. $5xy^2 + y^3 = c$ **17.** $2xy + x^3 = 5$ **19.** $2x = 2xy^2 - 15y$

Exercises 14–4, p. 490

1. $y = e^{-x}(x + c)$ **3.** $y = -\frac{1}{2}e^{-4x} + ce^{-2x}$ **5.** $y = -2 + ce^{2x}$ **7.** $y = x(3 \ln x + c)$

9. $y = \dfrac{8}{7}x^3 + \dfrac{c}{\sqrt{x}}$ **11.** $y = -\cot x + c \csc x$ **13.** $y = (x + c)\csc x$ **15.** $y = 3 + ce^{-x}$

17. $2y = e^{4x}(x^2 + c)$ **19.** $y = \frac{1}{4} + ce^{-x^4}$ **21.** $3y = x^4 - 6x^2 - 3 + cx$ **23.** $xy = (x^3 + c)e^{3x}$

25. $y = e^{-x}$ **27.** $y = x^2(3 + \sin x)$ **29.** $y = e^{\sqrt{x}} + (3e - e^2)e^{-\sqrt{x}}$

Exercises 14–5, p. 496

1. $y^2 = 2x^2 + 1$ **3.** $y = 2e^x - x - 1$ **5.** $y^2 = c - 2x$ **7.** $y^2 = c - 2 \sin x$

9. $N = N_0(0.5)^{t/2}, 0.354N_0$ **11.** 4130 years **13.** 12.9 min **15.** \$1083.29 **17.** $\lim\limits_{t \to \infty} \dfrac{E}{R}(1 - e^{-Rt/L}) = \dfrac{E}{R}$

19. 0.952 A **21.** $q = q_0 e^{-t/RC}$ **23.** 11.04 lb **25.** $v = 32(1 - e^{-t})$, 32 **27.** 10.0 ft/s

29. $x = t^2 + 1, y = \dfrac{1}{t^2 + 1}$ **31.** $p = 15.0(0.667)^{10^{-4}h}$ **33.** \$1260 **35.** $x = 4(1 + 2e^{-0.25t})$

Review Exercises for Chapter 14, p. 498

1. $2 \ln(x^2 + 1) - \dfrac{1}{2y^2} = c$ **3.** $y = e^{-2x}(x + c)$ **5.** $2x^2 + 4xy + y^4 = c$ **7.** $y = cx^3 - x^2$

9. $y = c(y + 2)e^{2x}$ **11.** $y = \frac{1}{2} + ce^{-4x}$ **13.** $y = \frac{1}{2}(c - x^2) \csc x$ **15.** $y = \dfrac{ce^{5/2}}{(x + 2)^{1/2}}$

17. $y = cxe^{x^3/3}$ **19.** $(x^2 + y^2)(4x + c) + 1 = 0$ **21.** $y = 2 \sin^2 x$ **23.** $y = 2x - 1 - e^{-2x}$

25. $x^2 + 2 \ln(1 - y) - 4 = 0$ **27.** $y = \frac{1}{4}x^2 \sin 4x$ **29.** $r = r_0 + kt$ **31.** 3.93 m/s

33. $N = 100e^{-0.026t}$ **35.** 14,400 **37.** $5y^2 + x^2 = c$ **39.** $q = c_1 e^{-t/RC} + EC$ **41.** $r = c \cos \theta$

Exercises 15–1, 15–2, p. 505

1. $y = c_1 e^{3x} + c_2 e^{-2x}$ **3.** $y = c_1 e^{-x} + c_2 e^{-3x}$ **5.** $y = c_1 + c_2 e^x$ **7.** $y = c_1 e^{6x} + c_2 e^{2x/3}$

9. $y = c_1 e^{x/3} + c_2 e^{-3x}$ **11.** $y = c_1 e^{x/3} + c_2 e^{-x}$ **13.** $y = e^x(c_1 e^{x\sqrt{2}/2} + c_2 e^{-x\sqrt{2}/2})$

15. $y = e^{3x/8}(c_1 e^{x\sqrt{41}/8} + c_2 e^{-x\sqrt{41}/8})$ **17.** $y = e^{3x/2}(c_1 e^{x\sqrt{13}/2} + c_2 e^{-x\sqrt{13}/2})$

19. $y = e^{-x/2}(c_1 e^{x\sqrt{33}/2} + c_2 e^{-x\sqrt{33}/2})$ **21.** $y = \frac{1}{5}(3e^{7x} + 7e^{-3x})$ **23.** $y = \dfrac{e^3}{e^7 - 1}(e^{4x} - e^{-3x})$

25. $y = c_1 + c_2 e^{-x} + c_3 e^{3x}$ **27.** $y = c_1 e^x + c_2 e^{2x} + c_3 e^{3x}$

Exercises 15–3, p. 510

1. $y = (c_1 + c_2 x)e^x$ **3.** $y = (c_1 + c_2 x)e^{-6x}$ **5.** $y = c_1 \sin 3x + c_2 \cos 3x$

7. $y = e^{-x/2}(c_1 \sin\frac{1}{2}\sqrt{7}x + c_2 \cos\frac{1}{2}\sqrt{7}x)$ **9.** $y = c_1 + c_2 x$ **11.** $y = c_1 \sin\frac{1}{2}x + c_2 \cos\frac{1}{2}x$

13. $y = (c_1 + c_2 x)e^{3x/4}$ **15.** $y = c_1 \sin\frac{1}{5}\sqrt{2}x + c_2 \cos\frac{1}{5}\sqrt{2}x$ **17.** $y = e^x(c_1 \cos\frac{1}{2}\sqrt{6}x + c_2 \sin\frac{1}{2}\sqrt{6}x)$

19. $y = (c_1 + c_2 x)e^{4x/5}$ **21.** $y = e^{3x/4}(c_1 e^{x\sqrt{17}/4} + c_2 e^{-x\sqrt{17}/4})$ **23.** $y = c_1 e^{x(-6 + \sqrt{42})/3} + c_2 e^{x(-6 - \sqrt{42})/3}$

25. $y = e^{(\pi/6 - 1 - x)}\sin 3x$ **27.** $y = (4 - 14x)e^{4x}$ **29.** $(D^2 - 9)y = 0$ **31.** $(D^2 + 9) = 0$

Exercises 15–4, p. 514

1. $y = c_1e^{2x} + c_2e^{-x} - 2$ **3.** $y = c_1\sin x + c_2\cos x + x^2 - 2$ **5.** $y = c_1e^{-x} + c_2e^{-3x} + \frac{1}{8}e^x + \frac{2}{3}$
7. $y = c_1 + c_2e^{3x} - \frac{3}{4}e^x - \frac{1}{2}xe^x$ **9.** $y = c_1e^x + c_2e^{-x} - \frac{1}{2}\sin x$
11. $y = c_1\sin 3x + c_2\cos 3x + x + 2\sin 2x + \cos 2x$ **13.** $y = c_1e^{-5x} + c_2e^{6x} - \frac{1}{3}$
15. $y = c_1e^{2x/3} + c_2e^{-5x} + \frac{1}{4}e^{3x}$ **17.** $y = c_1e^{2x} + c_2e^{-2x} - \frac{1}{5}\sin x - \frac{2}{5}\cos x$
19. $y = c_1\sin x + c_2\cos x - \frac{1}{3}\sin 2x + 4$ **21.** $y = c_1e^{-x} + c_2e^{-4x} - \frac{7}{100}e^x + \frac{1}{10}xe^x + 1$
23. $y = (c_1 + c_2x)e^{-3x} + \frac{1}{25}e^{2x} - e^{-2x}$ **25.** $y = \frac{1}{6}(11e^{3x} + 5e^{-2x} + e^x - 5)$
27. $y = -\frac{2}{3}\sin x + \pi\cos x + x - \frac{1}{3}\sin 2x$

Exercises 15–5, p. 521

1. $x = 4\cos 10t$ **3.** 20   **5.** $y = \frac{1}{4}\cos 16t$ **7.** $y = \frac{1}{4}\cos 16t + \frac{8}{63}\sin 2t - \frac{1}{63}\sin 16t$
9. $x = 0.150\cos 6.76\,t$ **11.** $q = 2.23 \times 10^{-4}e^{-20t}\sin 2240t$ **13.** $q = \frac{1}{100}(1 - \cos 316t)$
15. $e^{-10t}(c_1\sin 99.5t + c_2\cos 99.5t) - \frac{1}{2290}(2\cos 200t + 15\sin 200t)$ **17.** $\frac{20}{229}(2\sin 200t - 15\cos 200t)$
19. $-[(6.5 \times 10^{-4})\sin 400t + (8.5 \times 10^{-6})\cos 400t]$

Review Exercises for Chapter 15, p. 522

1. $y = c_1 + c_2e^{-x/2}$ **3.** $y = (c_1 + c_2x)e^{-x}$ **5.** $y = e^x(c_1\sin 2x + c_2\cos 2x)$ **7.** $y = c_1e^{0.2x} + c_2e^{-0.4x}$
9. $y = c_1e^{8x} + c_2e^{-7x} - \frac{1}{2}$ **11.** $y = (c_1 + c_2x)e^{-3x} + \frac{1}{3}x - \frac{2}{9}$
13. $y = c_1\sin\frac{1}{2}\sqrt{3}x + c_2\cos\frac{1}{2}\sqrt{3}x + x^2 - \frac{8}{3}$ **15.** $y = c_1e^{2x/3} + c_2e^{4x/3} + \frac{1}{2}x + \frac{25}{8}$
17. $y = e^{-x/2}(c_1e^{x\sqrt{5}/2} + c_2e^{-x\sqrt{5}/2}) + 2e^x$ **19.** $y = c_1\sin 3x + c_2\cos 3x + \frac{1}{8}\sin x$
21. $y = 2e^{-x/2}\sin(\sqrt{15}x/2)$ **23.** $y = \frac{1}{25}[16\sin x + 12\cos x - 3e^{-2x}(4 + 5x)]$
25. $y = (2 + 8t)e^{-4t}$; critically damped   **27.** $x = 0.668e^{-1.60t}\sin 6.74t$
29. $q = e^{-6t}(0.4\cos 8t + 0.3\sin 8t) - 0.4\cos 10t$ **31.** $i = 0$ **33.** $y = \dfrac{10}{3EI}[100x^3 - x^4 + xL^2(L - 100)]$

Exercises 16–1, p. 528

1.

x	0.0	0.2	0.4	0.6	0.8	1.0
y	1.00	1.20	1.44	1.72	2.04	2.40

$y = \dfrac{1}{2}x^2 + x + 1$

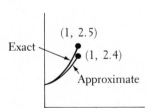

Exact   (1, 2.5)
(1, 2.4)
Approximate

3.

x	-0.2	-0.1	0.0	0.1	0.2	0.3	0.4	0.5
y	2.0000	2.1840	2.3937	2.6330	2.9069	3.2208	3.5815	3.9970

$y = 2.4233e^{0.2x^2 + x}$

Exact   (0.5, 4.2003)
(0.5, 3.9970)
$(-0.2, 2)$ Approximate

5.

x	0.0	0.1	0.2	0.3	0.4	0.5	0.6	0.7	0.8	0.9	1.0
y	1.00	1.10	1.21	1.33	1.46	1.60	1.75	1.91	2.08	2.26	2.45

7.

x	-0.2	0.0	0.2	0.4	0.5	(Not all values shown)
y	2.0000	2.4079	2.9436	3.6535	4.0935	

9.

x	0.0	0.2	0.4	0.6	0.8	1.0
y	0.0000	0.2010	0.4142	0.6573	0.9525	1.3303

(Not all values shown)

11.

x	0.0	0.2	0.4	0.6	0.8	1.0	1.2
y	0.0000	0.2000	0.4082	0.6436	0.9379	1.3614	2.1418

13.

x	0.0	0.2	0.4	0.6	0.8	1.0
y	1.5708	1.5608	1.5151	1.4415	1.3460	1.2334

(Not all values shown)

15.

x	0.0	0.1	0.2	0.3	0.4	0.5	0.6
y	1.0000	1.1082	1.2380	1.3917	1.5719	1.7812	2.0227

(Not all values shown)

17. 1.2076   **19.** $i_{approx} = 0.0804$ A, $i_{exact} = 0.0898$ A

Exercises 16–2, p. 533

1. $y_2 = 1 + x + x^2 + \frac{1}{3}x^3$; $y = \dfrac{1}{1-x}$; $y_2(0.1) = 1.1103$; $y(0.1) = 1.1111$

3. $y_3 = x^2 + \frac{1}{2}x^4 + \frac{1}{6}x^6$; $y = e^{x^2} - 1$; $y_3(1) = 1.67$; $y(1) = 1.72$

5. $y_3 = \frac{1}{2}x^2 + \frac{1}{20}x^5 + \frac{1}{160}x^8 + \frac{1}{4400}x^{11}$ **7.** $y_2 = \frac{3}{4} + x + \frac{1}{2}x^2 + \frac{1}{4}\cos 2x - \frac{1}{2}\sin 2x$

9. Maclaurin expansion: $y = 1 + x + x^2 + \frac{1}{3}x^3 + \frac{1}{12}x^4 + \cdots$

     Example A: $y_3 = 1 + x + x^2 + \frac{1}{3}x^3 + \frac{1}{24}x^4 + \cdots$

11. 0.58 ft/s

Exercises 16–3, p. 539

1. $F(s) = \displaystyle\int_0^\infty e^{-st}\, dt = -\frac{1}{s}e^{-st}\Big|_0^\infty = \frac{1}{s}$ **3.** $F(s) = \displaystyle\int_0^\infty e^{-st}\sin at\, dt = \frac{e^{-st}(-s\sin at - a\cos at)}{s^2 + a^2}\Big|_0^\infty = \frac{a}{s^2 + a^2}$

5. $\dfrac{1}{s-3}$ **7.** $\dfrac{6}{(s+2)^4}$ **9.** $\dfrac{s-2}{s^2+4}$ **11.** $\dfrac{3}{s} + \dfrac{2(s^2-9)}{(s^2+9)^2}$ **13.** $s^2 L(f) + sL(f)$

15. $(2s^2 - s + 1)L(f) - 2s + 1$ **17.** t^2 **19.** e^{-5t} **21.** $\frac{1}{2}t^2 e^{-t}$

23. $\frac{1}{54}(9t\sin 3t + 2\sin 3t - 6t\cos 3t)$

25. $1 - \frac{3}{2}e^{-t} + \frac{1}{2}e^t$ **27.** $\frac{1}{2}(2t - 1)e^t + \frac{3}{2}e^{-t}$ **29.** $3e^{-3t} + \sin 2t$

Exercises 16–4, p. 542

1. $y = e^{-t}$ **3.** $y = -e^{3t/2}$ **5.** $y = (1 + t)e^{-3t}$ **7.** $y = \frac{1}{2}\sin 2t$ **9.** $y = 1 - e^{-2t}$
11. $y = e^{2t}\cos t$ **13.** $y = 1 + \sin t$ **15.** $y = e^{-t}(\frac{1}{2}t^2 + 3t + 1)$ **17.** $v = 6(1 - e^{-t/2})$
19. $q = 1.6 \times 10^{-4}(1 - e^{-5 \times 10^{-3}t})$ **21.** $i = 5t \sin 50t$ **23.** $y = \sin 3t - 3t \cos 3t$

Review Exercises for Chapter 16, p. 543

1.

x	0.0	0.1	0.2	0.3	0.4	0.5
y	1.0000	0.9000	0.8200	0.7568	0.7085	0.6743

3.

x	0.0	0.4	0.8	1.2	1.6	2.0	(Not all values shown)
y	1.0000	1.4366	2.0115	2.7128	3.5330	4.4672	

5.

x	0.0	0.1	0.2	0.3	0.4	0.5	0.6	0.7
y	0.5000	0.5546	0.6266	0.7188	0.8359	0.9854	1.1843	1.4864

7.

x	0.0	0.1	0.2	0.3	0.4	0.5	(Not all values shown)
y	1.0000	0.9077	0.8329	0.7734	0.7282	0.6965	

9. $y_3 = 1 + x + \frac{1}{2}x^2 + \frac{1}{2}x^3 + \frac{1}{12}x^4 + \frac{1}{60}x^5$ **11.** $y_2 = -\frac{7}{36} + x - \frac{1}{9}x^3 + \frac{1}{4}x^4 + \frac{1}{18}x^6$
13. $y = e^{t/4}$ **15.** $y = \frac{1}{2}(e^{3t} - e^t)$ **17.** $y = -4 \sin t$ **19.** $y = \frac{1}{3}t - \frac{4}{9}\sin 3t$
21. $y_2 = \frac{7}{2} + x - (3 + x)e^{-x} + \frac{1}{2}e^{-2x}$; $y_2(0.5) = 2.0611$
23.

x	0.0	0.2	0.4	0.5	0.6	0.8	1.0	(Not all values shown)
y	1.0000	1.4085	1.8442	2.0678	2.2932	2.7442	3.1882	

25. $i = 12(1 - e^{-t/2})$; $i(0.3) = 1.67$ A   **27.** $q = 10^{-4}e^{-8t}(4.0 \cos 200t + 0.16 \sin 200t)$
29.

i	0.00	0.05	0.10	0.15	0.20	0.25	0.30
t	0.0000	0.3000	0.5988	0.8963	1.1925	1.4875	1.7813

31. $y = \frac{1}{4}t \sin 8t$ **33.** $y = \cos 3.16t$
35. $F(s) = \dfrac{b - a}{(s + a)(s + b)} = \dfrac{1}{s + a} - \dfrac{1}{s + b}$; $L^{-1}(F) = e^{-at} - e^{-bt}$

Index

Trigonometry

$$\sin \theta = \frac{y}{r} \qquad \cos \theta = \frac{x}{r} \qquad \tan \theta = \frac{y}{x} \qquad \cot \theta = \frac{x}{y} \qquad \sec \theta = \frac{r}{x} \qquad \csc \theta = \frac{r}{y}$$

Law of Sines: $\dfrac{a}{\sin A} = \dfrac{b}{\sin B} = \dfrac{c}{\sin C}$

Law of Cosines: $a^2 = b^2 + c^2 - 2bc \cos A$

Basic Identities

$$\sin \theta = \frac{1}{\csc \theta} \qquad \cos \theta = \frac{1}{\sec \theta} \qquad \tan \theta = \frac{1}{\cot \theta} \qquad \tan \theta = \frac{\sin \theta}{\cos \theta} \qquad \cot \theta = \frac{\cos \theta}{\sin \theta}$$

$$\sin^2\theta + \cos^2\theta = 1 \qquad 1 + \tan^2\theta = \sec^2\theta \qquad 1 + \cot^2\theta = \csc^2\theta$$

$$\sin 2\alpha = 2 \sin \alpha \cos \alpha \qquad \cos 2\alpha = \cos^2\alpha - \sin^2\alpha = 2 \cos^2\alpha - 1 = 1 - 2 \sin^2\alpha$$

$$\sin\frac{\alpha}{2} = \pm \sqrt{\frac{1 - \cos \alpha}{2}} \qquad \cos\frac{\alpha}{2} = \pm \sqrt{\frac{1 + \cos \alpha}{2}}$$

Other Relations

$$\sin (-\theta) = - \sin \theta \qquad -\frac{\pi}{2} \le \text{Arcsin } x \le \frac{\pi}{2}$$

$$\cos (-\theta) = \cos \theta \qquad 0 \le \text{Arccos } x \le \pi$$

$$\tan (-\theta) = - \tan \theta \qquad -\frac{\pi}{2} < \text{Arctan } x < \frac{\pi}{2}$$

$\pi \text{ rad} = 180°$

$s = \theta r$

$A = \frac{1}{2} \theta r^2$

Greek Alphabet

alpha	A α	eta	H η	nu	N ν	tau	T τ
beta	B β	theta	Θ θ	xi	Ξ ξ	upsilon	Y υ
gamma	Γ γ	iota	I ι	omicron	O o	phi	Φ ϕ
delta	Δ δ	kappa	K κ	pi	Π π	chi	X χ
epsilon	E ϵ	lambda	Λ λ	rho	P ρ	psi	Ψ ψ
zeta	Z ζ	mu	M μ	sigma	Σ σ	omega	Ω ω